PCR APPLICATIONS

PROTOCOLS FOR FUNCTIONAL GENOMICS

PCR APPLICATIONS

PROTOCOLS FOR FUNCTIONAL GENOMICS

Edited by

Michael A. Innis
Chiron Corporation, Emeryville, California

David H. Gelfand
Roche Diagnostic Systems, Alameda, California

John J. Sninsky
Roche Molecular Systems, Alameda, California

ACADEMIC PRESS
San Diego London Boston New York Sydney Tokyo Toronto

Front cover photograph (paperback edition only): Analysis of the human breast cancer cell line BT474 using comparative genomic hybridization (CGH). During CGH, green fluorescing tumor DNA and red fluorescing normal reference DNA samples were hybridized (along with excess unlabeled cot-1 DNA) to normal metaphase spreads. The chromosomes were then counterstained with DAPI. Chromosomal regions showing green are increased in copy number in BT474, while regions showing red are relatively decreased in copy number.

This book is printed on acid-free paper. ∞

Academic Press
a division of Harcourt Brace & Company
525 B Street, Suite 1900, San Diego, California 92101-4495, USA
http://www.apnet.com

Academic Press
24-28 Oval Road, London NW1 7DX, UK
http://www.hbuk.co.uk/ap/

International Standard Book Number: 0-12-372185-7 (hb)
International Standard Book Number: 0-12-372186-5 (pb)

PRINTED IN THE UNITED STATES OF AMERICA
99 00 01 02 03 04 EB 9 8 7 6 5 4 3 2 1

CONTENTS

Part Four
GENOMICS AND EXPRESSION PROFILING

CONTRIBUTORS

Numbers in parentheses indicate the pages on which the authors' contributions begin.

Richard D. Abramson (33), Applied Biosystems Division-Perkin Elmer, Foster City, California 94404

Ellen M. Beasley (55), Department of Genetics, Stanford Human Genome Center, Stanford University School of Medicine, Stanford, California 94305-5120

Nick Benson (341), Sidney Kimmel Cancer Center, San Diego, California 92121

Barry R. Bloom (49), Department of Microbiology and Immunology, Albert Einstein College of Medicine, Howard Hughes Medical Institute, Bronx, New York 10461

David Botstein (485), Department of Genetics, Stanford University School of Medicine, Stanford, California 94305

Bob Bonner (497), National Institute of Children's Health and Human Diseases, National Institutes of Health, Bethesda, Maryland 20892

Mary Ann D. Brow (537), Third Wave Technologies, Inc., Madison, Wisconsin 53719

Patrick Brown (485), Department of Biochemistry, Stanford University, Stanford, California 94305

Sheng-Yung P. Chang (285), Roche Molecular Systems, Alameda, California 94501

Koei Chin (473), University of California Cancer Center, San Francisco, California 94143-0808

L. A. Christel (105), Cepheid, Sunnyvale, California 94089

Rodrigo F. Chuaqui (497), Laboratory of Pathology, National Cancer Institute, National Institutes of Health, Bethesda, Maryland 20892

Kristina A. Cole (497), Laboratory of Pathology, National Cancer Institute, National Institutes of Health, Bethesda, Maryland 20892

David R. Cox (55), Department of Genetics, Stanford Human Genome Center, Stanford University School of Medicine, Stanford, California 94305-5120

Ronald W. Davis (445), Department of Biochemistry, Beckman Center, Stanford University Medical Center, Stanford, California 94305-5307

Alan D. Dean (95), Gene Check, Inc., Fort Collins, Colorado 80524

Peter A. Doris (231), Institute for Molecular Medicine, University of Texas Health Sciences Center, Houston, Texas 77025

Xiaozhu Duan (297), Chiron Corporation, Emeryville, California 94608-2916

Dwight B. DuBois (197), Cenetron Diagnostics, LLC, Austin, Texas, 78744

Barbara Dunn (485), Department of Genetics, Stanford University School of Medicine, Stanford, California 94305

Michael R. Emmert-Buck (497), Laboratory of Pathology, National Cancer Institute, National Institutes of Health, Bethesda, Maryland 20892

Tracy Ferea (485), Department of Genetics, Stanford University School of Medicine, Stanford, California 94305

David H. Gelfand (3), Roche Diagnostics Systems, Alameda, California 94501

Klaus Giese (297), Chiron Corporation, Emeryville, California 94608-2916

Joe W. Gray (473), University of California Cancer Center, San Francisco, California 94143-0808

Amanda L Hayward (231), Department of Biology, Yale University, New Haven, Connecticut 06520

Mark G. Herrmann (211), Department of Pathology, University of Utah Medical School, Salt Lake City, Utah 84132

Ralf Herwig (457), Max-Planck Institut für Molekulare Genetik, D-14195 Berlin, Germany

Russell Higuchi (263), Roche Molecular Systems, Alameda, California 94501

Cruz A. Hinojos (231), Institute for Molecular Medicine, University of Texas Health Sciences Center, Houston, Texas 77025

Michael J. Holland (429), Department of Biological Chemistry, University of California School of Medicine, Davis, California 95616

Rhonda Honeycutt (341), Phenotypics Corporation, San Diego, California 92121

Mary Huckabee (497), Laboratory of Pathology, National Cancer Institute, National Institute of Health, Bethesda, Maryland 20892

Jim R. Hully (169), PE Biosystems, Foster City, California 94044

Michael A. Innis (3), Chiron Corporation, Emeryville, California 94608-2916

Daniel B. Kainer (231), Institute for Molecular Medicine, University of Texas Health Sciences Center, Houston, Texas 77025

John J. Kang (429), Department of Biological Chemistry, University of California School of Medicine, Davis, California 95616

Steve Kaye (153), Department of Virology, University College London, London W1P 6DB, United Kingdom

Jenny M. Kelley (127), The Institute for Genomic Research, Rockville, Maryland 20850

Frank Kullmann (341), University of Regensburg, 93042 Regensburg, Germany

Laura C. Lazzeroni (55), Department of Genetics and Division of Biostatistics, Stanford University School of Medicine, Stanford, California 94305-5405

Jeffrey Y. Lee (497), Laboratory of Pathology, National Cancer Institute, National Institutes of Health, Bethesda, Maryland 20892

Hans Lehrach (457), Max-Planck Institut für Molekulare Genetik, D-14195 Berlin, Germany

Lance A. Liotta (497), Laboratory of Pathology, National Cancer Institute, National Institutes of Health, Bethesda, Maryland 20892

Emmanuel Liscum (505), Division of Biological Sciences, University of Missouri, Columbia, Missouri 65211

Jörgen Lonngren (521), Professional Genetics Laboratory AB, S-751 83 Uppsala, Sweden

Li Mao (355), Department of Thoracic/Head and Neck Medical Oncology, University of Texas M. D. Anderson Cancer Center, Houston, Texas 77030

Françoise Mathieu-Daudé (341), Sidney Kimmel Cancer Center, San Diego, California 92121

Michael McClelland (341), Sidney Kimmel Cancer Center, San Diego, California 92121

W. A. McMillan (105), Cepheid, Sunnyvale, California 94089

Sebastian Meier-Ewert (457), GPC-Aktiengesellschaft, D-82152 Martinsried, Germany

Catherine Moberg (521), Professional Genetics Laboratory AB, S-751 83 Uppsala, Sweden

Thomas W. Myers (141), Program in Core Research, Roche Molecular Systems, Alameda, California 94501

Richard M. Myers (55), Department of Genetics, Stanford Human Genome Center, Stanford University School of Medicine, Stanford, California 94305-5120

Peter S. Nelson (307), Department of Molecular Biotechnology and Medicine, University of Washington, Seattle, Washington 98195

M. A. Northrup (105), Cepheid, Sunnyvale, California 94089
Peter J. Oefner (231), Department of Biochemistry, Stanford University, Stanford, California 94305
Yun Oh (355), Department of Thoracic/Head and Neck Medical Oncology, University of Texas M. D. Anderson Cancer Center, Houston, Texas 77030
Brittan L. Pasloske (197), Ambion, Inc., Austin Texas 78744
Tomi Pastinen (521), Department of Human Molecular Genetics, National Public Health Institute, S-00300 Helsinki, Finland
K. Peterson (105), Cepheid, Sunnyvale, California 94089
F. Pourahmadi (105), Cepheid, Sunnyvale, California 94089
John Quackenbush (127), The Institute for Genomic Research, Rockville, Maryland 20850
Filippo Randazzo (393), Chiron Corporation, Emeryville, California 94608
R. Reynolds (73), Roche Molecular Systems, Alameda, California 94501
Anna-Louise Reysenbach (377), Environmental Sciences Department, Portland State University, Portland, Oregon 97207-0751
R. Saiki (73), Roche Molecular Systems, Alameda, California 94501
Mark Schena (445), Department of Biochemistry, Beckman Center, Stanford University Medical Center, Stanford, California 94305-5307
Armin O. Schmitt (457), metaGen Gesellschaft für Genomforschung, D-14195 Berlin, Germany
Todd Seeley (405), Chiron Corporation, Emeryville, California 94608
Chetan Seshadri (497), Laboratory of Pathology, National Cancer Institute, National Institutes of Health, Bethesda, Maryland 20892
Nicole L. Simone (497), Laboratory of Pathology, National Cancer Institute, National Institutes of Health, Bethesda, Maryland 20892
Gisela Sitbon (521), Professional Genetics Laboratory AB, S-751 83 Uppsala, Sweden
James Snider (329), PE Biosystems, Foster City, California 94404
John J. Sninsky (23), Roche Molecular Systems, Alameda, California 94501
James C. Stephans (297), Chiron Corporation, Emeryville, California 94608-2916
Ann-Christine Syvänen (521), Department of Human Molecular Genetics, National Public Health Institute, S-00300 Helsinki, Finland
Ayly L. Tucker (365), Genentech, Inc., San Francisco, California 94080

Costantino Vetriani (377), Institute of Marine and Coastal Sciences, Rutgers University, New Brunswick, New Jersey 08901-8521

Robert Wagner (95), Gene Check, Inc., Fort Collins, Colorado 80524

Cindy R. WalkerPeach (197), Cenetron Diagnostics, LLC, Austin, Texas, 78744

Robert Watson (263), Roche Molecular Systems, Alameda, California 94501

John Welsh (341), Sidney Kimmel Cancer Center, San Diego, California 92121

L. Western (105), Cepheid, Sunnyvale, California 94089

P. Mickey Williams (365), Genentech, Inc., San Francisco, California 94080

Matthew M. Winkler (197), Ambion, Inc., Austin Texas 78744

Carl T. Wittwer (211), Department of Pathology, University of Utah Medical School, Salt Lake City, Utah 84132

Hong Xin (297), Chiron Corporation, Emeryville, California 94608-2916

S. Young (105), Cepheid, Sunnyvale, California 94089

G. Zangenberg (73), Roche Molecular Systems, Alameda, California 94501

PREFACE

When approached about the possibility of editing another PCR volume, I thought the timing was appropriate given the explosion of PCR applications for mRNA quantitation, diagnosis, gene discovery, genomic analysis, and expression profiling. Also, although the newest crop of molecular biologists grew up using PCR routinely, I still find myself devoting significant time teaching first principles of PCR to help young investigators troubleshoot specific applications and to guide them through a confusing maze of choices as to which enzyme, buffer, cycling condition, etc., to use for which purpose.

Our objective in compiling *PCR Applications: Protocols for Functional Genomics* was to combine expert advice with contemporary protocols and perhaps a few personal reflections. *PCR Applications* makes a perfect companion to the earlier sourcebooks, *PCR Protocols* and *PCR Strategies,* because completely new chapters update the reader with the latest practical advice, theoretical information, and protocols without repeating the well-established PCR protocols in the earlier books.

This book, which follows the protocol format of the previous volumes, is divided into four sections. The first section, Key Concepts for PCR, contains chapters that provide practical and theoretical information needed to effectively navigate through the explosion of PCR-based techniques. The second section, Quantitative PCR, provides examples and protocols for real-time quantitative PCR and procedures for measuring gene expression. The last two sections, Gene Discovery, and Genomics and Expression Profiling, contain chapters of unique interest in the postgenomic scientific era.

We hope that, like *PCR Protocols* and *PCR Strategies, PCR Applications* will serve as a sourcebook and a practical tool for a wide range of research applications. If it causes you to think differently, then we will have succeeded.

The editors thank all the authors who contributed to this volume. We especially extend our thanks and appreciation to Emelyn

Eldredge, Acquisitions Editor at Academic Press, who inspired us to work together on *PCR Applications.* This book would not have been completed without her encouragement, dedication, and hard work.

Michael A. Innis (for the editors)

Part One

KEY CONCEPTS FOR PCR

1

OPTIMIZATION OF PCR: CONVERSATIONS BETWEEN MICHAEL AND DAVID

Michael Innis and David Gelfand

Our chapter titled Optimization of PCR, which was published in *PCR Protocols* 8 years ago, remains as valid today as it was then. Since that chapter was written, however, literally thousands of journal articles in which PCR was used have been published. While the basic PCR technique has not changed dramatically, predictably, numerous and diverse innovations have occurred, including the availability of multiple thermostable polymerases and proofreading polymerases; multiple hot start strategies; long-range PCR; real-time quantitative PCR; PCR-derived cDNA libraries, PCR strategies for generating normalized libraries, subtractive libraries, and representational difference analysis; buffer optimization, use of cosolvents, and mixtures of enzymes; and optimized primer design tools. In seeking to update the previously published chapter on PCR optimization for this book, we met for dinner at David's house and recorded our conversation. After reviewing the transcribed tape, we decided that the dialog might make an interesting and appropriate format for this chapter.

The efficiency of PCR is controlled by many parameters, such as polymerase type, buffer type, primer concentration and stability (T_m), dNTP concentration, cycling parameters, and complexity and con-

PCR Applications

centration of starting template. Because of the numerous applications and the assortment of polymerases available, we purposely omitted a standard PCR amplification protocol. The purpose of this chapter is to provide additional information concerning optimization of PCR to that which was published in *PCR Protocols.*

Hot Start Techniques

MI: In particular, when reaction mixtures are set up at room temperature, high background and/or low yield can occur, especially if amplifying a low-abundance target out of a complex mixture. "Hot start" minimizes the possibility of mispriming and misextension events by ensuring that the polymerase is either not added or not active during setup. What do you see as the relative advantages of the different hot start techniques, such as manual hot start, hot wax, AmpliTaq Gold, *Taq* antibodies, etc.?

DG: I think the least desirable of all of these choices is manual hot start for three or four reasons. Aside from burning your fingers, and things like that, you have these little tubes that you're taking out, you're cooling them down when you take them out, and you're adding something in the cycler in an open tube. That means that all of the tubes won't have the same history, and you have the possibility of cross-contamination of samples.

MI: If you are only analyzing one tube, it's not a problem.

DG: No, but if you're doing 20 or 30 or 96, it's a big problem.

MI: What about the hot wax method?

DG: It's OK. I think it's not as reproducible as AmpliTaq Gold, or apparently the antibody stuff. Oh, you also have two more options. First, the aptamers described by Sumeda Jayasena from NeXstar (Lin and Jayasena, 1997; Dang and Jayasena, 1996), but they are not yet commercially available. A second way of performing hot start is dUTP and UNG because the nonspecific extension products that are initiated during nonstringent setup conditions, when they incorporate dU, get trashed in the 50°, 2-minute UNG treatment step. Thus, those nonspecific primer extension products get degraded with UNG and can't be utilized either as primers or as templates. The dU-containing strands that are synthesized under setup conditions get degraded. Aside from its PCR product sterilization, false positive prevention feature, the dUTP-UNG is another way of doing hot start.

MI: Can you tell us what AmpliTaq Gold is?

DG: Yes. AmpliTaq Gold is reversibly chemically modified; the epsilon-amino groups of lysines are derivitized. The chemical modification is reversed at elevated temperature and low pH. The first U.S. patent issued in October 1997 and a second issued June 1998 (Birch *et al.*, 1997, 1998).

MI: What are some of the advantages of AmpliTaq Gold?

DG: With a preincubation of 95°C for 10 minutes, you reactivate about 40% of the enzyme activity, and you get continuing reactivation during subsequent PCR cycles.

MI: It's a kind of time-released enzyme.

DG: Time release, and the time release effect can be extended by not initially preactivating for as long at 95°C, and performing more cycles; that is, to get more enzyme when you need it, really in the later stages.

MI: We've had trouble using AmpliTaq Gold for RT-PCR.

DG: Sure, it's not compatible with RT-PCR, it's for DNA-PCR. It's compatible with RT-PCR if you're using MLV reverse transciptase plus AmpliTaq Gold, or AMV reverse transcripase plus AmpliTaq Gold. It's not compatible with r*Tth* for two reasons; the buffers, which give highest efficiency of RT-PCR and provide the broadest manganese window, are NOT 10 m*M* Tris, pH 8.3; they're bicine, tricine, and higher pH buffers. That is, the pH of the buffers that are used for single-buffer RT-PCR doesn't drop as much at 95°C as 10 m*M* Tris, pH 8.3 does. You need a pH below 7 at 95°C to reactivate AmpliTaq Gold. The metal ion buffers that are good for single-buffer RT-PCR with manganese don't drop sufficiently in pH to get good reactivation. And there's a second factor. If you're doing RNA RT-PCR, you don't want to go to 95°C before performing cDNA synthesis at 60, 65, or 70°C because, at 95°C in the presence of metal ion, the RNA will fragment from heavy metal-catalyzed chemical degradation. Thus, you don't get good, quantitative PCR with high sensitivity. It's variable.

It's also the case that in RT-PCR, you usually have much lower complexity in the template than you do when you have genomic DNA. A microgram of RNA is what's in 100,000 cells, and a microgram of DNA is what's in 150,000 cells. But the complexity of a microgram of RNA is far less than the complexity of a microgram of genomic DNA.

MI: Because only about 10% of the DNA is expressed?

DG: Right.

MI: One consideration is target copy number. If you have a low abundance copy number you want to use some form of a hot start, which you can effect by wax beads, by antibodies, by AmpliTaq Gold, or by aptamers.

DG: Yes.

MI: And one that is conveniently available and useful is the AmpliTaq Gold strategy. This is straightforward for DNA PCR. For RNA PCR you have to take into account something different.

Long-Range PCR

MI: The first consideration is target copy number and getting hot start. The next thing is the choice of enzyme for the length or product that you're going to want to make. In our second book there was good discussion about long-range PCRs; there have been a lot of developments in long-range PCR since then. Is there anything that you think is worth commenting on?

DG: The significant advance that sticks out in my mind as a major development beyond what Suzanne Cheng and Wayne Barnes found with regard to the buffers, and the enzyme blends (both being important to get long product), was the care and attention that's necessary to prepare intact template. If one is truly interested in genomic samples, and if your genomic DNA template is nicked, it's going to be very challenging to generate 20 and 30-kb PCR products from nicked templates. Thc standard way of analzying template integrity by either native agarose gel or pulse field gel does not tell you what is important. What is important is the nick distribution frequency, not the average size of double-stranded DNA. If your target site has a nick in it, you're not going to be able to amplify it. Suzanne Cheng and collaborators had a paper in *PCR Methods and Applications* that compared a half dozen or so different methods of preparing genomic DNA templates (Cheng *et al.*, 1995). There were a variety of methods that worked well and a whole variety of methods that worked poorly.

MI: But, I don't understand why long PCR wouldn't work by overlap PCR, even if you have nicks. If you have overlapping fragments, even these should go together in long-range PCR. So I don't understand unless the nicks are nonrandom.

G: I suspect some nicks are nonrandom while others are probably random.

MI: If the nicks are nonrandom then it's difficult, but if they're random it should be OK.

G: I gather you're saying that you'd make a primer extension product in one cycle that's then used to prime on another extension product. That requires, of course, that those two primer extension products find each other during the annealing time of a cycle, i.e., that they renature. In the original Alan Wilson paper on jumping PCR in the *Journal of Biological Chemistry* (Paabo *et al.*, 1990), very high concentrations of cloned template were used. Thus, partially extended primers could renature to generate the amplificable target sequence. With 0.1 μg of human genomic DNA, it may take an estimated 8000 hours for half of the initial primer extension products that are complementary to find each other.

MI: OK. So this is important with low starting concentrations.

G: With standard human genomic concentrations of 0.1 μg, you have

15,000 copies, not the 10^9 to 3×10^9 copies with 10 ng of starting plasmid template.

Magnesium Ion Concentration

MI: So, you've raised something that I think is interesting that's the metal ion concentration of the reaction. In our first book we discussed that it may be beneficial to optimize the magnesium concentration.

DG: I think that's still true.

MI: So, what do you think, because I've always thought that it's not so important to optimize the magnesium concentration. It's more important to minimize other divalent cations that are in your reaction. I found that a chelator in the buffer obviates the sharp dependence on magnesium concentration above the threshold that is required for polymerase activity. So I see no difference between the 2 and 5 m*M* if I've chelated my magnesium solution.

DG: Oh, OK. We don't do that so therefore I didn't know that.

MI: I've always made my magnesium solutions with EDTA (i.e., 1 m*M* $MgCl_2$, 0.1 m*M* EDTA).

DG: I think that's useful because there could always be cobalt, or manganese or zinc present.

MI: Right, heavy metal contaminants create truncated products at elevated temperature by strand cision.

DG: I do think high magnesium concentration favors several things that are contraindicated for high-specificity PCR. First, high magnesium will enhance the stability of mismatched primers. Second, high magnesium permits enzymes to bind to primer template when they otherwise wouldn't. That is, infidelity is enhanced and artifacts are enhanced at higher magnesium concentrations. Artifacts include infidelity, misextention, and plus one addition. Things that Mother Nature hasn't designed the enzyme to do. If one does a genetic fidelity assay over a magnesium titration, you'll have more infidelity at higher rather than lower magnesium concentration. You'll have better discrimination against misextension at low rather than high magnesium. "Plue one" addition, the nontemplate-directed addition of a single nucleotide at the 3′ end of a blunt-end duplex, is enhanced at higher magnesium concentration, minimized at lower magnesium concentration. If you want to have plus A addition for cloning with TA vectors, then you should choose the high end of the magnesium window. Or, if you want to do mapping studies with microsatellites and you don't want to have a mixture of blunt-end and plus one products, you may tend to use higher enzyme concentration than you otherwise would and higher magnesium concentration than you otherwise would. In addition, there's the trick from the Prostate Investigation Group at NCI of modifying the 5′ ends of the nonfluorescently labeled primer to favor plus one addition (Brownstein *et al.*, 1996). The ability of a nonproofreading enzyme to do plus one addition is controlled by enzyme concentration, magnesium ion concentration, as well as the sequence of the end of the duplex. The NCI group carried out many studies relating to the nature of the sequence at the end of the duplex to either favor or disfavor the plus one addition.

MI: Is plus one always A?

DG: It's almost always A, but the propensity or rate of addition of A is affected by magnesium concentration, cycling time, and the sequence at the end of the duplex. I don't remember the detail, but if for example, it's a C at the 5′-prime end of the opposite primer, so it's a CG then you are more likely to add plus one than if it's GC.

Reaction Components

MI: Are there any other components of the reaction that are critical?

DG: Well, there's been an increasing use of metal ion buffers whether it's for RT-PCR or DNA PCR; tricine and bicine. There is of course the single RT-PCR "EZ buffer." But also the "XL buffer" has a much lower $\Delta pH/\Delta t$. The drop in pH per degree temperature increase is much less in metal ion buffers than in Tris buffers. That minimizes, I believe, depurination at elevated temperature and subsequent fragmentation. That is, these buffers minimize damage to long primer extension products at high temperature by having the pH well above 7 at 95°C.

MI: So for long-range PCR, that's an important consideration, as well as for RT-PCR, for the same considerations.

DG: Well, actually no, I think for different considerations, because by the time you're heating to 95°C, you no longer have RNA around, and it doesn't matter. It's because these buffers are metal ion buffers as well as pH buffers, and don't require as tight a manganese window. There's a very narrow manganese window with Tris buffer. It becomes a very much broader manganese window with a bicine or a tricine buffer because they are metal ion buffers.

DG: It's not quite a joking mode, but for those companies that provide monoclonal antibodies to a thermostable enzyme, that's an additional component that goes into PCRs. There's been some suggestion but we haven't found it to be true, that thermostable pyrophosphatase is a benefit in PCR. The concentration of pyrophosphate that is required to inhibit standard PCR is far higher than the concentration of pyrophosphate that is generated in PCR, or generatable in PCR. Thus pyrophosphorolysis doesn't have a significant role to play in standard PCR. It may help some in very, very long product PCR. Where it is critical to minimize pyrophosphate concentration, of course, is in cycle sequencing, particularly with the new mutant sequencing enzymes which incorporate ddNTPs very well.

High-Fidelity PCR

DG: Let's talk about high fidelity PCR with proofreading enzymes.

MI: Yes, I think high fidelity is a great topic. Some of the considerations for high fidelity is the choice of enzymes and/or choice of conditions. So, why don't we talk about the choice of enzymes first and then go on to discuss optimal conditions for each enzyme?

DG: There are commercially available thermostable proofreading enzymes and thermostable nonproofreading enzymes, as well as blends of proofreading enzymes and nonproofreading enzymes primarily for extra-long PCR. Under appropriate conditions of use, the proofreading thermostable enzymes that are available from a variety of suppliers can certainly provide a high fidelity of cloned PCR product. The fidelity of cloned PCR product is important in those instances where you're cloning for expression, in molecular biology or you're cloning and sequencing to look at variation of related sequences from the target sample, where it may not be possible to derive the information that is needed by direct sequencing of the PCR product. In other cases, however, if you're interested in sequence information of a target sequence, the best thing to do is PCR it and sequence it directly. You really don't need to clone it. It's easy to directly cycle sequence PCR products. It's no longer a challenge. It's convention. I think that's a big change in the last 3 or 4 years for a variety of reasons.

Actually, we started doing that, as you know, on a G protein receptor project in '89.

Yes, we did do it in '89, but it wasn't "routine" then.

Now it's routine . . . preferred, I think because of the AmpliTaq FS enzyme (fluorescent sequencing enzyme) and the change in chemistry that makes it much easier to directly sequence PCR products. In those instances where you must clone and sequence or want high-fidelity replication of the target, there is *Pwo, Pfu, Kod,* Vent, and there's Deep Vent, and others, but they can really pose challenges in using them.

There's no free lunch!

I think we were very lucky when we first used *Taq* DNA polymearse. There was no proofreading activity. It would have been a lot harder to get PCR generally accepted if it had been as challenging initially as the proofreading enzymes are to work with. That's because proofreading enzymes degrade single-stranded DNA. That's how Mother Nature designed them to work, to remove mismatches. The primers are single-stranded DNA, and that means that the specificity that the user has designed into the primer can be changed during the course of the PCR. That is, having primers that are mismatched to the intended target sequences means they're not readily extendible.

I told Kary Mullis when he first presented the idea of PCR at the meeting in Monterey that I thought it was a great idea but on a practical basis it might not work if you were using Pol I, which is what he was proposing to do, because the 3′ to 5′ exonuclease proofreading activity would degrade the primers.

DG: There is a lot of nonspecificity in Klenow-mediated PCR, and I don't believe all of that nonspecificity comes from nonstringent 37°C annealing. That is, some of the nonspecificity also came from Klenow chewing back on the primers.

MI: Interestingly, Kary quotes that in a bad way now, saying that I told him that his idea would never work. That's not what I said, what I said was that you need an enzyme that won't degrade your primers to make it work.

DG: That is one problem that is partially addressable by chemical modification at the 3′ end of the primer, possibly with phosphorothioate or by methyl phosphonate modification of primer. However, that's not the only time that proofreading activity contributes negatively in PCR. At the end of each cycle, after the enzyme has completed extension and you have a completed duplex, the proofreading activity, of course, can remove the last few incorporated nucleotides and then because there's dNTP present, the polymerase activity repairs the nucleotides back in. It's idling in place. It's able to "repair in" on a template, but it is consuming dNTP and generating pyrophosphate. The problem comes at the end of extension when you heat to denature the duplex you've generated. You denature, and you now have a hyperthermophilic polymerase with an inherent proofreading activity and the ends of your PCR product have been denatured. Until it's cooled back down and primer has bound, that is during that temperature transition of up to 95°C and back to whatever the annealing temperature is, 50–60°C, the proofreading polymerase can be chewing at the 3′ end of the primer extension product which it last synthesized. Thus shortening it somewhat, an ill-defined number of nucleotides.

MI: That's why the yields are often lower with a proofreading enzyme.

DG: And why, if you go too many cycles, it can go away. PCR products can come up and then disappear. If the primer population is being shortened at it's 3′ end, and at each cycle you're nibbling some at 3′ end of the newly synthesized products, you may no longer have the 20 or 25 nucleotides of primer complementarity. However, proofreading DNA polymerase can provide a useful benefit if you have a clone you've sequenced and you want to do some molecular biology of assembling, recombining. You're starting with a nanogram of plasmid, and you only have to go 10 or 15 cycles to mutagenize or assemble the desired clone. The issue is important when you have something that is very low-copy number and you have to do 30, 35, or 40 cycles to get a cloneable product level because of very low copy number; it becomes a challenge.

Right, so for the reader here, in making your enzyme choice you really have to first determine what your application is going to be. If you need a short segment from a plasmid, you might want to use a high template and keep the cycle number low and then it's OK to use proofreading enzymes. But if you want to have product from genomic DNA, with a high efficiency, you should choose an enzyme that lacks the proofreading activity.

If you want to do long-range PCR, you use a mixture of proofreading and nonproofreading polymerases because when your reaction stalls you need to chew back and get another running start at it.

You have to remove the mismatches efficiently, then extend the primers. There's no choice, because the proofreading enzymes alone won't do it and nonproofreading enzymes alone won't do it. There is another of problem with the archae proofreading enzymes, features that are important in doing the same thing over and over. For example, if you're amplifying the same target sequence again and again in the laboratory and you're concerned about carryover. The archae enzymes are incompatible with dUTP. Thus, it's not possible to use dUTP-UNG for the product carryover sterilization or hot start features of dUTP-UNG.

Right, so while we're on dUTP-UNG why don't we just talk about it.

We've never discussed it in terms of an optimization, when would you choose to used dUTP in your PCR amplifications? Most people I know don't incorporate dUTP in their standard PCRs. So if they're not in the reference lab setting, is it important? What's your experience here?

Sure, it hadn't occurred to me back in that context but you're absolutely right. I very much think that it's an absolute requirement for all reference labs, or diagnostic labs, whether it's human (genomic), genetics diagnostic, or infectious diseaes diagnostics. With routine, repetitive PCRs or RT-PCRs, dUTP-UNG enormously mitigates the problem of PCR product carryover contamination. It still doesn't address the issue of sample mix-up. But that's a separate issue. If one's doing a one-off, or an occasional, intermittent, different PCR it's probably not necessary. If one is doing similar primer sets often, and doing low-copy, high-sensitivity amplification, one needs to seriously consider routinely employing either dUTP plus UNG or at least just dUTP. If one finds that they have a problem, they can recover from it. You say hey, well, OK you have a problem, fine, I will simply replace all my reagents. I'll get all new primers, I'll get new tubes of enzymes, I'll make new buffers, I'll make new master mix. And if I clean up the air and bench top with bleach and alcohol, I'll be OK. And change all my pipettes. Well, that's an awful lot of hassle to go through. While it's easy to make buffers, and it's easy to get another tube of enzyme, primers are sometimes a little more difficult. Because it's harder to check those as being the source of the contamination. You can always pull another tube of enzyme from someplace else, or one that's never been opened. It's a benefit for human genomic samples, particularly samples in microwell dishes. The inability to use dUTP-UNG is one of the reasons I don't favor MLV or AMV for RT-PCR. Since r*Tth* efficiently incorporates dUTP, you can fully utilize the benefits of a single enzyme, single tube, single buffer, and dUTP-UNG sterilization.

MI: Just on the record David, I think that nonproofreading enzymes under appropriate high-fidelity conditions give you maximum efficiency for every combination.

DG: I agree completely Michael. Two thumbs up.

MI: Yeah, great, but it's very difficult to convince other people of the logic of that when their superstitions and their lack of following directions for implementing high-fidelity conditions lead them to have different results. Why don't we briefly summarize the enzymes that you're familiar with and what the basic properties are and under what conditions you'd choose to use one versus another?

DG: The enzymes I am most familiar with are *Taq*, r*Tth*, Stoffel Fragment, AmpliTaq Gold, blends of r*Tth* and Vent as the r*Tth* XL product line, both for DNA templates and a separate product for long RNA templates. Those are the ones we use most often. But I know there's

Pfu from Stratagene, and *Pwo* from Boehringer-Mannheim, and Vent and Deep Vent from NEB. There are also blends of *Taq* from Boehringer-Mannheim, Stratagene has *Taq–Pfu,* and a Japanese company, Takara, I think, has *Taq, Pfu,* and *Taq–Kod* blends.

G: *Pwo* and *Pwo* plus X is Boehringer-Mannheim. I think they also have an AMV *Taq–Pwo* blend for RT-PCR. And LTI has some blends. They all provide protocols on how to use them, and I think they all work. However, I don't think you can use a blend or just a proofreading polymerase and assume that the product will have 100% fidelity. That is, when fidelity is important you still have to sequence your clone.

II: Good advice. Sequence your clones.

G: Think about what happens in the cell where not only are there proofreading polymerases, but there are also multicomponent mismatched repair systems. And there are still mutations. Nothing's always perfect. Your bias initially reflected that you prefer to use high-fidelity conditions with nonproofreading enzymes. The literature is replete with fidelity analyses of *Taq*-mediated PCRs. Some of those papers have reported error rates of 1/400 to 1/500 per nucleotide sequenced and others with no errors for 15,000 nucleotides sequenced and one error for 22,000 nucleotides sequenced. It's all using *Taq.* I don't think it was AmpliTaq from supplier A or supplier B that made the significant difference. It was the conditions under which it was used that made the difference.

II: I think we need to stress that because I think that you, and I and maybe five other people on this planet have an appreciation for that, outside of the company you work for. And for the rest, it is Black Magic and/or superstition. Or it's worked for Fred down the hall so it'll work for me.

G: I think one has to optimize conditions when using nonproofreading enzymes *Taq* or r*Tth,* in high-fidelity conditions to minimize errors. One uses low rather than high magnesium, and extension times that are short rather than long. Then, what one is doing is taking advantage of what may be termed misextension kinetics.

II: Also, low dNTP concentrations ($<50\ \mu M$) help with fidelity.

DG: Low balanced dNTP and minimal magnesium. For a nonproofreading enzyme to generate an error in a PCR product the enzyme has to make two mistakes. It has to misinsert and it has to misextend. One can control misextension by low dNTP, low magnesium, and minimizing extension times. The enzyme prefers to extend things that are matched and doesn't like to extend things that are mismatched as much. Rather than being overly permissive for extension time, you set your extension time to be one that is just sufficient for minimally necessary extension, like 95% of the primers get extended and the others don't, rather than extension times that are 5 or 50 times longer than necessary. You can even get higher sensitivity than when using a proofreading enzyme because Mother Nature didn't create proofreading enzymes with PCRologists in mind.

MI: Right. It's really hard to beat *Taq* for the combination of specificity, sensitivity, and yield. So, like you said earlier, we were incredibly lucky to have discovered the best enzyme first, and then to have enhanced it by adding other enzyme activities to it.

DG: A brief note on Stoffel Fragment. Stoffel Fragment excels for allele-selective PCR, for rare sequence PCR, and for rare mutation detection because it discriminates against misextention far better than *Taq* does. It's much easier to develop allele-selective PCR or mismatched selective PCR based on 3′-terminal mismatched primers. Stoffel Fragment has a more stringent requirement for a perfect match primer/template in the active site than is the case for full-length *Taq*. Finally, there are a few examples where the 5′ nuclease activity of wild-type *Taq* can cut the template. Those examples are ones in which the template assumes a secondary structure that looks like what we would call a TaqMan substrate or a 5′ nuclease substrate. When *Taq* clips the template in a region of secondary structure, you make a runoff product and block PCR. Well, Stoffel Fragments will not do that. AmpliTaq CS has a point mutation that eliminates 5′ nuclease activity; however, AmpliTaq CS is not commercially available except in a manual sequencing kit.

DG: Stoffel Fragment is the enzyme of choice for RAPDs or AP PCR. RAPD is from DuPont and AP PCR was developed by the group at the La Jolla Cancer Research Foundation. Stoffel Fragment is much better for those applications than full-length *Taq* enzyme; probably for two reasons. First, analysis is by size polymorphism. When your products are small you have better discrimination of size polymorphism than when your products are big. Stoffel Fragment tends to preferentially amplify smaller products and not longer products. Thus, you don't get smearing or background. Second, I think there's probably too much misextension under cycling conditions that are used in RAPDs and AP PCR (low temperature and long extension times). These cycling conditions are very nonstringent and probably there is too much misextension going on which masks the polymorphic differences that you'd like to see. Stoffel Fragment, because of its impaired ability to extend mismatches, cuts down the background so that the polymorphic differences are more readily detected.

II: What about AmpliTaq FS?

G: AmpliTaq FS is different from AmpliTaq in that it has two mutations. One mutation in the 5′ nuclease domain eliminates 5′ nuclease activity and the second mutation F667Y in the active site markedly enhances Kcat for incorporation of dideoxynucleotides. Thus, you only need an exceedingly low concentration of ddNTPs and you can raise the total dNTP concentration to be more favorable for primer extension. The second mutation was found in our Designer DNA polymerase screen, an *in situ* colony-based screen for mutants which incorporate ddNTPs much more efficiently than does the wild-type enzyme. Stan Tabor at Harvard found the same amino acid change by an entirely different route—splicing and substituting pieces of T_7 DNA polymerase into the Klenow Fragment of *E. coli* DNA polymerase.

Fluorescent Dye Labeling

II: What about a *Taq* that incorporates fluorescein dye-labeled nucleotides more efficiently?

G: We have obtained mutants of *Taq* with improved ability to incorporate fluorescein dye family analogs. Different fluorescent dye-labeled nucleotides are incorporated with markedly different efficiencies. The rhodamine dye family labeled nucleotides are generally incorporated more efficiently than normal nucleotides. In contrast the fluorescein dye family nucleotides, and the cyanine dye family nucleotides are usually incorporated much, much less efficiently than normal nucleotides. The discrimination is far more marked for pyrimidines than for purines. In our colony-based *in situ* genetic screen we looked for mutants of *Taq* with a markedly improved ability to incorporate the TET (fluorescein)-labeled dCTP. We found a mutant which was substantially improved in its ability to incorporate fluorescein dye family dideoxynucleotides for cycle sequencing. This mutation improved the relative discrimination against T_7 terminators. And, hence, greatly improved in the ability to utilize the T_7 terminators in a sequencing reaction. Fortuitously, it turns out that this mutation also improved in the uniformity of incorporation of rhodamine dye family terminators, although we were not screening for that. In addition, we found that this mutation significantly increased the extension rate about four–sixfold of F667Y *Taq*.

Primer Design Factors

MI: Good. I want to switch to the selection of primers, particularly in the context of choosing primers for real-time kinetic PCR. How do you optimize primer selection so that you maximize efficiency. What suggestions do you have on primers in general?

DG: I've by and large been using the program Oligo 5.0 to facilitate primer design, with the notion of trying to match the T_m/T_d of the primers to each other. If it's for multiplex PCR, all of the primers need to be the same, whether it's 65°C or 68°C, but have them all be the same. As well, avoid long runs of Gs in the primers, complementarity at the 3′ ends, and complementary at the 5′ end. In addition, Randy Saiki (unpublished data) found some time ago that one could minimize "primer dimer" in multiplex PCR and markedly improve sensitivity by having primers that end (3′) with AA.

MI: What annealing temperature should be used for primer design?

DG: 65°C or 68°C, let's say 65°C. That's a minimal estimate of what is permissible. What provides maximum specificity is always higher than what the Oligo program predicts.

MI: Do you use a program that calculates the effects of nearest neighbor bases in the analysis?

DG: Yes, nearest neighbor analysis, salt and glycerol concentration. We often use glycerol in our RT-PCRs (see Chapter 10).

MI: Do you have a hot start strategy for RT-PCR that allows you to have both primers present during the reverse transcription reaction? What are the considerations that one might undertake in optimizing for RT-PCR?

DG: We have several strategies for enhancing specificity through primer design. I just mentioned Randy Saiki's observation of, where possible, designing all primers to end AA. Steven Will and colleagues in RMS Chemistry Department have developed additional 3′ modifiers that markedly enhance specificity, reduce primer dimer, improve sensitivity, and are quite useful in KTC amplifications. These will be described in publications in a few months.

MI: Great, for me it's a major stumbling block if you want to set up a homogenous RT-PCR.

DG: We do them all the time. In addition, Kang and Holland at U.C. Davis (see Chapter 27) are doing experiments with yeast and 85% of all the primer pairs worked just great. Very few had background in KTC PCR using ethidium bromide single-tube, single-buffer, single-enzyme RT-PCR.

MI: We wrote in our first chapter about denaturation temperature, cycling time, extension time, the virtue of two vs. of three temperature PCR amplifications. What has been your experience with two temperature PCR? I generally have really good success when I design my own PCRs and I typically run them using just two temperatures (annealing/extension and denaturation). But others have had a lot of failures. When I suggest to other people to design their reactions according to first principles, some come back saying that it doesn't work the way I'd suggested. Usually, I discover that they have changed an important variable, which indicates to me that they didn't understand the principles. What is your opinion on cycling temperature parameters?

DG: I agree with your experience. My bias is that you'll always get maximum specificity at maximum permissible stringency. Using a lower temperature than necessary will favor extension from incorrectly primed templates. If you know the sequence of the target binding sites, and you know it is perfectly matched to your primer, use the highest possible stringency at the highest permissable temperature. If your target is not perfectly matched to the primer sequence, because the target sequence might be variable, then find out at what lower than maximally permissible temperature you need for maximum sensitivity. Then, after the first four or five cycles raise the temperature to match that calculated for the primers. Start out, for example, with three-temperature PCR if your primers aren't matched. After a few cycles you've incorporated your primers and your target binding site is now perfectly matched to the primer. It doesn't make sense to me to continue to use low stringencies that are far below that which is maximally selective. Related, is of course, if you're doing degenerate PCR priming you start out at low temperatures and wish to incorporate 8 to 10 nucleotides in the 5′ tail of your degenerate primer, now allowing you to use much higher stringency.

MI: What do you think of touchdown PCR?

DG: Not much. If you're starting at too high a stringency and you're not priming, you're doing all these cycles and nothing has happened. As you don't know when to stop dropping the temperature and as you continue to drop the temperature, you're not using the number of cycles appropriate for the target copy number. There are many ways to do PCR and there's not only one right way to do PCR. I can understand people who are making some new primers and are just going to do it once or twice and want to get a product and they'll do touchdown PCR. Fine. What really befuddles me is people who do touchdown PCR with primers that perfectly match sequences, such as amplifying inserts with vector-specific primers. This seems truly bizarre. And usually the same people perform far too many cycles of PCR. We know that an appropriately optimized PCR can detect a single copy of template sequence in a single PCR even if there's high-complexity genomic background in 40 to 42 cycles. You don't need even half that many if you are starting out with a ng of plasmid DNA.

MI: Another aspect of optimizing PCR is the denaturation temperature of the target DNA. Since our first chapter there's been a lot learned about the use of cosolvents in PCR.

DG: Yes, yes, the first chapter of the second book.

MI: I don't know if there's much to add to that chapter.

DG: I guess the only thing to reemphasize is that cosolvents are used primarily because they can affect strand separation temperature (T_{ss}), i.e., lower the strand separation temperature requirement to effect complete strand separation. Glycerol, formamide, and DMSO all lower T_m and T_{ss}; however, they do other things as well. They can inhibit enzyme activity, slow down extension rate, and adversely affect enzyme thermostability, or, conversely enhance enzyme thermostability. That is, glycerol enhances enzyme thermostability, and formamide and DMSO negatively impact enzyme thermostability. It makes most sense when amplifying GC-rich targets to start out using 5%, 10%, or 15% glycerol because not only will that lower the strand separation temperature, let's say $2\frac{1}{2}$ to 3° for 10% glycerol, but that concentration of glycerol will also markedly enhance thermostability. If 10% glycerol in insufficient, then titrate in either DMSO or formamide rather than starting out with 5% or 7% DMSO, which will certainly affect strand separation temperature maybe half a degree per percent, but simultaneously, adversely affect enzyme thermostability. A blend of 10% glycerol and a little bit of formamide or DMSO, as many papers have reported, gives a happy medium of adequate enzyme thermostability and the increased denaturation potential needed. Long PCR product formulations have organic cosolvents in them to assist in the denaturation of 20-kb products that may be quite GC rich.

DG: One thing that has always been an idea, a notion, that I think we should use. That is if you're studying a particular amplicon, it's probably worthwhile to determine what is the minimum necessary T_{den} for that particular amplicon. What is the temperature that you must reach to get complete strand separation? Let's say you routinely use 94°C, or 95°C, but if one wants to truly optimize the PCR for maximum specificity and you know the T_{den} of your product under the conditions you are using is 88°C, consider not going to 95°C. It's unnecessary. Not so much from the perspective of preserving enzyme activity, but rather if you make T_{den} 89°C or 90°C, many nonspecific, unintended products that you might otherwise generate which have a higher strand separation temperature requirement would fail to amplify. You use T_{den} control as a means of enhancing specificity.

MI: Right. This concept is expanded in Tom Myer's chapter in this book on RNA-specific PCRs (see Chapter 10).

MI: I hope we have encouraged people to experiment with different buffers, enzymes, and enzyme blends, as well as cycling parameters.

MI: Thank you, David, for dinner and for this discussion. Lets look at the status of PCR in another 5 years or so?

DG: OK

References

Birch, D. E., Laird, W. J., and Zoccoli, M. A. (1997, October 14). U.S. Patent No. 5,677,152. Nucleic acid amplification using a reversibly inactivated thermostable enzyme.

Birch, D. E., Laird, W. J., and Zoccoli, M. A. (1998, June 30). U.S. Patent No. 5,773,258. Nucleic acid amplification using a reversibly inactivated thermostable enzyme.

Brownstein, M. J., Smith, J. R., and Carpten, J. D. (1996). Modulation of non-templated nucleotide addition by *Taq* DNA polymerase: Primer modifications that facilitate genotyping. *Biotechniques* **21,** 1004–1010.

Cheng, S., Chen, Y., Monforte, J. A., Higuchi, R., and van Houten, B. (1995). Template integrity is essential for PCR amplifications of 20- to 30-kb sequences from genomic DNA. *PCR Methods Appl.* **4,** 294–298.

Dang, C., and Jayasena, S. D. (1996). Oligonucleotide inhibitors of *Taq* DNA polymerase facilitate detection of low copy number targets by PCR. *J. Mol. Biol.* **264,** 268–278.

Lin, Y., and Jayasena, S. D. (1997). Inhibition of multiple thermostable DNA polymerases by heterodimeric aptamer. *J. Mol. Biol.* **271,** 100–111.

Paabo, S., Irwin, D. M., and Wilson, A. C. (1990). DNA damage promotes jumping between templates during enzymatic amplification. *J. Biol. Chem.* **265**(8), 4718–4721.

2

THE CONVERGENCE OF PCR, COMPUTERS, AND THE HUMAN GENOME PROJECT: PAST, PRESENT, AND FUTURE

John J. Sninsky

Nineteen eighty-four was a memorable year. The polymerase chain reaction (PCR) was described for the first time at the Cetus Scientific meeting in Monterey, California; the Alta Summit meeting in Colorado was credited as the first significant event when the sequencing of the entire human genome was seriously contemplated; and the infamous commercial for the MacIntosh computer was aired. In a forshadowing of things to come, Russel Higuchi and coworkers reported the molecular cloning of fragments of the quagga genome, an extinct member of the horse family, using what had then become routine recombinant DNA methods; Luc Montagnier and Robert Gallo each reported the discovery of the virus thought to cause AIDS in that same year. Since then, PCR, the human genome project, and mouse-directed icon-display computers have become increasingly interwoven. One is hard pressed to imagine each without the other. The recruitment of PCR for numerous genome project tasks permitted the idea to be translated from a nearly unimaginable, prohibitively costly dream to an immense but tractable project. For example, it was recommended that sequence-tagged sites (STSs) be used to avoid the difficulty of comparing data from different laboratories and the logistics and expense of a huge plasmid database. Indeed, these authors referred to the STS map as "the centerpiece of the human physical mapping effort."

Fundamental observations and advances in one discipline when translated into another discipline lie at the heart of significant advances. The conception of PCR was based on the appreciation that the feedback loops of computers could be accomplished in molecular terms employing the repetitious synthetic capacity of DNA polymerase coupled to denaturation of double-stranded DNA and the targeting function of synthetic oligonucleotides or primers. The doubling of amplicons at each cycle in PCR and the doubling of transistors on computer chips, every year, now known as Moore's law, marks an interesting similarity. The translation of this simple but profound idea to the laboratory bench with meticulous and careful experimentation led to the PCR revolution.

PCR and Computers

Analogies can provide insight and clarity. By comparing similar features of two different systems or processes, individuals familiar with one may gain insight into the other discipline. This insight may catalyze future developments.

What do PCR and computers have in common? Amplification. The best known example of an active electronic amplifier is the transistor. The transistor was described first at a press conference in 1948. Integration of numerous transistors on a semiconducting silicon wafer is referred to as integrated circuit (now commonly referred to as "chip") and was first demonstrated in 1958.

In 1965, Gordon Moore wrote an article for *Electronics* magazine in which he predicted that the power and complexity of the integrated circuit would double every year. Moore's prediction has proven strikingly accurate. For example, Fig. 1 depicts the geometric growth of the number of transistors on Intel chips during the past approximately 20 years. Included in Fig. 1 is the exponential amplification of amplicons by PCR. Presenting both geometric processes on the same graph serves to emphasize their parallel evolution. PCR has proven to be a sophisticated molecular biological method, routinely carried out in labs to gather a wealth of data. The sequence informa-

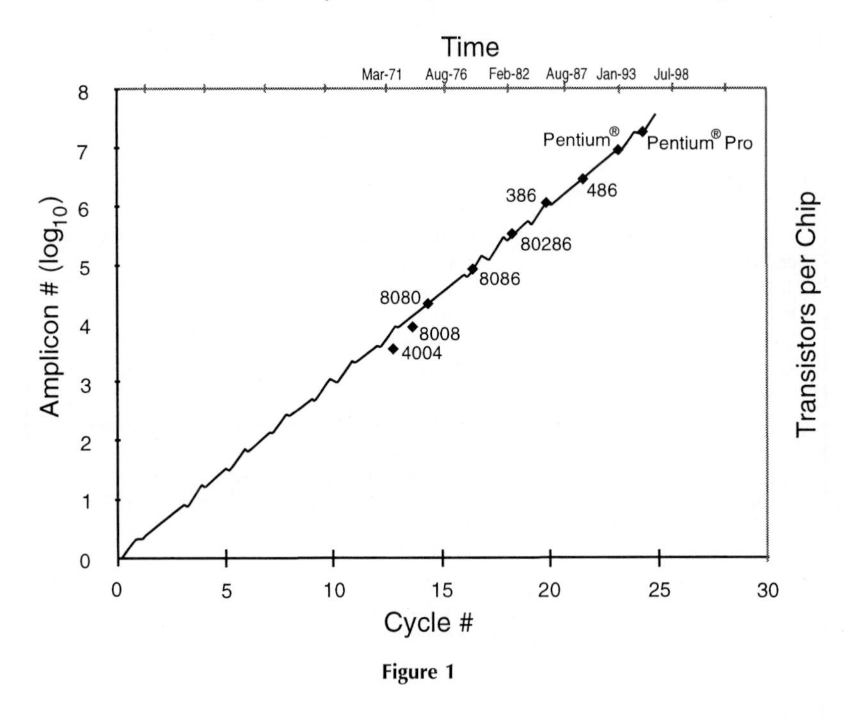

Figure 1

tion generated by PCR is rapidly providing the foundation for our understanding of genes and their functions.

Computers and PCR display marked analogies (Table 1), despite originating at different times in this century (1946 and 1984, respectively). Computers began as sophisticated devices appreciated and utilized by a select few. Over time, computer use expanded to the layperson and it is now in household use. PCR is following a similar path, having originated as an experimental exercise and progressed to a general workhorse found in nearly every molecular laboratory.

I now examine a few of the similarities between PCR and computers. Although the "engine" of a personal computer (PC) is its central processing unit (CPU), PCR is driven by DNA polymerase. Just as the versatility of the CPU is broadened with graphics and video accelerators (coprocessors), PCR versatility is extended with enzymes containing activities other than DNA polymerase, such as enzymes with different combinations of functional activities, mutated DNA polymerases, and the so-called "designer" DNA polymerases (David Gelfand). Recently, *Taq* DNA polymerase has been modi-

Table 1

Computer and PCR Analogies

Computer	PCR
Hardware	Thermal cycler MW plate readers and washers Direct optical monitoring
Feedback loops	Cycles
Clock speed (MHz)	Cycle times
Operating system	Buffer
CPU	DNA polymerase synthetic activity
Coprocessor	3′ nuclease activity (proofreading function—long PCR) 5′ nuclease activity (detection strategies) Uracil glycosylase
Bus	Optical fibers/CCD (video)
Applications	Targets (e.g., genes, pathogens)
Search algorithms (drivers)	Primer pair
Input	dNTPs
Output	Amplicons (signal)
Multitasking	Multiplexing (parallel processing)

fied to be inactive until thermal cycling begins and simultaneously titrate more enzyme into the reaction at later cycles without further reagent additions. Using this approach, near ideal amplification occurs in which only the required amount of reagent is active during the cycling process. Accessory replication proteins and new DNA polymerases will likely continue to aid our ability to refine PCR.

One of the most important functions in a PC is the driver. The correlating activity in PCR are the primers. Just as electronic search engines vary in their effectiveness, so do primers for PCR. By necessity, PCR reactions contain primer in excess to the target sequence. This results in occasional inappropriate pairing and erroneous amplification. Appropriate primer design can sometimes mean the difference between a successful and an unsuccessful amplification reaction. Despite numerous investigators having postulated primer design rules or guidelines, important advances continue to result in new design suggestions.

One of the greatest advances for the PC was the advent of multitasking, or the ability to execute simultaneous applications. The corresponding breakthrough for PCR was multiplexing, in which distinct segments of DNA or RNA can simultaneously be amplified in a single reaction. Microsoft provides examples of two fundamentally distinct types of multitasking: cooperative and preemptive. In the case of cooperative multitasking (Windows 3.1), the operating system relinquishes control to an application. With preemptive multitasking (Windows '95), the operating system is always in control. Cooperative multiplex PCR occurs when the targets are at the same copy number and have equally efficient primer pair systems. The result it simultaneous amplification of targeted gene segments. On the other hand, preemptive multiplexing occurs when the targets exist at markedly different copy numbers. Amplification of the highest copy number target overwhelms the reaction and may terminate when reagents are exhausted. Thus, amplification of the low-number template is precluded.

The size of computers has been reduced from the room-sized ENIAC in 1946 to handheld digital assistants so popular today. Miniaturization of microprocessors and computers has spurred widespread use of these devices. Likewise, not surprisingly, miniaturization of PCR has and will continue to catalyze the expanded use of this powerful procedure. Miniaturization of PCR was approached in two fundamentally different ways. The first feat was reduction of the reaction volume. Original amplification reactions were carried out in 100-μl volumes. An interest in assembling large parallel reactions in limited space resulted in reaction volumes as small as 5–10 μl. The second feat of miniaturization was the modification of thermal cyclers. Once a lab bench behemoth, there now exist battery-operated thermal cycling devices for a small number of reaction vessels.

We have now come full circle. Just as the computer programming loop sparked the conception of PCR, PCR has sparked the harnessing of PCR for computation. Only 10 years after its creation, Adelman described how PCR, thought to be restricted to biological sciences, could be used for computational calculations. This insightful paper was rapidly followed by other studies employing PCR to confirm and extend such an approach. The unfolding of the role of PCR in molecular computation will bruly be fascinating.

Information Space

Even though PCR is a biochemical procedure, it is best thought of as an information (generator) tool; the true potential of PCR and the resulting information will be realized only when this information is assembled in large interactive relational databases. These databases will likely be established by multiple sources. The resulting compendium of cross-referenced information should be accessible to the eventual user of the information regardless of the user's training and background experience. Ideally, the resulting reports will be fully hypertext linked to facilitate examination and validation of the data.

Sequence Space

Victor McKusick's *Mendelian Inheritance of Man* has been replaced with an online version that is imminently more interrogatable. Figure 2 shows a graph of the number of genes identified during the past 20 years; it shows an apparent bimodal distribution. Cursory analysis suggests two different rates of human gene identification with the shift occurring around 1988. One can question whether the Human Genome Project, which began in 1990, has contributed to

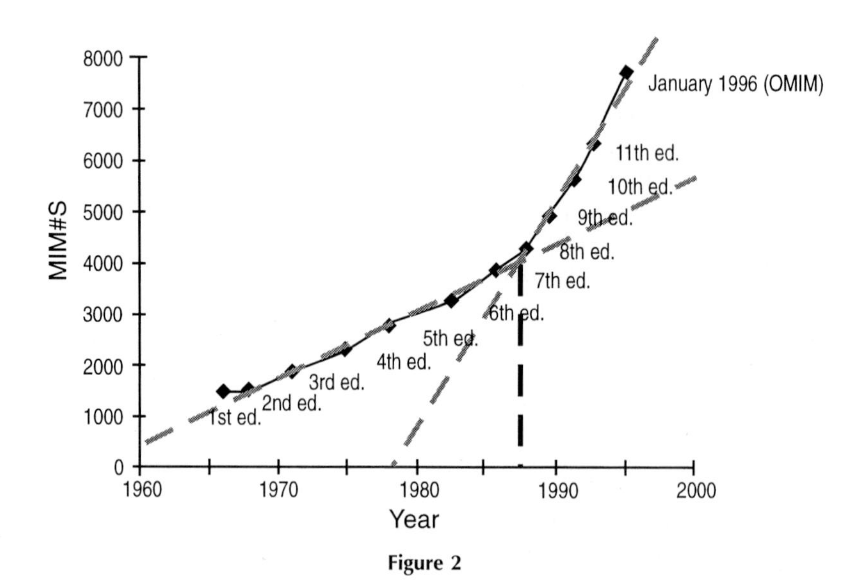

Figure 2

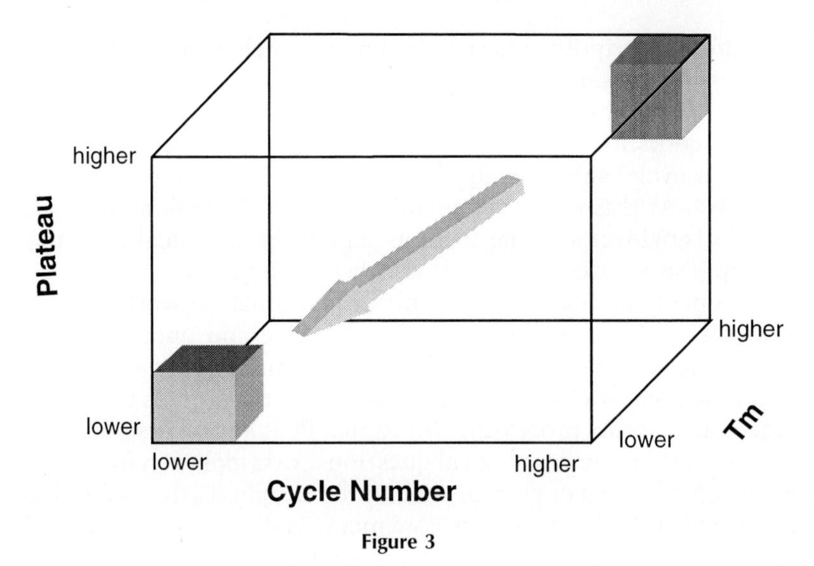

Figure 3

the enhanced rate of gene discovery. I contend that the increased use of PCR—applied toward the Human Genome Project—is the basis for this rapid increase in the identification of human genes.

In a prophetic statement in 1945, Vanavar Bush stated that "The physician, puzzled by patient's reactions, strikes the trail established in studying an earlier similar case, and runs rapidly through analogous case histories, with side references to the classics for the pertinent anatomy and histology." In time, this associative retrieval path will have at its core the nucleotide sequences gathered by PCR. These searches will be trails through sequence space. The ease of PCR and its reduction to a procedure capable of being carried out by a vast number of laboratories highlights the number of investigators inputting data but not assessing this information. The tide will soon turn from the gathering of sequence information to its distribution for diagnosis and application toward cures.

Genomics

We have embarked on a new era in the biological sciences—an era in which the unit of study has changed from the gene to the genome, from the transcription and translation of a single or a small handfull of genes to thousands of mRNAs and proteins simultaneously, and from individual biochemical pathways or circuits to cellular regulatory networks. We each recall rote memorization of the Kreb's cycle (the TCA cycle) and have already linked this simple circuit to other pathways. At this point, the metabolic networks look more like an electrical engineer's wiring diagram than the biochemical conversion of simple substrates by a small number of enzymes.

Our understanding of large multidimensional networks will undoubtedly shed valuable light on pathogenic and oncogenic pathways. Indeed, the catalyst for this fundamentally new way to consider biological studies rests on the development of tools that permit large-scale parallel processing. Multiplex PCR and arrays are critical to probing these new biological questions. Extrapolation from levels of mRNA to levels of protein will not be simple. Early studies have uncovered mRNA–protein discordances. Will the extrapolation be based on single mRNA–protein comparisions or on information from hundreds of mRNA–protein comparisons?

It is interesting to note, given the computer–PCR analogy I discussed, that photolithographic procedures employed in the semiconductor industry are currently being used to generate arrays of *in situ* synthesized DNA chips. These DNA chips have a capacity of hundreds of thousands of probes that can be used for sequence scanning, mutation detection, resequencing, and gene expression analysis.

Gene expression, or mRNA profiling, will help us to understand how genotype translates into phenotype. Key in the multidimensional networks will be those genes that reverberate through the entire network versus those that perturb limited circuitry. These differences will be representative of the biological organism's capacity to accommodate or tolerate environmental and genetic insults.

The Next Millenium

What does the future of PCR look like? Our ability to precisely predict how PCR will unfold is confounded by the continued evolution of this powerful technology. The future of PCR likely holds further advances in (i) multiplexing, (ii) kinetic PCR, (iii) mRNA profiling, (iv) designer DNA polymerases and accessory proteins,

(v) applications for handheld PCR, and (vi) applications for multidisciplinary PCR.

The initial report of PCR described a single amplification of a 110-base pair fragment. Since this report, the length of an amplicon has been extended to 22 kb and the number of simultaneously amplified segments has increased to more than 30 individual segments of nucleic acid. Will we soon be able to amplify hundreds of segments that are all thousands of basepairs long? If so, we will optimize extraction of genetic information in a single reaction. This is particularly heartening because we will be kept busy for several years unveiling the genetic polymorphisms and their associated haplotypes so critical to individuality and disease predisposition.

Deciphering the entire genome will provide only the primary genetic blueprint of human life. This blueprint will provide the foundation on which we will need to build circuits and eventually interactions of the circuits into networks that lie at the heart of cell biology.

This first decade of PCR will be recognized by our use of an incredibly powerful but relatively crude tool. Our vision of PCR in the future is compromised by our past experience and biases. Indeed, the inherent skepticism that precluded the conception of PCR prior to 1984 is likely to continue to dog us. In the future, weaving of PCR, computers, and the Human Genome Project more tightly together into the fabric of science will occur.

3

THERMOSTABLE DNA POLYMERASES: AN UPDATE

Richard D. Abramson

The discovery and subsequent commercialization of the PCR method of DNA amplification (Mullis and Faloona, 1987; Saiki *et al.*, 1988) has focused considerable interest on the DNA polymerases of thermophilic organisms. Many of these DNA polymerases are both thermoactive and thermostable, capable of catalyzing polymerization at the high temperatures required for stringent and specific DNA amplification, as well as withstanding the even higher temperatures necessary for DNA strand separation. An extensive review of the enzymatic properties of thermostable DNA polymerases characterized at the time was published in 1995 (Abramson, 1995). This review provides an update of the field of thermostable DNA polymerase enzymology, with a focus on those recent discoveries that have made a major impact on the PCR technique and are of the most significant importance to the practical researcher.

Recently Characterized Thermostable DNA Polymerases

Several new DNA polymerase enzymes have been identified and characterized from thermophilic eubacteria in the past few years.

PCR Applications

Until recently, although conserved throughout mesophilic eubacteria, multimeric DNA polymerases similar to *Escherichia coli* Pol III holoenzyme replicase had not been identified in thermophilic eubacteria. DNA polymerase III comprises the catalytic core of the *E. coli* replicase, distinguished from other polymerases in the ability of the catalytic α subunit to interact with other replication proteins at the replication fork to form the holoenzyme. McHenry *et al.* (1997) were the first to report the existence of a 130-kDa DNA polymerase purified from *Thermus thermophilus* (*Th*) which was distinct from the 94-kDa DNA polymerase I protein previously identified and cross-reacted specifically with *E. coli* anti-pol III α monoclonal antibodies. The polymerase preparation has a specific activity of 8,000,000 units/mg at 60°C, and when corrected for the presence of contaminating/associated proteins, it is estimated to have an approximate specific activity of 2×10^6, comparable to that of purified *E. coli* pol III. Two additional proteins, tightly associated with the *Tth* α subunit, were found to be homologous to *E. coli* γ and τ subunits of DNA polymerase III holoenzyme, both products of the *DnaX* gene (McHenry *et al.*, 1997). Similarly, Yurieva *et al.* (1997) cloned the *Tth DnaX* gene and, like their *E. coli* counterparts, verified that the two recombinant proteins possessed a DNA-stimulated ATPase activity. These studies suggest that similar to mesophilic eubacteria, a conventional holoenzyme replicase also exists in thermophiles.

A recent report identified two DNA polymerase enzymes from the thermoacidophilic eubacterium *Bacillus acidocaldarius* (De Falco *et al.*, 1998). The single-subunit enzymes, with molecular weights of 117 and 103 kDa, respectively, differ in a variety of properties, including thermostability, thermoactivity, and processivity. In addition, the lower molecular weight protein lacks 5′ to 3′ exonuclease activity, which is present in the 117-kDa enzyme.

Similarly, several new DNA polymerases have been identified in thermophilic Archaea. A new DNA polymerase has been identified in *Pyrococcus furiosus* (*Pfu*), distinct from the previously identified α-like (family B) 90-kDa enzyme. This novel DNA polymerase is composed of two protein subunits, with molecular weights of 69,294 and 143,161 respectively, encoded by tandem genes and transcribed as part of a single operon (Uemori *et al.*, 1997). The DNA polymerase has a strong 3′ to 5′ exonuclease. A homologous DNA polymerase was identified in the methanogenic archaeon *Methanococcus jannaschii* (Ishino *et al.*, 1998). New Archaea DNA polymerases of the family B type have also been identified recently from *Pyrococcus* sp. strain KOD1 (Takagi *et al.*, 1997), *Thermococcus* sp. TY (Niehaus *et al.*, 1997), and *Thermococcus* sp 9°N-7 (Southworth *et al.*, 1996). These proteins all have an approximate molecular weight of 90 kDa, and all possess 3′ to 5′ exonuclease activity. Two family B DNA polymerase genes were also cloned from hyperthermophilic archaeon *Pyrodictium occultum* (Uemori *et al.*, 1995). Additionally, a third family B DNA polymerase gene was identified in *Sulfolobus solfataricus* (*Sso*), encoding for a protein highly divergent at the amino acid level from the two *Sso* family B polymerases previously described (Edgell *et al.*, 1997).

Recent Progress in Thermostable DNA Polymerase Enzymology

Structure Analysis: The Power of the Crystal

The crystallization and structure determination of full-length *Thermus aquaticus* (*Taq*) DNA polymerase (Kim *et al.*, 1995; Urs *et al.*, 1995; Eom *et al.*, 1996) as well as a Klenow-like 5′ to 3′-deficient protein fragment (Korolev *et al.*, 1995; Li *et al.*, 1998) have been reported. The structures of the polymerase domains of *Taq* DNA polymerase and of the Klenow fragment of *E. coli* Pol I are almost identical, whereas the structure of the corresponding 3′ to 5′ exonuclease domain is dramatically altered in *Taq* DNA polymerase, resulting in the absence of this activity. The 5′ to 3′ exonuclease active site of *Taq* DNA polymerase is positioned approximately 70 Å from the polymerase active site in a separate domain, although it is suggested that this is not the position in solution, in which presumably the 5′ nuclease domain is located much closer to the central mass of the polymerase domain (Kim *et al.*, 1995). Crystals prepared in the presence of a blunt-end DNA duplex (Eom *et al.*, 1996) or dNTPs (Li *et al.*, 1998) identified those amino acid residues involved in substrate binding and active site catalysis.

The crystal structure of another large fragment, Klenow-like thermostable DNA polymerase, *Bacillus stearothermophilus* (*Bst*), has also been determined (Kiefer *et al.*, 1997). The structure of the polymerase domain is very similar to the Klenow fragment of *E. coli* Pol I; however, as with *Taq* DNA polymerase, the putative 3′ to 5′ exonuclease domain is highly divergent in sequence retaining only a common fold. Changes in the catalytic residues known to coordinate the binding of the divalent metal ions required for exonuclease activity and in loops connecting homologous structure elements result in the absence of 3′ to 5′ exonuclease activity in *Bst* DNA polymerase. No obvious specific interactions contributing to thermostability were identified within the structure, suggesting that thermostability is a result of more subtle interactions throughout the protein.

Mutational Analysis: Designer Enzymes

Mutational analyses of a number of thermostable DNA polymerases have contributed to a greater understanding of DNA polymerase enzymology and resulted in several new mutant enzymes of practical importance. Mutational analyses of the residues comprising the 5′ to 3′ exonuclease domain of *Taq* DNA polymerase have identified several amino acids which alter 5′ nuclease activity. Merkens *et al.* (1995) identified two residues, Arg25 and Arg74, that when mutated resulted in significantly reduced 5′ to 3′ exonuclease activity while exhibiting wild-type polymerase activity. Similarly, Kim *et al.* (1997) identified Arg74, Lys82, and Arg84 as residues that when mutated to alanine resulted in an 80–90% reduction in 5′ to 3′ exonuclease activity. A similar mutational analysis of the 5′ to 3′ exonuclease activity of *Bacillus caldotenax* DNA polymerase characterized two mutants corresponding to two known *E. coli* DNA polymerase I 5′ nuclease mutations, Gly184 to aspartic acid and Gly192 to aspartic acid, as deficient in 5′ nuclease activity (Ishino *et al.*, 1995). Comparison of those residues shown to be important by mutational analysis to the crystal structure of the 5′ nuclease domain of *Taq* DNA polymerase reveals that most of these amino acid residues do not lie within the proposed active site and, in fact, Arg74, Lys82, and Arg84 lie within a completely disordered loop. Further refinement of the crystal structure of the nuclease domain complexed with either substrate or product is needed to fully reconcile the mutational data with the structure.

One particular *Taq* DNA polymerase mutant, Gly46 to aspartic acid, has been characterized as having wild-type polymerase activity, extension rate, and processivity while exhibiting a more than 1000-fold reduction in 5′ to 3′ exonuclease activity (Abramson *et al.*, 1994). The Gly46 residue, as determined by crystal structure, lies in the middle of an α-helix, apparently well removed from the active site (Kim *et al.*, 1995); however, the residue is highly conserved among DNA polymerase I enzymes (R. D. Abramson, unpublished data). This mutant enzyme (commercially available as AmpliTaq DNA Polymerase, CS from PE Applied Biosystems, Foster City, CA) has an enhanced utility in DNA cycle sequencing, in which optimal substrates for the exonuclease activity are known to form during cycled extension, resulting in the cleavage of both the DNA template and the newly synthesized terminated strand. The combination of wild-type polymerase function (as opposed to the altered polymerase characteristics seen with deletion mutants such as Stoffel fragment) with substantially decreased 5′ nuclease activity results in superior sequencing ladders with high accuracy, even intensity, and low background when compared to wild-type *Taq* DNA polymerase (Abramson *et al.*, 1994).

Random mutagenesis of 39 residues comprising the O-helix substrate binding site of the polymerase domain of *Taq* DNA polymerase identified a number of active mutants as well as those amino acid residues that were either immutable or tolerant of only highly conservative replacements (Suzuki *et al.*, 1996). Two residues found to be immutable (Arg659 and Lys663) and two found to be tolerant only of conservative substitutions (Phe667 and Tyr671) are located by crystal structure analysis on the side of the O-helix facing the incoming dNTP, confirming the functionally active configuration of the polymerase domain. Preliminary evidence indicates that Arg659 and Lys663 interact with the phosphate groups of the incoming dNTP, Phe667 interacts with the deoxyribose moiety of the dNTP, and Tyr671 binds to the template primer.

Many of the active O-helix mutants were characterized and found to have decreased fidelity while exhibiting near wild-type specific activity (Suzuki *et al.*, 1997). Two of these mutants yielded mutation frequencies at least 7- and 25-fold greater, respectively, than that of the wild-type enzyme. Additionally, a preliminary report indicates that at least some of these mutants exhibited increased fidelity, showing greater discrimination against 3′-mismatched extensions compared to wild-type polymerase (Newcomb *et al.*, 1997). Thermostable DNA polymerases with reduced fidelity may have utility in molecular PCR applications requiring increased mutagenesis, such as random mutagenesis and applied molecular evolution. Mutants with increased discrimination against extending primers with 3′ mismatches have utility in the amplification of rare mutant alleles in a background of wild-type sequence as required in the analysis of somatic mutations associated with cancer as well as in those PCR applications generally requiring highly specific primer extension and high-fidelity replication. These mutants may also prove to be useful in PCR DNA fingerprinting applications, such as RAPD PCR, in which fewer but possibly more informative sequences will be amplified with greater reproducibility.

The Phe667 residue of *Taq* DNA polymerase, shown to be tolerant of only conservative substitutions (Suzuki *et al.*, 1996) and proposed to interact with the deoxyribose moiety of the dNTP within the substrate binding site, has been mutated to a tyrosine, resulting in a mutant enzyme with decreased discrimination for dideoxyribonucleotides (Tabor and Richardson, 1995; Kalman *et al.*, 1995). Whereas wild-type *Taq* DNA polymerase has an average incorporation rate for dNTPs that is 3000 times that of ddNTPs, the mutant polymerase actually has a twofold preference for ddNTPs over dNTPs. Conversely, wild-type T7 DNA polymerase, which has a tyrosine residue at the homologous site in the protein, efficiently incorporates ddNTPs, whereas a mutant containing phenylalanine at that position has an average incorporation rate for dNTPs that is 8000 times that of ddNTPs (Tabor and Richardson, 1995). Discrimination against ddNTPs by wild-type *Taq* DNA polymerase has been shown to be the result of a slow rate of phosphodiester bond formation and not initial substrate binding (Brandis *et al.*, 1996). Whether a hydroxyl moiety (provided either by the dNTP or the tyrosine residue) is required to stabilize a catalytically productive orientation or is required for binding an essential ligand such as a metal ion has yet to be determined.

The Tyr667 mutant of *Taq* DNA polymerase has great utility as a DNA sequencing enzyme. Efficient incorporation of ddNTPs improves the uniformity of terminations since discrimination against dideoxynucleotides is sequence specific and reduces the requirement for high concentrations of ddNTPs in the sequencing reaction. This is especially important for automated DNA sequencing using fluorescent dideoxy-terminators, in which unincorporated nucleotides lead to a high background. A deletion mutant of *Taq* DNA polymerase lacking the 5′ to 3′ exonuclease domain and containing the Phe667 to Tyr point mutation (commercially available as Thermo Sequenase DNA polymerase from Amersham Life Science, Cleveland, OH) demonstrated more even terminations and improved base calling with both fluorescent and radioactive cycle sequencing (Reeve and Fuller, 1995; Vander Horn *et al.*, 1997). A similar mutant enzyme containing the Gly46 to Asp 5′ nuclease point mutation and the Phe667 to Tyr mutation (commercially available as AmpliTaq DNA Polymerase, FS from PE Applied Biosystems) showed improved accuracy and termination uniformity in automated fluorescent cycle sequencing and resulted in sequence data of sufficiently high quality as to allow for reproducible heterozygous polymorphism detection (Chadwick *et al.*, 1996; Parker *et al.*, 1996).

Long-Distance PCR: Two Enzymes Are Better than One

Limitations to the PCR process typically prohibited reliable amplification of sequences more than 5–10 kilobases (kb) in length. DNA polymerases lacking 3′ to 5′ proofreading exonuclease typically have a misincorporation rate such that the probability of the polymerase incorporating a mismatched nucleotide that effectively prohibits further extension in the majority of the amplification products becomes very likely in extensions beyond 10 kb. Alternatively, those thermostable enzymes possessing proofreading activity, although demonstrating lower misincorporation rates, are limited in their ability to perform long extensions due to the long extension times necessary for long amplifications, thereby exposing primers and template to extensive 3′ degradation. Extensive secondary structure of the template may also limit extension with proofreading DNA polymerases due to the ability of the enzyme to idle between polymerase and exonuclease mode rather than performing strand displacement synthesis.

Recent reports demonstrating amplification products of up to 42 kb in length take advantage of the properties of both proofreading and nonproofreading enzymes using an enzyme blend containing primarily a nonproofreading thermostable DNA polymerase with a very small amount of proofreading polymerase, contributing a small amount of thermostable 3′ to 5′ exonuclease activity (Barnes, 1994; Cheng *et al.*, 1994; Cheng, 1995). Cheng *et al.* (1994), using an enzyme blend of *Tth* DNA polymerase and *Thermococcus litoralis* DNA polymerase in a 45 : 1 ratio (based on polymerase units), were able to amplify a 22-kb fragment of the β-globin gene cluster from genomic DNA, and in a 125 : 1 ratio they were able to amplify a 39-kb fragment from λ bacteriophage DNA. Similarly, Barnes (1994) was able to amplify a 35-kb fragment from λ bacteriophage DNA using a mixture of a deletion mutant of *Taq* DNA polymerase (Klentaq) and *Pfu* DNA polymerase in a ratio ranging from 160 : 1 to 640 : 1 (based on polymerase units).

The fidelity of these polymerase blends has been analyzed. Barnes (1994) demonstrated an approximately 13-fold increase in the fidelity of a Klentaq/*Pfu* DNA polymerase blend over *Taq* DNA polymerase alone and an almost 2-fold increase in the fidelity of a Klentaq/*Pfu* DNA polymerase blend over *Pfu* DNA polymerase alone. Cline *et al.* (1996) reported a mean error rate of a *Taq*/*Pfu* DNA polymerase mixture (16 : 1 ratio based on polymerase units) that was 30% lower than the mean error rate of *Taq* DNA polymerase but 6-fold higher than that with *Pfu* DNA polymerase alone, whereas a Klentaq/*Pfu* DNA polymerase blend had an error rate that was 3-fold higher than that of *Pfu* DNA polymerase but 70% lower than that seen with the *Taq*/*Pfu* DNA polymerase blend. The improved fidelity and ability to amplify long sequences makes these polymerase mixtures quite useful for a variety of PCR applications. Many polymerase blends are commercially available, including r*Tth* DNA Polymerase, XL (PE Applied Biosystems), Expand DNA Polymerase (Boehringer–Mannheim, Indianapolis, IN), *Taq Plus* DNA Polymerase (Stratagene, La Jolla, CA), Elongase DNA polymerase (Life Technologies, Gaithersburg, MD), and LA *Taq* DNA Polymerase (Takara Biomedical, Otsu, Hsiga, Japan).

Hot Start PCR: Thermally Activated Enzyme Preparations

Improvements in PCR sensitivity, specificity, and product yield were afforded by the introduction of the hot start technique for performing PCR in 1990 (Faloona *et al.*, 1990). Hot start PCR is a simple modification of the original PCR process, whereby the amplification reaction is initiated at an elevated temperature. In conventional PCR amplifications, in which active reaction components are subjected to temperatures lower than the desired annealing temperature, nonintended priming can occur, resulting in nonspecific product formation and primer oligomerization. Once low-temperature nonspecific priming occurs, either during PCR set up or in the initial temperature increase which occurs in the first amplification cycle, these products will be efficiently amplified throughout the remaining PCR cycles. Amplification of nonintended product often results in poor yields of the desired product(s). This in turn can reduce the sensitivity of the experiment either by decreasing the desired amplification signal or by obscuring it with high background. When nonspecific products are avoided the polymerase activity is directed to the desired target only, thereby increasing sensitivity and yield and decreasing nonspecific background amplification (Fig. 1).

Hot start PCR was initially performed manually by adding an essential component of the reaction to the reaction mixture subset only after those reactants were heated to an elevated temperature. However, many researchers found this approach cumbersome and time consuming and the technique was not widely adopted. Chou *et al.* (1992) developed a semiautomated means of performing hot start PCR using a wax bead that when first melted and then cooled, formed a solid wax barrier used to separate an essential reaction component from the bulk of the reagents. The wax layer subsequently melts upon heating during the first PCR cycle and convectively mixes the two aqueous layers, resulting in an fully active reaction, greatly increasing the specificity, yield, and precision of amplifying low copies of target DNA and permitting routine amplification of a single target molecules (Chou *et al.*, 1992).

Alternate methods of automating hot start PCR have also been developed. Sharkey *et al.* (1994) demonstrated the utility of *Taq* DNA polymerase antibodies as thermolabile inhibitors of enzyme activity. At low temperatures (20–40°C), antibody–antigen interaction results in potent inhibition of DNA polymerase activity. As the temperature is raised upon beginning the thermal cycling process, the antibodies are heat denatured, thereby releasing fully active *Taq* DNA polymerase. Preincubation of *Taq* DNA polymerase with such antibodies yielded highly specific PCR reactions with increased sensitivity (Sharkey *et al.*, 1994; Kellogg *et al.*, 1994). Commercially available *Taq* DNA polymerase antibody is offered by many vendors, including Clontech Laboratories (Palo Alto, CA), Life Technologies, and Sigma Chemicals (St. Louis, MO).

A similar method of hot start PCR has been described by Dang and Jayasena (1996) using high-affinity oligonucleotide ligands (aptamers) rather than protein antibodies to selectively inhibit *Taq* DNA polymerase activity at low temperatures. In the presence of ligand, *Taq* DNA polymerase is no longer active at room temperature but retains its full activity at temperatures above 40°C. Unlike antibody inhibitors, oligonucleotide aptamer inhibition is thermally reversible since the ligand does not become irreversibly denatured at high temperatures. The addition of the aptamer to a PCR amplification eliminated the need for manual hot start and improved the efficiency of detection of low copy number targets. Aptamers that

Conventional PCR

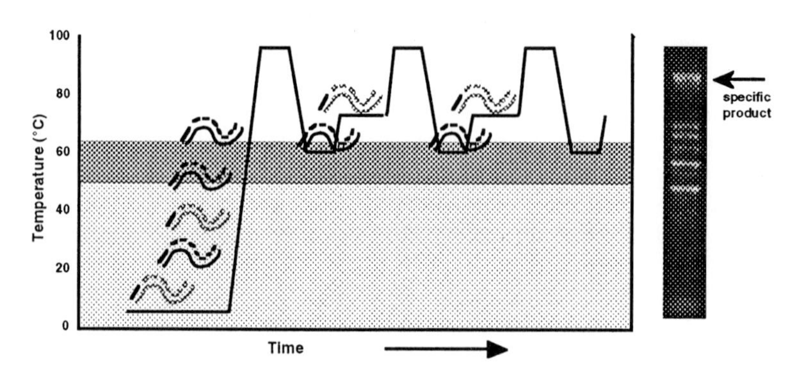

Hot Start PCR

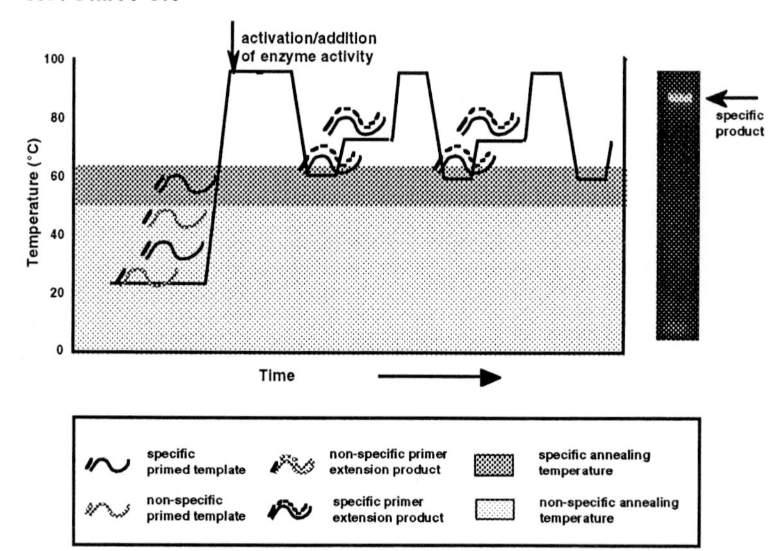

Figure 1 Schematic of PCR setup and initial cycle. The dynamics of PCR setup and initial cycle vary, depending on the sample temperature and the amount of active enzyme present. In conventional PCR, mispriming events that occur on the initial upramp can lead to amplification of nonspecific products because active enzyme is present throughout the entire process. With hot start PCR, active enzyme is present only at temperatures above the specific annealing temperature, which helps minimize extension of misprimed templates.

inhibit other thermostable DNA polymerases, including *Tth* DNA polymerase and the Stoffel fragment of *Taq* DNA polymerase, have also been constructed (Dang and Jayasena, 1996; Lin and Jayasena, 1997).

A modified form of *Taq* DNA polymerase requiring thermal activation has recently been described (Birch *et al.*, 1996). AmpliTaq Gold DNA Polymerase (commercially available from PE Applied Biosystems) is a chemically modified enzyme, such that the modification renders the enzyme inactive. Upon thermal activation the modifier is permanently released, regenerating active enzyme. The resultant hot start reaction provides increased sensitivity, specificity, and yield and has proved especially useful when performing multiplex PCR, in which multiple primer pairs are used to simultaneously amplify multiple sequences in a single reaction (Birch *et al.*, 1996; Zimmermann and Mannhalter, 1998). AmpliTaq Gold enzyme is also useful in RNA PCR, in which Moloney murine leukemia virus reverse transcriptase is used to perform the reverse transcription reaction in the presence of inactive DNA polymerase followed by a heat activation step to allow for active DNA polymerase enzyme to perform the PCR amplification. Incorporating hot start into the reaction allowed for a two-order of magnitude increase in sensitivity, with increased specificity and product yield (Fig. 2).

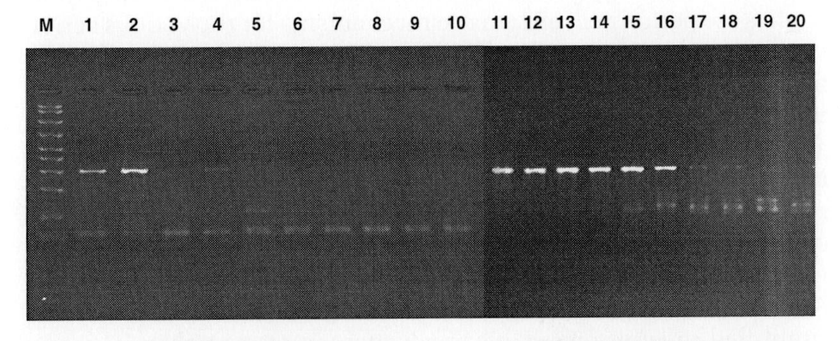

Figure 2 Hot start RNA PCR with AmpliTaq Gold DNA polymerase. Reverse transcription was performed with 15 U of MuLV RT in the presence of either 2.5 U of *Taq* DNA polymerase (lanes 1–10) or 2.5 U of AmpliTaq Gold DNA polymerase (lanes 11–20) in a 50-μl reaction. A 12-min RT reaction is followed by 43 cycles of PCR, amplifying a 308-bp target from pAW109 cRNA. Lanes 1–2 and 11–12, 5×10^4 starting RNA copies; lanes 3–4 and 13–14, 5×10^3 copies; lanes 5–6 and 15–16, 5×10^2 copies; lanes 7–8 and 17–18, 50 copies; lanes 9–10 and 19–20, negative control (0 copies of RNA).

Unlike other methods of performing hot start PCR, activation of AmpliTaq Gold DNA Polymerase can be precisely modulated to slowly release active enzyme over time, allowing enzyme activity to increase with cycle number as template increases. An initial high-temperature pre-PCR incubation step activates approximately 40% of the available enzyme activity, limiting the amount of active enzyme at the beginning of the amplification reaction when low amounts of primer template substrate molecules are present. With each additional cycle, as template increases, enzyme activity increases as more modified enzyme is activated during each high-temperature denaturation phase of the PCR. This "time-release" characteristic of thermal activation results in increased specificity compared to alternate hot start methods in which 100% of the enzyme activity is available at a point in the reaction when substrate is most limiting.

The Future

In recent years, considerable progress has been made toward understanding the mechanism of action of thermostable DNA polymerases. As the crystal structure of these enzymes in the presence of their various substrates/products is refined, and additional mutagenesis studies are performed, enzymatic mechanisms for the various activities these proteins possess will be proposed and experimentally verified. The potential for constructing mutant thermostable DNA polymerases subtly enhanced for any specific activity or research application is limitless. The ability to reconstruct a replicative thermostable DNA polymerase holoenzyme system *in vitro,* once the rest of the replication enzymes from a thermophile are cloned and characterized, may also have enormous utility in applications in which rapid, highly processive, accurate DNA synthesis is desired. Additionally, as previously developed methods continue to become simplified and automated due to advances in reagent chemistry, the enormous power of PCR will continue to expand.

References

Abramson, R. D. (1995). Thermostable DNA polymerases. In *PCR Strategies* (M. A. Innis, D. H. Gelfand, and J. J. Sninsky, Eds.), pp. 39–57. Academic Press, San Diego.

Abramson, R. D., Reichert, F. L., Starron, A., and Akers, J. (1994). Improved cycle sequencing using a new mutant enzyme. AmpliTaq DNA Polymerase, CS. *Clin. Chem.* **40,** 2339.

Barnes, W. M. (1994). PCR amplification of up to 35-kb DNA with high fidelity and high yield from λ bacteriophage templates. *Proc. Natl. Acad. Sci. U.S.A.* **91,** 2216–2220.

Birch, D. E., Kolmodin, L., Wong, J., Zangenberg, G. A., Zoccoli, M. A., McKinney, N., Young, K. K. Y., and Laird, W. J. (1996). Simplified hot start PCR. *Nature* **381,** 445–446.

Brandis, J. W., Edwards, S. G., and Johnson, K. A. (1996). Slow rate of phosphodiester bond formation accounts for the strong bias that *Taq* DNA polymerase shows against 2′,3′-dideoxynucleotide terminators. *Biochemistry* **35,** 2189–2200.

Chadwick, R. B., Conrad, M. P., McGinnis, M. D., Johnston-Dow, L., Spurgeon, S. L., and Kronick, M. N. (1996). Heterozygote and mutation detection by direct automated fluorescent DNA sequencing using a mutant *Taq* DNA polymerase. *BioTechniques* **20,** 676–683.

Cheng, S. (1995). Longer PCR amplifications. In *PCR Strategies* (M. A. Innis, D. H. Gelfand, and J. J. Sninsky, Eds.), pp. 313–324. Academic Press, San Diego.

Cheng, S., Fockler, C., Barnes, W. M., and Higuchi, R. (1994). Effective amplification of long targets from cloned inserts and human genomic DNA. *Proc. Natl. Acad. Sci. U.S.A.* **91,** 5695–5699.

Chou, Q., Russell, M., Birch, D. E., Raymond, J., and Bloch, W. (1992). Prevention of pre-PCR mis-priming and primer dimerization and improves low-copy-number amplifications. *Nucleic Acids Res.* **20,** 1717–1723.

Cline, J., Braman, J. C., and Hogrefe, H. H. (1996). PCR fidelity of *Pfu* DNA polymerase and other thermostable DNA polymerases. *Nucleic Acids Res.* **24,** 3546–3551.

Dang, C., and Jayasena, S. D. (1996). Oligonucleotide inhibitors of *Taq* DNA polymerase facilitates detection of low copy number targets by PCR. *J. Mol. Biol.* **264,** 268–278.

De Falco, M., Grippo, P., Rossi, M., and Orlando, P. (1998). Multiple forms of DNA polymerase from the thermo-acidophilic eubacterium *Bacillus acidocaldarius:* Purification, biochemical characterization and possible biological role. *Biochem. J.* **329,** 303–312.

Edgell, D. R., Klenk, H. P., and Doolittle, W. F. (1997). Gene duplications in evolution of archaeal family B DNA polymerases. *J. Bacteriol.* **179,** 2632–2640.

Eom, S. H., Wang, J., and Steitz, T. A. (1996). Structure of *Taq* polymerase with DNA at the polymerase site. *Nature* **382,** 278–281.

Faloona, F., Weiss, S., Ferre, F., and Mullis, K. (1990). Direct detection of HIV sequences in blood: High gain polymerase chain reaction. Paper presented at the 6th International Conference on AIDS, San Francisco. [Abstract No. 1019].

Ishino, Y., Takahashi-Fujii, A., Uemori, T., Imamura, M., Kato, I., and Doi, H. (1995). The amino acid sequence required for 5′ → 3′ exonuclease activity of *Bacillus caldotenax* DNA polymerase. *Protein Eng.* **8,** 1171–1175.

Ishino, Y., Komori, K., Cann, I. K., and Koga, Y. (1998). A novel DNA polymerase family found in Archaea. *J. Bacteriol.* **180,** 2232–2236.

Kalman, L. V., Abramson, R. D., and Gelfand, D. H. (1995). Thermostable DNA polymerases with altered discrimination properties. *Genome Sci. Technol.* **1,** 42.

Kellogg, D. E., Rybalkin, I., Chen, S., Mukhamedova, N., Vlasik, T., Siebert, P. D., and Chenchik, A. (1994). TaqStart antibody: "Hot Start" PCR facilitated by a neutralizing monoclonal antibody directed against *Taq* DNA polymerase. *BioTechniques* **16,** 1134–1137.

Kiefer, J. R., Mao, C., Hansen, C. J., Basehore, S. L., Hogrefe, H. H., Braman, J. C., and Beese, L. S. (1997). Crystal structure of a thermostable *Bacillus* DNA polymerase I large fragment at 2.1 Å resolution. *Structure* **5,** 95–108.

Kim, Y., Eom, S. H., Wang, J., Lee, D.-S., Suh, S. W., and Steitz, T. A. (1995). Crystal structure of *Thermus aquaticus* DNA polymerase. *Nature* **376,** 612–616.

Kim, Y., Kim, J. S., Park, Y., Chang, C. S., Suh, S. W., and Lee, D. S. (1997). Mutagenesis of the positively charged conserved residues in the 5′ exonuclease domain of *Taq* DNA polymerase. *Mol. Cells* **7,** 468–472.

Korolev, S., Nayal, M., Barnes, W. M., Di Cera, E., and Waksman, G. (1995). Crystal structure of the large fragment of *Thermus aquaticus* DNA polymerase I at 2.5 Å resolution: Structural basis for thermostability. *Proc. Natl. Acad. Sci. U.S.A.* **92,** 9264–9268.

Li, Y., Kong, Y., Korolev, S., and Waksman, G. (1998). Crystal structure of the Klenow fragment of *Thermus aquaticus* DNA polymerase I complexed with deoxyribonucleoside triphosphates. *Protein Sci.* **7,** 1116–1123.

Lin, Y., and Jayasena, S. D. (1997). Inhibition of multiple thermostable DNA polymerases by a heterodimeric aptamer. *J. Mol. Biol.* **271,** 100–111.

McHenry, C. S., Seville, M., and Cull, M. G. (1997). A DNA polymerase III holoenzyme-like subassembly from an extreme thermophilic eubacterium. *J. Mol. Biol.* **272,** 178–189.

Merkens, L. S., Bryan, S. K., and Moses, R. E. (1995). Inactivation of the 5′-3′ exonuclease of *Thermus aquaticus* DNA polymerase. *Biochim. Biophys. Acta* **1264,** 243–248.

Mullis, K. B., and Faloona, F. (1987). Specific synthesis of DNA *in vitro* via a polymerase-catalyzed chain reaction. *Methods Enzymol.* **155,** 335–350.

Newcomb, T. G., Suzuki, M., Jackson, A. L., and Loeb, L. A. (1997). High fidelity *Taq* polymerases for mutation detection. *FASEB J.* **11,** A1249.

Niehaus, F., Frey, B., and Antranikian, G. (1997). Cloning and characterisation of a thermostable alpha-DNA polymerase from the hyperthermophilic archaeon *Thermococcus* sp. TY. *Gene* **204,** 153–158.

Parker, L. T., Zakeri, H., Deng, Q., Spurgeon, S., Kwok, P. Y., and Nickerson, D. A. (1996). AmpliTaq DNA polymerase, FS dye-terminator sequencing: Analysis of peak height patterns. *BioTechniques* **21,** 694–699.

Reeve, M. A., and Fuller, C. W. (1995). A novel thermostable polymerase for DNA sequencing. *Nature* **376,** 796–797.

Saiki, R. K., Gelfand, D. H., Stoffel, S., Scharf, S., Higuchi, R., Horn, G. T., Mullis, K. B., and Ehrlich, H. A. (1988). Primer-directed enzymatic amplification of DNA with a thermostable DNA polymerase. *Science* **239,** 487–491.

Sharkey, D. J., Scalice, E. R., Christy, K. G., Jr., Atwood, S. M., and Daiss, J. L. (1994). Antibodies as thermolabile switches: High temperature triggering for the polymerase chain reaction. *Bio/Technology* **12,** 506–509.

Southworth, M. W., Kong, H., Kucera, R. B., Ware, J., Jannasch, H. W., and Perler, F. B. (1996). Cloning of thermostable DNA polymerases from hyperthermophilic marine Archaea with emphasis on *Thermococcus* sp. 9°N-7 and mutations affecting 3′-5′ exonuclease activity. *Proc. Natl. Acad. Sci. U.S.A.* **93,** 5281–5285.

Suzuki, M., Baskin, D., Hood, L., and Loeb, L. A. (1996). Random mutagenesis of *Thermus aquaticus* DNA polymerase I: Concordance of immutable sites *in vivo* with the crystal structure. *Proc. Natl. Acad. Sci. U.S.A.* **93,** 9670–9675.

Suzuki, M., Avicola, A. K., Hood, L., and Loeb, L. A. (1997). Low fidelity mutants in the O-helix of *Thermus aquaticus* DNA polymerase I. *J. Biol. Chem.* **272,** 11,228–11,235.

Tabor, S., and Richardson, C. C. (1995). A single residue in DNA polymerases of the *Escherichia coli* DNA polymerase I family is critical for distinguishing between deoxy- and dideoxyribonucleotides. *Proc. Natl. Acad. Sci. U.S.A.* **92,** 6339–6343.

Takagi, M., Nishioka, M., Kakihara, H., Kitabayashi, M., Inoue, H., Kawakami, B., Oka, M., and Imanaka, T. (1997). Characterization of DNA polymerase from *Pyrococcus* sp. KOD1 and its application to PCR. *Appl. Environ. Microbiol.* **63,** 4504–4510.

Uemori, T., Ishino, Y., Doi, H., and Kato, I. (1995). The hyperthermophilic archaeon *Pyrodictium occultum* has two alpha-like DNA polymerases. *J. Bacteriol.* **177,** 2164–2177.

Uemori, T., Sato, Y., Kato, I., Doi, H., and Ishino, Y. (1997). A novel DNA polymerase in the hyperthermophilic archaeon, *Pyrococcus furiosus:* Gene cloning, expression, and characterization. *Genes Cells* **2,** 499–512.

Urs, U. K., Sharkey, D. J., Peat, T. S., Hendrickson, W. A., and Murthy, K. (1995). Characterization of crystals of the thermostable DNA polymerase I from *Thermus aquaticus. Proteins Struct. Funct. Genet.* **23,** 111–114.

Vander Horn, P. B., Davis, M. C., Cunniff, J. J., Ruan, C., McArdle, B. F., Samols, S. B., Szasz, J., Hu, G., Hujer, K. M., Domke, S. T., Brummet, S. R., Moffett, R. B., and Fuller, C. W. (1997). Thermo sequenase DNA polymerase and *T. acidophilum* pyrophosphatase: New thermostable enzymes for DNA sequencing. *BioTechniques* **22,** 758–762.

Yurieva, O., Skangalis, M., Kuriyan, J., and O'Donnell, M. (1997). *Thermus thermophilus* dnaX homolog encoding gamma- and tau-like proteins of the chromosomal replicase. *J. Biol. Chem.* **272,** 27,131–27,139.

Zimmermann, K., and Mannhalter, J. W. (1998). Comparable sensitivity and specificity of nested PCR and single-stage PCR using a thermally activated DNA polymerase. *BioTechniques* **24,** 222–224.

4

MUSINGS ON MICROBIAL GENOMES[1]

Barry R. Bloom

One of the curiosities of the Human Genome Project, and the choice of the initial six organisms whose DNA sequences were to be obtained, is that none was a human pathogen. Despite the fact that infectious diseases represent the largest cause of death in the world—not cardiovascular disease or cancer—pathogen genomes were ignored in the initial formulation of genomes that would be of scientific interest. (*Escherichia coli* was included, but the K-12 strain of *E. coli* chosen for the genome project was totally debilitated and nonpathogenic.) For someone concerned with infectious diseases and vaccines and the humane uses of science, perhaps temporary solace can be obtained by recalling Tallyrand's wonderful reflection upon the assignation of the Duc de Broglie: "It was worse than a crime, it was a blunder!" Due to public concern about emerging infections and increasing drug resistance of microbial pathogens, that blunder is now being rectified. A major international effort is under way, at the National Institutes of Health, (NIH), academic institutions in the United States, Europe, and Japan, and in industry, to obtain the

[1] This chapter was adapted by permission from *Nature* **378,** 236, copyright 1995 Macmillan Magazines Ltd.

DNA sequences of microbial genomes, including the major pathogens of man and animals.

While it may be somewhat surprising, the first genome sequence that was to have been completed was that of *Mycobacterium leprae,* the causative agent of leprosy, although by a curious turn of fate it was not. Almost 15 years ago, the World Health Organization gave a promising biotech company a start-up grant to work on the genome, which, along with cosmids from Bill Jacobs and my lab, enabled the company to obtain initial sequences of the high G + C DNA to obtain NIH grants for the *M. leprae* genome. A wonderful circle of history was to have been closed since *M. leprae* was the first bacterial pathogen of man ever identified—by the Norwegian G. Armauer Hanson in 1873, 9 years before Koch discovered the tubercle bacillus. This most mysterious of pathogens has eluded the best efforts of science for more than a century to cultivate it in the laboratory; it can be grown to large numbers only in the armadillo. Alas, halfway through the sequencing of the *M. leprae* genome, the aforementioned biotech company had the revelation that there was no money in leprosy, stopped work on the genome, and diverted its attention to the sequence of *Helicobacter pylori,* the principal cause of ulcers and probably stomach cancer. As a result, *H. pylori* and not *M. leprae* was probably the first microbial genome sequenced. Alas, we shall probably never know because this momentous scientific achievement was not published but rather announced in 1995 only in a press conference. The DNA sequence was apparently sold to a major pharmaceutical company with the stipulation that it not be made publicly available: thus the inauspicious launching of the pathogen genome project.

The first complete and published sequence of a microbial genome on the public record, carried out by The Institute for Genome Research (TIGR) in collaboration with university collaborators, was that of *Hemophilus influenzae Rd* (Fleischmann *et al.*, 1995; Smith *et al.*, 1995). It was indeed a monumental achievement, and the sequence of each of the 16 microbial genomes completed subsequently has contributed a unique story and added some intriguing mysteries. The genomes of some of the major human pathogens have been completed, including *Borrelia bergorferi, Treponema pallidum, Mycoplasma pneumoniae, Neisseria gonorrhoae,* and *Streptococcus pyogenes.* Also, genomes of other major pathogens are in the process of being completed, including *M. tuberculosis, S. typhimurium, V. cholerae, P. aeruginosa, S. aureus, Legionella pneumophila,* and *C. trachomatis.* Even some parasite genomes are in progress, including leishmania, trypanosomes, plasmodium causing malaria, and schistosome worms. It is particularly gratifying that the sequence of *H. pylori* was recently completed and placed in the public domain by TIGR, which has been exemplary in this regard. Also, that of *M. leprae* is currently under way at TIGR. To be fair to history, one must acknowledge that these are by no means the first complete genomes to be sequenced—the first being the phage, ϕX174 (5386bp), by Sanger in 1977 and the most complex was cytomegalovirus (229 kb)—but they do represent some of the most complex genomes to date.

What have we learned? I focus on a few lessons from the first two complete bacterial genomes. *Hemophilus influenzae* Rd has a relatively small genome (1.8 Mb), a base composition similar to that of human DNA (38% G + C), and it serves well as a model bacterium. *Mycoplasma genitalium* was chosen because it has the smallest genome known for a self-replicating organism (580 kb) that might provide insight into the minimal functional gene set for a living organism (Fraser *et al.*, 1995). The completed genomes represented a staggering achievement for several reasons. Physical maps did not exist for either genome. Consequently, they were sequenced by a random shotgun strategy applied to the whole genome using three- to nine-fold coverage and high throughput and analyzed with sophisticated assembly and alignment software (Fleischmann *et al.*, 1995; Smith *et al.*, 1995). For *H. influenzae* Rd, 28,000 sequencing reactions were performed; for *M. genitalium*, the entire genome was sequenced on eight sequencing machines by five individuals in 2 months. Gaps were closed by sequencing overlapping fragments obtained from independent libraries. The first story told by these genomes is that the shotgun approach is remarkably rapid, error-free, and cost-effective. The error rate was estimated to be between 1 base in 5,000–10,000. An equal amount of time was required for analysis of the sequence data, including searches and comparison of existing bacterial sequence data, identification of putative genes, and annotation of the genetic map, including putative operons, regulatory regions, starts, and stops. The sequencing cost at this efficiency of scale was 30¢ per base, or about $200,000 for the complete sequence of the mycoplasma genome. Each microbial genome represents a one-time investment with knowledge that can be used for all time—a terrific bargain!

What has sequencing taught us about the genomes? In the case of *H. influenzae* Rd, 1743 coding regions were predicted, and as the result of the analysis and annotation process, 1007 of these were assigned to functional roles in the cells, including amino acid metabolism, biosynthesis of cofactors, cell envelope, intermediary and energy metabolism, lipid metabolism, nucleotide synthesis, replication, transcription, translation, and transport. The distribution of these open reading frames (ORFs) by predicted function in *H. influenzae* Rd and subsequent genomes published by TIGR was graphically revealed in color-coded maps which set a standard for providing a useful picture of the genome organization (see *http://www.-tigr.com*). For example, it indicates at a glance the amount of DNA that a typical bacterium would invest in various functions (e.g., 10% for energy metabolism, 17% for transcription and translation, 12% for transport, and 8% for cell envelope proteins). In the case of the 470 predicted coding regions of *M. genitalium,* an organism which lives in association with mammalian cells, the investment in other functions was quite different. For example, while *H. influenzae* had 68 genes for amino acid biosynthesis, *M. genitalium* has only 1. It has no cytochromes or genes of the tricarboxylic acid cycle. Since *M. genitalium* is by no means the earliest eubacterium, but has learned to simplify its life by its association with mammalian hosts, it was fascinating to learn what it could afford to discard and still survive. However, it has committed almost 5% of its genome to repeated elements encoding an adhesin that presumably allows it to stick to the cells that nurture it and which could be varied by recombination to allow antigenic variants to elude immune responses.

This leads to some mysteries. Foremost is the fact that despite the wealth of information in the sequence databases from microorganisms, one-third of the ORFs of each of the genomes sequenced predict sequences that cannot be assigned biological functions. This suggests that each organism has evolved some very specialized proteins, not common to others, to carry out important or unique functions or possibly that insufficient information is available to account for the species diversity. What do the 90 genes found in *M. genitalium,* but not in *H. influenzae* Rd, actually do? Finally, how does *M. genitalium* survive without a transcription factor (σ) for the stress response. Perhaps in simplifying its lifestyle, it has learned how to avoid stress; this achievement can possibly serve as a lesson for us all.

What does all this mean for microbiologists? At best, it means that they will be freed to do more biology rather than molecular biology. The need for random mutagenesis and screening—very inefficient approaches to defining gene functions even when transposable elements are available and overwhelming when not—will be superseded by PCR amplification of specific genes of interest, of known or unknown function, that can readily be mutagenized, precisely deleted from or inserted into the chromosome more rapidly to approach the mysteries of uniqueness and diversity. For pathogens, it means defining the genes that control virulence and that are necessary for protection. It means that the complete information on these should be available to scientists everywhere, not only in print but also in useable databases, to allow scientific imaginations all over the world to make use of this knowledge. Also, it means that the scientific community must be responsible and insist that *all* sequenced genomes be publicly available.

What of the future? The next advance will surely be the study not merely of the biological potential of the static genomes but also of the patterns of gene expression in various environmental circumstances. Screening for gene expression in pathogens following treatment with antibiotics, in different tissues of the body or in different hosts, will allow new insights into pathogenesis and open up new possibilities for diagnosis and prevention and treatment. Clearly, the new chip technologies are going to make this possible (Schena *et al.*, 1996). The immediate dilemma most ordinary scientists face will be what to do when outgunned by the few with access to these fantastic new technologies. In the immortal words of the waiter at Ferrara's, a favorite Italian pastry shop in New York, when asked by my wife how they made a wonderful iced expresso dessert called *granita de caffe*: "Lady, you need a machine. You don't got a machine!" Alas, we all are going to need the machines. When we have them, I hope we can settle in to the long and serious task ahead—understanding microbe biology.

References

Fleischmann, R. D., Adams, M. D., White, O., *et al.* (1995). *Science* **269,** 496–512.

Fraser, C. M., Gocayne, J. D., White, O., *et al.* (1995). *Science* **270,** 5235, 397–403.

Schena, M., Sholan, D., Heller, R., Chai, A., Brown, P., and Davis, R. W. (1996). *Proc. Natl. Acad. Sci. USA* **93,** 10614–10619.

Smith, H. O., Tomb, J.-F., Dougherty, B. A., Fleischmann, R. D., and Venter, J. C. (1995). *Science* **269,** 538–540.

5

STATISTICAL REFINEMENT OF PRIMER DESIGN PARAMETERS

Ellen M. Beasley, Richard M. Myers, David R. Cox, and Laura C. Lazzeroni

In the course of developing markers to build high-resolution radiation hybrid maps of the human genome, we have developed tens of thousands of primer pairs for amplifying sequences from complex, mammalian, whole genome DNA templates. The scale of this development gives us a unique opportunity to examine the parameters we use for designing successful oligonucleotide pairs for PCR, using statistical modeling on large data sets. In this chapter, we present the results of such a statistical analysis and show that modifying primer design parameters significantly reduces marker development failures. We identified three primer parameters—length, GC content, and, specifically, the GC content of the 3′ half of the primer—that were strongly associated with sequence-tagged site (STS) development failure. By modifying the primer selection parameters that we specified, we dramatically improved our marker development success. Our statistical analysis demonstrates the potential of exploratory data analysis to optimize primer parameters, or other quantifiable PCR variables, to maximize experimental success. While our focus is on primers designed for marker development, the results of this analysis should apply whenever high specificity of priming with a complex template is required.

PCR Applications

Radiation Hybrid Mapping: A Primer Design Problem

Modern maps of complex genomes are built by localizing unique STSs (Olson *et al.*, 1989) in radiation hybrid cell lines (Cox *et al.*, 1990; Hudson *et al.*, 1995; Gyapay *et al.*, 1996; Schuler *et al.*, 1996; Stewart *et al.*, 1997). An STS is a pair of oligonucleotides that gives a unique product after amplification on the target genome. Radiation hybrids for mapping the human genome have been constructed by fusing irradiated human cells with a hamster cell line host (Gyapay *et al.*, 1996; Stewart *et al.*, 1997). Amplification of the STS is performed by using genomic DNA prepared from the hybrid cell lines. The presence of the entire hamster genome in the template DNA imposes the condition that the STS not give a PCR product on hamster DNA that is the same size as or that interferes with scoring of the product from the human DNA.

In order to maximize the number and usefulness of our markers, we design STSs on both expressed sequence tag (EST; single-read sequence of cDNA clones; Adams *et al.*, 1991) and genomic sequences. When designing STSs on EST sequences we focus on the 3′ untranslated region to minimize the problem of homologies between humans and hamsters in protein coding regions. Most of the sequence we use for STS development is from a single sequencing read less than 500 bases in length. As expected, our STS development success is highly influenced by sequence quality.

STS Development and Assay

STS Design

We design STSs on human sequence data from several sources. Before primer selection, we first remove or mask sequences that are inappropriate primer binding sites. These sequences include human repeat sequences and cloning vector and *Escherichia coli* sequences that can be present in a clone. We use a modification of the program RepeatMasker, developed at the University of Washington, Seattle (A. F. A. Smit, and P. Green, RepeatMasker at *http://ftp.genome.*

washington.edu/RM/RepeatMasker.html) to screen against a file with common vector sequences, the entire *E. coli* genome, and all identified human repeat sequences—Repbase. Repbase is a useful, regularly updated source for human repeat sequences; it is copyrighted by the Genetic Information Research Institute and available at *http://www.grinst.org/~server/repbase.html.* Once these sequence elements have been masked, we use Primer3, developed at the Whitehead Institute for Biomedical Research by Steve Rozen and Helen J. Skaletsky (1996, 1997), to select primers. Primer3 code can be obtained from *http://www-genome.wi.mit.edu/genome_software/other/primer3.html* or a web-base server can be used at *http://www-genome.wi.mit.edu/cgi-bin/primer/primer3.cgi.*

Primer3 is a primer design program that allows the user to specify many primer and PCR product characteristics. We use only a subset of the allowable variables (Table 1). Both primer length and primer melting temperature (Tm) are represented by three values: an optimal value and minimum and maximum values that define the allowable range. The primer Tm is calculated by the nearest neighbor method of Breslauer *et al.* (1986) with the calculation method of Rychlik and associates (1990). An accurate calculation of Tm requires a value for the salt concentration (m*M* concentration of KCl). The reader should refer to the Primer3 web site for a full explanation of the allowable parameter variables. While the user may choose to specify only some of the variable values, it is important to note that Primer3 has preset default values for some parameters.

STS Assay Protocol

We develop at least 300 STSs every week; therefore, our goal is to design STSs that give a high initial success rate under uniform assay conditions. While it is possible that many apparent design failures could be made to work by customizing assay conditions for each primer pair, we do not reassay failed STSs because of the cost of optimization for individual STSs. Instead, we have optimized our assay protocol to maximize overall STS development success, with STSs developed in-house and at other genome centers.

Our standard 10-μl PCR assay contains 25 ng template DNA (Research Genetics), 0.8 μM each oligonucleotide primer, 200 μM dNTP mix, 10 m*M* Tris–HCl (pH 8.3), 50 m*M* KCl, 2.5 m*M* $MgCl_2$, and 0.35 unit AmpliTaq Gold (PE Applied Biosystems). Duplicate assays are run in 96-well cycle plates. The following is the cycle program:

Table 1

Original Primer3 Preferences for STS Design

Original Primer3 Setting	Primer Parameter Description
PRIMER_OPT_SIZE = 20	Optimal primer length (bases)
PRIMER_MIN_SIZE = 18	Minimum primer length
PRIMER_MAX_SIZE = 25	Maximum primer length
PRIMER_NUM_NS_ACCEPTED = 0	Number of indeterminant bases accepted
PRIMER_PRODUCT_SIZE_RANGE = 221-350, 90-220	Product size range; Primer3 will use the first range entered preferentially
PRIMER_OPT_TM = 62	Optimal primer melting temperature
PRIMER_MIN_TM = 59	Minimum primer melting temperature
PRIMER_MAX_TM = 65	Maximum primer melting temperature
PRIMER_MIN_GC = 20	Minimum %GC
PRIMER_MAX_GC = 80	Maximum %GC
PRIMER_SALT_CONC = 50	Salt concentration (m*M* KCl) in reaction
PRIMER_SELF_ANY = 8	Maximum number of contiguous bases able to self-anneal anywhere in the primers
PRIMER_SELF_END = 3	Maximum number of contiguous bases able to self-anneal at the 3′ end of the primers
PRIMER_GC_CLAMP = 0	GC clamp at 3′ ends of primers

10 min at 95°C to denature the template and activate the enzyme, 30 cycles of 30 sec at 94°C, 30 sec at 60°C, and 23 sec at 72°C, followed by a final 3.5-min elongation at 72°C. Reactions are then held at 4°C until detection by ethidium bromide staining of reactions run on 3.0% SeaPlaque (FMC Corp.) agarose gels. Further details on our current assay protocol, and any future changes, can be found at the Stanford Human Genome Center web site (*http://www-shgc.stanford.edu*).

Defining STS Design Failure

Our initial assay with a newly designed primer pair consists of two reactions—one with human genomic DNA and the other with hamster genomic DNA as the template. A successful STS is one that gives a unique product of the correct size with the human template and has no product near that size in the hamster sample. The three types of STS development failures, in order of frequency, are background products in the hamster sample that would interfere with scoring the STS, the appearance of a product of the expected size in both the human and the hamster samples, and the absence of PCR product in the human sample. There are several reasons why STSs fail, including the following:

1. Homologous sequences are present in the hamster genome.
2. Homologous sequences are present in the human, i.e., the STS sequence is not unique.
3. The sequence used to design the STS is not an accurate genomic sequence. This includes hidden introns in cDNA sequences and incorrect sequence.
4. Primer characteristics can result in false priming or failure to amplify.

We chose to focus on primer properties associated with STS failure to minimize failures caused by this easily controlled variable.

Statistical Examination of Marker Development Failure

The Data Set

Many STS design failures are due to problems with the DNA sequence underlying the STS design. To focus on issues that are more likely to relate to primer design, we analyzed a group of STSs developed on the basis of high-quality, genomic sequence from BAC ends (BAC end sequences were kindly provided by M. D. Adams and J. C. Venter, The Institute for Genomic Research). By studying the success rate of STS development with this group of sequences, we minimized the effect of sequence quality and eliminated problems due to possible disagreement between cDNA and genomic sequence. This relatively uniform, high-quality set of human DNA sequence made it feasible to examine the primer characteristics in this group of STSs because other factors contributing to STS failure were minimized. We examined the primer and sequence characteristics of a set of 392 STSs, including 146 failures (37%), developed by using our original primer design parameters (Table 1). Our goal was to generate a model containing a short list of apparently important predictors that, if modified, might reduce STS failure.

Statistical Methods

Variable Screening Strategy We wanted to identify STS and primer-pair characteristics associated with STS failure using as few preconceived hypotheses as possible. Variable selection was an important component of the analysis because there are an unlimited number of ways in which such potential characteristics can be defined. We began by defining candidate variables for STS primer pairs and PCR products within each of six logical categories of sequence characteristics: length, GC content, trend in GC concentration within a primer, homologies, repeats, and balanced representation of all four bases in the sequence. Within each class, we defined multiple variables for both the primer pair and the entire STS using alternative formulations of the characteristic. For example, for a given primer pair, we considered the sum of the primer lengths and the minimum and maximum of the two lengths. The minimum and/or maximum would be a better predictor if primer failure results when either primer exceeds an effective size limitation, whereas the sum of the lengths would be better if adding a base to either primer had the same effect on failure whatever the initial length of that primer. See Table 2 for a complete list of candidate variables within all six categories.

We used the statistical tools described later to first develop and test variables within each logical category, while ignoring the effect of variables in the remaining five categories. Using this process, we identified one to four variables from each category that were both statistically significant and represented more or less conceptually distinct characteristics. Simultaneous screening of these remaining candidates led to a final statistical model containing seven failure-associated variables. Lastly, we further refined the definitions of some of the remaining variables. For example, we obtained a slightly better fitting model by replacing STS sequence length, in bases, by a two-level categorical variable for short and long STSs.

As we considered many alternative formulations for the explanatory variables, we thought it wise to limit statistical modeling of the

Table 2

Variables Tested: Primer and STS Properties Tested in the Model

- Length
  - PCR product length
  - Primer 1 length
  - Primer 2 length
  - Minimum primer length
  - Maximum primer length
  - Total primer length, sum of lengths
- GC content
  - Fraction of sequence represented by GC
  - Fraction of primer 1 represented by GC
  - Fraction of primer 2 represented by GC
  - Fraction of both primers represented by GC
  - Maximum GC concentration in either primer
  - Minimum GC concentration in either primer
- Balanced representation of bases
  - Fraction of sequence represented by most frequent base
  - Fraction of sequence represented by two most frequent bases
  - Fraction of sequence represented by three most frequent bases
  - Same three variables examined for primer 1
  - Same three variables examined for primer 2
- Trend in GC content from 5′ to 3′ end
  - Fraction of GC in primer 1 occurring in 3′ half of the primer
  - Fraction of GC in primer 2 occurring in 3′ half of the primer
  - Average of trend in primer 1 and primer 2
  - Maximum trend in either primer
  - Minimum trend in either primer
- Simple repeats
  - Maximum run of single repeat in sequence
  - Maximum run of single repeat in primer 1
  - Maximum run of single repeat in primer 2
  - Maximum run of single repeat in either primer
  - Same four variables for two-base repeats
  - Same four variables for three-base repeats
  - An indicator for any extreme value of repeat elements in sequence
  - An indicator for any extreme value of repeat elements in either primer
- Sequence complements (including reverse complements)
  - Match between primers 1 and 2 (>5)
  - Match between primer 1 and itself (>5)
  - Match between primer 2 and itself (>5)
  - Match between the sequence and itself (>11)
  - Match between the entire STS sequence and primer 1 (>8)
  - Match between the entire STS sequence and primer 2 (>8)

relationship between these variables and STS failure to the logistic regression model described later. Simple assessment of the logistic model suggested that its fit to the data was satisfactory. We also did not consider potential variable interactions in which the presence of changes in two variables can be vastly different than the individual effects of changes in each of the two variables.

Statistical Tools During the screening process described previously, we used several statistical methods in the statistical software package S-Plus (MathSoft; *http://www.mathsoft.com*) to evaluate candidate explanatory variables. Spector (1994) provides a useful introduction to this software. The variables that comprise our final model consistently demonstrated a significant effect upon STS failure under all statistical approaches.

We tested the association of each categorical variable for association with STS failure using standard Chi-square tests. For example, a variable for the presence or absence of a short sequence homology (>5 bases) between the two members of a primer pair yielded a Chi-square statistic of 4.54 ($p = 0.033$). This indicates that the presence of a between-primer homology is marginally statistically significant for predicting failure when considered in the absence of other potentially explanatory variables. We tested for a difference in the mean value of each quantitative variable between successful and failed STSs using standard t tests. For example, the t statistic for a mean difference in the maximum primer GC content (the higher of the two primer GC content levels) was 8.04 ($p < 0.001$). This statistically significant result ignores the role of alternative explanatory variables. Details for implementing these procedures can be found in any introductory statistics text.

The tests described previously do not address the simultaneous effects of explanatory variables upon STS failure. For example, the t statistic for maximum primer length is -6.80 ($p < 0.001$) and is also statistically significant. This may reflect a causative effect of primer length on STS failure. However, maximum primer length is negatively correlated (-0.53) with maximum primer GC content. Such correlation can lead to significant test results for both variables when, in fact, only one variable has a direct effect on the outcome. Single-variable t tests do not distinguish between these possibilities.

We used logistic regression to model the simultaneous effect of multiple explanatory variables on STS failure. For a given STS, let x_i be the value for a given STS of the ith of d explanatory variables in the model. Let p be the failure probability for that STS. Under the logistic regression model, the odds of failure are

$$\frac{p}{1-p} = exp\,(\beta_0 + \beta_1 x_1 + \beta_2 x_2 + \ldots + \beta_d x_d)$$

This model is also written in terms of the log odds of failure as

$$\log_e \frac{p}{1-p} = \beta_0 + \beta_1 x_1 + \beta_2 x_2 + \ldots + \beta_d x_d$$

The baseline failure probability is determined by the unknown value of β_0. The unknown effect of the ith variable upon STS failure, which is estimated from the data, is denoted by β_i. An introduction to logistic regression can be found in Agresti (1990).

To complete logistic regression models containing alternative sets of explanatory variables, we used the residual deviance, which is analogous to the residual sum of squares in linear regression. When comparing models in which the smaller model contains a subset of the variables in the larger model, differences in the residual deviance can be interpreted as a likelihood ratio statistic (twice the difference in the log of the maximized likelihoods). Akaike's information criterion (AIC, 1974), which includes a variable in the final model only if it reduces the residual deviance by at least 2, provides a useful guideline for variable selection.

We tried both forward and backward variable selection strategies. Forward selection begins with a model containing no explanatory variables and, one at a time, adds the best available meeting the AIC criterion. The forward selection algorithm stops when no more such variables are available. Backward selection begins with a large number of variables in the model. Variables that are least important according to the criterion are removed one at a time. The backward selection algorithm stops when removing any of the remaining variables would increase the residual deviance by 2 or more. Our final model was based on a subjective assessment of the results under all of the approaches described previously and included those variables that were consistently significant under any statistical method.

Results of STS Design Failure Analysis

The statistical analysis identified three primer characteristics that were strongly predictive of STS failure: shorter primer length (represented by two variables), higher primer GC content, and an increasing AT to GC trend toward the 3′ end of the primers. These variables demonstrated strong effects upon STS failure under all methods of statistical analysis. Weaker effects that also remained in the final model include the presence of a short sequence homology between the two primers, the length of the longest run of repeats of length 3 within either primer, and whether the product length was long (>220) as opposed to short (<187). The choice to include some of these weaker effects depended on the strategy used to build the model but does not notably change the estimated effects of the stronger variables upon STS failure.

The final model is shown in Table 3. For each variable, the second column gives the estimate of β_i and its standard error. The third column shows the change in the residual deviance caused by the removal of the corresponding variable from the model. The associated p value corresponds to the likelihood ratio test for that variable based on the deviance statistic. In this context, the p value should be interpreted as a measure of the relative amount of evidence for a given variable. The testing paradigm is not appropriate because variables were selected based on the presence of a detectable association with failure. Consequently, we have made no attempt to adjust the p values for multiple tests. The exploratory nature of this analyusis means that true significance levels cannot be attached to the results, which we have instead confirmed experimentally. An alter-

Table 3

Results of Logistic Regression Analysis of STS Development Failure

	Estimate ± SE	Deviance Statistic (p value)	Effect (change in x_i)
β_0	7.6 ± 10.2		
Minimum primer length	−0.8 ± 0.5	3.72 (0.054)	2.3 (−1)
Maximum primer length	−0.3 ± 0.1	8.28 (0.004)	1.3 (−1)
Primer match	0.7 ± 0.4	4.05 (0.044)	2.0 (yes)
Maximum three-base repeats in primer	0.7 ± 0.2	8.28 (0.005)	2.0 (+1)
Larger product size	0.3 ± 0.2	2.00 (0.157)	1.3 (yes)
Maximum %GC in primer	18.1 ± 3.5	30.55 (<0.001)	6.1 (+.1)
Maximum AT–GC trend	8.0 ± 1.6	29.31 (<0.001)	2.2 (+.1)

native approach would have been to select variables based on a random subset of the data and to use the remaining data to validate the model. The fourth column represents the estimated effect of each variable in terms of the odds of failure. For example, if maximum primer GC content is increased by 10%, the odds of failure is multiplied by 6.1. The graphs shown in Fig. 1 illustrate the effect of some of these variables. Failure rates for GC content, AT-GC trend, and sequence length are based on interval-grouped data.

This analysis does not suggest that the other variables examined are not important determinants of success; rather, it suggests that the settings for other important variables may already be close to optimal. The preferences used for designing the STSs in this data set are quite standard for the human genome mapping community and give reasonable success rates.

Revising Primer Design Parameters: The Experimental Results

We were able to modify the primer selection preferences within Primer3 to change failure-associated variables identified by the statistical analysis. The original and revised preferences are shown in Table 4. In the data analysis, we defined a varible, trend, that measures the fraction of primer GC content in the 3′ half of the primer. This variable, while significantly predictive of failure, cannot be explicitly controlled in Primer3. We reasoned that trend is a significant predictor of failure because increasing the GC content of the 3′ end of the primer will stabilize false priming at the 3′ end and will increase background and decrease specificity. To reduce the stability of 3′-end annealing, we instead adjusted the Maximum End Stability, which is the maximum allowable value for the ΔG of duplex disruption (Breslauer *et al.*, 1989) for the terminal five 3′ bases, to a value of 8, slightly more stringent than the 9 that Rychlik (1993) recomends. We chose not to modify the product size value for experimental reasons.

We have analyzed the STS development results for 166 sequences with STSs developed with both the original and the revised primer design preferences. The summary of results for this sequence is shown in Table 5. Among STSs that failed with the original preferences, 62 (93%) passsed when using the revised preferences. Among

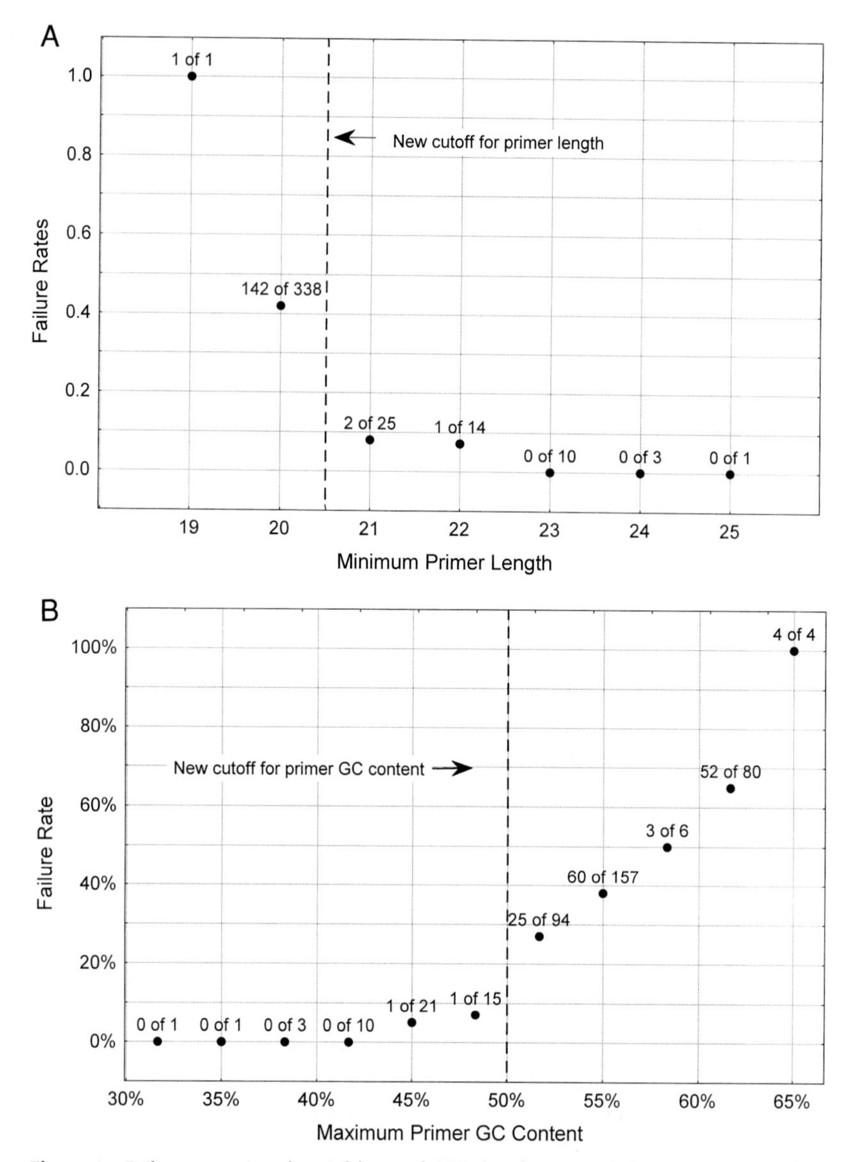

Figure 1 Failure-associated variables and STS development failure rates. A–D show the relationship between four failure-associated variables and the failure rate. The data in B and D are interval grouped. In all cases, the numbers of successful STSs of the total number of STSs in the group are shown above the data point. Figure 1A shows the results for the minimum primer length, the shortest primer of the pair. Figure 1B illustrates the relationship between the primer with highest primer GC content in the STS and development failure. Figure 1C presents the result for the AT to GC trend variable—the percentage of GC that is in the 3′ half of the primer. Figure 1D shows failure rates for STSs grouped by their PCR product length. Only the variables illustrated in A and B were directly used to modify the primer-selection parameters; the new limit is shown with a vertical dashed line.

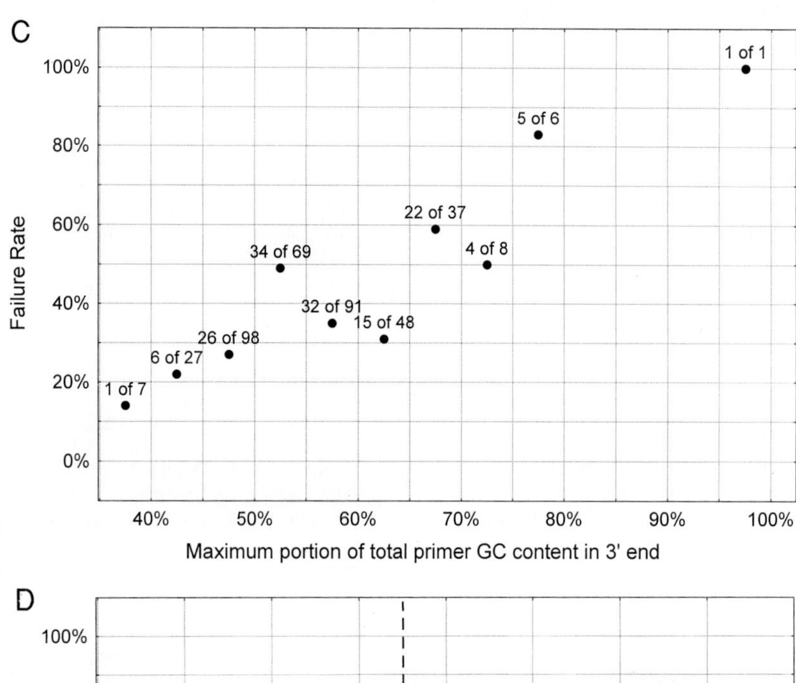

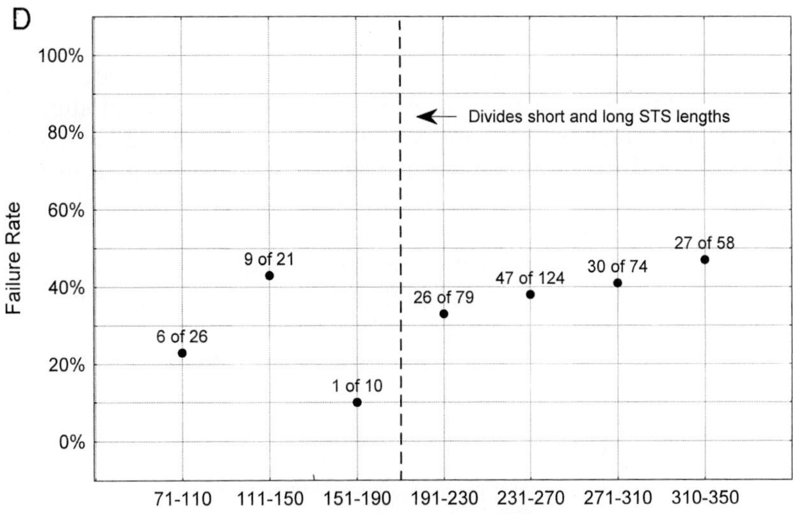

Figure 1 (*Continued*)

previously successful STSs, 96 (97%) also passed with the new preferences. This supports a causative relationship between the primer design parameters and marker development success. As shown in

Table 4

Revised Primer3 Preferences

Original Primer3 Setting	Revised Primer3 Setting
PRIMER_OPT_SIZE = 20	PRIMER_OPT_SIZE = 23
PRIMER_MIN_SIZE = 18	PRIMER_MIN_SIZE = 21
PRIMER_MAX_SIZE = 25	PRIMER_MAX_SIZE = 26
PRIMER_NUM_NS_ACCEPTED = 0	PRIMER_NUM_NS_ACCEPTED = 0
PRIMER_PRODUCT_SIZE_RANGE = 221-350, 90-220	PRIMER_PRODUCT_SIZE_RANGE = 221-350, 90-220
PRIMER_OPT_TM = 62	PRIMER_OPT_TM = 62
PRIMER_MIN_TM = 59	PRIMER_MIN_TM = 59
PRIMER_MAX_TM = 65	PRIMER_MAX_TM = 65
PRIMER_MIN_GC = 20	PRIMER_MIN_GC = 20
PRIMER_MAX_GC = 80	PRIMER_MAX_GC = 50
PRIMER_SALT_CONC = 50	PRIMER_SALT_CONC = 50
PRIMER_SELF_ANY = 8	PRIMER_SELF_ANY = 8
PRIMER_SELF_END = 3	PRIMER_SELF_END = 3
PRIMER_GC_CLAMP = 0	PRIMER_GC_CLAMP = 0
PRIMER_MAX_END_STABILITY = 100 (default setting)	PRIMER_MAX_END_STABILITY = 8

Table 5, the observed difference between the overall success rates of the two methods was 35%. A 95% confidence interval suggests that, with more data, the improvement will be between 28 and 43%. This is a startling improvement in development success and suggests that primer design optimization can dramatically improve assay results.

Because our new primer preferences were developed in response to failures under our original preferences among this same set of STSs, it could be suggested that the new preferences are optimal for only these STSs. However, the fact that the new success rates among

Table 5

STS Development Success Improves with Revised Primer3 Parameters

	Old Preferences	Revised Preferences
Failed	67	8
Passed	99	158
Success rate	60%	95%

previous failures (93%) and among previous successes (97%) are both superior to the previous overall success rate (63%) argues against this conclusion. The slightly lower success rate among previous failures vs that among previous successes might suggest the presence of other failure-associated characteristics in these STSs.

A comparison between the assay results for 12 sequences is shown in Fig. 2. The major consistent difference in the results with the new preferences is that the background smear or multiple bands observed in many hamster samples is eliminated. The two major failure classes, failures due to background or an identical band, both showed significant improvement with the revised parameters. Both of these improvements can probably be attributed to the decrease in the local duplex stability with the revised primer design parameters. The background smear is likely to be caused by false priming, which is reduced by decreasing the overall GC content and specifically decreasing the duplex stability of duplex formation at the 3′ end. While the identical band in the hamster band probably reflects sequence conservation, the decrease in local duplex stability must increase specificity by increasing sensitivity to any sequence mismatches.

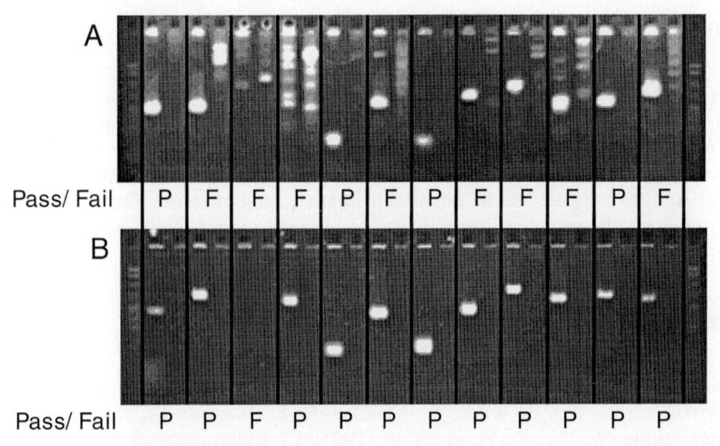

Figure 2 PCR results with original and revised Primer3 parameters. The results of STS development assays are shown with the original (A) and the revised (B) primer-picking preferences. In all cases the PCR reactions are loaded in pairs with human DNA template on the left and hamster on the right. The STSs in each column of data, delineated by black vertical lines, are designed on the same sequence. The score, pass (P) or fail (F), is shown underneath each pair of reactions.

Conclusions

We have determined primer selection criteria that are quite stringent and give excellent results in our hands. Our particular application allows this level of stringency because we are willing to allow some sequences (1–3%) to fail in the primer design step and we allow primers to be picked anywhere in the sequence. Obviously, for many applications it will be critical to be able to pick primers for every sequence and the user may want to determine precisely the sequence that is amplified. The public availability of Priimer3, with its easily modified preferences, allows the user to start with stringent, optimial primer characteristics and to gradually relax these standards, over subsequent rounds of selection, until a primer pair is found. For amplification of a specific sequence from a complex template, the user may have much better success by performing two rounds of PCR: the first with these optimized primer parameters to enhance the chance of success, and the second using the product of the first PCR reaction as the template and selecting primers to refine the product to the user's specifications.

Acknowledgments

This study was supported by Grant HG00206 from the National Human Genomic Research Institute. We thank Mark D. Adams and J. Craig Venter for the BAC end sequences that were used in this study. We acknowledge Steve Rozen and Helen J. Skaletsky of the Whitehead Institute for Biomedical Research for their development, support, and distribution of the Primer3 primer-picking program. We thank Christopher Piercy for oligonucleotide syntheses and Wei-Lin Sun STS development assays.

References

Adams, M. D., Kelley, J. M., Gocayne, J. D., Dubnick, M., Polymeropoulos, M. H., Merril, C. R., Wu, A., Olde, B., Moreno, R. F., Moreno, R., Kerlavage, A. R., McCombie, W. R., and Venter, J. C. (1991). Complementary DNA sequencing: "Expressed Sequence Tags" and the Human Genome Project. *Science* **252,** 1651–1656.

Agresti, A. (1990). *Catagorical Data Analysis.* Wiley, New York.

Akaike, H. (1974). A new look at statistical model identification. *IEEE Trans. Automatic Control* **19,** 716–723.

Breslauer, K. J., Frank, R., Blocker, H., and Marky, L. A. (1986). Predicting DNA duplex stability from the base sequence. *Proc. Natl. Acad. Sci. USA* **83,** 3746–3750.

Cox, D. R., Burmeister, M., Price, E. R., Kim, S., and Myers, R. M. (1990). Radiation hybrid mapping: A somatic cell genetic method for constructing high-resolution maps of mammalian chromosomes. *Science* **250,** 245–250.

Gyapay, G., Schmitt, K., Fizames, C., Jones, H., Vega-Czarny, N., Spillett, D., Muselet, D., Prud'Homme, J. F., Auffray, C., Morissette, J., Weissenbach, J., and Goodfellow, P. N. (1996). A radiation hybrid map of the human genome. *Hum. Mol. Genet.* **5,** 339–346.

Hudson, T. J., Stein, L. D., Gerety, S. S., Ma, J., Castle, A. B., Silva, J., Slonim, D. K., Baptista, R., Kruglyak, L., Xu, S.-H., Hu, X., Colbert, A. M. E., Rosenberg, C., Reeve-Daly, M. P., Rozen, S., Hui, L., Wu, X., Vestergaard, C., Wilson, K. M., Bae, J. S., Maitra, S., Ganiatsas, S., Evans, C. A., DeAngelis, M. M., Ingalls, K. A., Nahf, R. W., Horton, L. T., Jr., Anderson, M. O., Collymore, A. J., Ye, W., Kouyoumjian, V., Zemsteva, I. S., Tam, J., Devine, R., Courtney, D. F., Renaud, M. R., Nguyen, J., O'Connor, T. J., Fizames, C., Fauré, S., Gyapay, G., Dib, C., Morissette, J., Orlin, J. B., Birren, B. W., Goodman, N., Weissenbach, J., Hawkins, T. L., Foote, S., Page, D. C., and Lander, E. S. (1995). An STS-based map of the human genome. *Science* **270,** 1945–1954.

Olson, M., Hood, L., Cantor, C., and Botstein, D. (1989). A common language for physical mapping of the human genome. *Science* **245,** 1434–1435.

Rychlik, W. (1993). In *Methods in Molecular Biology, Vol. 15: PCR Protocols: Current Methods and Applications* (B. A. White, Ed.), pp. 31–40. Humana Press, Totowa, NJ.

Rychlik, W., Spencer, W. J., and Rhoads, R. E. (1990). Optimization of the annealing temperature for DNA amplification in vitro. *Nucleic Acids Res.* **18,** 6409–6412.

Schuler, G. D., Boguski, M. S., Stewart, E. A., Stein, L. D., Gyapay, G., Rice, K., White, R. E., Rodriguez-Tomé, P., Aggarwal, A., Bajorek, E., Bentolila, S., Birren, B. B., Butler, A., Castle, A. B., Chiannilkulchai, N., Chu, A., Clee, C., Cowles, S., Day, P. J. R., Dibling, T., Drouot, N., Dunham, I., Harris, M., Harrison, P., Brady, S., Hicks, A., Holloway, E., Hui, L., Hussain, S., Louis-Dit-Sully, C., Ma, J., MacGilvery, A., Mader, C., Maratukulam, A., Matise, T. C., McKusick, K. B., Morissette, J., Mungall, A., Muselet, D., Nusbaum, H. C., Page, D. C., Peck, A., Perkins, S., Piercy, M., Qin, F., Quackenbush, J., Ranby, S., Reif, T., Rozen, S., Sanders, C., She, X., Silva, J., Slonim, D. K., Soderlund, C., Sun, W.-L., Tabar, P., Thangarajah, T., Vega-Czarny, N., Vollrath, D., Voyticky, S., Wilmer, T., Wu, X., Adams, M. D., Auffray, C., Walter, N. A. R., Brandon, R., Dehejia, A., Goodfellow, P. N., Houlgatte, R., Hudson, J. R., Jr., Ide, S. E., Iorio, K. R., Lee, W. Y., Seki, N., Nagase, T., Ishikawa, K., Nomura, N., Phillips, C., Polymeropoulos, M. H., Sandusky, M., Schmitt, K., Berry, R., Swanson, K., Torres, R., Venter, J. C., Sikela, J. M., Beckmann, J. S., Weissenback, J., Myers, R. M., Cox, D. R., James, M. R., Bentley, D., Deloukas, P., Lander, E. S., and Hudson, T. J. (1996). A gene map of the human genome. *Science* **274,** 540–546.

Spector, P. (1994). *An Introduction to S and S-Plus.* Duxbury, Belmont, CA.

Stewart, E. A., McKusick, K. B., Aggarwal, A., Bajorek, E., Brady, S., Chu, A., Fang, N., Hadley, D., Harris, M., Hussain, S., Lee, R., Maratukulam, A., O'Connor, K., Perkins, S., Piercy, M., Qin, F., Reif, T., Sanders, C., She, X., Sun, W.-L., Tabor, P., Voyticky, S., Cowles, S., Fan, J.-B., Mader, C., Quackenbush, J., Myers, R. M., and Cox, D. R. (1997). An STS-based radiation hybrid map of the human genome. *Genome Res.* **7,** 422–433.

6

MULTIPLEX PCR: OPTIMIZATION GUIDELINES

G. Zangenberg, R. K. Saiki, and R. Reynolds

Many research and diagnostic assays involve the analysis of multiple regions within a gene (e.g., CFTR) or the analysis of multiple loci (e.g., for human identification). Rather than perform individual PCR amplification reactions for each region or locus, it is often desirable to amplify all sequences of interest simultaneously in a "multiplex" reaction. Multiplex PCR also offers a significant time and cost saving advantage, especially when a large number of individuals need to be analyzed. Another benefit of multiplex PCR is that only a single aliquot of DNA or RNA is required rather than an aliquot for each marker to be analyzed. This aspect is particularly important for forensic applications, prenatal diagnosis, and some clinical applications in which tissue and DNA samples are frequently limited. Reducing the number of tubes to which aliquots of DNA need to be added also minimizes the possibility of contamination and sample mix-up during reaction setup.

One of the first multiplex PCR systems was designed for the detection of mutations in the dystrophin gene (Chamberlain *et al.*, 1988). Nine PCR products were amplified simultaneously and analyzed by gel electrophoresis. Deletions within the gene could be readily detected by mobility shifts in the gels. The first commercially avail-

PCR Applications

able multiplex PCR kit (AmpliType PM) was designed for the forensic science community (Budowle *et al.*, 1995). Polymorphic regions of six genetic markers located on five chromosomes are amplified in a single tube using as little as 1 ng DNA extracted from samples collected at crime scenes and from victims. Alleles are identified by hybridization of PCR product to immobilized sequence-specific oligonucleotide probes which detect polymorphisms within the amplified regions. Several commercial kits are available for simultaneous amplification and detection of multiple markers containing length polymorphisms [e.g., short tandem repeat (STR) loci]. These systems use primers for different loci which amplify fragments whose length variants do not overlap when analyzed on agarose or polyacrylamide gels (Lins *et al.*, 1996; Kimpton *et al.*, 1994). The number of STR loci analyzed simultaneously can be increased by using multiple fluorescent dye-labeled primer pairs. For research applications, new reagents and protocols that greatly improve the specificity of PCR (single target and multiplex) have been introduced and will be discussed in detail in the following sections.

Multiplex PCR Optimization Strategies

Most of the guidelines and strategies used to optimize single-target PCR (Saiki, 1989) are similarly helpful for optimization of multiplex PCR systems. However, rapid optimization of a specific and efficient multiplex PCR assay requires that attention be paid to some additional critical factors. The principal challenge in developing a multiplex PCR system is overcoming primer dimer formation. Primer dimers are short products that are usually formed prior to the first PCR cycle and are typically caused by primers that have 3′ complementary overlaps of two or more bases (Chou *et al.*, 1992). Although it is unlikely that such short overlaps will form at the elevated temperatures normally employed for PCR, room temperature setup of a reaction containing these primers will permit transient annealing of the 3′ termini to occur. Under these conditions, there is also sufficient polymerase activity to begin extending the annealed primers to create the short products. Once formed, these dimers are very efficiently amplified and effectively compete with amplification of the desired target. As the number of targets in multiplex reactions increases, the probability that at least two primers will have 3′

complementary overlaps is significantly increased. As the number of primers that can anneal at their 3′ ends increases, so does primer dimer accumulation.

An approach that has been used with single-target reactions to minimize primer dimer formation is to limit the reaction for one of the components, typically either the DNA polymerase or the $MgCl_2$, until the temperature of the reaction solution is higher than the annealing temperature (Chou *et al.*, 1992). Under these "hot start" PCR conditions the primers cannot anneal to each other and be extended to form primer dimer. Hot start PCR can be set up manually but it is labor-intensive if a large number of samples need to be amplified. Furthermore, each time the reaction tube is opened, the possibility of contamination is increased. Simpler methods for hot start PCR and a new guideline for primer design that greatly reduces primer dimer formation will be discussed later. The combination of well-designed primers and hot start PCR eliminates nearly all primer dimer formation. It can further be controlled by careful optimization of the remaining reaction components.

Other significant challenges to optimizing multiplex systems are maintaining specific amplification and obtaining comparable yields of each amplicon. The recommendations regarding primer design, enzyme choice, reaction component optimization, and thermal cycling parameters discussed in the following sections will minimize primer dimer formation and nonspecific template-dependent amplification and favor equivalent yields of all amplified targets.

Primer Design

As with single-target PCR systems, primers with similar melting temperatures and without stable secondary structure (i.e., hairpins) should be chosen. Primer length can vary between 16 and 28 nucleotides, depending on GC content. Potential sequences can be screened quickly using a primer design software package (e.g., Oligo). In addition, a sequence homology search (BLAST) may help avoid nonspecific primer annealing to pseudogenes or partially homologous DNA sequences. For most multiplex PCR applications, particularly those involving the analysis of multiple length polymorphisms (i.e., VNTRs), it is advantageous to design primers such that the PCR products can be readily distinguished by gel electrophoresis. Gel-resolvable amplicons are convenient to assess the relative amplification efficiencies of each product during optimization. To avoid preferential amplification of the shorter amplicons, the size range between the smallest and largest products should not exceed ~300 bp whenever possible. These two guidelines, while helpful, do place some restrictions on primer design because they limit which sequences can be selected. Furthermore, when analyzing a large number of single-nucleotide polymorphisms, the number of targets may exceed the ability to distinguish individual bands on gels. For these multiplex systems, designing primers to amplify particular sizes of products is no longer useful. It is also likely that a more robust multiplex reaction will be achieved if all the fragments are approximately the same size, thus avoiding preferential amplification due to length differences.

Other authors who address multiplex PCR (Kebelmann-Betzing *et al.*, 1998; Henegariu *et al.*, 1997; Chamberlain and Chamberlain, 1994) recommend similar rules for primer design as those that are used for single-target PCR but do not address the difficulty of selecting a group of primers without occurrence of complementary 3′ overlap. As primers are designed and added to the multiplex set, it becomes increasingly complicated to track the 3′ sequences already in use and to avoid complementary termini. While this problem should be amenable to a computerized solution, such software has not yet become widely available. One relatively simple technique that can be used to create primers without these terminal overlaps is to design the oligonucleotides so they all share a common, nonpalindromic 3′ end. For example, if all multiplex primers are positioned on their target sequences so that they each have the dinucleotide "AA" at their 3′ termini, they will be unable to form complementary two-base overlaps. Similarly, other dinucleotide common termini could be used as long as they are nonpalindromic (e.g., "GC" termini would be a poor choice). Furthermore, by choosing a common terminal sequence at the 3′ ends of all primers in the multiplex it is substantially easier to identify problematic sequences that can lead to the formation of stable overlaps. For example, if the AA dinucleotide is chosen to be the common 3′ end for a set of multiplex primers, it is obvious that primers should not be positioned such that they will end in "TTAA" because stable four-base overlaps will form.

This strategy was used to develop a cystic fibrosis multiplex PCR system in which 20 products, from 129 to 598 bp in length, were coamplified using AmpliTaq DNA Polymerase without significant primer dimer accumulation. Each of the 40 primers in this system ends in a common AA dinucleotide (R. Saiki, unpublished data).

While the use of common terminal dinucleotides can be very helpful, it is not always possible to position a primer to have the desired two bases at its 3′ end. For example, the product size may be too big relative to the other PCR products in the system, the resulting PCR product may overlap another target in the multiplex, or the 5′ region adjacent to the terminal AA may be too A-T or G-C rich to obtain the appropriate melting temperature for the primer. In these situations, designing primers with a single A (or other common nucleotide) at the 3′ end, coupled with a hot start step, is usually sufficient to eliminate primer dimer (R. Reynolds, unpublished data).

Enzyme Choices

The enzyme used most often for single-target and multiplex PCR is the recombinant thermostable DNA polymerase from *Thermus aquaticus* (e.g., AmpliTaq DNA Polymerase). However, the Stoffel Fragment of *Taq* DNA polymerase may often prove to be better suited to multiplex PCR than the holoenzyme. Stoffel Fragment is more thermostable and exhibits optimal activity over a broader range of magnesium ion concentrations (Lawyer *et al.*, 1993), which is beneficial to multiplex PCR. In addition, Stoffel Fragment lacks the intrinsic 5′ to 3′ exonuclease activity of *Taq* DNA polymerase. This activity may degrade the template DNA in cases in which regions of the single-stranded template fold to form hairpin structures (Abramson, 1995). Stoffel Fragment will extend through these regions by strand displacement, thereby avoiding exonucleotlytic digestion of the template.

Another enzyme option is the chemically modified version of AmpliTaq DNA Polymerase (AmpliTaq Gold). This enzyme is particularly valuable for multiplex PCR systems because it is inactive until it is heated at 95°C, providing an effective hot start. At this temperature, primer dimer cannot form and compete with amplification of the specific targets. Birch *et al.* (1996) describe in detail this and other benefits of using AmpliTaq Gold-mediated hot start PCR for both single-target and multiplex PCR. TaqStart antibody has also been used to temporarily inactivate the polymerase and improve multiplex PCR (Kellogg *et al.*, 1994). These hot start PCR methods are easier and more reproducible than the approaches involving manual addition of key reagents described previously.

Stoffel Fragment can also be derivatized to be inactive until it is heated at 95°C by the same chemical modification used for AmpliTaq

Gold. Several of the examples presented later show that the chemically modified version of Stoffel Fragment (Stoffel CM) is superior to AmpliTaq Gold. In particular, Stoffel Fragment CM improves amplification of GC-rich targets that may have very complex secondary structures. Due to Stoffel CM's high thermostability, the denaturation temperature during cycling can be raised to achieve complete melting of the DNA target. Another study revealed a striking difference between the AmpliTaq Gold and Stoffel CM enzymes: Stoffel CM incorporates dUTP, a part of the UNG-mediated PCR product carryover prevention technique (Longo *et al.*, 1990), much less efficiently than AmpliTaq Gold. This difference was also observed in the unmodified version of the enzymes. Suggestions to compensate for this difference are described in Example 5.

Due to the different properties of the AmpliTaq and Stoffel Fragment DNA polymerases, these polymerases have different requirements for reaction components and thermal cycling parameters. The adjustments that need to be made are described under *Reaction Component Optimization* and *Thermal Cycling Parameters.*

Reaction Component Optimization

Each of the components in a PCR amplification contributes to the success or failure of the reaction. Therefore, the concentration of each component needs to be considered when optimizing the multiplex reaction. In general, the components are optimized one at a time, while holding the starting concentration of the other components constant. All the individually optimized conditions are tested together and additional modifications can be made if necessary. Recommended starting concentrations are provided in the following sections.

Since the concentration of $MgCl_2$ and the annealing temperature have the greatest effect on specificity and yield, it is efficient to begin the optimization process by performing a $MgCl_2$ titration experiment (e.g., 4, 6, and 8 m*M*) at three annealing temperatures (e.g., ±2°C window around predicted optimal temperature). The test windows for $MgCl_2$ concentration and annealing temperature can then be narrowed and the best combination can be used for subsequent reaction component and thermal cycling optimization experiments.

Buffer and Salt The optimal buffer concentration and pH, as well as the KCl concentration, depend on the DNA polymerase used in the reaction. This chapter is limited to a discussion of the hot start polymerases AmpliTaq Gold and Stoffel CM. Their performance in a multiplex reaction is superior to the unmodified forms of these enzymes and to manual hot start techniques. AmpliTaq Gold works very efficiently in reactions containing 15 m*M* Tris–HCl (pH 8.0) and 50 m*M* KCl. In contrast, Stoffel CM has optimal activity in reactions containing 10 m*M* Tris–HCl buffer and 10 m*M* KCl.

$MgCl_2$ Using suboptimal concentrations of $MgCl_2$ can result in high levels of nonspecific amplification and reduced product yield. For most single-target amplification reactions, 1.5 m*M* $MgCl_2$ is a suitable starting point for assay optimization. This concentration is also useful for screening candidate primer pairs, including those that may become part of a multiplex system.

In general, it appears that higher order multiplex amplifications benefit from increased $MgCl_2$ concentration on the order of 3–10 m*M*. This concentration range for multiplex PCR is significantly higher than the optimal $MgCl_2$ concentration determined for the individual primer pairs which may vary between 1 and 3 m*M*. Once candidate multiplex primer pairs are identified, they should be tested together in reactions containing higher concentrations of $MgCl_2$, preferably at several annealing temperatures as recommended previously. Up to 36 different target sequences, varying in length from 94 to 600 bp, have been amplified simultaneously using 8 m*M* $MgCl_2$ (G. Zangenberg, unpublished data).

Another consideration for optimizing $MgCl_2$ concentration is total dNTP concentration. Since dNTPs quantitatively bind divalent cation, substantial changes to nucleotide concentrations may affect optimal $MgCl_2$ concentrations. For example, a standard amplification reaction with 1.5 m*M* $MgCl_2$ and 200 μM of each dNTP would have an effective concentration of 0.7 m*M* $MgCl_2$ because the DNA precursors would bind 800 μM Mg^{2+}.

DNA Polymerase As described under *Enzyme Choices*, AmpliTaq Gold and Stoffel CM are the preferred enzymes for multiplex PCR because they essentially eliminate primer dimer formation and mispriming. Although multiplex systems can be optimized using AmpliTaq DNA Polymerase (e.g., AmpliType PM kits coamplify six targets) or other sources of *Taq* DNA polymerase, more extensive primer testing and fine-tuning of the reaction component concentrations and cycling parameters are required.

As a starting point for optimizing enzyme concentration and screening candidate primers, 5–10 units per 100-μl reaction can be used. Generally, as the number of targets to be amplified in the multiplex PCR increases, the units of enzyme per reaction will have to be increased to obtain acceptable yields of all targets.

Deoxynucleoside Triphosphates The standard starting point for deoxynucleoside triphosphate (dNTP) concentration optimization is 200 μ*M* for each dNTP, which is sufficient to produce approximately 25 μg DNA in a 100-μl reaction. For those applications in which UNG sterilization will be used to prevent PCR product carryover contamination, the TTP in the reaction must be replaced with dUTP (Longo *et al.*, 1990). In these reactions, replace the TTP with two to three times the amount of dUTP (i.e., 400–600 μ*M*). In addition to increasing the concentration of dUTP in the reaction, it may be necessary to adjust reaction conditions depending on the chosen DNA polymerase because these enzymes incorporate dUTP with different efficiencies (see Example 5). Further adjustments may be necessary if the target sequence is unusually AT rich.

Primers For initial testing of candidate primer pairs in single-target and multiplex reactions, 200–250 n*M* of each primer can be used. Primer concentration generally ranges from 100 to 400 n*M* (each primer) and may vary between targets due to differences in priming efficiencies. The optimal primer concentration for each marker needs to be determined empirically under multiplex conditions to achieve equivalent yields of all PCR products. Also, it is important to use primers that are ≥90% pure when possible.

Input DNA For initial multiplex PCR optimization experiments, use no more than 10–100 ng human genomic DNA (3,000–30,000 single-copy genes). A common error made when performing PCR is to add too much extracted DNA to the reaction (e.g., >100 ng). Adding excess DNA to a reaction can lead not only to increased nonspecific template-dependent amplification but also to weak or no amplification if the extracted sample contains a DNA polymerase inhibitor. As discussed under Example 4, an optimized multiplex system can tolerate several hundred nanograms of input DNA but there is no benefit from adding such an excessive amount, unless the DNA sample is severely degraded.

A multiplex PCR system can also be optimized to produce sufficient PCR product for subsequent typing or detection from as little as 250 pg DNA. This is an important feature for multiplex PCR applications in which the sample is limited (e.g., analysis of forensic samples). When working with limited sample amounts, product yield may be increased by increasing the number of cycles or, to some extent, increasing the amount of enzyme and/or the length of the pre-PCR heat-activation step of AmpliTaq Gold or Stoffel CM.

Reaction Volume Successful amplification reactions from as little as 10 μl have been reported and the upper limit is usually 100 μl. Choice of reaction volume depends in part on the type of thermal cycler used and the method of PCR product analysis. Another factor affecting choice of reaction volume is cost of reagents consumed, most notably primers and DNA polymerase.

Thermal Cycling Parameters

To optimize multiplex PCR amplification reactions in which three or more primer pairs are being tested together, use the guidelines discussed in the following sections for choosing parameters.

Pre-PCR Activation of Chemically Modified Enzymes Generally, the pre-PCR activation step is performed at 95°C for 12–15 min. This temperature should be used for activation of the enzyme even when a different temperature is used for denaturation during the PCR cycles. At 95°C, maximum enzyme activation is achieved within 15 min (Birch *et al.*, 1996). For some applications, this step will be shorter so that enzyme is continuously activated during subsequent PCR cycles. Under these "time-release" conditions, increased specificity can be obtained but up to 50 cycles may be required to produce enough PCR product for analysis (Kebelmann-Betzing *et al.*, 1998).

Denaturation For most systems, denaturation can be carried out for 30–60 sec at 95°C. In some thermal cyclers, the time can be reduced to 15 sec. Frequently, PCR cycling protocols include a "predenaturation" step but this is not necessary when using the chemically modified version of the enzymes.

Primer Annealing The choice of annealing temperature depends on the melting temperature chosen for the primers. Higher annealing temperatures result in more specific amplification but may also lead to loss of longer target amplification and overall reduction of PCR product yield. Lower annealing temperatures often result in an overall increase in nonspecific amplification but also increase specific target amplification. Typically, annealing times can be kept short (e.g., 20–45 sec) but these may vary between thermal cyclers. The best balance between specific amplification and high product yield can be determined by a series of amplifications at annealing temperatures that differ by a single degree.

Extension With both AmpliTaq Gold and Stoffel CM, the extension step is performed at 72°C. The extension time can be varied between 20 and 60 sec, depending on the thermal cycler used and product size. Stoffel CM or Stoffel Fragment will require more time to complete strand synthesis because these enzymes are less processive than the *Taq* DNA polymerases. Longer PCR products will also require longer extension times than short ones.

Final Extension A final strand extension step is added after cycling is completed. This step is carried out at 72°C for 7–10 min. Many protocols include a "hold" step in the thermal cycler at 4–15°C following the final extension. The hold step is a convenience if the cycling is completed after the end of the workday. However, if the reaction contains UNG and dUTP, the PCR products should be analyzed or transferred to a −20°C freezer as soon as possible after the final extension step. The UNG enzyme retains some activity after cycling and will degrade the PCR product after extended incubation at 4–15°C.

Cycle Number The cycle number is determined by considering the amount of input DNA and the length of the pre-PCR activation step. Using the starting conditions recommended previously (i.e., 10–100 ng DNA and 12- to 15-min activation at 95°C), 32–34 cycles should be sufficient to produce an acceptable amount of multiplex PCR products.

Examples

The following examples illustrate some issues that are important for optimization of multiplex PCR systems and general use of the chemically modified DNA polymerases. In each of the examples, a multiplex PCR containing 15 genetic markers is used. A 19-plex system is also used in the first example for comparison of different DNA polymerases. The sizes of the PCR products and the primer concentrations are listed in Table 1 for each multiplex system. As target sequence allowed, 3′ AA and 3′ A were chosen.

Either the Perkin–Elmer DNA Thermal Cycler 480 or GeneAmp PCR System 2400 instrument was used to amplify the multiplex systems. The cycling conditions are listed in Table 2. The concentrations of the reaction components are listed within each example. Electrophoresis was performed using a 3.5% MetaPhor agarose gel, cast and run in 0.5× Maniatis TBE buffer containing 0.5 μg/ml ethidium bromide. Most efficient separation is achieved when gel box, gel, and running buffer are chilled at 4°C and electrophoresis

Table 1

Primer Concentrations and PCR Product Length for the Markers Included in the Multiplex Amplification Reactions

15-Plex Multiplex			19-Plex Multiplex		
Locus	Primer Concentration (n*M*)	Length (bp)	Locus	Primer Concetration (n*M*)	Length (bp)
TNF-α	200	313	PROS1	300	232
TNF-β	200	275	DQA1	250	222/225
TNF-β	200	244	LDLR	150	214
PROS1	300	232	ALAD	200	197
CCR2	200	214	GYPA	150	190
ALAD	200	197	HBGG	175	172
CCR5	100	184	ABO/7	400	165
ABO/7	400	165	ABO/6	200	158/159
ABO/6	200	158/159	B7S8	175	151
FcεRi-β	200	145	PON	200	144
HTR2	200	132	HTR2	200	132
Eotaxin	200	125	ZFX/ZFY	200	126
CAII	225	113	ILIB	225	117
SRD	250	102	CAII	225	113
IL-4	200	95	CST5	225	108
			GC	225	105
			SRD	250	102
			ANT1	300	98
			LIPC	300	94

Table 2

Multiplex PCR Cycling Conditions

Step	TC480	GAPS2400
Enzyme activation	95°C, 8 min	95°C, 12 min
Denaturation	95°C, 1 min	95°C, 45 sec
Annealing	60°C, 30 sec	60°C, 45 sec
Extension	72°C, 30 sec	72°C, 45 sec
Final extension	72°C, 7 min	72°C, 7 min
Hold	15°C, forever	15°C, forever
Number of cycles	32	32

is carried out in a cold room or refrigerator. The gel should be run at about 16 V/cm for 2 to 3 hr depending on the fragment sizes to be separated, including a length standard to indicate fragment sizes. For all the examples, the GibcoBRL 25-bp DNA Ladder was run in the lanes labled "M." The 125-bp band is approximately two to three times brighter than the other ladder bands. For gel electrophoretic separation of smaller size fragments differing by less than 3 or 4 bp from each other, gel matrixes such as GeneAmp Detection Gel (Perkin–Elmer) or polyacrylamide will allow sufficient band resolution.

Example 1: Comparison of Chemically Modified and Unmodified AmpliTaq DNA Polymerase and AmpliTaq DNA Polymerase, Stoffel Fragment

The yield and specificity of multiplex reactions (100 μl) containing approximately equivalent amounts of AmpliTaq DNA Polymerase, AmpliTaq Gold (both 5 U), Stoffel Fragment (8 U), or Stoffel CM (20 U) are compared in Fig. 1. Lanes 2 (15 plex) and 7 (19 plex) contain AmpliTaq DNA Polymerase; lanes 3 (15 plex) and 8 (19 plex) contain AmpliTaq Gold DNA Polymerase. The reactions containing AmpliTaq Gold do not have primer dimer or template-dependent background. Also, several products that are not amplified with AmpliTaq are clearly visible in the AmpliTaq Gold reactions. Similar effects are seen when the chemically modified Stoffel Fragment (Stoffel CM) is used (lanes 5 and 9) in place of Stoffel Fragment (lanes 4 and 10).

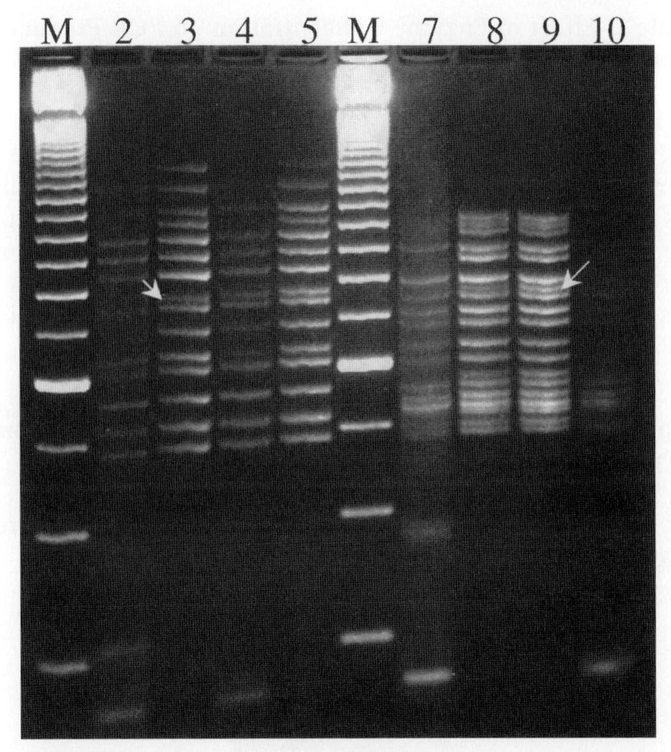

Figure 1 Comparison of chemically modified and unmodified AmpliTaq DNA Polymerase and AmpliTaq DNA Polymerase, Stoffel Fragment. See Table 3 for specific reaction conditions.

In addition to these differences between the modified and unmodified versions of the same enzyme, differences between Stoffel Fragment and AmpliTaq also were observed. Reactions containing Stoffel Fragment have less primer dimer and nonspecific background than reactions containing AmpliTaq DNA Polymerase (Fig. 1, lanes 4 and 10 vs lanes 2 and 7, respectively). Another difference observed is that use of Stoffel CM results in greater yield of the ABO/exon 7 PCR product than AmpliTaq Gold (Fig. 1, lanes 5 and 9 vs lanes 3 and 8, respectively; see Fig. 1, arrow). This PCR product is very GC rich and may have a complex secondary structure in which the template's 5′ end tends to fold back to form a hairpin structure. In such a case, *Taq* DNA polymerase's 5′ exonuclease activity may cause template degradation, whereas Stoffel Fragment, lacking the 5′ exonuclease activity, will perform strand displacement synthesis.

Example 2: Effect of Enzyme Concentration on PCR Product Yield and Specificity

Multiplex reactions containing ~10 U (Fig. 2, lane 2), ~20 U (lane 3), and ~30 U (lane 4) of Stoffel CM per 100 μl were amplified under the conditions described in Tables 2 and 3. The reaction run in lane 2 does not contain sufficient enzyme to amplify the largest targets in the multiplex. In contrast, increasing the amount of enzyme by two- or threefold results in complete amplification of all DNA targets (lanes 3 and 4) but a small amount of primer dimer is now visible.

Example 3: Effect of Buffer pH and Enzyme Concentration on PCR Product Yield

AmpliTaq Gold and Stoffel CM enzymes are activated more efficiently when the Tris–HCl buffer pH is lowered from 8.3 to 8.0,

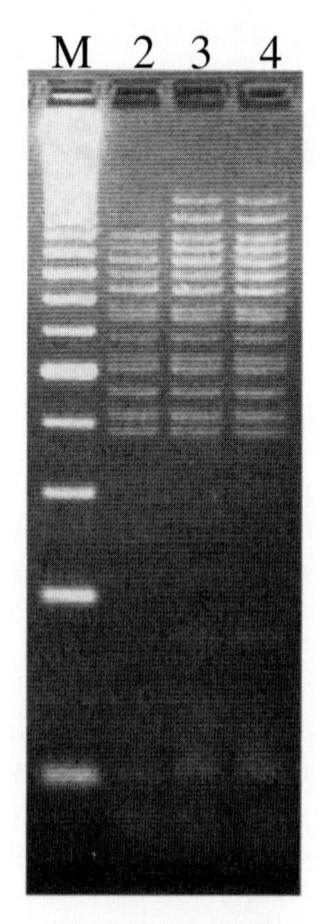

Figure 2 Effect of enzyme concentration on PCR product yield and specificity.

resulting in increased product yield and greater amplification of longer targets. Figure 3 shows the effect of reducing the buffer pH from 8.3 (lane 2) to 8.0 (lane 3) on Stoffel CM amplification efficiency. The reaction conditions used were described in Table 3 but the buffer pH and enzyme concentrations were varied. Reducing the buffer pH to 8.0 and increasing the amount of Stoffel CM twofold results in even more efficient amplification, particularly for the longer targets (lane 4).

Example 4: Effect of Input DNA Concentration on PCR Product Yield and Specificity

As discussed under *Reaction Component Optimization,* adding too much extracted DNA to a reaction can result in increased nonspecific amplification, especially when the multiplex PCR conditions have not been carefully optimized. If the quantity of DNA is determined

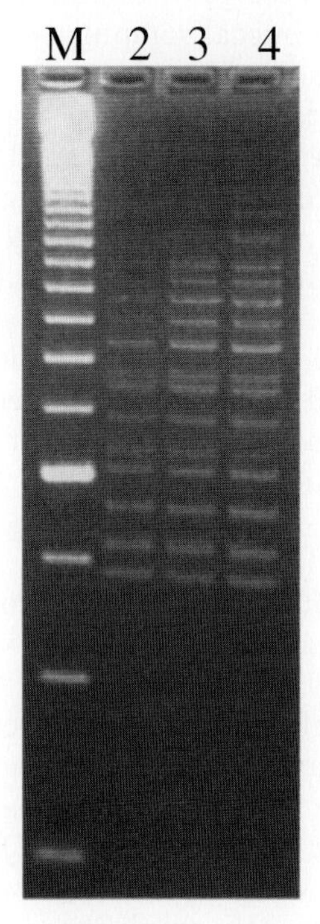

Figure 3 Effect of buffer pH and enzyme concentration on PCR product yield.

Table 3

PCR Reaction Conditions (100 μl) for Example 1

AmpliTaq DNA Polymerase/ AmpliTaq Gold	AmpliTaq DNA Polymerase, Stoffel Fragment/Stoffel CM
15 m*M* Tris–HCl, pH 8.0	10 m*M* Tris–HCl, pH 8.0
50 m*M* KCl	10 m*M* KCl
4 m*M* $MgCl_2$	4 m*M* $MgCl_2$
200 μ*M* dNTP each	200 μ*M* dNTP each
5 U AmpliTaq/AmpliTaq Gold	8 U Stoffel/10–30 U Stoffel CM[a]
10 ng DNA	10 ng DNA
Primer concentrations in Table 1	Primer concentrations in Table 1

[a]Twenty units of Stoffel CM are equivalent to 5 U AmpliTaq Gold.

routinely prior to amplification, then an amount of template that is known to give adequate yield and specific products can be added. However, for many high-throughput sample screening procedures, the DNA is not quantitated prior to amplification and the amount of DNA extracted from buccal scrapings and blood samples can be highly variable. For these applications, it is important to ensure that the multiplex system is optimized so that both small and large amounts of input DNA can produce reliable results.

Figure 4 shows the results of amplifying 10 ng (lane 2), 100 ng (lane 3), 500 ng (lane 4), and 1 μg (lane 5) DNA in the 15-plex system using Stoffel CM and the conditions described in Table 3. A well-optimized system can tolerate input DNA amounts that range over two orders of magnitude. However, even under optimized conditions, as the amount of input DNA is increased, amplification of the smaller products increases relative to that of the longer products and primer dimer products become visible.

Example 5: Efficiency of TTP vs dUTP Incorporation by Stoffel CM

Figure 5 shows the results of multiplex reactions containing Stoffel CM and either TTP (lane 2) or dUTP (lane 3). In the presence of dUTP, the longer target sequences are not amplified and one of the smaller products is not amplified either. One explanation may be that very AT-rich gene targets, such as the *PROS1* locus (Fig. 5, arrow), will be less efficiently amplified.

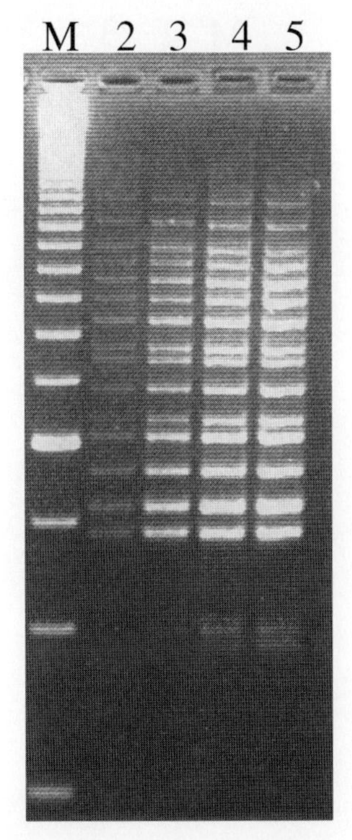

Figure 4 Effect of input DNA concentration on PCR product yield and specificity.

This difference between TTP and dUTP incorporation has not been observed with either AmpliTaq or AmpliTaq Gold DNA polymerases (data not shown). Increasing the amount of enzyme and adding a portion of TTP to the reaction in addition to dUTP (e.g., at a ratio of 1 : 4) may help increase amplification efficiency when using the UNG carryover prevention procedure with Stoffel Fragment or Stoffel CM.

Multiplex PCR Troubleshooting

For troubleshooting, we assume that single-target PCR has been carried out successfully but the multiplex reaction does not have the desired specificity and/or yield. Problems commonly encountered when developing a multiplex PCR system are listed in Table 4 along with possible causes and recommended actions.

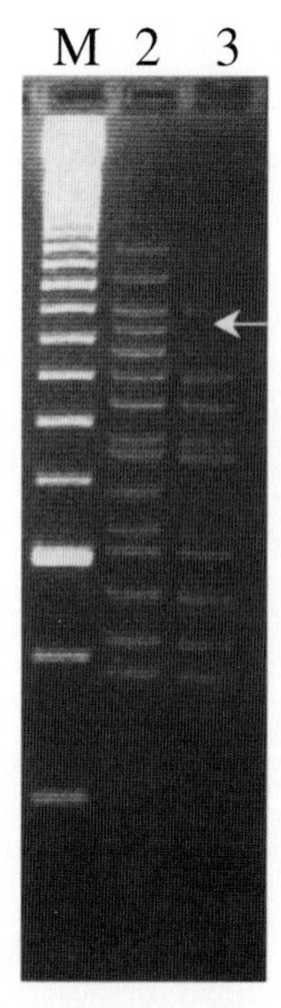

Figure 5 Efficiency of TTP vs dUTP incorporation by Stoffel CM.

Conclusions

This chapter described several strategies that should expedite the development of reliable multiplex amplification reactions. Of particular importance are those suggestions involving primer design and hot start polymerases that inhibit the formation of primer dimer artifact. By preventing dimer accumulation, one of the main impediments to successful multiplex PCR is removed. These guidelines are generally applicable in any situation in which the various target sequences are initially present at approximately equal concentrations. For those cases in which there may be large differences among target copy numbers, such as a panel for several infectious disease pathogens, the overriding challenge becomes one of dependably amplifying a low copy target (pathogen A) that might be present in the same specimen as a high copy target (pathogen B). These techniques have not been discussed in this chapter.

Table 4

Multiplex PCR Troubleshooting

Problem	Possible cause	Recommended action
No PCR products at all	Enzyme activation not complete	Check cycling parameters and pre-PCR heat activation of chemically modified enzymes
	Component left out of the reaction	Prepare new reaction mix
	Inhibitors in sample	Try alternative extraction procedure
	Product degradation due to UNG activity	Freeze PCR product immediately after PCR
Low yield of all products	Incomplete amplification	Increase pre-PCR enzyme activation time
		Increase cycle number
		Increase enzyme concentration
		Increase $MgCl_2$ concentration
	Tris–HCl buffer pH too high (suboptimal activation of the chemically modified enzymes)	Decrease buffer pH to approximately 8.0
	Primer concentration too low	Increase primer concentration
	Annealing temperature too high	Decrease annealing temperature
	Insufficient or degraded DNA sample	Check input DNA concentration and DNA quality
Low yield of one product	Primer concentration too low	Increase primer concentration of the low-yield PCR product
	Poor primer design	Redesign primers if necessary (see *Primer Design*)
	Secondary structure of template	Try alternative enzymes (e.g., Stoffel Fragment)
	Inefficient incorporation of dUTP	Add some TTP to the reaction
		Increase enzyme concentration
	Annealing temperature too high	Decrease annealing temperature

(*continues*)

Table 4 (*Continued*)

Problem	Possible cause	Recommended action
Low yield or loss of one or more large fragments	Preferential amplification of smaller products	Increase primer concentration of large products
	Enzyme concentration too low	Increase enzyme concentration
	Extension time too short	Increase extension time
	Degraded DNA sample	Check DNA quality
	DNA concentration too low	Check input DNA concentration
High background, nonspecific PCR products	Enzyme concentration too high	Decrease enzyme concentration
	Too much enzyme at early cycles	Use hot start methods (manual hot start or simplified hot start methods; e.g., AmpliTaq Gold, Stoffel CM, or TaqAntibodies)
	Pre-PCR heat cycle too long	Decrease pre-PCR heat activation of chemically modified enzymes
	$MgCl_2$ concentration too high	Decrease $MgCl_2$ concentration
	Poor primer design	Redesign primers (see *Primer Design*)
	Suboptimal amount of DNA	Check input DNA concentration
	Annealing temperature too low	Increase annealing temperature

The maximum number of DNA fragments that can be routinely amplified in a multiplex PCR has not been determined, nor is it clear that such a limitation exists. As a practical matter, current methods for analyzing PCR products (e.g., probe capture hybridization) do not require more than 30–40 amplicons. On the horizon, however, are high-density probe microarrays, the so-called DNA chips, which will require vastly more complicated multiplex reactions. Some of these chips, designed for genotyping applications, may eventually contain probes for thousands of single-nucleotide polymorphisms (Wang *et al.*, 1998). The new technologies that will be needed to accommodate these "kiloplex" reactions remain to be discovered.

Acknowledgments

Special thanks to David E. Birch for chemically modifying Stoffel Fragment, Keith Bauer for providing unmodified Stoffel Fragment, David E. Birch, Walter Laird, and Lori Kolmodin for helpful discussions about AmpliTaq Gold and Stoffel CM, and the DNA Synthesis Group for providing oligonucleotides. The authors also thank Tom White and Karen Walker for their comments.

References

Abramson, R. D. (1995). Thermostable DNA polymerases. In *PCR Strategies* (M. A. Innis, D. H. Gelfand, and J. J. Sninsky, Eds.), pp. 39–57. Academic Press, San Diego.

Birch, D. E., Kolmodin, L., Laird, W. J., McKinney, N., Wong, J., Young, K. K. Y., Zangenberg, G. A., and Zoccoli, M. A. (1996). Simplified hot start PCR. *Nature* **381,** 445–446.

Budowle, B., Lindsey, J. A., DeCou, J. A., Koons, B. W., Giusti, A. M., and Comey, C. T. (1995). Validation and population studies of the loci LDLR, GYPA, HBGG, D7S8, and Gc (PM loci), and HLA-DQα using a multiplex amplification and typing procedure. *J. Forensic Sci.* **40**(1), 45–54.

Chamberlain, J. S., and Chamberlain, J. R. (1994). Optimization of multiplex PCRs. In *The Polymerase Chain Reaction* (K. B. Mullis, F. Ferré, and R. A. Gibbs, Eds.). Birkhäuser, Boston.

Chamberlain, J. S., Gibbs, R. A., Ranier, J. E., Nguyen, P. N., and Caskey, C. T. (1988). Deletion screening of the Duchenne muscular dystrophy locus via multiplex DNA amplification. *Nucleic Acids Res.* **16,** 11141–11156.

Chou, Q., Russell, M., Birch, D. E., Raymond, J., and Bloch, W. (1992). Prevention of pre-PCR mis-priming and primer dimerization improves low-copy-number amplifications. *Nucleic Acids Res.* **20**(7), 1717–1723.

Henegariu, O., Heerema, N. A., Dlouhy, S. R., Vance, G. H., and Vogt, P. H. (1997). Multiplex PCR: Critical parameters and step-by-step protocol. *BioTechniques* **23,** 504–511.

Kebelmann-Betzing, C., Seeger, K., Dragon, S., Schmitt, G., Möricke, A., Schild, T. A., Henze, G., and Beyermann, B. (1998). Advantages of a new *Taq* DNA polyerase in multiplex PCR and time-release PCR. *BioTechniques* **24,** 154–158.
Kellogg, D. E., Rybalkin, I., Chen, S., Mukhamedova, N., Vlasik, T., Siebert, P. D., and Chenchik, A. (1994). TaqStart antibody: Hot start PCR facilitated by a neutralizing monoclonal antibody directed against *Taq* DNA polymerase. *BioTechniques* **16,** 1134–1137.
Kimpton, C. P., Fisher, D., Watson, S., Adams, M., Urquhart, A., Lygo, J., and Gill, P. (1994). Evaluation of an automated DNA profiling system employing multiplex amplification of four tetrameric STR loci. *Int. J. Legal Med.* **106,** 302–311.
Lawyer, F. C., Stoffel, S., Saiki, R. K., Chang, S. Y., Landre, P. A., Abramson, R. D., and Gelfand, D. H. (1993). High-level expression, purification, and enzymatic characterization of full-length thermus aquaticus DNA polymerase and a truncated form deficient in 5′ to 3′ exonuclease activity. *PCR Methods Appl.* **2,** 275–287.
Lins, A. M., Sprecher, C. J., Puers, C., and Schumm, J. W. (1996). Multiplex sets for the amplification of polymorphic short tandem repeat loci-silver stain and fluorescence detection. *BioTechniques* **20,** 882–889.
Longo, M. C., Berninger, M. S., and Hartley, J. L. (1990). Use of uracil DNA glycosylase to control carry-over contamination in polymerase chain reactions. *Gene* **93**(1), 125–128.
Saiki, R. K. (1989). The design and optimization of the PCR. In *PCR Technology, Principles and Applications for DNA Amplification,* pp. 7–16. Stockton Press, New York.
Saiki, R. K., Gelfand, D. J., Stoffel, S., Scharf, S. S., Higuchi, R., Horn, G. R., Mullis, K. B., and Erlich, H. A. (1988). Primer directed enzymatic amplification of DNA with a thermostable DNA polymerase. *Science* **239,** 487.
Wang, D. G., Fan, J. B., Siao, C. J., Berno, A., Young, P., Sapolsky, R., Ghandour, G., Perkins, N., Winchester, E., Spencer, J., Kruglyak, L., Stein, L., Hsie, L., Topaloglou, T., Hubbell, E., Robinson, E., Mittmann, M., Morris, M. S., Shen, N., Kilburn, D., Rioux, J., Nusbaum, C., Rozen, S., Hudson, T. J., Lipshutz, R., Chee, M., and Lander, E. S. (1998). Large-scale identification, mapping, and genotyping of single-nucleotide polymorphisms in the human genome. *Science* **280,** 1077–1082.

7

THE USE OF IMMOBILIZED MISMATCH BINDING PROTEIN FOR THE OPTIMIZATION OF PCR FIDELITY

Robert Wagner and Alan D. Dean

PCR has revolutionized molecular biology. However, PCR error rates (insertion of noncomplementary nucleotides) are commonly on the order of 10^{-5} to 10^{-6}, in contrast to *in vivo* DNA synthesis in which error rates are on the order of 10^{-10} (i.e., one error per 10 billion bases replicated). Furthermore, the multicycle nature of PCR allows errors to multiply and accumulate rapidly. Errors made in the first round of PCR are duplicated in each of the subsequent rounds, while new errors are being created at a fairly constant rate per nucleotide, such that a large fraction of product molecules may carry one or more sequence differences from the starting template. Nonetheless, for many applications this high error rate poses little or no problem. For example, when sequencing is the goal, an error rate in the product molecules as high as 1% per position will have no effect on the results or will, at worst, produce small background peaks under the peaks for the correct bases. However, for some applications, such as cloning or mutation detection, high error rates can be a major problem.

PCR also suffers from problems of mispriming, wherein products, single or double stranded, are occasionally produced from nontarget

PCR Applications

sites. Given that PCR depends on a pair of 16- to 25-base oligonucleotide primers finding their perfectly complementary sequence among the 6×10^9 bases in a haploid human genome, it is perhaps remarkable that mispriming does not overwhelm the process.

Determining the fidelity of PCR amplification is currently either relatively imprecise (in the case of mispriming) or extremely laborious (in the case of misincorporation). Mispriming is generally measured by gel electrophoresis, wherein mispriming is assumed to be absent or rare if the amplicon produces a single band. However, it may be that mispriming is generally asymmetric, such that most of the products are single stranded and any individual product is relatively rare. Alternately, mispriming may produce some double-stranded products, but at much lower frequency than the target amplicon. Thus, mispriming may be relatively common but the products virtually invisible in gel electrophoresis.

Misincorporation is generally detected in assays involving reversion of known mutations or forward mutations of cloned genes (Lundberg *et al.,* 1991). In these assays, the amplicons are cloned and individually transformed into bacteria in which the alterations and their frequency can be detected and measured. These assays commonly take a week or more to complete.

Immobilized mismatch binding protein (IMBP)-based assays provide a rapid, robust, and inexpensive means to measure PCR fidelity (both mispriming and misincorporation) and to facilitate PCR optimization (Debbie *et al.,* 1997). IMBP specifically binds mismatch-containing double-stranded DNA or double-stranded DNA containing one to four contiguous unpaired bases (Wagner *et al.,* 1995).

Mismatch binding proteins function *in vivo* as the first step in mismatch repair, namely, mismatch recognition. Perhaps best characterized of the mismatch binding proteins is the *Escherichia coli* protein, MutS (Su and Modrich, 1986; Su *et al.,* 1988; Parker and Marinus, 1992). Both *in vivo* and *in vitro,* MutS does not recognize all mismatches with equal efficiency but, somewhat counterintuitively, recognizes best those mismatches which most resemble base pairs (Radman and Wagner, 1986; Modrich, 1991). Nuclear magnetic resonance studies have revealed that well-repaired mismatches are invariably intrahelical and have some stacking and pairing interactions (Fazakerley *et al.,* 1986; Hunter *et al.,* 1986). Unpaired bases are recognized by virtue of their ability to exist stacked in an intrahelical configuration. In addition, MutS recognizes mismatches best when they are in GC-rich regions (Jones *et al.,* 1987), presumably because of the increased helical stability in such regions. Thus, the MutS recognition spectrum perfectly complements the polymerase error spectrum, i.e., polymerases tend to make errors wherein the mispairs most resemble base pairs, to make most errors in GC-rich regions (Petruska and Goodman, 1985), and to make small additions or deletions, particularly in regions of repeating bases (Kroutil *et al.*, 1996).

Because of its recognition spectrum, MutS would seem to be a logical choice for a mutation detection system. By using DNA of known sequence and annealing it to test DNA, it is possible to form heteroduplexes which will contain mispaired or unpaired bases when the test sequence differs from the known sequence by a single base substitution or by a small addition or deletion. These heteroduplexes will be substrates for MutS binding. Early attempts to use MutS in mutation detection assays produced less than satisfactory results. The assays used MutS in filter binding, nuclease protection or gel shift assays (Ellis *et al.*, 1994; Jiricny *et al.*, 1988; Lishanski *et al.*, 1994; Wagner and Radman, 1995) and the results generally suffered from poor binding of all but the best repaired mismatches and frequently gave high backgrounds. The use of mismatch binding proteins in routine mutation detection became practical with the discovery that specific immobilization of the protein improved both the binding spectrum and the discrimination between mismatched and perfectly paired molecules (Wagner *et al.*, 1995). Early work with IMBP employed immobilization on nitrocellulose and magnetic beads. Recently, IMBP 96-well plates have been developed.

IMBP assays, by virtue of their ability to detect all the common polymerase errors in PCR, including mispriming, provide an ideal means to monitor PCR fidelity and determine optimum conditions for high-fidelity PCR amplification (Debbie *et al.*, 1997).

Protocols

In general, it is most desirable to optimize PCR conditions for precisely the fragment and primer pair that will be used experimentally because optimum conditions have been found to vary with fragment, primer pair, and even the form of the template (e.g., genomic DNA or plasmid). However, the variations in optimum conditions for a given enzyme with different fragments or primers are frequently less significant than variations between enzymes.

Using IMBP, PCR fidelity can be measured directly or indirectly. For direct measurement, PCR amplification is performed with at least one of the primers labeled (the most commonly used label is 5′-biotin) and the level of background signal is measured for a variety of conditions. For indirect measurement, amplification is performed with unlabeled primers and the amplicon is used to compete with a standard mismatched substrate for IMBP binding. Indirect measurement may be somewhat less sensitive than direct measurement but has the major advantage of not requiring a labeled primer.

Direct Measurement of PCR Fidelity

PCR amplification should be performed using primers of the exact sequence as will be used in the final application except that one or both primers should be 5′-biotin labeled. It is generally necessary to do several different amplifications in which one or more conditions of amplification are varied (e.g., pH, $MgCl_2$ concentration, and annealing temperature). To increase detection of polymerase errors it may be helpful to amplify for additional cycles beyond the number which will be used to produce the desired product in the final application. Once optimum conditions have been determined for a given enzyme and a particular primer pair, it may be sufficient to check only one or two conditions when using the same enzyme with a different primer pair.

The amplicons should be run on gels to confirm the presence of product and quantitated, either by OD_{260} or by comparison to samples of comparable size and known concentration on polyacrylamide gels. Three or four different amounts of amplicon (generally in the range of 5–50 ng) should be tested in an IMBP assay.

IMBP assays can be performed in any of the currently available formats (nitrocellulose, magnetic beads, and microtiter plates) although the simplest and most cost-effective method is almost certainly microtiter plates. IMBP plates and plate kits are available from Gene Check, Inc. (Fort Collins, CO). Plates should be prepared by washing according to the manufacturers protocol and samples added in volumes of 20–40 μl. The entire IMBP plate protocol requires less than 1 hr to perform. IMBP plate kits include positive and negative controls, which should be included with each set of assays. In direct measurement assays, higher fidelity corresponds to lower signal. In other words, the more errors in an amplicon, including misincorporation and mispriming, the greater the amount of labeled IMBP substrate and the higher the signal in the assay.

Indirect Measurement of PCR Fidelity

PCR fidelity can be measured indirectly by using a competitive IMBP assay in which a labeled substrate for IMBP binding is mixed with various amounts of PCR amplicon or amplicons prepared under various conditions. In indirect measurement assays, higher signal corresponds to higher fidelity since PCR errors produce competing molecules which will reduce signal. The major advantage of indirect fidelity determination is that the amplicon can be produced with unlabeled primers.

Assays are performed exactly as described for direct measurement except that a competitive substrate (Gene Check, Inc.) is added immediately prior to applying the samples to the IMBP plate. It is important when using indirect measurement to add a range of sample quantities so as to ensure that the assay is being performed in a concentration range in which competition is easily detectable. Concentrations of PCR products which are too high will give excessive competition and make it difficult or impossible to distinguish the effects of different conditions. Similarly, too low a concentration of PCR products may not reveal any competition, independent of PCR conditions.

Results and Discussion

Effects of Buffer Composition and pH on PCR Fidelity

Several different buffer conditions have been examined for *Pfu* amplification. In these experiments, *Pfu* (Stratagene, La Jolla, CA) was used to amplify a 190-bp fragment of the sheep prion gene. Sheep genomic DNA and a pCRII plasmid (Invitrogen, San Diego, CA) are the templates into which the 190-bp fragment has been cloned. The primers for this fragment have been described (Debbie *et al.*, 1997). Conditions were selected using Stratagene's Optiprime kit. All amplification reactions were examined by gel electrophoresis for the presence of products. Those giving products were tested in IMBP assays with both direct and indirect measurement of PCR fidelity. The results are presented in Figs. 1 and 2. Clearly, buffer conditions have an enormous influence both on the success of PCR and on its fidelity. Several products which give single bands in polyacrylamide gels give widely varying results in IMBP assays. Similar results for different polymerases have been reported (Debbie *et al.*, 1997). It

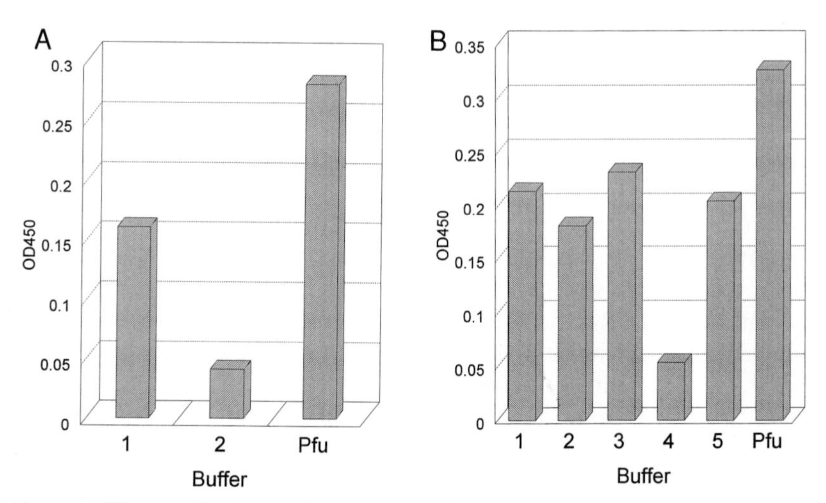

Figure 1 Effects of buffer conditions on PCR fidelity by direct IMBP assays. A region of the sheep prion gene was amplified from genomic (A) or plasmid (B) DNA templates using *Pfu* polymerase (Stratagene). Buffers all contained 0.3 μM primers (reverse primer 5′ biotin labeled), 0.2 m*M* dNTPs, and 25 m*M* KCl, but they varied in pH and $MgCl_2$ concentration (Optiprime System, Stratagene). 1, 10 m*M* Tris (pH 8.3), 1.5 m*M* $MgCl_2$; 2, 10 m*M* Tris (pH 8.3), 3.5 m*M* $MgCl_2$; 3, 10 m*M* Tris (pH 8.8), 1.5 m*M* $MgCl_2$; 4, 10 m*M* Tris (pH 8.8), 3.5 m*M* $MgCl_2$; 5, 10 m*M* Tris (pH 9.3), 1.5 m*M* MgCl. The reaction buffer provided by Stratagene (*Pfu*) is 10 m*M* KCl, 10 m*M* $(NH_4)_2SO_4$, 20 m*M* Tris (pH 8.75), 2 m*M* $MgSO_4$, 1% Triton X-100, and 10 μM BSA. Cycling conditions were four cycles of 94°C for 30 sec, 66°C for 30 sec, 72°C for 30 sec; four cycles of 94°C for 30 sec, 64°C for 30 sec, 72°C for 30 sec; four cycles of 94°C for 30 sec, 62°C for 30 sec, 72°C for 30 sec; four cycles of 94°C for 30 sec, 60°C for 30 sec, 72°C for 30 sec; 25 cycles of 94°C for 30 sec, 58°C for 30 sec, and 72°C for 30 sec; and a final extension for 5 min at 72°C. PCR products were examined by polyacrylamide gel electrophoresis (only with buffers 1, 2, and *Pfu* was the genomic template amplified). To test for differences in fidelity, biotin-labeled amplicons were denatured for 3 min at 94°C and allowed to reanneal for 30 min at 75°C before adding to the IMBP assay (40 ng/well in a total volume of 30 μl). Color was developed following the manufacturer's protocol and measured at OD_{450}. Data are the average of two measurements and are representative of at least two experiments.

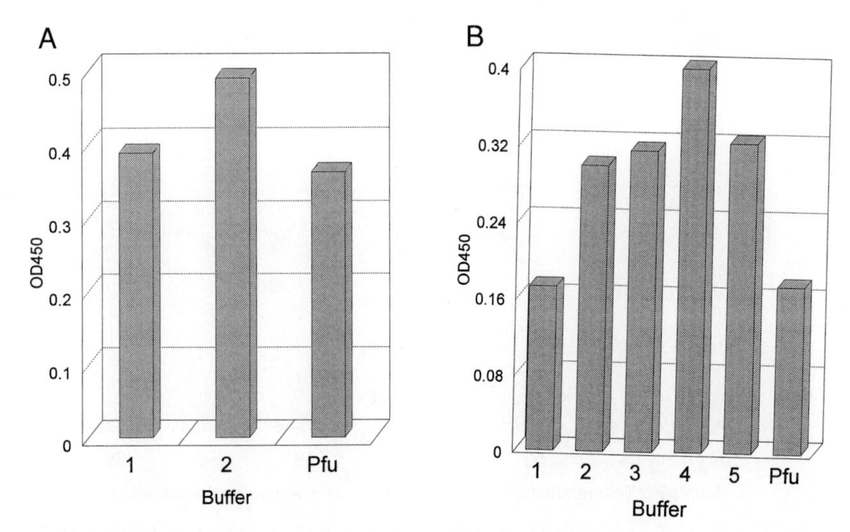

Figure 2 Effects of buffer conditions on PCR fidelity by indirect IMBP assays. Genomic (A) or plasmid (B) DNA templates were amplified using the same series of buffers described for measuring fidelity directly (Fig. 1) except that the amplicons were not labeled with biotin. To measure fidelity, samples (100 ng) were denatured, reannealed, and mixed with biotin-labeled competitive substrate (Gene Check, Inc.; 10 ng/well). Total volume of sample plus competitive fragment was 30 μl. The mixture was added to an IMBP assay plate.

is interesting to note that the optimal conditions for high-fidelity amplification of genomic DNA were not precisely the same as the conditions for high-fidelity amplification of the same fragment when it was cloned into a plasmid. Similar differences have been observed with amplifications involving different primer pairs (data not shown).

Effects of Annealing Temperature on PCR Fidelity

It is generally recognized that higher annealing temperatures produce fewer mispriming errors and protocols such as hot start and touchdown have been developed to minimize the effects of mispriming. Figure 3 shows the effect of varying annealing temperature on direct measurement PCR fidelity IMBP assays. Templates were either sheep genomic DNA or plasmid DNA as described previously. The results indicate that PCR fidelity decreases (i.e., signal increases) with decreasing annealing temperature. Similar results are obtained

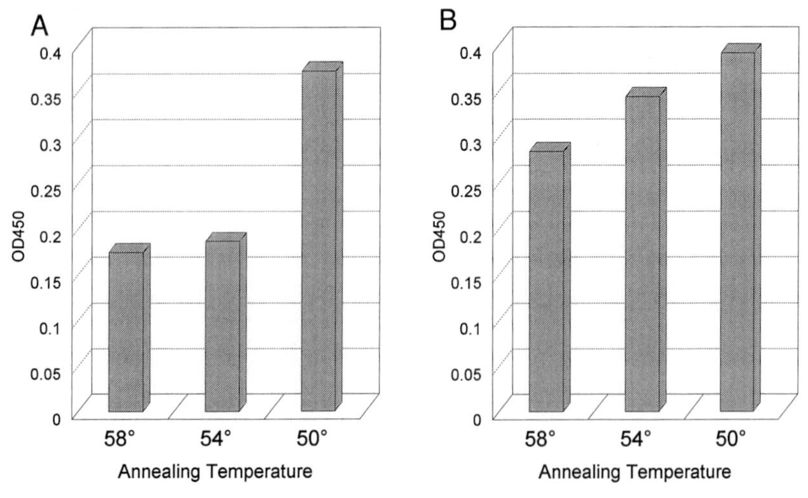

Figure 3 Effect of annealing temperature on PCR fidelity by direct IMBP assays. Genomic (A) and plasmid (B) DNA template amplification conditions were optimized for fidelity as described in the legends to Figs. 1 and 2 [i.e., four plasmid and two genomic buffers were used). Cycling conditions were 35 cycles of 94°C for 1 min, annealing temperatures 58°, 54°, or 50°C for 1 min, 72°C for 1 min, and a final extension at 72°C for 10 min. To measure PCR fidelity, the biotin-labeled samples were denatured, reannealed, and added to the IMBP assay (20 ng/well in a total volume of 30 μl).

with indirect measurement (Fig. 4) in which signal decreases with decreasing annealing temperature. Note that the changes observed with plasmid template are much less than those observed with equal amounts of amplicon from genomic template. This difference presumably reflects a difference in amounts of mispriming observed with plasmids and genomic DNA templates. The much higher target to nontarget DNA ratio in plasmid template relative to genomic template would be expected to result in less mispriming with plasmid template. Misincorporation rates should be comparable. Therefore, it appears that changes in temperature of annealing result primarily in changes in mispriming.

Conclusion

Optimizing PCR conditions involves determining the set of conditions, including annealing temperature, buffer pH, and $MgCl_2$ con-

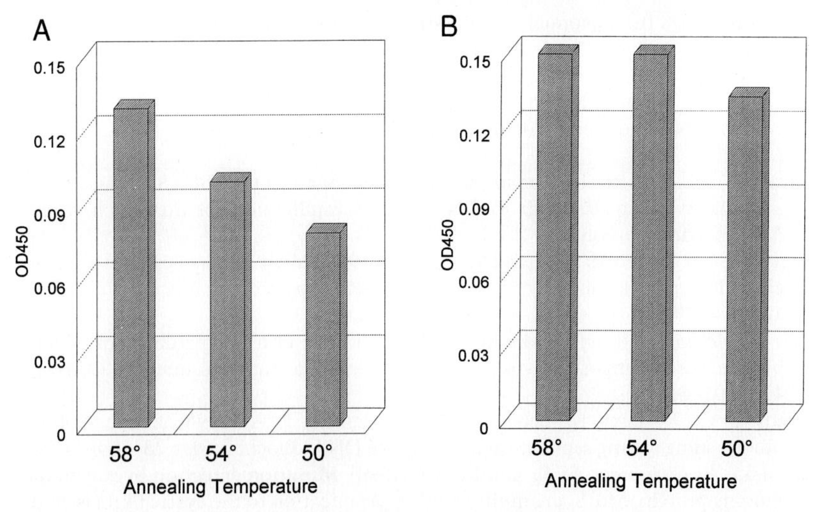

Figure 4 Effects of annealing temperature on PCR fidelity by indirect IMBP assays. Genomic (A) or plasmid (B) DNA templates were amplified as described in the legend to Fig. 3 except that the amplicons were not biotin labeled. To measure fidelity, samples (20 ng) were denatured, reannealed, and mixed with competitive fragments (10 ng/well). Total volume sample plus competitive fragment was 30 μl. The mixture was added to an IMBP assay plate.

centration, which produce sufficient product for the intended use and produce that product with maximum fidelity. IMBP assays provide a convenient, inexpensive, and rapid method to determine PCR fidelity. They are also useful for comparing polymerases and for determining lot to lot variation in polymerase fidelity. For such quality control assays it is possible to use a plasmid (or similar) template with high target to nontarget ratio to minimize the effects of mispriming and concentrate primarily on misincorporation differences. The use of IMBP for optimization and fidelity determination can dramatically increase the utility and consistency of PCR amplification.

References

Debbie, P., Young, K., Pooler, L., Lamp, C., Marietta, P., and Wagner, R. (1997). Allele identification using immobilized mismatch binding protein: Detection and identification of antibiotic-resistant bacteria and determination of sheep susceptibility to scrapie. *Nucleic Acids Res.* **25,** 4825–4829.

Ellis, L. A., Taylor, G. R., Banks, R., and Baumberg, S. (1994). MutS binding protects heteroduplex from exonuclease digestion *in vitro*: A simple method for detecting mutations. *Nucleic Acids Res.* **22,** 2710–2711.

Fazakerley, G. V., Quignard, E., Woisard, A., Guschlbauer, W., van der Marel, G. A., van Boom, J., Jones, M., and Radman, M. (1986). Structures of mismatched base pairs in DNA and their recognition by the *E. coli* mismatch repair system. *EMBO J.* **5,** 3697–3703.

Hunter, W. N., Brown, T., Anand, N. N., and Kennard, O. (1986). Structure of an adenine–cytosine base pair in DNA and its implication for mismatch repair. *Nature* **320,** 552–555.

Jiricny, J., Su, S.-S., Wood, S. G., and Modrich, P. (1988). Mismatch-containing oligonucleotide duplexes bound by the *E. coli mutS*-encoded protein. *Nucleic Acids Res.* **16,** 7843–7853.

Jones, M., Wagner, R., and Radman, M. (1987). Repair of a mismatch is influenced by the base composition of the surrounding nucleotide sequence. *Genetics* **115,** 605–610.

Kroutil, L. C., Register, K., Bebenek, K., and Kunkel, T. A. (1996). Exonucleolytic proofreading during replication of repetitive DNA. *Biochemistry* **23,** 1046–1053.

Lishanski, A., Ostrander, E. A., and Rine, J. (1994). Mutation detection by mismatch binding protein, MutS, in amplified DNA: Application to the cystic fibrosis gene. *Proc. Natl. Acad. Sci. U.S.A.* **91,** 2674–2678.

Lundberg, K. S., Shoemaker, D. D., Adams, M. W. W., Short, J. M., Sorge, J. A., and Mathur, E. J. (1991). High-fidelity amplification using a thermostable DNA polymerase isolated from *Pyrococcus furiosus. Gene* **108,** 1–6.

Modrich, P. (1991). Mechanisms and biological effects of mismatch repair. *Annu. Rev. Genet.* **25,** 229–253.

Parker, B. O., and Marinus, M. G. (1992). Repair of DNA heteroduplexes containing small heterologous sequences in *Escherichia coli. Proc. Natl. Acad. Sci. U.S.A.* **89,** 1730–1734.

Petruska, J., and Goodman, M. (1985). Influence of neighboring bases on DNA polymerase insertion and proofreading fidelity. *J. Biol. Chem.* **260,** 7533–7539.

Radman, M., and Wagner, R. (1986). Mismatch repair in *Escherichia coli. Annu. Rev. Genet.* **20,** 523–538.

Su, S.-S., and Modrich, P. (1986). *Escherichia coli mutS*-encoded protein binds to mismatched DNA base pairs. *Proc. Natl. Acad. Sci. U.S.A.* **83,** 5057–5061.

Su, S.-S., Lahue, R. S., Au, K. G., and Modrich, P. (1988). Mispair specificity of methyl-directed DNA mismatch correction *in vitro. J. Biol. Chem.* **263,** 6829–6835.

Wagner, R., and Radman, M. (1995). Mismatch binding protein-based mutation detection systems. In *Methods: A Companion to Methods in Enzymology,* Vol. 7, pp. 199–203. Academic Press, San Diego.

Wagner, R., Debbie, P., and Radman, M. (1995). Mutation detection using immobilized mismatch binding protein (MutS). *Nucleic Acids Res.* **23,** 3944–3948.

8

A NEW GENERATION OF PCR INSTRUMENTS AND NUCLEIC ACID CONCENTRATION SYSTEMS

M. A. Northrup, L. A. Christel, W. A. McMillan, K. Petersen, F. Pourahmadi, L. Western, and S. Young

Current commercial instruments for amplification of nucleic acids via PCR have satisfied the technical requirements to prove the viability and potential of this powerful biological analysis technique. However, there are significant opportunities to advance the instrumentation technology to allow for the extension of the PCR technique beyond the laboratory and make it low cost, rapid, flexible, and automated. Similar improvement opportunities in sample purification and processing techniques also exist.

Speeding up the ramping rates between denature, extension, and anneal temperatures has been shown to augment the PCR process (Wittwer *et al.*, 1990), and this concept has developed into a commercial product (Wittwer *et al.*, 1997). In addition, the ability to combine homogenous detection methodologies with PCR has also been incorporated into commercially available products (Idaho Technology LightCycler and Perkin–Elmer/Applied Biosystems Model 7700). Recent advances in fluorescent detection chemistries, such as the TaqMan (Livak *et al.*, 1995) and molecular beacons (Tyagi and Kramer, 1996; Tyagi *et al.*, 1998), have been exploited with these new commercial thermal cycling products allowing product quantitation capabilities. Despite these advances in instrument technologies, there are still limitations in functionality, and therefore there are opportunities for improvements. Some of these new commercial opportunities for technological improvements include independent control of the thermal cycling and optical systems at each reaction site, rapid thermal cycling of the larger volumes (100 μl or greater) required for clinical analyses of infectious diseases, low power consumption, use of solid-state optical components (which can lower cost), modularity, ease of serviceability, and portability.

Although it has been shown that quantitative, homogenous PCR assays can be performed in both large bench-top instruments and a portable format (Northrup *et al.*, 1998), pre-PCR sample preparation remains the significant bottleneck in terms of the need for human intervention, complexity, and lack of automation. Sample preparation protocols range from a simple dilution of potential PCR inhibitors to complex filtering, multiple centrifugations, lysing, mixing, solid phase extraction, etc. Automation of some of these procedures has typically taken the form of robots replacing human-mediated steps and several solid phase extraction techniques. As with the analytical thermal cyclers for PCR, opportunities exist for improvements that will augment the distribution and applicability of the PCR technique. Specific examples of these technological improvements include replacing human-mediated fluid handling steps with microfluidic devices that are self-contained and automated, combining process steps in flowthrough disposable cartridge formats and speeding up the steps with micromachined, electronically controlled "chips."

In this chapter, we describe new instrument systems and devices that take advantage of the improvements afforded by these technological advances.

Historical Perspective on Miniaturized PCR Devices: "PCR on a Chip"

In addition to the commercial development of thermal cyclers for PCR, there have been a number of alternative, research-based developments of innovative devices for performing PCR. In particular, silicon micromachining or microelectromechanical fabrication techniques have been applied to make several versions of miniaturized thermal cyclers, or what has been referred to as PCR on a chip. The relative utility and functionality of these devices vary, and a brief overview is presented.

The first PCR results from a silicon reactor with integrated heaters were obtained in 1993 (Northrup *et al.*, 1993). The results from that device included the rapid (20-sec cycles) and low-power (1 or 2 W) amplification of a 142-bp target (gag region of HIV) cloned into MS-2 bacteriophage. That work showed the ability to significantly decrease thermal mass and provide direct, integrated, and thermally isolated heating to each reaction site (50-μl reaction volume). Performing PCR directly on silicon has certain biocompatibility problems, however. Wilding *et al.* (1994) published PCR results from a silicon device without heaters. That device was thermally cycled with the use of an external "hot plate." The PCR results were marginal, with large "primer dimer" bands probably due to the slow ramping rates (~0.3°C/sec) or possibly due to inhibition by the silicon surfaces as pointed out later by the same group (Cheng *et al.*, 1996). The latter results showed an improvement in performance but still required external heating sources. Another group (Burns *et al.*, 1996) performed PCR on a micromachined device with integrated heaters that was part of a larger integrated structure. Those results also showed variations in productivity, possibly due to the thermal sinking effect of the large substrate or inhibition by the surfaces.

Performing PCR directly in the silicon device was abandoned by the original group that performed the work in 1993. Instead, the silicon reaction chambers were used as sleeve-type reactors with disposable, thin-walled, polypropylene liners (Northrup *et al.*, 1995, 1996a,b, 1997, 1998). The use of disposable liners suggests many advantages, including reduced cost per assay, the removal of inhibitors, and possible cross-contamination, and it allows for the direct coupling of the PCR reactor to a microelectrophoresis channel (Woolley *et al.*, 1996). Woolley *et al.* showed rapid thermal cycling (30 cycles in 15 min), direct, hands-off transfer of PCR product into a microelectrophoresis channel, and a 70-sec electrophoretic separation. The total time of analysis was 20 min. A micromachined electrochemiluminescence cell for possible PCR product detection has also been devised. Results from the detection of potential DNA labels (ruthenium) have been provided but not in conjunction with PCR (Hsueh *et al.*, 1996).

Several recent articles have been published with new ideas for performing PCR on a chip. Waters *et al.* (1998) reported results on a combination chip (a glass slide with etched channels and chambers) that performs cell lysis, PCR, and electrophoresis. The micromachined device was not fully integrated because external heaters, electrodes, and detectors were provided to perform the required functions. The PCR chamber was a small part of a large substrate and the heating was provided by an external hot plate thermal cycler (MJ Research, Inc.). As a result, the thermal cycling was extremely slow, requiring more than 4 hr to perform 24 cycles (9 min per cycle in dwell times and an additional 35 min for ramping between temperatures). Products visualized by agarose gel electrophoresis indicate that a significant amount of primer-dimer artifacts were generated. These results might be explained by slow thermal cycling, the unusually low annealing temperature (37°C), or possible inhibitors present.

Recently, two groups have attempted to design and fabricate micromachined devices for flowthrough PCR. The concept has been demonstrated on a macro scale, but two recent publications describe attempts to miniaturize the process. Kohler *et al.* (1998) fabricated a sophisticated silicon/glass chip with integrated heaters for performing flowthrough PCR. This work included temperature and flow rate data, but PCR was not demonstrated. Kopp *et al.* (1998) showed PCR results from an etched glass microchannel device in which the fluid path meandered over three copper heating blocks (5 W each). These authors provided results of a sample containing 10^8 molecules following 20 amplification cycles. The sensitivity of the system appeared compromised as flow rates were increased from 5.8 to 72.9 nl/sec, corresponding to total reaction times ranging from 18.8 to 1.5 min needed to complete 20 cycles. Although these results appear promising, one must keep in mind the relatively small reaction volumes (10 μl), the high-input copy numbers added per sample (10^9), the lack of integrated functionality (i.e., heaters or detectors), and the results from varying flow rates suggesting reduced sensitivity as individual cycle times drop below 20 sec.

Clearly, exciting microdevices are being devised to perform PCR. However, until they meet the performance requirements established by larger systems and offer a unique advantage in terms of instrumentation design, they will remain in research laboratories. Some of the principles learned during these exercises can provide important lessons for the development of future DNA analysis tools. The sleeve-type reactor system described previously (Northrup *et al.*, 1998) is currently being applied to a commercial system called the SmartCycler. Some of the principles from that work, such as the integration of heaters, low thermal mass reaction sleeves, disposable reaction inserts, independent control at each reaction site, and the incorporation of solid-state optical components, are being implemented. The commerical version of these technological advancements, as incorporated into the SmartCycler, are described in the following section.

Independent Control and Real-Time Fluorescent Detection of Each PCR Reaction Site: The SmartCycler

Conventional PCR thermal cyclers, including the versions that have optical detection capabilities, rely on uniform heating and cooling of all reaction sites. Typically the reaction sites are machined metal blocks with a variety of holes for insertion of the reaction tubes or, in the case of the LightCycler, are small volume glass capillary tubes arranged within an oven with a single heat source and cooling apparatus. The high surface to volume ratio of the small glass capillaries provides a very rapid thermal response for small reaction volumes as a carousel moves the tube into an optical read station for fluorescence detection. In the larger analytical thermal cycler (PE/ABI 7700), fiber optic cables provide the optical path to the heated metal block platform from a laser-based illumination source and back to the detector.

The Cepheid SmartCycler is based on the I-CORE (Intelligent Cooling/heating Optics Reaction) module. The I-CORE module (Fig. 1 and Fig. 2a) contains low thermal mass, high thermal conductivity ceramic (A1N) heat exchanging devices that have embedded thin-film resistive heaters. The I-CORE is based on a patented chemical reaction system (Northrup *et al.*, 1996a). Also within the I-CORE module are two miniature optics blocks that provide four-color excitation from light-emitting diodes and four detection channels based on silicon photodetectors. Optical wavelength discrimination is provided by a series of embedded reflective, cut-off, and band pass filters similar to those used in much larger optical bench systems. An eight-

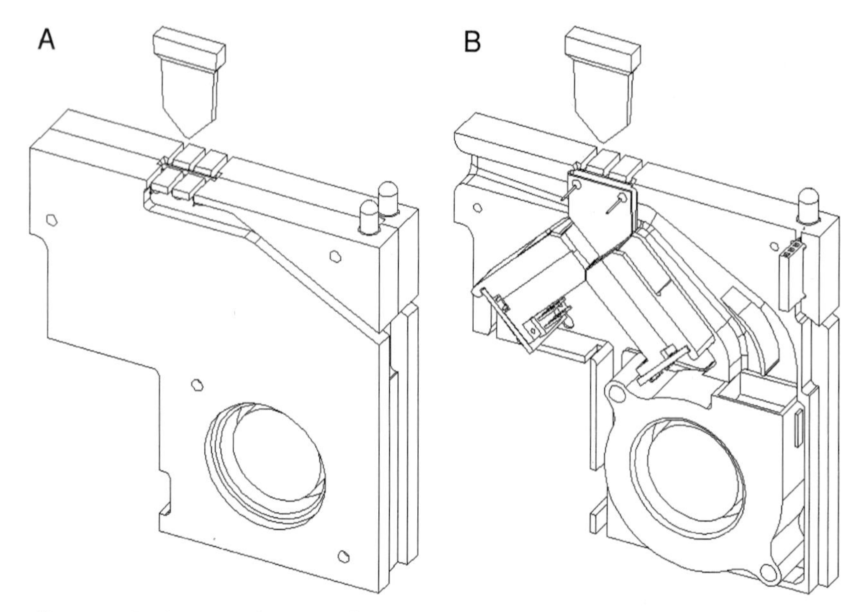

Figure 1 Engineering drawings showing the I-CORE module with the covers on (a) and with one side of the cover removed (b). The large-volume, fast thermal response reaction tube is shown, along with the internal heat-exchanging chambers, cooling fan, and dual (four-color excitation and four-color detection) optical interrogation devices.

site SmartCycler is shown in Fig. 2b and 2c. Results showing four independent thermal cycles are displayed in Fig. 3.

Instrument Control and Data Analysis Software

One of the opportunities for improvments of PCR instrumentation is in control of the reaction chambers and optical detection systems. The information obtained from these systems can be used as feedback to the instrument for optimization of the reaction parameters. Also, significant improvements in the user interface can improve the utility, ease of use, and productivity of the instrument. The SmartCycler incorporates these new features. A brief description of some of the key user interface features is presented.

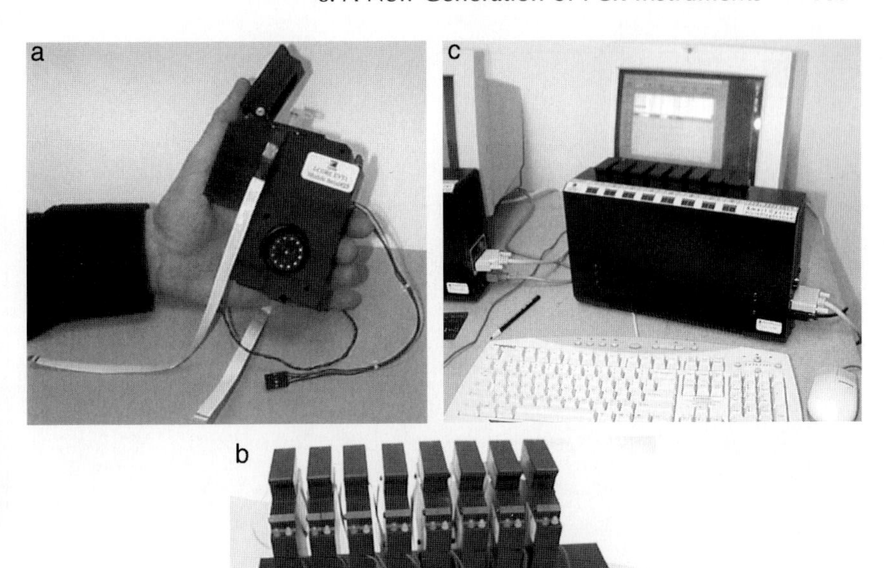

Figure 2 Photographs of the ICORE components, assembly, and complete SmartCycler instrument. Figures show (a) one ICORE module, (b) how they are assembled within the instrument, and (c) the complete instruments along with the computer.

Instrument Program Menu

New programs (a series of heating and cooling steps) are created from a graphical user interface screen (Fig. 4). The template provided can be used to create a specific user-defined program. Most of the functions are self-explanatory to someone familiar with running PCR reactions. Within the four programming areas, any combination of holds and cycling profiles can be generated by entering temperatures and times (steps). Then the user can select one or more sites to run the selected program, or independent sites can be separately programmed by performing the same procedure outlined previously. All information is associated and stored with each particular run in the Results section.

Instrument Menu Screen

This screen shown in Fig. 5 displays and monitors the actual thermal cycling status and optical output at each reaction site. Each window

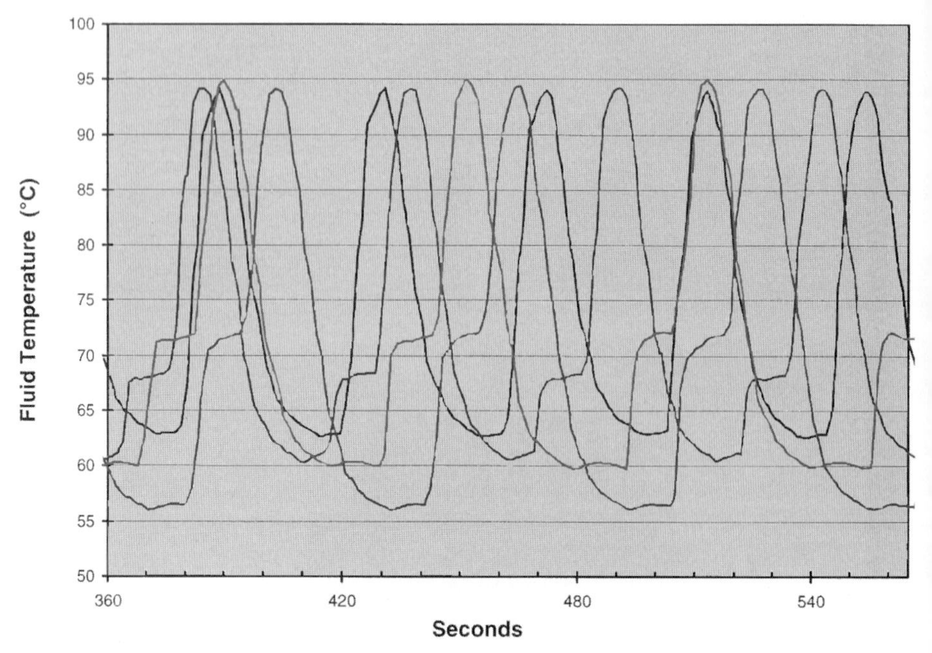

Figure 3 Example of a series of simultaneous thermal profiles corresponding to four different PCR assays (*Bacillus subtilus* bacterium, cftr exon 10 genetic marker, lambda virus control, and MS2 virus) run simultaneously on the SmartCycler.

corresponds to one of the multiple reaction sites as labeled on the instrument. Specific program information is displayed in real time by selecting the site of interest (e.g., site 4 is selected in Fig. 5).

Results Menu Screen

The results of a particular run are displayed by program name, date, and operator in the Results menu screen shown in Fig. 6. Other information displayed includes time program started and finished, the reaction site used, added "tags," and the final program status (completed, failed, or stopped by user). A graphical display of the entire run can be viewed by selecting temperature chart (Fig. 6) or optical results (Fig. 7).

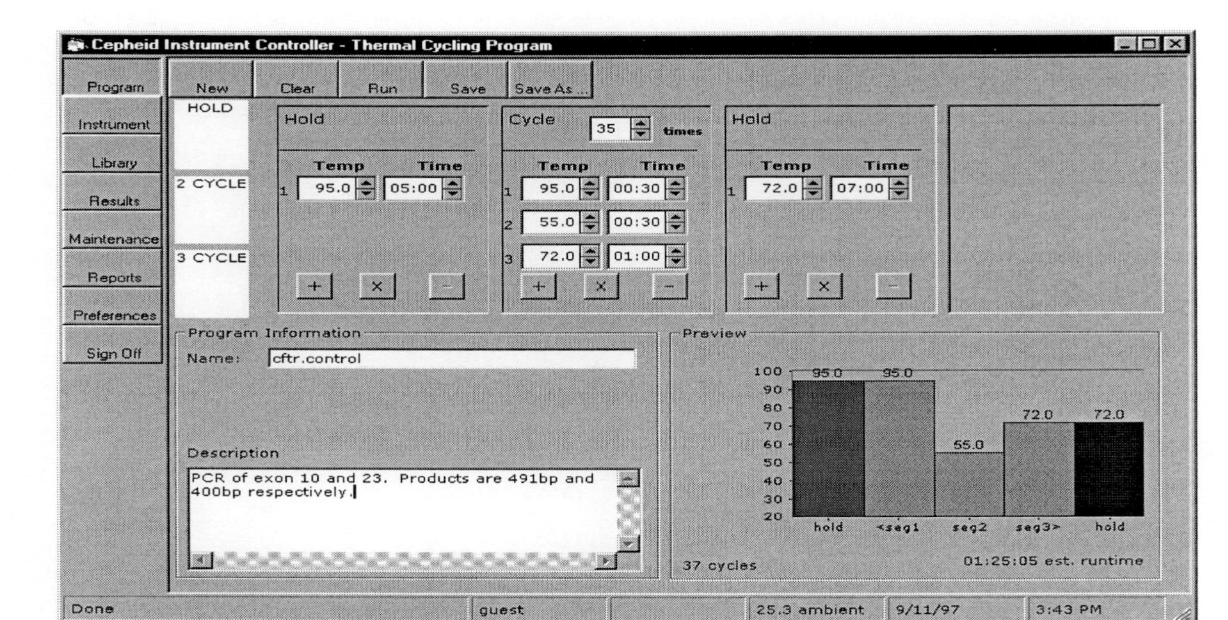

Figure 4 Example of an instrument programming screen from the SmartCycler user interface. This screen shows the use of graphics to easily and simply program one or more reaction sites.

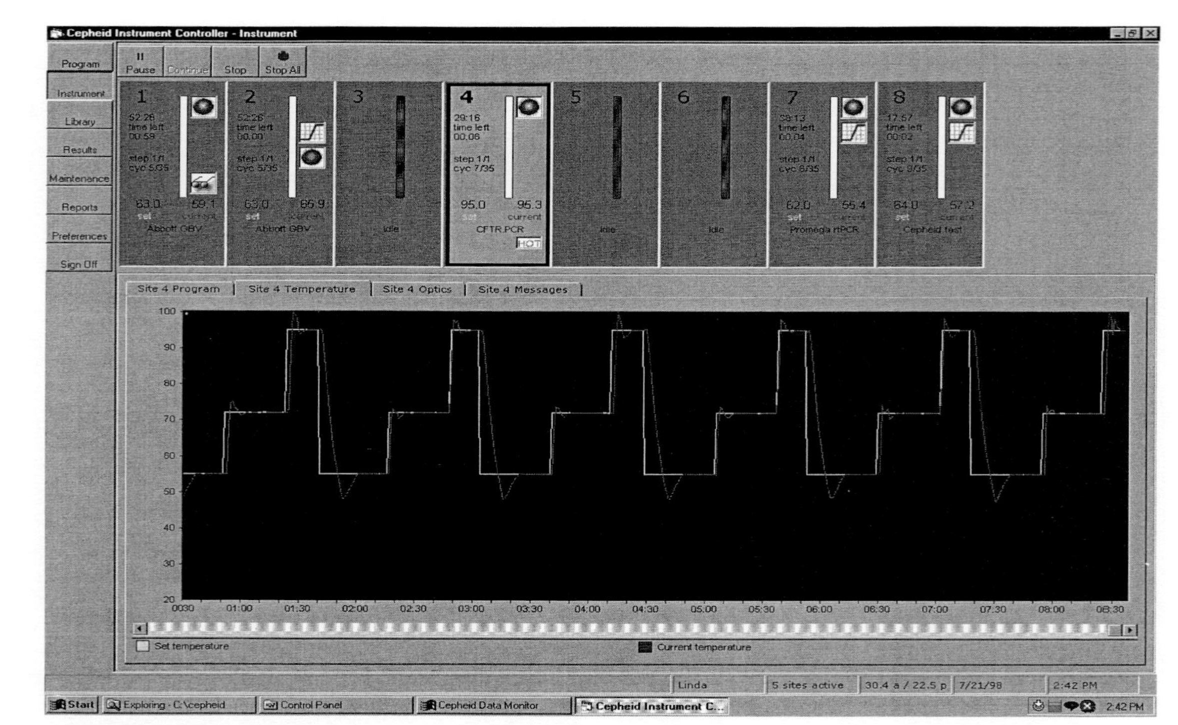

Figure 5 Example of an instrument menu screen from the SmartCycler user interface. Included are the real-time thermal data from the reaction site highlighted. This screen emphasizes the user's ability to completely and independently program individual reaction sites.

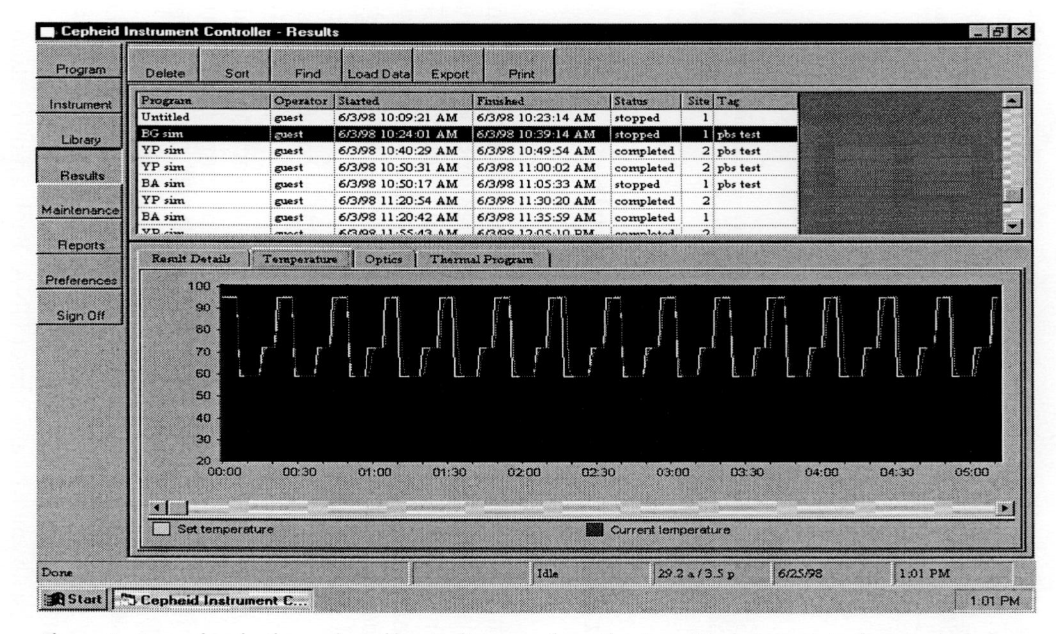

Figure 6 Example of a thermal profile result screen from the SmartCycler user interface. Included are some results from a variety of files (top) and the specific thermal profile from the particular program highlighted (bottom). Optical results from a particular assay are shown in Fig. 7.

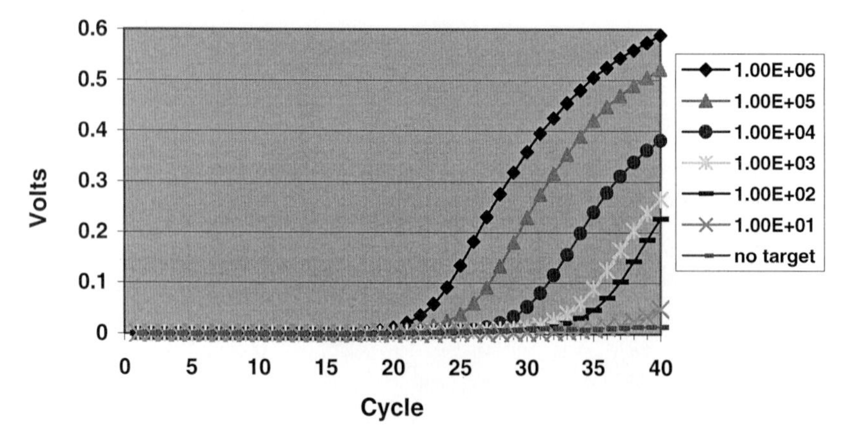

Figure 7 Serial dilution series (10–1,000,000 starting copies) of real-time optical detection of the β-actin gene by the TaqMan technique on the SmartCycler. See text for assay details.

TaqMan β-Actin Detection Results

Real-time monitoring of PCR reactions was carried out using components from the TaqMan β-actin Detection Reagents kit purchased from Perkin–Elmer (Catalog No. 401846) and Ready-To-Go PCR Beads purchased from Pharmacia Biotech (Catalog No. 27-9555-01). Human genomic DNA was purchased from Promega (Catalog No. G3041) and used as the target nucleic acid in the following experiments. The 150-μl reactions contained 300 n*M* each of forward and reverse primers and 200 n*M* of fluorescent probe (FAM/TAMRA). The reconstituted PCR beads (six beads/150-μl reaction) provided a buffer containing 10 m*M* Tris–HCl (pH 9.0 at room temperature), 50 m*M* KCl, 1.5 m*M* $MgCl_2$, 200 μM each dNTP, *Taq* polymerase at 0.6 U/μl, and stabilizers, including BSA. Additional $MgCl_2$ was added to a final concentration of 3.5 m*M*. Human genomic DNA was serially diluted from 3 μg to 30 pg, representing an input target number ranging from 1×10^6 to 10 molecules; a negative control was run in the absence of any target DNA. Each amplification cycle consisted of 95°C for 15 sec and 60°C for 60 sec, preceded by a one-time hold at 96°C for 3 min, as recommended by the manufacturer (Perkin–Elmer). The cycles were repeated 40 times. The magnitude of the fluorescence at each target concentration (Fig. 7) correlated strongly with the amount of the amplification product (294 bp) generated, as determined by subsequent agarose gel electrophoresis and ethidium bromide staining.

Nucleic Acid Extraction Cartridges

Figure 8 shows a photograph of a prototype disposable nucleic acid purification cartridge with integrated reaction chamber for PCR and real-time optical interrogation. Within this cartridge is a nucleic acid capture chip, reagent and sample input ports, reagent mixing and waste chambers, fluid channels, and flow sensors. This particular cartridge is designed to perform simple nucleic acid capture, washing, and controlled elution into the integrated 100-μl reaction/detection chamber. External fluid sources and motive sources provide automated, computer-controlled processing within the cartridge. The PCR reaction vessel fits directly into the SmartCycler instrument.

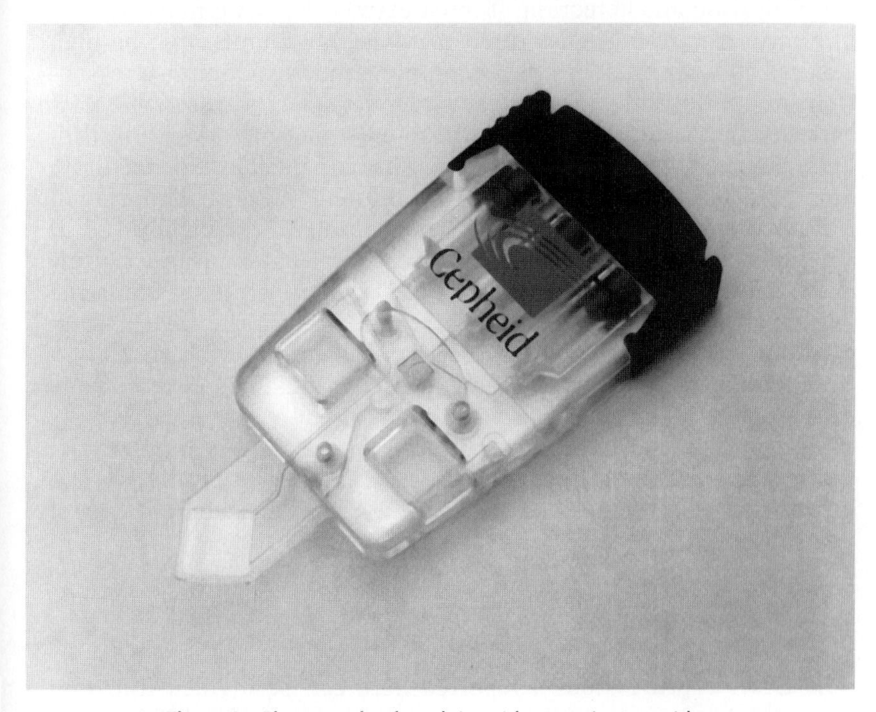

Figure 8 Photograph of nucleic acid extraction cartridge.

Therefore, all the processing after sample input is automated, and the user is not required to perform any pipetting or sample handling. The high surface area chip is used as a device with a silicate (SiO_2) surface for controlled nucleic acid capture and elution as described later.

Many NA purification kits are available commercially (Biorad Labs, Hercules, CA; Promega, Madison, WI; and Qiagen, Inc., Santa Clarita, CA). Typically, these kits utilize some form of silica gel, glass matrix, or membrane as the capture medium since it is known that NA will bind to glass or other silica-type surfaces under the proper chemical conditions (Boom *et al.*, 1993). New micromachining techniques and innovations both simplify and improve NA extraction and concentration. We have fabricated high surface area NA capture surfaces of oxidized silicon using deep reactive ion etching (DRIE) (Klaassen *et al.*, 1995; Bhardwaj and Ashraf, 1995) and have shown that such structures can be used to controllably capture and release DNA from test solutions. Several parameters are of primary importance to system performance. The chip must have sufficient binding capacity to retain the quantity of NA required for subsequent amplification and detection. It must extract the NA efficiently, i.e., the ratio of target NA captured to target NA input must be high (exceed 50%). It must be able to complete the extraction in a reasonable amount of time, i.e., the capture must occur at reasonable flow rates, so that clinically realistic volumes (sometimes several milliliters) can be processed expeditiously. The chip must also concentrate the input NA, allowing the NA to be eluted (released) into a small volume for PCR amplification. These volumes are typically 100 μl or less to conserve expensive PCR reagents. The chip must allow wash solutions to be efficiently passed through while retaining the NA so that PCR inhibitors can be excluded from the final elution without requiring large volumes of wash solution typically associated with commerical kits. Finally, the chip must have a reasonable cost in order to be used in a disposable format.

The process used to produce the NA capture chips is simpler than competing technologies such as LIGA (Marques *et al.*, 1997). The fabrication of these devices has been described elsewhere (Christel *et al.*, 1998).

DNA Capture Studies

The DNA capture studies reported here can be divided into two types: high concentration inputs (on the order of 100–1000 ng/ml) and low concentration inputs (at or below 10^5 copies of target DNA). These two regimes have relevance to different clinical situations. For low concentration studies, chip output was processed through a standard PCR protocol using a Perkin–Elmer 9700 thermal cycler. Reference samples spanning several orders of magnitude in target starting copy number were run in parallel. PCR product was then processed through gel electrophoresis, typically with 1% agarose gels and ethidium bromide staining. Photographs of the resulting gel bands were compared visually to the reference standards to estimate the starting copy number in the chip elutions.

High Concentration Results

For high concentration studies, a solution of plasmid digest DNA was used as the starting material (Bio-Rad, Product No. 170-3123). This solution consists of a mix of DNA with an average length of about 500 bp. The material is tagged with fluorescein, a fluorescent marker, allowing *in situ* observation of the DNA during sample processing.

The chip elutions were analyzed using a technique similar to that of Rye *et al.* (1993) and modified by Christel *et al.* (1998). The ranges of experimental parameters are given in Table 1. Note that the internal volume of the chip is about 0.2 μl, leading to a residence time of 200 msec at a flow rate of 1.0 μl/sec.

The total binding capacity of high aspect ratio chips was investigated using an input of 400 ng of DNA (400 μl of a 1000 ng/ml solution) followed by 400 μl of wash and then elution. All flow rates

Table 1

Experimental Conditions for High Concentration DNA Capture Experiments

Parameter	Range
DNA concentration	100–1000 ng/ml
DNA volume	400 μl
DNA total dose	40–400 ng
DNA flow rate	0.1–5.0 μl/sec
Wash volume	400 μl
Wash flow rate	0.5–5.0 μl/sec
Elution flow rate	0.5 μl/sec

were 0.5 μl/sec. By comparing each elution signal to the standards and summing them, we concluded that 11 or 12 ng of DNA was captured and then eluted from the chip. It has been reported (Vogelstein and Gillespie, 1979) that the maximum binding capacity of glass is approximately 40 ng/cm^2. Since the internal surface area of the chip is approximately 0.36 cm^2, it is evident that in this experiment, an amount of DNA consistent with the maximum binding capacity of glass has been captured. As a comparison, the same protocol was run utilizing a "flat" chip without the enhanced surface area. This chip had an internal surface area of about 0.06 cm^2. In this case, about 2.5 ng of DNA was captured and eluted, with nearly one-sixth of it captured on the DRIE chip (Fig. 9). This is also consistent with binding at the capacity limit of glass.

Low Concentration Results

For the low concentration studies λ DNA (48,000 bp) (Pharmacia, Bridgewater, NJ) was used as a bacteria simulant. Starting solutions were again prepared by dilution of λ stock with chaotropic salt solutions. Starting copy number for the first set of experiments was 5×10^4 copies in 500 μl of solution. As a control, the experimental protocol was first run using a chaotropic solution without DNA. Standards of known copy number (10^4 and 5×10^4) were also prepared using both water that had passed through the chip test system and pure water.

The protocol was similar to that of the high concentration studies. Figure 10 shows the gel photograph from one such experiment. The standards prepared with system water are shown first, followed by the run with blank solution and the run with DNA solution. By comparing the first 25-μl chip elution to the 10^4 and 5×10^4 standards in system water, it is estimated that the first chip elution contains about 2.5×10^4 copies. This implies a 10$\times$ concentration effect and a 50% capture efficiency.

There are several advantages to this approach to the purification of nucleic acids: Continuous-flow operation allows the processing

Figure 9 SEM of the nucleic acid extraction chip (a) and a close-up of the same chip (b) showing the details of the geometry. This silicon chip was fabricated using the DRIE process.

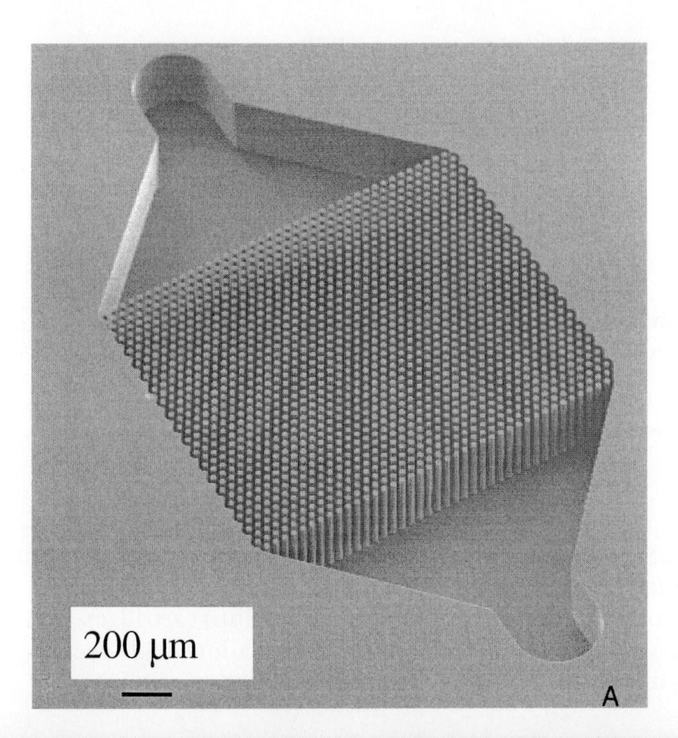
200 µm
A

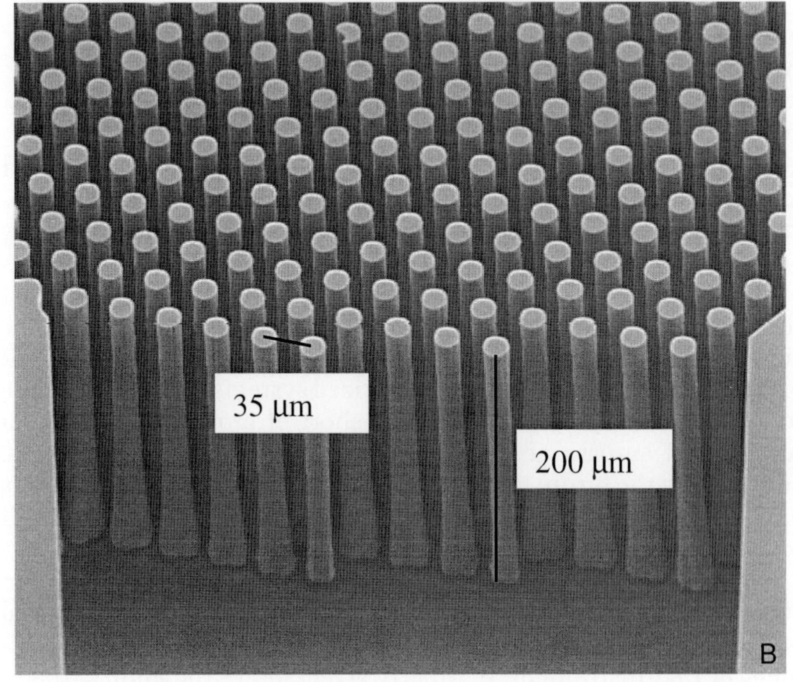
35 µm
200 µm
B

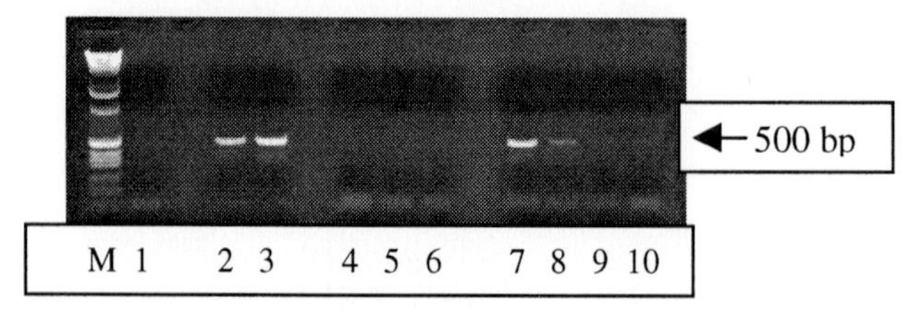

Figure 10 Photograph of electrophoretic gel showing DNA chip capture. M, molecular weight standard; 1, Neg Ctl; 2, 10^4 standard; 3, 5×10^4 standard; 4–6, control run with no DNA; 7–10, chip elutions from DNA run. By comparing lane 7, obtained from the first 25-μl chip elution, to standard lanes 2 and 3, it is estimated that a capture efficiency of 50% was acheived, with a concentration factor of about 10X in the first elution.

of large volumes of sample containing extremely low concentrations of target DNA or RNA; application of local heat increases elution efficiency; the device is easily integrated into cartridges; and the low internal volume and high surface area allow high recovery rates, particularly for low concentration samples.

Summary

The PCR technique has clearly evolved into an important tool for researchers and clinicians. This has been afforded by the commercialization of robust and dependable instruments for thermal cycling and, recently, with homogenous fluorescence detection. The state-of-the-art instruments that include real-time, homogeneous, fast thermal cycling, and quantitative detection capabilities still leave significant opportunities for improvements. Efforts to develop PCR on a chip or micromachined/miniaturized systems have shown some interesting capabilities but still fall short of providing the type of results that surpass or even equal those of commercial systems. However, in the future the development of new nucleic acid systems based on some of the principles from such research devices will probably occur.

This chapter describes the extension of previous work based on silicon micromachining that has shown equivalent and improved performance over commercial systems. The commercial embodiment of that instrument, or SmartCycler, shows the ability to perform thermal cycling with real-time homogeneous product detection using low-cost solid-state components. Other improvements over commercial systems include new graphical user interface, independent control of each reaction site, modularity, and rapid thermal cycling of large volumes.

In the areas of pre-PCR sample processing, often the bottleneck of analyses, silicon micromachining has been applied to the development of flowthrough sample processing cartridges. Ultimately, all the processing and homogenous quantitative detection will take place in one low-cost disposable, integrated system. The PCR user community anxiously awaits these changes to bring PCR to the next level of utility.

References

Belgrader, P., Smith, J. K., Weedn, V. W., and Northrup, M. A. (1998). The application of a portable, miniature thermal cycler for human identification. *J. Forensic Sci.* **43**(3), 315–119.

Bhardwaj, J. K., and Ashraf, H. (1995). Advanced silicon etching using high density plasmas. In *Proceedings of the Society of Photo-Optical Instrumentation Engineers, Micromachining and Fabrication Technology*, Bellingham, WA, Vol. 2639, pp. 224–233.

Boom, W. R., Adriaanes, H. M. A., Kievits, T., and Lens, P. F. (1993). Process for isolating nucleic acid, U.S. Patent No. 5,234,809.

Burns, M. A., Mastrangelo, C. H., Sammarco, T. S., Man, F. P., Webster, J. R., Foerster, B. N., Jones, D., Fields, Y., Kaiser, A. R., and Burke, D. T. (1996). Microfabricated structures for integrated DNA analysis. *Proc. Natl. Acad. Sci. USA* **93,** 5556–5561.

Christel, L. A., Petersen, K., McMillan, W., and Northrup, M. A. (1998). Automated nucleic acid probe assays using silicon microstructures for nucleic acid concentration. Submitted for publication.

Cheng, J., Shoffner, M. A., Hvichia, G. E., Kricka, L. J., and Wilding, P. (1996). Chip PCR II. Investigations of different PCR amplification systems in microfabricated silicon-glass chips. *Nucleic Acid Res.* **24,** 380–385.

Hsueh, Y.-T., Smith, R. L., and Northrup, M. A. (1996). A microfabricated electrochemiluminescence cell for the detection of amplified DNA. *Sensors Actuators* **B33,** 110–114.

Ibrahim, M. S., Lofts, R. S., Henchal, E. A., Jahrling, P., Weedn, V. W., Northrup, M. A., and Belgrader, P. (1998). Real-time microchip PCR for detecting single-base differences in viral and human DNA. *Anal. Chem.* **70**(90), 2013–2017.

Klaassen, E. H., Petersen, K., Noworolski, J. M., Logan, J., Maluf, N. I., Brown, J., Storment, C., McCulley, W., and Kovacs, G. T. A. (1995). Silicon fusion bonding and deep reactive ion etching: A new technology for microstructures. In *Proceedings of the 8th International Conference on Solid-State Sensors and Acutators*, Stockholm, pp. 556.

Kohler, J. M., Dillner, U., Mokansky, Poser, S., and Schultz, T. (1998). Micro channel reactors for fast thermocycling. Paper presented at Process Miniaturization: 2nd

International Conference on Micoreaction Technology, AIChE National Meeting, March 9–12, New Orleans.

Kopp, M. U., de Mello, A. J., and Manz, A. (1998). Chemical amplification: Continuous-flow PCR on a chip. *Science* **280,** 1046–1048.

Livak, K. J., Flood, S. J. A., Marmaro, J., Giusti, W., and Deetz, K. (1995). *PCR Methods Appl.* **4**(3), 357–362.

Marques, C., Desta, Y. M., Rogers, J., Murphy, M. C., and Kelly, K. (1997). Fabrication of high-aspect-ratio microstructures on planar and nonplanar surfaces using a modified LIGA process. *J. Micro Electro-Mechanical Systems* **6**(4), 329–336.

Northrup, M. A., Ching, M. T., White, R. M., and Watson, R. T. (1993). DNA amplification in a microfabricated reaction chamber. In *Transducers '93, Seventh International Conference on Solid Sensors and Actuators,* pp. 924–927. IEEE Proceedings, Yokohama, Japan.

Northrup, M. A., Hills, R. F., Landre, P., Lehew, S., Hadley, D., and Watson, R. (1995). A MEMS-based DNA analysis system. In *Transducers '95, Eighth International Conference on Solid State Sensors and Actuators. pp. 764–767.* IEEE Proceedings, Stockholm, Sweden.

Northrup, M. A., Mariella, R. P., Carrano, T. V., and Balch, J. W. (1996a). Silicon-based sleeve devices for chemical reaction, U.S. Patent No. 5,589,136.

Northrup, M. A., Beeman, B., Hills, R. F., Hadley, D., Landre, P., and Lehew, S. (1996b). Integrated miniature DNA-based analytical instrument. In *Analytical Methods and Instrumentation, Special Issue on MicroTAS* (H. M. Widmer, Ed.), pp. 153–157. Ciba Geigy, Basel.

Northrup, M. A., Beeman, B., Hadley, D., Landre, P., and Lehew, S. (1997). *Development of a micromachined chemical reaction chamber: Application to the polymerase chain reaction.* In *Automation Technologists for Genome Characterization* (T. J. Beugelsdijk, Ed.). Wiley, New York.

Northrup, M. A., Hadley, D., Landre, P., Lehew, S., Richards, J., and Stratton, P. (1998). A miniature DNA-based analytical instrument based on micromachined silicon reaction chambers. *Analy. Chem.* **70**(5), 918–922.

Rye, H. S., Dabora, J. M., Quesada, M. A., Mathies, R. A., and Glazer, A. N. (1993). Fluorometric assay using dimeric dyes for double- and single-stranded DNA and RNA with picogram sensitivity. *Anal. Biochem.* **208,** 144–150.

Tyagi, S., and Kramer, F. R. (1996). Molecular beacons: Probes that fluoresce upon hybridization. *Nature Biotechnol.* **14,** 303–308.

Tyagi, S., Bratu, D. P., and Kramer, F. R. (1998). Multicolor molecular beacons for allele discrimination. *Nature Biotechnol.* **16,** 303–308

van Driëenhuizen, B. P., Maluf, N. I., Opris, I. E., and Kovacs, G. T. A. (1997). Force-balanced accelerometer with mG resolution, fabricated using silicon fusion bonding and deep reactive ion etching. In *Proceedings of the 9th International Conference on Solid-State Sensors and Actuators. "Tranducers 97,"* Chicago, pp. 1229–1230.

Vogelstein, B., and Gillespie, D. (1979). Preparative and analytical purification of DNA from agarose. *Proc. Natl. Acad. Sci. USA* **76**(2), 615–619.

Waters, W. C., Jacobson, S. C., Kroutchinina, Khandurina, J., Foote, R. S., and Ramsey, J. M. (1998). Microchip device for cell lysis, multiplex PCR amplification, and electrophoretic sizing. *Anal. Chem.* **70**(1), 158–162.

Wilding, P., Shoffner, M. A., and Kricka, L. J. (1994). *Clin. Chem.* **40,** 1815–1818.

Wittwer, C. T., Fillmore, G. C., and Garling, D. J. (1990). Minimizing the time required for DNA amplification by efficient heat transfer to small samples. *Anal. Biochem.* **186,** 328–331.

Wittwer, C. T., Ririe, K. M., Andrew, D. A., David, D. A., Gundry, R. A., and Balis, U. J. (1997). The LightCycler™: A microvolume multisample fluorimeter with rapid temperature control. *BioTechniques* **22,** 176–181.

Woolley, A. T., Hadley, D., Landre, P., deMello, A. J., Mathies, R. A., and Northrup, M. A. (1996). Functional integration of PCR amplification and capillary electrophoresis in a microfabricated DNA analysis device. *Anal. Chem.* **68,** 4081–4086.

9

SEQUENCING PCR PRODUCTS

Jenny M. Kelley and John Quackenbush

PCR (Saiki *et al.*, 1985; Mullis *et al.*, 1986; Mullis and Faloona, 1987) has proven invaluable because it allows the amplification of specific DNA segments without having to first clone them in a microbial host. This has facilitated a wide range of applications, including mutation detection for disease diagnosis and genetic linkage studies (Saiki *et al.*, 1985; Weissenbach *et al.*, 1992), amplification of ribosomal genes for evolutionary studies (Sogin *et al.*, 1989; Woese, 1987), and environmental sampling (Bej *et al.*, 1990), as well as the completion of genome sequencing projects (Fleischmann *et al.*, 1995). Of particular importance to these applications is that PCR generates DNA in quantities sufficient for sequencing (Sanger and Coulson, 1975); the direct sequencing of PCR products allows the generation of DNA sequence data directly from patient and environmental samples and from genomic regions that are difficult to clone.

The flexibility of PCR has allowed a variety of primer design strategies to be used for both DNA amplification and sequencing (Fig. 1). Nested PCR and sequencing primers can be used to amplify common segments from a mixed population; these can then be distinguished by the use of a specific internal sequencing primer. Sam-

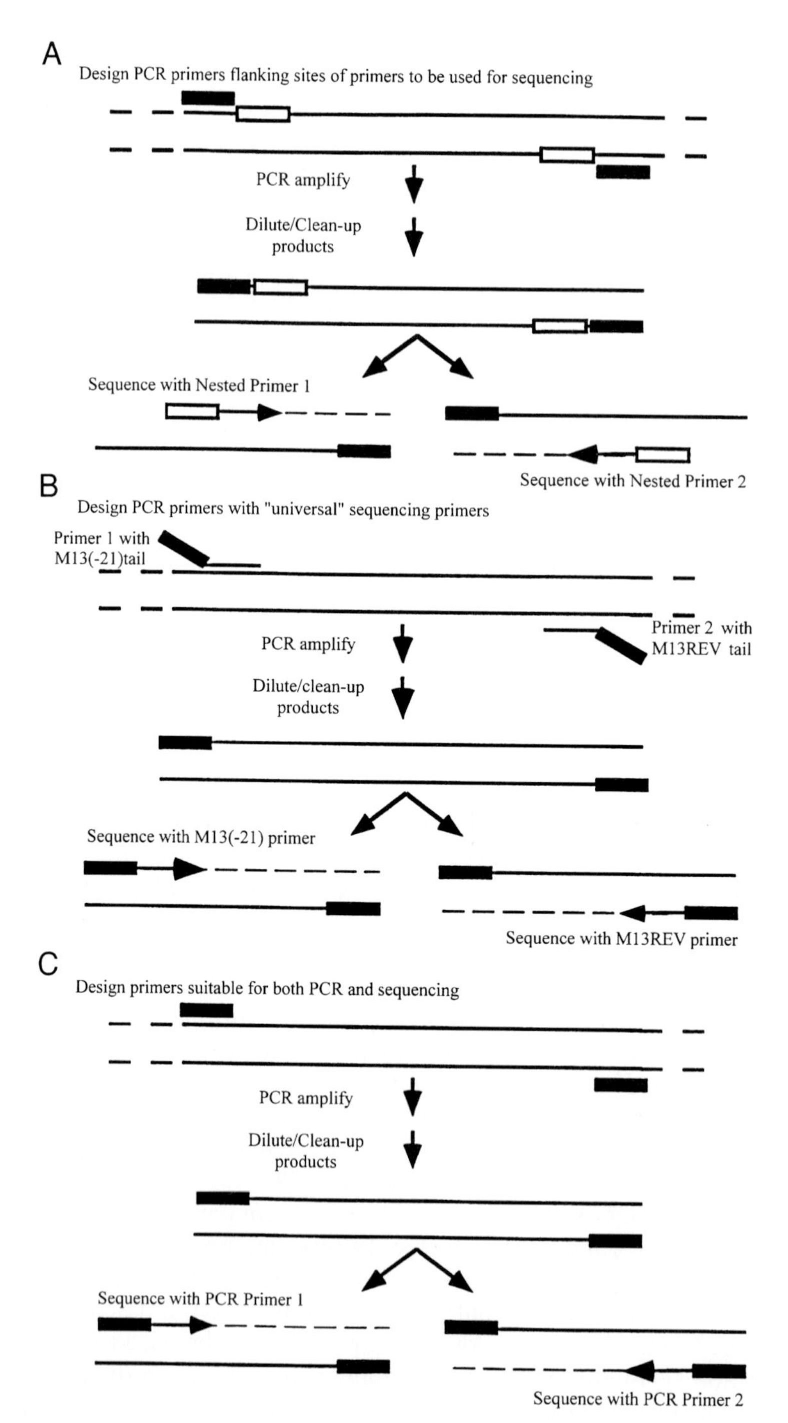

Figure 1 Primer design strategies for amplification and sequencing using (A) nested primers, (B) primers with "universal" tails, and (C) the same primers for both reactions.

ples amplified with specific primers can be sequenced using a universal sequencing primer through the addition of a 5′ "tail" to one or both of the amplification primers. This approach is particularly attractive for sequencing using dye-labeled primers. The amplification primers can also be used as sequencing primers, reducing the cost of primer synthesis. Each of these approaches should be evaluated in the context of the specific application, but all can yield high-quality DNA sequence data.

Regardless of the primer design strategy, there are two elements that are crucial for successful sequencing of PCR products. First, primer selection and amplification conditions must be optimized to yield sufficient quantities of the specific target without producing significant background. The use of a nested sequencing primer can help overcome some of the problems associated with the presence of nonspecific amplification products, but successful sequence determination can still be compromised by contaminating DNAs. While optimization of the amplification reaction is important for sequencing of PCR products, it is discussed elsewhere in this book and consequently will not be addressed in this chapter.

Second, unused PCR primers and nucleotides must be eliminated prior to sequencing. Radioactive and dye-terminator sequencing are particularly sensitive to the presence of excess amplification primers. Both forward and reverse PCR primers can act as sequencing primers, resulting in the generation of labeled sequence ladders. These additional fragments cannot be distinguished from the intended target, making it impossible to interpret the sequence data. Even in dye-primer sequencing, in which only those DNA segments complementary to the sequencing primer generate labeled sequence ladders, presequencing cleanup can be important for the generation of high-quality data. The excess primer and nucleotides carried over from PCR can alter the ratios of reagents in the sequencing reaction sufficiently to compromise both the fluorescent signal and the reaction fidelity.

A variety of approaches have been developed for the purification and sequencing of PCR products, including dilution for direct sequencing, column purification, gel purification, and exonuclease I/shrimp alkaline phosphatase (Exo I/SAP) treatment. Each of these has certain advantages and disadvantages, and each should be evaluated based on the requirements of the specific application.

Dilution and direct sequencing: If the PCR reaction has been rigorously optimized to use minimal concentrations of both primers and nucleotides without interfering with amplification, the reaction products can simply be diluted in water, usually at 1:5 or 1:10 dilutions, and used directly as sequencing templates (Trower *et al.*, 1995). This has the advantage of requiring minimal processing prior to sequencing but the disadvantage that optimization and characterization of both PCR and sequencing reactions must be performed so that the protocol delivers sufficient quality sequence data. This requires not only that the PCR reactions be optimized with regard to primer and nucleotide concentrations but also that any secondary PCR products that could interfere with sequencing be eliminated. The investment can be worthwhile, however, if the same product is to be sequenced from a large number of sources.

Column purification: There are a number of commercially available products for PCR purification. The Qiagen QIAquick PCR Purification Kit (Product No. 28104) uses a proprietary silica gel membrane that absorbs DNA and allows it to be eluted in 50 μl or less of water or Tris buffer. The columns separate oligonucleotides of 40 bases or fewer from the PCR product DNA, allowing primers and nucleotides to be removed. Ultrafiltration columns allow separation based on molecular weight. Centricon 100 columns (Amicon, Product No. 4212) have a 100-kDa molecular weight cutoff and consequently can be used to separate the desired PCR products from single-stranded DNA less than approximately 300 bases and double-stranded DNA less than 125 bases in length. Other commercially available spin columns use gel filtration chromatography media such as Sephacryl (Amersham Pharmacia Biotech) to preferentially bind small DNA segments allowing the elution of larger products for sequencing. The advantage of column purification products is that they are quick and relatively simple to use and they tend to provide highly reproducible results. The disadvantage is that any contaminating PCR products are not removed and these may interfere with subsequent sequencing. Furthermore, while primers and primer dimers are eliminated, primer oligomers may not be eliminated. These can generate contaminating extension products; therefore, care must be taken to ensure that the PCR primers do not produce such products. Finally, since recovery from columns can vary, DNA quantity and quality should be checked by agarose gel electrophoresis prior to sequencing.

Gel purification: Agarose gel electrophoresis separates PCR reaction products and by-products by size, at which point the desired band can be excised from the gel and extracted using any of a variety of methods, including commercial products such as the QIAquick Gel Extraction Kit (Qiagen, Product No. 28704) or other simple techniques (Heery *et al.*, 1990; Zhen and Swank, 1993). The primary advantage of gel purification is that it specifically separates the target fragment from contaminating PCR products and other amplification artifacts, primers, and nucleotides. There are a number of disadvantages, however. First, gel purification is not a rapid technique nor is it amenable to automation. Second, ethidium bromide staining and UV illumination are used to identify the reaction products in the gel and this can nick the PCR template, introducing artificial stops in radioactive and dye-primer sequencing approaches. Third, gel electrophoresis is not perfectly efficient in separating DNA fragments based on size, so some residual contamination of primers and other products may remain; dilution of the extracted products can effectively remove these. Finally, some agarose products contain small molecules that can interfere with enzymatic reactions, including sequencing, so care should be taken in selecting both the agarose and the extraction technique to eliminate any such artifacts.

Exo I/SAP: Following PCR, treatment with Exo I (Amersham, Product No. E 70073Z) degrades single-stranded DNA, including unincorporated primers, and SAP (Amersham Pharmacia Biotech, P/N E 70092Y) treatment dephosphorylates the residual oligonucleotides, inactivating them. Following treatment, the enzymes are heat inactivated and the products can then be used in sequencing reactions (Werle *et al.*, 1994; Hanke and Wink, 1994). This approach has the advantage of being relatively simple and cost-effective as well as amenable to high-throughput applications. A disadvantage is that secondary products are not removed by this process. Consequently, the PCR must still be optimized to prevent their formation.

Of these techniques, we have found the Exo I/SAP method to be consistently both cost-effective and highly reliable independent of primer design strategy. Consequently, we use this approach to PCR sequencing almost exclusively. As a demonstration of the application of this technique, we describe the PCR amplification and sequencing of two *Arabidopsis thaliana* plasmid clones, ATDSB57 and ATDSB59, each containing an insert of approximately 1.8 kb.

Protocols

PCR Amplification of Plasmid DNA

PRIMERS FOR AMPLIFICATION OF PUC18 INSERTS

pUC18pcr6	TGT GAG TTA GCT CAC TCA TTA GGC AC
pUC18pcr5	GCT GCA AGG CGA TTA AGT TGG GTA A
pUC18pcr5.T7	TAA TAC GAC TCA CTA TAG GGG CTG CAA GGC GAT TAA GTT GGG TAA

These primers were designed to flank the M13(−21) (TGT AAA ACG ACG GCC AGT) and M13REV (CAG GAA ACA GCT ATG ACC) priming sites in the pUC18 vector and thus allow these to be used as nested sequencing primers. The pUC18pcr5.T7 primer is identical to the pUC18pcr5 primer at its 3′ terminus, but it contains a 5′ tail consisting of the T7 promoter primer sequence commonly used in cycle sequencing; the T7 primer sequence does not occur in the pUC18 cloning vector.

Amplification Protocol

1. DNA templates were prepared for *A. thaliana* plasmid clones ATDSB57 and ATDSB59 using standard alkaline lysis techniques (Sambrook *et al.*, 1989). PCR reactions using both the pUC18pcr6/pUC18pcr5 and pUC18pcr6/pUC18pcr5.T7 primer pairs were used to amplify the inserts from these clones. Each 100-μl PCR reaction contained the following:

Sterile H_2O	52.5 μl
Perkin–Elmer PCR buffer II	10.0 μl
$MgCl_2$	8.0 μl (25 m*M*)
GeneAmp dNTP mixture	8.0 μl (2 m*M* each dNTP)
Perkin–Elmer *Taq* polymerase	0.5 μl

Primer-pUC18pcr6	10.0 μl (0.2 μM)
Primer-pUC18pcr5 (or pUC18pcr5.T7)	10.0 μl (0.2 μM)
Template DNA	1.0 μl (approximately 100 ng)
Total volume	100 μl

Amplification reaction mixtures were prepared in 200-μl Perkin–Elmer MicroAmp PCR tubes, capped, and loaded into a MJ Research PT-100 thermocycler with a heated lid. The plasmid insert DNA was then amplified using the following cycling profile:

Thermal profile	No. cycles
95°C, 2 min	1
95°C, 15 sec; 69°C, 30 sec; 72°C, 1 min	3
95°C, 15 sec; 67°C, 30 sec; 72°C, 1 min	3
95°C, 15 sec; 65°C, 30 sec; 72°C, 2 min	3
95°C, 15 sec; 65°C, 30 sec; 72°C, 4 min	3
72°C, 5 min	1
4°C, hold	

2. Following PCR, amplification products were checked by agarose gel electrophoresis. Ten percent (10 μl) of the amplified product was combined with 2 μl of loading buffer and loaded onto a 1% agarose gel in 1× TAE (plus ethidium bromide), run at 44 mV for 3 hr, and imaged by UV transillumination. The results are shown in Fig. 2.

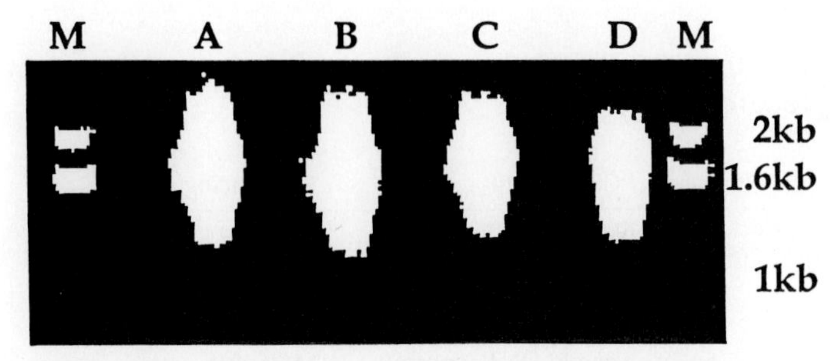

Figure 2 PCR amplified clone inserts using PCR primers both without (left) and with (right) the T7 primer tail described in the text. M, Gibco/BRL 1-kb ladder; A, clone ATDSB57; B, clone ATDSB59; C, ATDSB57 (with tailing primer); D, ATDSB59 (with tailing primer).

Exo I/SAP Treatment Protocol

1. For each sample to be sequenced, the following Exo I/SAP reaction mixture was prepared:

SAP (1 U/μl)	0.5 μl
Exo I (10 U/μl)	0.4 μl
10× SAP buffer	0.5 μl
Sterile H_2O	8.9 μl
Total volume	10.3 μl

2. For each PCR product to be sequenced, 8 μl of the reaction products was added to 10 μl of the Exo I/SAP mixture in a 200-μl MicroAmp tube and incubated in an MJ Research PT-100 thermocycler with a heated lid at 37°C for 1 hr, followed by 72°C for 15 min (to inactivate the Exo I and SAP). Reactions were held at 4°C until used for sequencing.

Dye Terminator Sequencing Protocol

1. Sequencing reactions were prepared using Perkin–Elmer Applied Biosystems FS+ Dye Terminator Ready Reaction Sequencing reagents (Product No. 402080) using the protocol supplied by the manufacturer. Each sequencing reaction contained the following:

Sterile H_2O	8.5 μl
ABI FS+ Dye Terminator sequencing mix	6.0 μl
Sequencing primer (0.2 μM)	1.5 μl
Treated PCR product or template DNA	2.0 μl
Total volume	20 μl

For each of the *A. thaliana* plasmid clones, five sequencing reactions were prepared:

Reaction	PCR primers	Sequencing primer
1	pUC18pcr6/pUC18pcr5	M13(−21)
2	pUC18pcr6/pUC18pcr5	pUC18pcr5
3	pUC18pcr6/pUC18pcr5.T7	M13(−21)
4	pUC18pcr6/pUC18pcr5.T7	T7
5	Plasmid DNA (control)	M13(−21)

In reactions 1 and 3, the M13(−21) primer (TGT AAA ACG ACG GCC AGT) serves as an internal, "nested" sequencing primer. For reaction 2, the sequencing primer pUC18pcr5 is the same primer that was used in the PCR amplification reaction. For reaction 4, the T7 primer (AAG GCG ATT AAG TTG GGT AA) is the tail added to the pUC18pcr5.T7 primer used for PCR amplification. Reaction 5 used DNA template rather than the Exo I/SAP-treated PCR products for the appropriate clone to serve as a control.

2. Sequencing reactions were prepared in 200-μl MicroAmp tubes, capped, loaded into a MJ Research PT-100 thermocycler with a heated lid, and processed using the following thermal profile:

Thermal profile	No. cycles
95°C, 2 min	1 cycle
94°C, 10 sec; 50°C, 10 sec; 60°C, 2 min	30 cycles
4°C, hold	

3. Mini-spin columns were prepared for removal of unincorporated dye terminators. Sephadex G-50 Fine (Amersham, Product No. 17-0573-02) was rehydrated in sterile H_2O (approximately 15 ml per gram) and autoclaved, after which excess water was removed. Approximately 350 μl of the Sephadex slurry was added to each well of a Pall Silent Monitor 96-well 0.45μm loprodyne filter plate (Pall Europe Ltd., Portsmouth, UK, Product No. SMO45LP) after which the plate was centrifuged for 2 min at 1500*g* to remove excess water. This was repeated a second time to completely fill the minicolumns.
   The sequencing reaction products were then loaded onto the minicolumns. The Silent Monitor plate was placed on top of a Perkin–Elmer 9600 reaction plate and centrifuged for 2 min at 750*g*. The Sephadex columns were discarded and the reaction products that passed through the columns were dried in a vacuum desiccator.
4. Sequencing reactions were resuspended in formamide/EDTA/blue dextran loading buffer, run for 10 hr on a 48-lane ABI 377 Automated DNA Sequencer (Perkin–Elmer) using the 48-cm run module with a 4.5% LongRanger gel (FMC, Product No. 50615).
5. Sequence data were analyzed using the ABI200 base calling module and reanalyzed using phred, an improved base caller written by Phil Green of the University of Washington (see *http://genome.washington.edu* or contact Phil Green via e-mail at *phg@u.washington.edu*).

6. The phred-called sequences for each sample were then collected, trimmed to remove low-quality and vector sequence, and assembled using TIGR Assembler (Sutton *et al.*, 1995) to provide sequence alignments for analysis.

Results and Conclusions

Each of the sequencing reactions produced more than 650 bases of high-quality DNA sequence data as shown in the following table, and a representative alignment of the sequences and the associated chromatograms are shown in Fig. 3. The various amplification/sequencing primer combinations described under Protocols are reflected in the suffix appended to the sequence names.

Sequence name	Read length
ATDSB57.1	665
ATDSB57.2	771
ATDSB57.3	662
ATDSB57.4	693
ATDSB57.5	757
ATDSB59.1	660
ATDSB59.2	757
ATDSB59.3	676
ATDSB59.4	784
ATDSB59.5	770

Both the read lengths provided in the table and a visual inspection of the sequence chromatograms demonstrate the quality of the sequence data that can be obtained by sequencing Exo I/SAP-treated PCR products. The DNA sequence read lengths and sequence quality are comparable to those obtained by sequencing the plasmid template directly. The sequence from the PCR products does not suffer from any significant background, nor does it demonstrate any peculiar artifacts that might interfere with its interpretation. In fact, for each of the clones, there are no discrepancies between the sequences obtained from the PCR products and that from the plasmid control.

One may note that the PCR products sequenced here were amplified from a plasmid clone and consequently it may be argued that

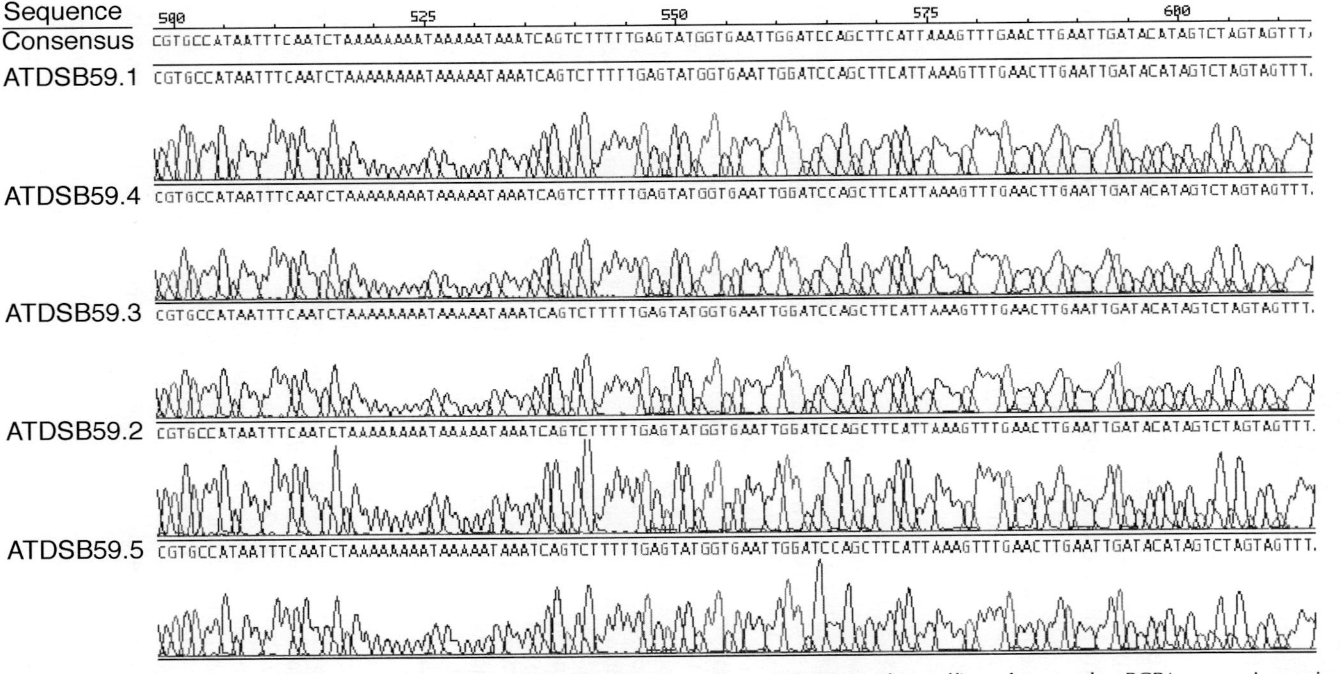

Figure 3 Sequence trace data for the five samples prepared from clone ATDSB59. The suffix refers to the PCR/sequencing primer combinations described in the text. ATDSB59.1 and ATDSB59.3 were sequenced using a nested M13(−21) primer, ATDSB59.2 used one of the amplification primers as a sequencing primer, ATDSB59.4 was sequenced using a T7 primer added as a tail to amplification primers, and ATDSB59.5 was a plasmid control sequenced using the M13(−21) primer. All sequencing reactions yielded high-quality sequence data spanning more than 650 bases of the target plasmid.

this would not represent the results to be expected from products amplified from genomic DNA. While PCR from genomic templates requires, in general, greater optimization than does plasmid amplification, PCR sequencing can still yield high-quality sequence data. Indeed, the sequencing of PCR products is currently being used to catalog single nucleotide polymorphisms by identifying overlapping peaks in the background of otherwise high-quality sequence (Nickerson *et al.*, 1997). Clearly, if such sensitive determinations can be made by sequencing PCR products, the results obtained should be sufficient for most general sequencing applications.

References

Bej, A. K., Steffan, R. J., DiCesare, J., Haff, L., Atlas, R. M. (1990). Detection of coliform bacteria in water by polymerase chain reaction and gene probes. *Appl. Environ. Microbiol.* **156,** 307–314.

Fleischmann, R. D., Adams, M. D., White, O., Clayton, R. A., Kirkness, E. F., Kerlavage, A. R., Bult, C. J., Tomb, J.-F., Dougherty, B. A., Merrick, J. M., McKenney, K., Sutton, G., FitzHugh, W., Fields, C., Gocayne, J. D., Scott, J., Shirley, R., Liu, L.-I., Glodek, A., Kelley, J. M., Weidman, J. F., Phillips, C. A., Spriggs, T., Hedlom, E., Cotton, M. D., Utterback, T. R., Hanna, M. C., Nguyen, D. T., Saudek, D. M., Brandon, R. C., Fine, L. D., Fritchman, J. L., Fuhrmann, J. L., Geoghagen, N. S. M., Gnehm, L. A., McDonald, L. A., Small, K. V., Fraser, C. M., Smith, H. O., and Venter, J. C. (1995). Whole-genome random shotgun sequencing and assembly of *Haemophilus influenzae* Rd. *Science* **269,** 496–512.

Hanke and Wink (1994). Direct DNA sequencing of PCR-amplified factor inserts following enzymatic degradation of primer and dNTPs. *BioTechnique* **17,** 858.

Heery, D. M., Gannon, F., and Powell, R. (1990). A simple method for subcloning DNA fragments from gel slices. *Trends Genet.* **6,** 173.

Mullis, K. B., and Faloona, F. A. (1987). Specific synthesis of DNA *in vitro* via a polymerase-catalyzed chain reaction. *Methods Enzymol.* **155,** 335–350.

Mullis, K. B., Faloona, F. A., Scharf, S. J., Saiki, R. K., Horn, G. T., and Erlich, H. A. (1986). Specific enzymatic amplification of DNA sequences *in vitro*: The polymerase chain reaction. *Cold Spring Harbor Symp. Quant. Biol.* **51,** 263–273.

Nickerson D. A., Tobe, V. O., and Taylor, S. L. (1997). PolyPhred: Automating the detection and genotyping of single nucleotide substitutions using fluorescence-based resequencing. *Nucleic Acids Res.* **25**(14), 2745–2751.

Saiki, R. S., Scharf, S., Faloona, F., Mullis, K. B., Horn, G. T., Erlich, H. A., and Arnheim, N. (1985). Enzymatic amplification of β-globin genomic sequences and restriction site analysis for diagnosis of sickle cell anemia. *Science* **230,** 1350–1354.

Sambrook, J., Fritsch, E. F., and Maniatis, T. (1989). *Molecular Cloning: A Laboratory Manual.* Cold Spring Harbor Laboratory Press, Cold Spring Harbor, NY.

Sanger, F., and Coulson, A. R. (1975). A rapid method for determining sequences in DNA by primed synthesis with DNA polymerase. *J. Mol. Biol.* **94,** 441–448.

Sogin, M. L., Gunderson, J. H., Elwood, H. J., Alonso, R. A., and Peattie, D. A. (1989).

Phylogenetic meaning of the kingdom concept: An unusual ribosomal RNA from *Giardia lamblia*. *Science* **243,** 75–77.

Sutton, G. G., White, O., Adams, M. D., and Kerlavage, A. R. (1995). TIGR assembler: A new tool for assembling large shotgun sequencing projects. *Gen. Sci. Tech.* **1,** 9–19.

Trower, M. K., Burt, D., Purvis, I. J., Dykes, C. W., and Christodoulou, C. (1995). Fluorescent dye-primer cycle sequencing using unpurified PCR products as templates: Development of a protocol amenable to high-throughput DNA sequencing. *Nucleic Acids Res.* **23,** 2348–2349.

Weissenbach, J., Gyapay, G., Dib, C., Vignal, A., Morissette, J., Millasseau, P., Vaysseix, G., and Lathrop, M. (1992). A second-generation linkage map of the human genome. *Nature* **359,** 794–801.

Werle, E., Schneider, C., Renner, M., Volker, M., and Fiehn, W. (1994). Convenient, single-step, one tube purification of PCR products for direct sequencing. *Nucleic Acids Res.* **22,** 4354.

Woese, C. R. (1987). Bacterial evolution. *Microbiol. Rev.* **51,** 221.

Zhen, L., and Swank, R. T. (1993). A simple and high yield method of recovering DNA from agarose gels. *BioTechniques* **14,** 894.

10

RECENT ADVANCES IN HIGH-TEMPERATURE REVERSE TRANSCRIPTION AND PCR

Thomas W. Myers

A recombinant DNA polymerase from the thermophilic eubacterium *Thermus thermophilus* (r*Tth* pol) was found to possess efficient reverse transcriptase activity in the presence of Mn^{2+} (Myers and Gelfand, 1991). This high-temperature reverse transciptase (RT) reaction can be followed by PCR amplification resulting in the amplification of the newly synthesized cDNA in a two-step coupled process requiring only the addition of a single (thermostable) enzyme. The original procedure was based on performing the RT step in the presence of Mn^{2+}, followed by chelation of the Mn^{2+} with ethylenebis(oxyethylenenitrilo)tetraacetic acid (EGTA), and then activating the PCR amplification with Mg^{2+}. Subsequently, reaction conditions allowing r*Tth* pol to efficiently carry out both RT and DNA amplification in a single buffer containing Mn^{2+} were determined (Myers and Sigua, 1995). This protocol has formed the basis of many virus detection assays including HCV (Young *et al.*, 1993) and HIV (Mulder *et al.*, 1994).

Prior to the use of r*Tth* pol, reverse transcription was generally performed at lower reaction temperatures using either AMV or MuLV reverse transcriptases. A common problem encountered when

PCR Applications

using mesophilic reverse transcriptases is the inability of the enzymes to synthesize cDNA through stable RNA secondary structures. Since many of the potential RNA secondary structures will likely be unstable at the higher reaction temperatures used with r*Tth* pol, long cDNA synthesis can be achieved. An additional advantage gained by performing high-temperature RT is the increase in specificity of primer hybridization and subsequent extension by the r*Tth* pol. Furthermore, the intrinsic RNase H activity of the native retroviral reverse transcriptases has been suggested to be detrimental to the production of full-length cDNA during reverse transcription (Kotewicz *et al.*, 1988). The r*Tth* pol does possess an inherent 5′ → 3′ exonuclease/RNase H activity similar to *Escherichia coli* DNA polymerase I (Auer *et al.*, 1995). However, r*Tth* pol does not cleave RNA–DNA hybrids endonucleolytically, as is the case with retroviral reverse transcriptases.

Selective Amplification of RNA

We have utilized the 5′ → 3′ exonuclease/RNase H activity of r*Tth* pol, in conjunction with the incorporation of nucleotide analogs that lower the strand separation temperature requirements for PCR amplification, to selectively amplify RNA in the presence of genomic DNA of analogous sequence (Auer *et al.*, 1996). Following first-strand cDNA synthesis by r*Tth* pol, the RNA strand of the newly formed RNA–DNA hybrid is digested by the RNase H activity of r*Tth* pol, allowing the PCR primer to hybridize and initiate second-strand cDNA synthesis. The incorporation of dIMP residues during first- and second-strand cDNA synthesis, as well as subsequent PCR cycles, reduces the strand separation temperature of the resultant DNA–DNA duplex and allows for a strand separation temperature that is below that required for denaturation of genomic duplex DNA containing conventional nucleotides. Thus, by using a lower strand separation temperature than normal, the genomic DNA is not denatured and cannot be amplified. This strategy does not require the presence of introns and is amenable to applications in which pseudogenes are present or when differentiating between unspliced retroviral RNA and integrated proviral DNA.

As an example demonstrating our RNA selective amplification strategy, we preferentially amplified intracellular tumor necrosis factor-α (TNFα) mRNA in the presence of analogous genomic DNA from cultured cells (Auer *et al.*, 1996). The primer set used in the experiment allowed for differentiation of the RT-PCR product based upon amplicon size, dependent on which nucleic acid had been utilized as the target template–RNA or genomic DNA (Fig. 1). The primer binding sites spanned intron 3 and the resultant amplicon was 255 bp if mRNA was the template, whereas a product of 556 bp was observed if genomic DNA was the template. When dITP was used and a restrictive strand separation temperature of 83°C was employed, only product derived from RNA was observed following RT-PCR (Fig. 1, lane 2). No product was observed at the 83°C strand separation temperature in the absence of the RT step (lane 3) or when dGTP was substituted for dITP (lane 4). At a nonrestrictive strand separation temperature of 95°C, both RNA and DNA were used as targets for the RT-PCR (lane 5).

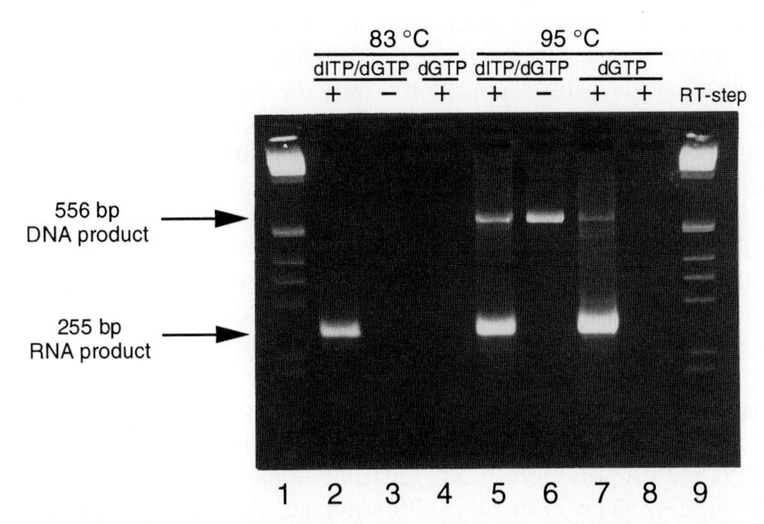

Figure 1 RNA selective amplification of TNF-α mRNA from cultured cells. HL-60 cells (~500) in 250 ng/μl poly(rA) were added directly to the reactions. The TNF-α mRNA was reverse transcribed and the cDNA or genomic DNA was amplified by PCR using a T_{den} of 83°C (lanes 2–4) or 95°C (lanes 5–8) as indicated. Oligonucleotides AW113 and TNF-α_2 were added to each reaction. Control reactions contained 300 μM dGTP (lanes 4, 7, and 8). HL-60 cells were omitted in the negative control reaction (lane 8). All reactions were incubated at 60°C for 60 min (RT step), except for the negative RT step reactions, which were kept on ice (lanes 3 and 6). Reproduced by permission of Oxford University Press from Auer *et al.* (1996). Selective amplification of RNA utilizing the nucleotide analog dITP and Thermus thermophilus DNA polymerase. *Nucleic Acids Res.* **24,** 5021–5025.

Long RT-PCR

The amplification of DNA targets longer than 20 kb has become practical and specific protocols and reviews of the technology have been published (Barnes, 1994; Cheng, 1995; Cheng *et al.*, 1994; Foord and Rose, 1994). However, routine cDNA synthesis and DNA amplification has been limited to approximately 4 kb, resulting primarily from the inability of the mesophilic reverse transcriptases to efficiently synthesize "full-length" cDNA. Long cDNA synthesis has become increasingly important for many molecular cloning applications. The protocol described here focuses on the process of performing high-temperature RT of long RNA targets and the subsequent DNA amplification of the cDNA. The technique described here (Hirose and Myers, 1997) is an extension of the original long-target PCR processes (Barnes, 1994; Cheng *et al.*, 1994). We have generated cDNA >15 kb using measles virus as a template for RT-PCR (Fig. 2). Reverse transcription was accomplished using a single specific oligonucleotide primer complementary to the 3′-terminus of the measles virus RNA. Subsequent PCR amplifications were achieved using specific oligonucleotide primer pairs at intervals progressively 3′ to the resultant first-strand cDNA.

Similar to the amplification of long DNA targets, the successful RT-PCR of long RNA targets depends on a number of variables, including sample integrity, reaction buffer pH, monovalent and divalent salts, cosolvents, primer design, sufficient extension time during reverse transcription and PCR steps, and the polymerase and nucleolytic proofreading activities of the enzymes. The inclusion of a secondary thermostable DNA polymerase possessing a $3' \rightarrow 5'$ proofreading exonuclease activity (Vent$_R$ DNA Polymerase) in the r*Tth* pol (r*Tth* DNA polymerase, XL) greatly enhances the long-target RT-PCR process. The r*Tth* DNA Polymerase, XL not only promotes efficient reverse transcription and DNA-dependent DNA synthesis but also reduces the number of nucleotide misincorporations that might otherwise prematurely terminate synthesis or lead to the accumulation of mutations present in the PCR product (C. Sigua, N. Moonsamy, and T. Myers, unpublished data). The inclusion of a proofreading exonuclease into both the RT and PCR amplification steps results in enhanced fidelity. However, it cannot be overemphasized that reaction conditions are extremely important in the overall fidelity achieved. Enhanced fidelity may be achieved by using the

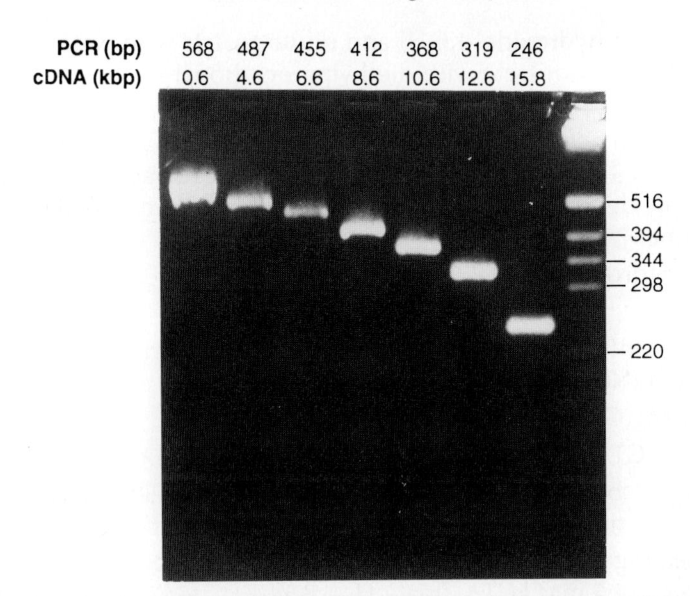

Figure 2 Two-buffer r*Tth* DNA Polymerase, XL RT-PCR of measles virus RNA isolated from tissue culture cells infected with measles virus. Reverse transcription was performed with the oligonucleotide MZLA1 as the "downstream" RT primer complementary to the 3′ terminus of the RNA. Subsequent PCR amplifications were achieved using specific oligonucleotide primer pairs at intervals progressively 3′ to the resultant first-strand cDNA. The reverse transcription step was performed at 60°C for 120 min, followed by a 1-min predenaturation step at 95°C then 40 cycles of 95°C for 15 sec, 65°C for 30 sec for each primer pair. Electrophoretic analysis of PCR products was performed with 5% of the PCR amplification on a 3% (w/v) NuSieve 1% (w/v) SeaKem GTG agarose gel stained with ethidium bromide.

minimal extension time and highest extension temperature possible, using low dNTP and divalent metal ion concentrations, and amplifying for as few cycles as necessary (Eckert and Kunkel, 1991; Goodman, 1995).

Materials and Methods

Reagents and Supplies

Glycerol

1*M* bicine–KOH (pH 8.3): Adjust an approximately 1 *M* solution of *N*,*N*-bis(2-hydroxyethyl)glycine (bicine) to pH 8.3 with 45%

potassium hydroxide (KOH) and dilute to 1 *M* with water. Filter sterilize, do not treat with diethylpyrocarbonate.

3 *M* KOAc (pH 7.5): Adjust an approximately 3 *M* solution of potassium acetate (KOAc) to pH 7.5 with glacial acetic acid and dilute to 3 *M* with water. Note: Very little acetic acid is required and pH decreases upon dilution, so adjust pH near final volume.

N-[tris(hydroxymethyl)glycine (tricine)

KOAc

100 m*M* $Mn(OAc)_2$: Manganese(II)acetate tetrahydrate from Aldrich (No. 22,977–6)

25 m*M* $Mg(OAc)_2$ (Perkin–Elmer)

500 m*M* EGTA (pH 8.0)

5X long RT buffer: 150 m*M* tricine, 375 m*M* KOAc, 50% (v/v) glycerol, final pH = 8.5

5X chelating buffer: 220 m*M* tricine, 470 m*M* KOAc, 47% (v/v) glycerol, 1.5 m*M* EGTA, final pH = 8.5

Deoxynucleoside triphosphates: Neutralized 10 m*M* solutions were from Perkin–Elmer, except for dITP, which was obtained from Pharmacia.

Primers: MZLA1 (5′-CAGGATTAGGGTTCCGGAGTTCAACC-3′) and TNFα2 (5′-GAACCCCGAGTGACAAGCCTG-3′) were synthesized on an Applied Biosystem 394 DNA synthesizer. Oligodeoxynucleotide AW 113 was obtained from Perkin–Elmer.

Measles virus RNA: Measles virus (ATCC VR-24-Edmonston strain) was obtained from the American Type Culture Collection and total cellular RNA was isolated as described by Shimizu *et al.* (1993). Measles virus RNA was stored in 50 ng/μl *E. coli* rRNA.

TNF-α-induced HL-60 cells were as described in Auer *et al.* (1996).

Escherichia coli rRNA: 16S and 23S rRNA from Boehringer-Mannheim was phenol extracted, ethanol precipitated, and adjusted to 126 ng/μl in water.

Poly (rA): Pharmacia

r*Tth* DNA Polymerase, XL: 2 U/μl from Perkin–Elmer

r*Tth* DNA Polymerase: 2.5 U/μl from Perkin–Elmer

Uracil-*N*-glycosylase (UNG): 1 U/μl from Perkin–Elmer

Thin-walled Gene Amp tubes (Perkin–Elmer)

GeneAmp PCR System 9600 (Perkin–Elmer)

SeaKem GTG agarose (FMC)

1-kb DNA ladder (GIBCO BRL)
5 μg/μl ethidium bromide

Protocols

RNA Selective RT-PCR

Combined RT and PCR Amplification (50 μl)

50 m*M* bicine–KOH (pH 8.3)

100 m*M* KOAc (pH 7.5)

13% glycerol (v/v) (this 13% is in addition to the 2% contributed by the r*Tth* pol)

300 μ*M* each of dATP, dCTP, and dTTP (or dUTP)

250 μ*M* dGTP

350 μ*M* dITP

0.25 μ*M* RT "downstream" primer

0.25 μ*M* PCR "upstream" primer

5 U r*Tth* pol

3 m*M* $Mn(OAc)_2$

500 ng poly(rA)

1 U UNG (if using dUTP)

The RT reaction and PCR were then performed consecutively in a GeneAmp PCR System 9600 as follows:

30 min at 60°C
2 min at 78°C for 1 cycle
10 sec at 78°C, 30 sec at 50°C, and 50 sec at 65°C for 40 cycles

Long RT-PCR

Reverse Transcription Reaction (20 μl)

1X long RT buffer
200 μ*M* each of dATP, dCTP, dGTP, and dTTP
0.75 μ*M* RT downstream primer
1.1 m*M* $Mn(OAc)_2$

5 U r*Tth* DNA Polymerase, XL
RNA (≤250 ng)
Incubate for 15–120 min at 60°C

PCR (100 μl)

Add 80 μl of the following solution to the RT action:

1X chelating buffer
1.0 m*M* Mg $(OAc)_2$
0.19 μ*M* PCR upstream primer

The reactions are then amplified in a GeneAmp PCR System 9600 as follows:

1 min at 95°C for 1 cycle
15 sec at 95°C and 1–6 min at 65°C for 20 cycles
15 sec at 95°C and 1–6 min at 65°C for 20 cycles with 15-sec autoextension (for longer targets)
7 min at 60°C
Hold at 15°C

Discussion

The RNA selective amplification scheme described here is sensitive and robust. The method avoids multiple reaction processes and the reaction is carried out in a single tube with a single buffer utilizing the enzymological properties of a single enzyme. However, when utilizing the RNA selective amplification strategy, the incorporation of a nucleotide analog that lowers the strand separation temperature necessitates the readjustment of the annealing temperature of the PCR amplification to a lower temperature. The use of various mixtures of dITP and dGTP allows for modulation of the strand separation temperature and the annealing temperature of the PCR. Another important precaution involves the integrity of any DNA present during RNA selective RT-PCR. Lesions in the DNA, such as nicks and gaps, would allow for nick translation synthesis with concomitant incorporation of dIMP into the DNA and a possible decrease of the strand separation temperature. We have not observed DNA integrity to be a serious challenge since the DNA damage would likely be randomly distributed and nick translation synthesis is considerably slower than cDNA synthesis. However, precaution is advised, especially when contemplating particular sample preparation methodologies.

High-temperature reverse transcription using r*Tth* pol helps to alleviate many of the problems typically encountered when amplifying long RNA targets. However, increased metal ion catalyzed hydrolysis of template RNA is observed at the elevated temperatures (Brown, 1974) used with r*Tth* pol. Therefore, factors such as increased enzyme activity and reaction specificity, as well as the reduction of RNA secondary structure, must be balanced with target degradation. The generation of long cDNA is improved using somewhat lower reaction temperatures (<70°C) to minimize hydrolysis of the template and by modifying reaction parameters to optimize for longer products. The use of oligonucleotide primers with relatively high melting temperatures is important for the strategy presented for long RT-PCR due to the high glycerol concentration present in the two-buffer RT-PCR (approximately 16 and 11% for the RT and PCR steps, respectively). The increased glycerol concentration effectively lowers the T_m of the primers and thus needs to be considered in the design of the primers. The use of sequence-specific oligonucleotide primers is preferred, especially for long RT-PCR in which reaction specificity is critical.

The efficiencies of the reverse transcription and DNA amplification steps decrease as the length of the target increases. Consequently, the greater the sensitivity required, the more care one needs to ensure optimal reaction conditions are used. Determining the optimal deoxynucleoside triphosphate and divalent metal ion concentrations for both the RT and PCR steps is important. Improved product yield is often achieved by increasing the dNTP concentration to 100 μM each dNTP and raising the final Mg $(OAc)_2$ concentration to 1.0 mM during the PCR amplification. However, increased product yield must be weighted against a decrease in fidelity (C. Sigua, N. Moonsamy, and T. Myers, unpublished data). Although not addressed in this technique, the preparation of full-length starting template RNA is equally important.

When performing long cDNA synthesis for RT-PCR, the two-buffer protocol is the method of choice since the RT and PCR steps can be performed at their own optimal conditions. In addition to achieving increased enzymatic activity for the two independent reactions, conditions known to influence the fidelity of DNA polymerases can be altered much more readily than they can with a single-buffer method (Myers and Sigua, 1995). The negative effect of Mn^{2+} on the fidelity of DNA synthesis is well-known for *E. coli* pol I and *Thermus aquaticus* DNA polymerase (Beckman *et al.*, 1985; Leung *et al.*, 1989). The addition of EGTA to chelate preferentially the Mn^{2+} in the two-buffer procedure is expected to decrease any effects that the Mn^{2+} may have on the fidelity of r*Tth* pol during the PCR amplification. It is worth nothing that under reaction conditions generally used for RT (high concentrations of $Mg^{2+}$and dNTPs) with retroviral reverse transcriptases, the MuLV reverse transcriptase was determined to have a lower fidelity than the Mn^{2+} activated r*Tth* pol during reverse transcription (Myers, *et al.*, 1995). The low dNTP (160 μM total) and Mg^{2+} (0.8 mM) concentrations suggested for the two-buffer method also may be conducive to higher fidelity synthesis during the PCR since similar conditions were demonstrated to increase the fidelity of *Taq* pol (Eckert and Kunkel, 1990).

Finally, it is critical to realize the enormous amplification of product produced during PCR. Low levels of DNA contamination, especially from previous PCR amplification reactions, samples with high DNA levels, or positive control templates, can result in product formation, even in the absence of purposefully added template DNA. The utilization of the dUTP/UNG carryover prevention system (Longo *et al.*, 1990) allowed for in the selective amplification of RNA strategy is not possible in the procedure described here for long RT-PCR due to the composition of the r*Tth* DNA Polymerase, XL (Slupphaug *et al.*, 1993). Accordingly, procedures must be followed to minimize carryover of amplified DNA (Higuchi and Kwok, 1989).

References

Auer, T., Landre P. A., and Myers, T. W. (1995). Properties of the 5′ → 3′ exonuclease/ ribonuclease H activity of *Thermus thermophilus* DNA polymerase. *Biochemistry* **34,** 4994–5002.

Auer, T., Sninsky, J. J., Gelfand, D. H., and Myers, T. W. (1996). Selective amplification of RNA utilizing the nucleotide analog dITP and *Thermus thermophilus* DNA polymerase. *Nucleic Acids Res.* **24,** 5021–5025.

Barnes, W. M. (1994). PCR amplification of up to 35-kb DNA with high fidelity and high yield from λ bacteriophage templates. *Proc. Natl. Acad. Sci. U.S.A.* **91,** 2216–2220.

Beckman, R. A., Mildvan, A. S., and Loeb, L. A. (1985). On the fidelity of DNA replication: Manganese mutagenesis *in vitro. Biochemistry* **24,** 5810–5817.

Brown, D. M. (1974). Chemical reactions of polynucleotides and nucleic acids. In *Basic Principles in Nucleic Acid Chemistry* (P. O. P. Ts'o, Ed.), pp. 43–44. Academic Press, New York.

Cheng, S. (1995). Longer PCR. In *PCR Strategies* (M. A. Innis, D. H. Gelfand, and J. J. Sninsky, Eds.), pp. 313–324. Academic Press, San Diego.

Cheng, S., Fockler, C., Barnes, W. M., and Higuchi, R. (1994). Effective amplification of long targets from cloned inserts and human genomic DNA. *Proc. Natl. Acad. Sci. U.S.A.* **91,** 5695–5699.

Eckert, K. A., and Kunkel, T. A. (1990). High fidelity DNA synthesis by the *Thermus aquaticus* DNA polymerase. *Nucleic Acids Res.* **18,** 3739–3744.

Foord, O. S., and Rose, E. A. (1994). Long-distance PCR. *PCR Methods Appl.* **3,** S149–S161.

Goodman, M. F. (1995). DNA polymerase fidelity: Misinsertions and mismatched extensions. In *PCR Strategies* (M. A. Innis, D. H. Gelfand, and J. J. Sninksy Eds.), pp. 17–31. Academic Press, San Diego.

Higuchi, R., and Kwok, S. (1989). Avoiding false positives with PCR. *Nature* **339,** 237–238.

Hirose, T., and Myers, T. W. (1997). PCR and RT/PCR amplification of long DNA and RNA. *Exp. Med.* **15,** 20–23.

Kotewicz, M. L., Sampson, C. M., D'Alessio, J. M., and Gerard, G. F. (1988). Isolation of cloned Moloney murine leukemia virus reverse transcriptase lacking ribonuclease H activity. *Nucleic Acids Res.* **16,** 265–277.

Leung, D. W., Chen, E., and Goeddel, D. V. (1989). A method for random mutagenesis of a defined DNA segment using a modified polymerase chain reaction. *Technique* **1,** 11–15.

Longo, M. C., Berninger, M. S., and Hartley, J. L. (1990). Use of uracil DNA glycosylase to control carry-over contamination in polymerase chain reactions. *Gene* **93,** 125–128.

Mulder, J., McKinney, N., Christopherson, C., Sninksy, J., Greenfield, L., and Kwok, S. (1994). Rapid and simple PCR assay for quantitation of human immunodeficiency virus type 1 RNA in plasma: Application to acute retroviral infection. *J. Clin. Microbiol.* **32,** 292–300.

Myers, T. W., and Gelfand, D. H. (1991). Reverse transcription and DNA amplification by a *Thermus thermophilus* DNA polymerase. *Biochemistry* **30,** 7661–7666.

Myers, T. W., and Sigua, C. L. (1995). Amplification of RNA: High temperatures reverse transcription and DNA amplification with *Thermus thermophilus* DNA polymerase. In *PCR Strategies* (M. A. Innis, D. H. Gelfand, and J. J. Sninsky, Eds.), pp. 58–68. Academic Press, San Diego.

Myers, T. W., Sigua, C. L., Lawyer, F. C., and Gelfand, D. H. (1995). Fidelity of MMLV reverse transcriptase and *Thermus thermophilus* DNA polymerase during reverse transcription and DNA amplification. *FASEB J.* **9,** A1336.

Shimizu, H., McCarthy, C. A., Smaron, M. F., and Burns, J. C. (1993). Polymerase chain reaction for detection of measles virus in clinical samples. *J. Clin. Microbiol.* **31,** 1034–1039.

Slupphaug, G., Alseth, I., Eftedal, I., Volden, G., and Krokan, H. E. (1993). Low incorporation of dUMP by some thermostable DNA polymerases may limit their use in PCR amplifications. *Anal. Biochem.* **211,** 164–169.

Young, K. K. Y., Resnick, R. M., and Myers, T. W. (1993). Detection of hepatitis C virus RNA by a combined reverse transcription-polymerase chain reaction assay. *J. Clin. Microbiol.* **31,** 882–886.

11

VIRAL GENOTYPING BY A QUANTITATIVE POINT MUTATION ASSAY: APPLICATION TO HIV-1 DRUG RESISTANCE

Steve Kaye

The association of point mutations in the genome of HIV-1 with resistance to antiretroviral drugs was first described for resistance to the nucleoside analog 3′-azidodeoxythymidine (AZT, zidovudine; Larder and Kemp, 1989). Four mutations in the reverse transcriptase (RT) gene at codons 67, 70, 215, and 219 were described, and later a fifth mutation at codon 41 was associated with high-level resistance (Kellam *et al.*, 1992). Introduction of the mutations, both singly and in combination, into wild-type, drug-sensitive infectious molecular clones of HIV-1 by site-directed mutagenesis confirmed their contributions to AZT resistance. Subsequently, mutations coding for resistance to all currently licensed antiretroviral drugs have been described (Schinazi *et al.*, 1996).

Although the final arbiter of drug resistance remains the phenotypic assay, which is the concentration of drug required to inhibit virus replication in culture, the use of genotypic assays for detection and quantification of resistance mutations has become widespread in monitoring patients undergoing therapy. In contrast to culture-based methods, genotypic assays are less hazardous, less expensive,

PCR Applications

and have a higher sample throughput. Furthermore, PCR-based genotypic assays allow a greater proportion of samples to be analyzed since PCR isolation rates for viral genomes are higher than rates for isolation by culture, and genotypic assays do not perturb the viral genome population, whereas virus culture may itself impose a selective pressure on the virus population.

The point mutation assay (PMA) developed in our laboratory is a PCR-based microtiter-format assay which allows the detection and quantification of any mutation within the amplified region (Kaye *et al.*, 1992). The assay is quantitative, measuring the proportions of wild-type and mutant sequence at a given point in the viral genome population amplified from the sample. Thus, the assay will give information on the rates of evolution of the mutations under the influence of the selective pressure exerted by the drug treatment.

The assay described is for quantification of resistance-associated mutations in a PCR-amplified region of the 5′ end of the HIV-1 RT gene but can be adapted to the assay of any mutation in any PCR product simply by changing the PCR primer and probe sequences.

Principle of the Assay

Figure 1 shows a schematic diagram of the assay. The following steps are shown:

A. Generation of a biotinylated PCR product in a nested PCR reaction, encompassing the region of the genome containing the mutation site of interest (X).
B. Capture of the biotinylated product on four streptavidin-coated microtiter wells. After capture of the products, the other components of the PCR are washed away.
C. Denaturation of the captured products using 0.15 *M* NaOH and washing away of the nonbiotinylated DNA strand.
D. Annealing of an oligonucleotide probe to the captured single strand. The probe is designed such that its 3′ terminus is one base short of the point being assayed.
E. Addition of a single ^{35}S-labeled dNTP to each of the four wells, together with Klenow DNA polymerase. If the base at the point of interest is complementary to the added dNTP the enzyme will extend the annealed oligonucleotide probe with a

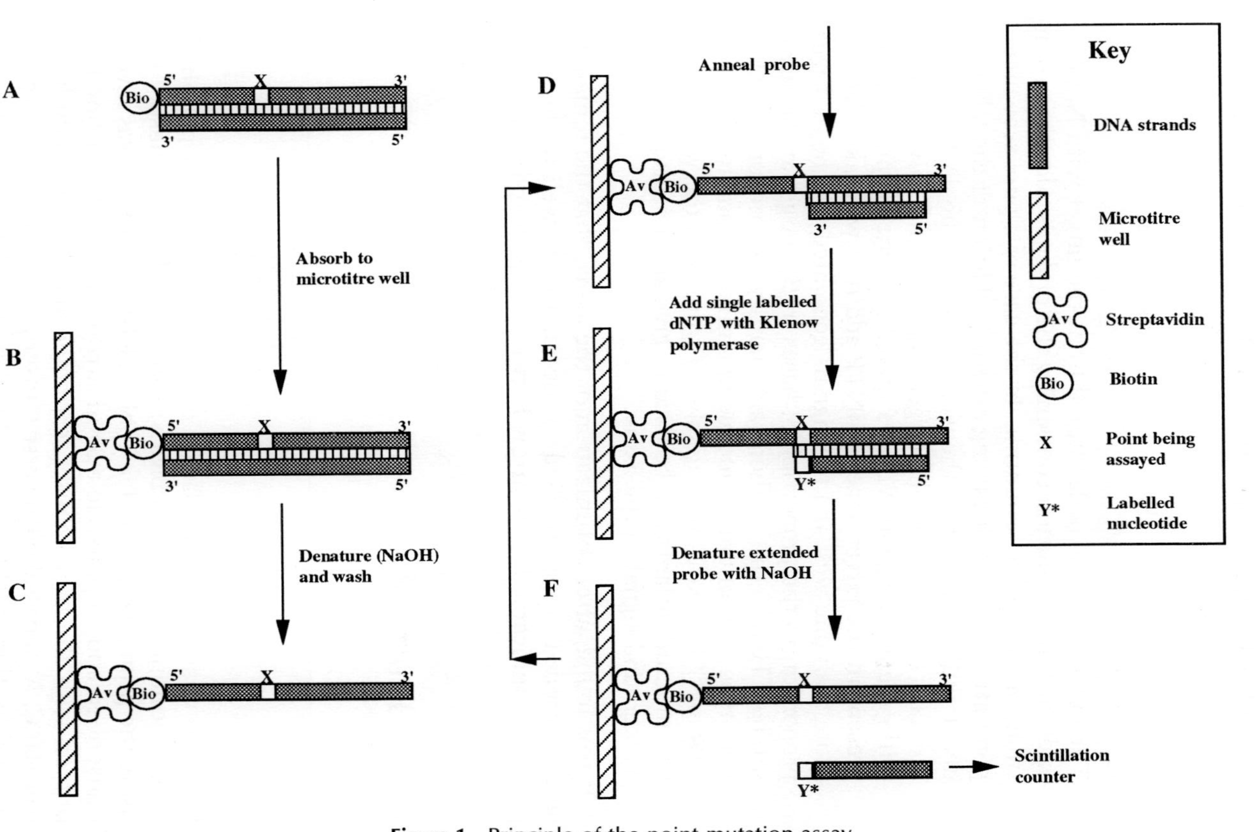

Figure 1 Principle of the point mutation assay.

single labeled dNTP (or with more than one where a run of identical bases occurs in the target). Ideally, labeled ddNTPs would be used so that multiple additions could not occur, but a suitable source of labeled ddNTPs was not available for this project.

F. The extended probe is denatured from the target with NaOH, and the NaOH solution containing the probe is added to a scintillation cocktail for counting. In the case of a pure sequence population at the point being assayed, addition of the labeled dNTP will occur only in the well containing the dNTP complementary to the base in the target sequence, and only this well will generate a signal. Where a mixed sequence population is present in the target, labeled dNTP addition will take place in more than one well and the signals generated will be in direct proportion to the proportion of bases in the target at the site of interest. Hence, the method can quantify the proportions of wild-type and mutant sequence in a mixed population of viral genomes. The bases that are absent from both the wild-type and mutant variants can be used to give an indication of the background signal in the assay.

It was found, after optimization of the assay, that the remaining single-stranded target attached to the microtiter well could be reprobed, as indicated by the arrow between steps F and D in Fig. 1.

Assay Protocol

PCR

Cell-free viral RNA from plasma or proviral DNA from peripheral blood mononuclear cells (PBMCs) were amplified in a nested PCR using the primers described in Table 1. RT-PCR for RNA was reverse transcribed and amplified in a 50-μl single-tube reaction using r*Tth*, primed with the outer antisense primer SPP8. Cycling conditions were 60°C, 30 min to allow reverse transcription followed by 35 cycles of amplification at 94°C, 1 min; 55°C, 1 min; and 72°C, 1 min.

DNA was amplified in a 50-μl reaction with *Taq* polymerase using the same amplification conditions. Second-round amplifications

Table 1
Primers Used for Amplifying the RT Gene of HIV-1 for Analysis by PMA

Primer	Sequence	Size	Location[a]
Outer set			
SPP1A (sense)	GTAGGACCTACACCTGTCAACATAA	25 mer	397–421
SPP8 (antisense)	GACTTGCCCAATTCAATTTTCCCAC	25 mer	1246–1270
Inner set			
SPP2A (sense)[b]	TGTTGACTCAGATTGGTTGCACTTTA	26 mer	434–459
SPP6A (antisense)[c]	TTCTGTATGTCTTAGACAGTCCAGCT	26 mer	1216–1241

[a] Location in RT gene of strain HXB2.
[b] SPP 2A biotinylated for use with probes ARP1, ARP3, ARP4C, and ARP5D.
[c] SPP6A biotinylated for use with probe ARP2B.

(RNA or DNA) were carried out in 50-μl reactions after transfer of 1 μl of first-round product. Cycling conditions were 30 cycles at 94°C, 1 min; 55°C, 1 min; and 72°C, 1 min. To ensure amplification, 5 μl of second-round product was analyzed on an ethidium bromide-stained agarose gel, and 40 μl was used in the PMA.

Streptavidin Coating of Microtiter Wells

Microtiter wells for use in PMA (Nunc, Maxisorb U-well) were coated with 25 μl/well streptavidin coating solution (0.01 *M* Tris–HCl, pH 7.6, containing 2.5 mg/ml streptavidin) and the wells were incubated overnight at room temperature in a moist box. The wells were washed twice with PBS and filled with blocking solution (PBS containing 1% BSA and 0.1% sodium azide, 0.35 ml/well) and incubated at room temperature for 1 hr. The wells were stored, for up to 6 months, at 4°C with the blocking solution in the wells.

PMA Method

The method described is for the analysis of 24 samples (22 test samples and 2 controls) in a complete 96-well plate. Smaller numbers were analyzed in strips of microtiter wells, reducing the quantities of reagents proportionately.

The storage buffer was aspirated from the stored streptavidin-coated microtiter wells and 15 μl/well TTA buffer was added. Ten microliters of each sample of second-round PCR product was added to four wells and mixed with the TTA diluent. The wells were incubated at room temperature for 15 min and washed three times in TTA wash buffer. Captured PCR products were denatured by the addition of 40 μ/well 0.15 *M* NaOH and the plate was incubated at room temperature for 5 min. The wells were washed four times in TTA buffer. After the addition of 25 μl/well anneal mix, the wells were sealed and incubated at 63°C for 3 min by floating in a shallow water bath. The plate was allowed to cool slowly to room temperature for 30 min by placing it on a metal block heated to 63°C and left to cool. Probe sequences for AZT resistance-associated mutations are shown in Table 2. Labeling mix (9.8 μl) was added to each well (labeling mix A, G, C, or T to one well for each sample) and the plate was incubated at room temperature for 2 min. The wells were washed six times with TTA with a 1-min soak between the fourth and fifth washes. Due to the short incubation time with the labeling mix, the wells were washed sequentially in the order in which the label was added (i.e., A, G, C, and then T). The labeled probles were denatured from the captured strands by the addition of 40 μl/well 0.15 *M* NaOH and the plate was incubated at room temperature for 5 min. The 40 μl of NaOH solution was mixed with 100 μl Microscint 40 scintillation cocktail (Canberra Packard) in a white microtiter plate and counted for 1 min/well on a Topcount microtiter scintillation reader (Canberra Packard).

Table 2

Point Mutation Assay Probes for ZDV Resistance-Associated Mutations

Codon[a]	Probe	Sequence	Wild type[b]	Mutant[c]
41	ARP5D	AAATTTTCCCTTCCTTTTCCA	T × 3	A or G
67	ARP1	TTTTCTCCATTTAGTACTGT	C	T × 7
70	ARP2B	AAAGAAAAAAGACAGTACTA	A × 2	G
215	ARP4C	CTGATGTTTTTTGTCTGGTGTG	G	A × 2 (Phe) T (Tyr)
219	ARP3	AAGTTCTTTCTGATGTTTTT	T	G × 2

[a] RT gene of HIV-1.
[b] Bases added to probe in PMA by wild-type sequence.
[c] Bases added to probe in PMA by mutant sequence.

Reprobing Modification of PMA

After the final removal of the now labeled probe with 0.15 *M* NaOH for scintillation counting, the single-stranded target remains attached to the microtiter well and can therefore be reprobed for another mutation.

After denaturation of the probe/target hybrid with 0.15 *M* NaOH and transfer of the labeled probe to the counting plate the wells were washed four times with TTA wash fluid and aspirated. Note that the denaturation step was limited to 5 min. to prevent extended NaOH treatment weakening the binding of the target to the streptavidin-coated well, either by weakening of the biotin/streptavidin bond or the streptavidin/microtiter well bond. The PMA protocol was resumed with the addition of a new anneal mix and annealing of the new probe. If necessary, after washing the wells were stored prior to reprobing by leaving the wells filled with TTA wash fluid and storing the wells at 4°C in a moist box for up to 3 days.

Calculation of PMA Results

The readings for the following bases were divided as follows:

Codon 41:	Divide the T signal by 3
Codon 67:	Divide the T signal by 7
Codon 70:	Divide the A signal by 2
Codon 215:	Divide the A signal by 2
Codon 219:	Divide the G signal by 2

This was required because at the point at which these bases occur in the sample being analyzed an identical base or bases follow and the polymerase adds more than one labeled base. The mean of the bases which do not occur at the point being analyzed was calculated as follows:

Codon 41:	C only
Codon 67:	Mean of A and G
Codon 70:	Mean of C and T
Codon 215:	C only
Codon 219:	Mean of A and C

This background signal was subtracted from the wild-type (W/T) and mutant signals:

Codon 41	W/T = T	Mutant = A or G
Codon 67	W/T = C	Mutant = T
Codon 70	W/T = A	Mutant = G
Codon 215	W/T = G	Mutant = A or T
Codon 219	W/T = T	Mutant = G

The percentage of wild type or mutant (M) was calculated as follows:

$$\%M = \frac{\text{corrected M signal}}{\text{corrected M signal} + \text{corrected W/T signal}} \times 100$$

$$\%W/T = \frac{\text{corrected W/T signal}}{\text{corrected W/T signal} + \text{corrected M signal}} \times 100$$

An example of the calculation used for the codon 70 mutation is shown in Table 3. To facilitate calculation, the data from the Topcount microtiter scintillation counter were downloaded directly to an Excel spreadsheet.

Controls

Wild-type and mutant plasmid controls were run in parallel with the test samples. The plasmid samples used were HXB2 and RTMC, a mutant derived from HXB2 by site-directed mutagenesis.

PMA Solutions

Wash buffer/TTA diluent (10×)	100 m*M* Tris–HCl containing 0.5% Tween 20 1.0% Na azide (Stored as 10× concentrate at room temperature)

Table 3

Example of PMA Calculation–Codon 70 Mutation (Probe ARP2B)

Sample	Base	Signal	A signal: 2	Mean C + T	Signal minus mean C + T	% Wild type (W) or mutant (M)
1	A	9027	4513	175	4338	W, 100
	G	183			8	M, 0
2	A	1046	523	24	490	W, 10
	G	4415			4391	M, 90

PMA diluent	40 m*M* Tris–HCl, pH 7.6, containing 20 m*M* $MgCl_2$ 50 m*M* NaCl (Stored at 4°C)	
Anneal mix (for 96 wells)	PMA diluent	2475 μl
	Probe stock,	25 μl
	(Made up immediately before use)	
Labeling mix (for 24 wells)	Klenow polymerase stock	48 μl
	0.1 *M* DTT	24 μl
	[^{35}S]dNTP dilution	24 μl
	Water	144 μl
	(Made up immediately before use)	
Probe stock	10 μ*M*	
Klenow polymerase stock	2 μl Klenow polymerase 200 μl PMA diluent (Made up immediately before use)	
[^{35}S]dNTP dilution	1:10 in water of 1000–1500 Ci/mmol (NEN Dupont) These labels were all purchased at the same time to have the same activity; the 1:10 dilution was stored at −20°C.	

Validation of PMA

Construction of Standard Curve

Figure 2 shows a standard curve for the AZT-resistance mutation at codon 215 (threonine to tyrosine) of the RT gene of HIV-1. The curve was constructed by mixing known proportions of wild-type (HXB2) and mutant (RTMC) plasmids (approximately 1000 copies total input). Assays were performed in triplicate to demonstrate the assay reproducibility.

Comparison with Selective PCR, Sequencing, and Phenotyping

Further validation of PMA was made using RNA samples (tissue culture supernatants) from sequential isolates from a patient on AZT monotherapy. These isolates had previously been assayed for phenotypic AZT resistance and genotypic AZT resistance by selec-

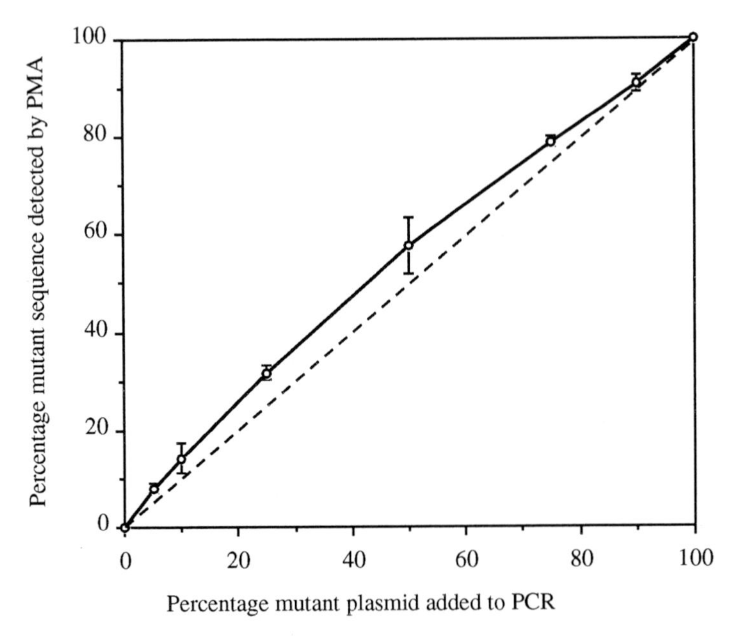

Figure 2 PMA standard curve for mutation at RT codon 215 (Thr to Tyr). Dashed line indicates equivalence. Each point was assayed in triplicate.

tive PCR at codons 67, 70, 215, and 219 only (Boucher *et al.*, 1992). The results of the phenotypic drug-resistance assays and genotypic resistance assayed by PMA and selective PCR are given in Table 4.

Of 24 points assayed only one discordance was observed (sample 105/F, codon 219), which was designated pure mutant by selective PCR and 94% mutant by PMA.

Application to Clinical Samples

Example 1: Monitoring of Genotypic Drug Resistance in Treated Patients

Figure 3 shows an example of the application of the assay to the monitoring of genotypic drug resistance in a patient treated with AZT monotherapy. Mutations were assayed in cell-free viral RNA from plasma at codons 41, 67, 70, 215, and 219 of the RT gene. The

Table 4

Comparative Analysis of ZDV Resistance-Associated Mutations in Sequential RNA Samples from a Treated Patient by PMA and Selective PCR

			Codon 67, D67N		Codon 70, K70R		Codon 215, T215Y/F		Codon 219, K219 Q	
Sample No.	Weeks on drug	AZT ID_{50} (μM)	PMA (%)[a]	PCR	PMA (%)	PCR	PMA (%)	PCR	PMA (%)	PCR
105/A	0	0.015	0	W	0	W	1	W	0	W
105/B	24	0.16	0	W	0	W	98	M	0	W
105/C	60	0.5	8	X	42	X	97	M	0	W
105/D	86	0.63	98	X	96	X	96	M	0	W
105/E	115	0.7	1	W	4	X	97	M	0	W
105/F	136	3.5	99	M	100	M	99	M	94	M

Note: Abbreviations used: M, mutant; W, wild type; X, mixed wild type and mutant.

[a] PMA result = percentage mutant sequence.

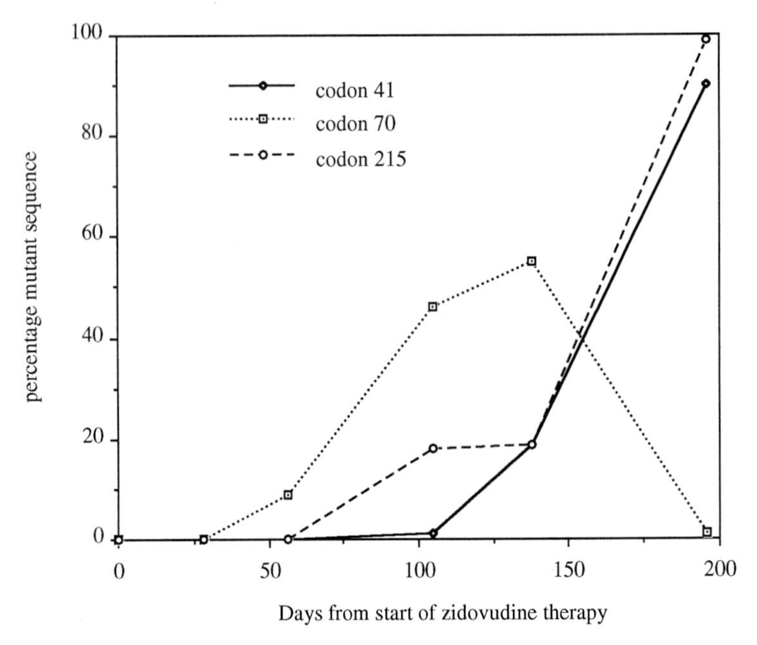

Figure 3 Changes in proportions of mutant sequence during zidovudine monotherapy.

changes in proportions of mutant sequence seen in this patient are typical of the mutation patterns commonly seen with AZT monotherapy. Mutation at codon 70 is the first to be observed and is subsequently displaced by a virus containing codon 41 and 215 mutations in combination. The mutations at codons 67 and 219 which usually appear later in the course of therapy were not seen in the patient shown in this example. The mutation at codon 70 confers a low level of drug resistance (approximately 8-fold), whereas the combined mutations at codons 41 and 215 give a high level of resistance (approximately 60-fold).

Example 2: Study of HIV-1 Replication Dynamics *in Vivo*

The quantitative data provided by PMA has allowed us to study the dynamics of HIV-1 infection by quantifying the changes in proportions of mutant sequence in cell-free virus RNA and proviral DNA from PBMCs (Kaye *et al.,* 1995).

An example of the pattern seen is shown in Fig. 4. The acquisition of the codon 70 mutation in RNA anticipates its acquisition in DNA. Similarly, the loss of the mutation beyond 300 days is seen in RNA before it is seen in DNA. These results suggest that the viral sequences represented in proviral DNA have a slower rate of turnover than the sequences represented in RNA. Therefore, quantifying the lag between RNA and DNA can be used to determine the length of the latent phase of the virus life cycle in various infected cell types. Such data can be used to determine the time needed to continue suppression of virus replication with highly active combination therapies in order to eliminate infected cell populations from the patient.

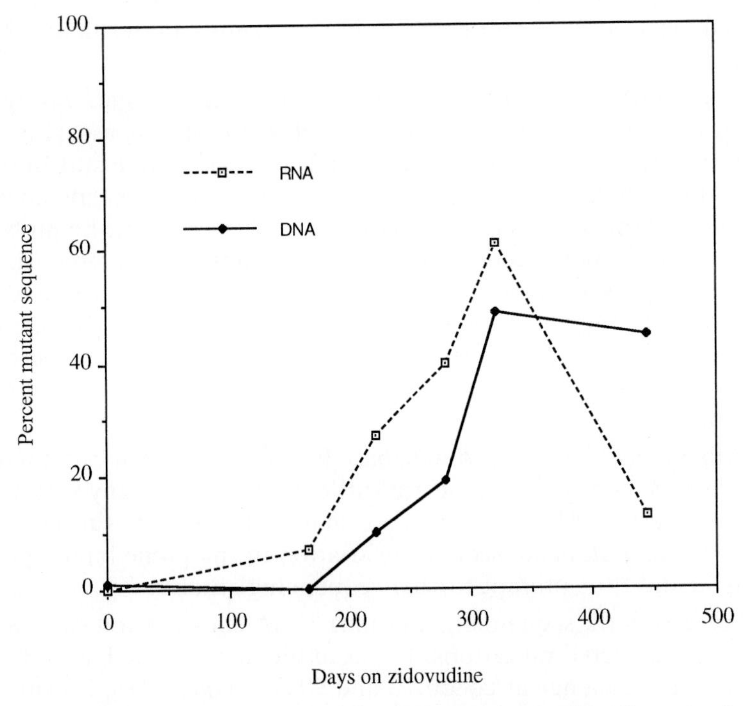

Figure 4 Changes in proportions of mutant sequence at HIV-1 RT codon 70 in plasma virus RNA and PBMC proviral DNA during zidovudine monotherapy.

Problems

The assay is generally robust and easily adaptable to any mutation of interest. Minor problems that have been encountered during the development and application of the assay include the following:

1. As with any quantitative assay the accuracy and reproducibility of the results is dependent on the sample size. As the number of RNA or DNA copies being added to the PCR reaction decreases below 10, sampling error inevitably leads to inaccuracies in the final quantitative result.
2. Occasionally, the probe may weakly anneal to other sequences within the same amplified sample, giving rise to weak background signals. For example, the probe used for detecting codon 41 (probe ARP5D in Table 2) gives a constant signal of 5% mutant sequence, even with pure wild-type plasmid input. Use of wild-type and mutant controls in assay runs can overcome this problem.
3. Long runs of identical bases at the point being assayed can result in a proportionately lower signal. For example, assaying with the codon 67 probe (ARP1 in Table 2) should result in the addition of seven T residues in the mutant sequence, and consequently the assay signal is divided by seven. In fact, the probe is probably not extended by the full amount and the proportion of mutant sequence in a mixed population is underestimated. However, the deviation of the actual result from the expected result in a standard curve is only minor (<10% from expected result) and the accuracy is probably sufficient for most applications.
4. Mismatches of even a single base between the probe and target at the 3′ end of the probe are sufficient to prevent any reaction in the assay. The absence of a signal in the assay is sufficient in itself to indicate sequence variation in the probe binding region and thus prompt further sequencing studies.
5. As with bulk sequencing methods, PMA does not indicate linkage of detected mutations. For example, if a sample has 50% mutant sequence at codon 70 and 50% at codon 215, it cannot be determined whether both mutations are present on the same virus genome.

Future Developments

Since the basic protocol is applicable to any mutation in any PCR product, the method can be applied to any mutation of interest in any organism. Clearly, there are probably more convenient ways of detecting simple allelic mutations in which the only options are 100% wild type or mutant or 50% heterozygous. The assay is best suited to applications in which populations of sequences need to be monitored for the acquisition or loss of a point mutation in serial samples.

As an alternative to the use of ^{35}S-labeled dNTPs, fluorescein-labeled ddNTPs can be used. It may be possible to detect the signal directly with newly developed high-sensitivity fluorometers, or the signal can be generated with an antifluorescein–enzyme conjugate and a colorimetric or chemiluminescent substrate (Ballard and Boxall, 1997). The combination of the PMA protocol with fluorescent-labeled ddNTPs and fluorescence-based automated sequencers can be used in a single-reaction multiprobe version of the assay—site-specific sequencing (Martinez-Picado *et al.*, 1997). Multiple probes of differing lengths are added to a single reaction and the extended labeled oligonucleotides can then be separated and quantified on an automated fluorescence-based sequencer such as the Applied Biosystems Prism 377.

Conclusions

The quantitative point mutation assay described allows the detection and quantification of single base changes in a mixed population of wild-type and mutant sequence. The assay is easily adapted to the assay of any mutation in any PCR amplified product. No modifications to the basic protocol are required to quantify a new mutation of interest, other than the design of new PCR primers and detection probe. The microtiter format of the assay gives the assay a high throughput and the potential for automation. Because the assay is quantitative, it can be applied not only to the detection of mutations in a mixed population of wild-type and mutant sequence but also to studies of the rates of acquisition or loss of a mutation over time in serial samples.

Acknowledgments

I acknowledge the excellent technical assistance of Julie Bennett, Liz Comber, and Shrenee Nesaratnam and the advice and guidance of Professor Clive Loveday and Professor Richard Tedder (all from the Department of Virology, University College London). The plasmids used in the construction of the standard curve were a generous gift from Dr. Brendan Larder, Glaxo-Wellcome Research, Stevenage, UK. Development of the assay was supported by grants from the Medical Research Council and the Trustees of University College London Hospitals.

References

Ballard, A. L., and Boxall, E. H. (1997). Colourimetric point mutation assay: For detection of precore mutants of hepatitis B. *J. Virol Methods* **67**, 143–152.

Boucher, C. A. B., O'Sullivan, E., Mulder, J. W., Ramaurtarsing, C., Kellam, P., Darby, G., Lange, J. M. A., Goudsmit, J., and Larder, B. A. (1992). Ordered appearance of zidovudine resistance mutations during treatment of 18 human immunodeficiency virus-positive subjects. *J. Infect. Dis.* **165,** 105–110.

Kaye, S., Loveday, C., and Tedder, R. S. (1992). A microtitre format point mutation assay: Application to the detection of drug resistance in human immunodeficiency virus type-1 infected patients treated with zidovudine. *J. Med. Virol.* **4,** 214–246.

Kaye, S., Comber, E., Tenant-Flowers, M., and Loveday, C. (1995). The appearance of drug resistance-associated point mutations in HIV type-1 plasma virus RNA precedes their appearance in proviral DNA. *AIDS Res. Hum. Retrovirus.* **11,** 1221–1225.

Kellam, P., Boucher, C. A. B., and Larder, B. A. (1992). Fifth mutation in human immunodeficiency virus type-1 reverse transcriptase contributes to the development of high-level resistance to zidovudine. *Proc. Natl. Acad. Sci. U.S.A.* **89,** 1934–1938.

Larder, B. A., and Kemp, S. D. (1989). Multiple mutations in HIV-1 reverse transcriptase confer high-level resistance to zidovudine (AZT). *Science* **246,** 1155–1158.

Martinez-Picado, J., Sutton, L., Helfant, A. H., Savara, A., Kaplan, J., and D'Aquila, R. T. (1997). Novel clonal analyses of resistance to HIV-1 RT and protease inhibitors. International Workshop on HIV Drug Resistance Treatment Strategies and Eradication, St. Petersburg, FL, June. [Abstract No. 45]

Schinazi, R. F., Larder, B. A., and Mellors, J. W. (1996). Mutations in retroviral genes associated with drug resistance. *Int. Antiviral News* **4,** 95–107.

12

IN SITU PCR

Jim R. Hully

This chapter presents a comprehensive overview of the *in situ* PCR technique with emphasis on current recommendations and protocols. Fixation, permeablization, and diffusion, important issues that continue to dissuade researchers from adopting this technique, are discussed in more detail. Finally, this review is intended as a resource for directing the reader to more detailed treatises on all the relevant aspects of *in situ* amplification. This review is not intended to be exhaustive because there are more than 500 citations for *in situ* amplification and it has been the topic of a major publication (Herrington and O'Leary, 1998).

In situ polymerase chain reaction (IS-PCR) is a collective term used to describe primer-driven amplification of a DNA or RNA (IS-RTPCR) template by PCR and its subsequent detection within the confines of a histological tissue section or cell preparation. Many different names and acronyms exist in the literature to describe minor technical variations: *in situ* PCR, PCR *in situ*, PCR *in situ* hybridization, in-cell PCR, direct and indirect ISPCR, PCR-driven ISH, localized *in situ* amplification, PCR-ISH and RTPCR-ISH, direct *in situ* single-copy, in-cell RT-PCR, labeled primer *in situ* polymerase chain reaction, PI-PCR, and G-PCR. Unfortunately, *in situ* PCR or IS-PCR are not recognized subject headings in the databases of either the National Institute of Health or the American Chemical Society.

Since its first description in the scientific literature by Ashley Haase and colleagues in 1990, IS-PCR has attracted a certain notoriety. In principle, it is relatively simple although technically laborious, but it quickly became apparent that it was challenging to optimize and without careful controls could readily generate false positives due to nonspecific amplification or diffusion (Komminoth *et al.*, 1992; O'Leary *et al.*, 1994a). Unfortunately, the plethora of papers that appeared in the early years failed to address these important caveats, suffered from poor reproducibility, and presented widely divergent and even contradictory protocols. Currently, there is still a belief that performing IS-PCR is part art and part science. Such a dogma does little credit to those who have dedicated long hours solving these problems or to the results available through IS-PCR.

The technique has been the subject of many reviews—some critical (Komminoth and Long, 1993; Teo and Shaunak, 1995b; Lewis, 1996c; O'Leary *et al.*, 1996; Long and Komminoth, 1997) and others more expansive and that include detailed protocols (Bagasra *et al.*, 1994; Gu, 1994; Hybaid, 1994; Nuovo, 1994; Lewis, 1996a; Bagasra and Hansen, 1997; Herrington and O'Leary, 1998; Muro-Cacho, 1997; O'Leary, 1998a).

Why perform PCR *in situ*? Where conventional PCR can answer the question of whether tissues are positive or not for the sequence of interest, it can only convey limited information about the cellular phenotypes and architecture of the sample. Such spatial information is available using conventional *in situ* hybridization but it lacks the sensitivity inherent to PCR. The general consensus is that nonisotopic *in situ* hybridization cannot reliably detect <20 copies per nucleus. This quickly becomes a limiting factor when one appreciates that most species of mRNA are of low abundance and the vase majority of functional genes are single copy. The integration of PCR and *in situ* hybridization combines exquisite sensitivity with anatomical definition enabling the investigator to look for extremely rare transcripts or single-copy genes at the cellular or subcellular levels. Direct comparisons of sensitivity between *in situ* hybridization and IS-PCR are limited but Jin *et al.* (1995) clearly showed the increased sensitivity gained by using IS-PCR when compared to *in situ* hybridization.

The DNA and mRNA targets for *in situ* amplification are not limited to eukaryotes (Hodson *et al.*, 1995; Tolker-Nielsen *et al.*, 1997) or zoological specimens (Goad *et al.*, 1991; Johansen, 1997), and have extended to the electron microscope (Kareem *et al.*, 1997). The primary applications have been the studies of infectious diseases, in particular viruses, genetic disorders, gene expression, genetic manipulations, and cancer. The potential exists for IS-PCR to be used in studying minimal residual disease, toxicology, forensics, and perhaps as a diagnostic tool in the clinic.

Protocols

It is important to emphasize that no single IS-PCR protocol is universally applicable nor should results be dependent on the vagaries of a single protocol. The most important criteria for embracing any protocol is that it be robust and reproducible. Too many published protocols are incapable of migrating beyond the laboratories, or perhaps the water supplies of their creators. As with any scientific endeavor, results can ony be measured by the controls, an truism often forgotten by those performing IS-PCR. The exquisite sensitivity of PCR combined with a lack of understanding of fixation has generated significant problems. Consequently, each and every experiment must be accompanied by a set of key controls designed to eliminate the possibility of false positives or negatives.

The protocol for IS-PCR is conveniently divided into three stages: (i) preparation of sample for PCR, which includes the key steps of fixation and permeablization; (ii) the actual PCR; and (iii) the detection of the amplified signal. Although arbitrary, such a division serves to emphasize in descending order, the relative importance of each of these steps to successful IS-PCR. Appendix 1 describes a robust and reproducible IS-PCR protocol that can be adapted to most samples and targets.

Fixation

Control of all aspects of fixation is paramount to achieving unequivocal and reproducible results with IS-PCR (O'Leary *et al.*, 1994b). Unfortunately, most clinical or experimental samples are fixed with routine histological diagnosis in mind and little regard for any future applications or developments. Furthermore, how the sample is prepared is also conducive to obtaining superior and consistent results. However, most "routine" fixatives are compatible with PCR and IS-PCR, although there are some exceptions (Tokuda *et al.*, 1990; An and Fleming, 1991; Ben-Ezra *et al.*, 1991; Greer *et al.*, 1991a,b; O'Leary *et al.*, 1994b). Cross-linking fixatives such as "formalin" or (para)formaldehyde as well as precipitating fixatives such as simple alcohols and acetone can give excellent IS-PCR results. When comparing these two major classes of fixatives, cross-linking agents produce better morphology but usually require longer pretreatment to expose the template. Precipitating fixatives are less damaging to nucleic acids but are not as capable at maintaining cellular integrity.

For consistent results the cross-linking fixative should have a neutral pH and be adequately buffered if it is not prepared fresh, the reagents should be of the highest quality, and the length of fixation should not exceed 24 hr. An excellent fixative is 4% formaldehyde (this is incorrectly but better known as 10% formalin) in a phosphate buffer, pH 7.0–7.4, prepared within 24 hr of use. Prolonged fixation offers no advantages and serves only to introduce unwarranted template damage as well as extend the permeablization steps (Tokuda *et al.*, 1990). For a detailed review and history of formaldehyde fixation, consult Fox *et al.* (1985). Precipitating fixatives such as methanol alone or together with acetone in a 3 : 1 ratio for 2–8 hr are excellent alternatives, particularly if using cultured or isolated preparations of cells.

There are a number of special fixatives [e.g., Permeafix (OrthoDiagnostics, Raritan, NJ, 800-322-6374) and Streck's Tissue Fixative (Streck Laboratories, Inc., Omaha, NE, 800-228-6090)] designed to be more compatible with immunocytochemistry and *in situ* hybridization. Although they are costly, there is evidence that they are better for cell preparations because they reduce or eliminate the need for permeablization (Patterson *et al.*, 1996; Uhlmann *et al.*, 1998). Concerns over the hazards of prolonged exposure to formalin fumes or use of flammable fixatives has prompted the development of alternative fixatives. To date, there are no data regarding their compatibility with IS-PCR. Glutaraldehyde should be avoided for IS-PCR because of the extensive cross-linking effected by its dual-reactive sites. In addition, unless these sites are sequestered, cross-links will continue to form long after the reagent has been removed. Other fixative which are not recommended are those that contain inhibitors of PCR such as heavy metals or significant amounts of acids, such as Bouin's or Zenker's (Jackson *et al.*, 1990; Greer *et al.*, 1991b).

Tissue samples should be collected as quickly as possible from the cadaver, trimmed into manageable blocks, and, if necessary, washed free of red blood cells before immersion in the fixative. The smaller the block, the more reliable and consistent will be fixation. As a general rule at least one of the block's dimensions should not exceed 5 mm. After fixation of 18–24 hr they should be processed promptly into paraffin wax or stored in 70% ethanol for up to a few days until processing can occur. Cultured cells can be prepared by a number of different methods. The quickest method is to harvest the cells, either enzymatically or mechanically, wash off culture media, and cytospin directly onto the silanized microscope slide. However, most if not all cytospins can only place one specimen per slide, thereby limiting the number of samples. To increase the number of available samples per slide, cells can be grown directly on the slide and subdivided using a hydrophobic marker (Zhang and Wadler, 1997). The cells should be fixed for 1–12 hr, washed, and air-dried. If the slides are to be stored, they should be kept desiccated at −70°C. For more consistent results, resuspend the harvested cells in fixative for up to 24 hr, wash, and process as a pellet into paraffin wax as described below (O'Leary *et al.*, 1994b; Levin *et al.*, 1996).

Conventional reagents and protocols can be employed for dehydrating and infiltrating the samples. Once embedded, the blocks can be stored for many years without any significant deterioration to the DNA. Use of low-melting temperature paraffin wax may reduce the template damage associated with exposure to high temperatures but it must be balanced against poor handling qualities and increased cost. Conventional 5- or 6-μm sections are mounted onto silanized slides. To minimize contamination the contents of the water bath should be changed and the microtome knife cleaned between blocks.

There are different methods for increasing the adherence of tissue sections or cells to microscope slides to minimize their loss during the harsh preparatory steps of IS-PCR (Dyanov and Dzitoeva, 1995). The most effective method has been to silanize the slides with aminopropyltriethoxysilane (APES). For protocols on applying APES see Lewis (1996a) and Bagasra and Hansen (1997). However, Teo and Shaunak (1995a) reported that preincubation of the PCR mixture on APES-coated slides rendered the reagents inactive, which could be reversed using lecithin or fetal calf serum. It should be noted that exposed glass surfaces will greatly affect PCR efficiency presumably due to sequestration of reagents. To increase the adherence most protocols call for an additional heating step. Paraffin-embedded samples should be baked horizontally at 60°C (or hot enough to melt the wax) for 36 hr for DNA to firmly attach the specimen to the slide as well as reduce nonspecific staining (Lewis, 1996a). Alternatively, Bagasra and Hansen (1997) recommend a heat-stabilization treatment in which the cells or sections are dry baked at 105°C for 5–120 sec followed by prolonged formalin fixation. Sections from cryopreserved tissue should be mounted on silanized slides, air-dried, and fixed in one of the aforementioned fixatives for 30–60 min prior to IS-PCR (Kovalenko *et al.*, 1997).

A slightly different approach consists of attaching the samples to fragments of slides or coverslips which can fit into conventional PCR tubes (Spann *et al.*, 1991; Ertsey and Scavo, 1998). The advantage is the use of regular thermal cyclers but the same pretreatments are required and sample size will be limited.

Permeabilization

Rarely will a section of tissue or cells be suitable for PCR, except perhaps those fixed with the newer molecular biology-orientated fixatives as previously noted. Invariably, the nucleic acids are unavailable as a consequence of fixation and/or biological barriers such as membranes and binding proteins. Such samples need to be "permeabilized," which involves denuding the nucleic acids of associated proteins, facilitating entry of the PCR reagents while maintaining structural integrity, and avoiding diffusion of the target or amplicon from the site of origin. Such requirements are almost mutually exclusive and impossible to fulfill when one considers how little is known about fixation or about creating and measuring the "holes" in the cellular architecture.

The majority of published IS-PCR protocols use some form of proteolytic digestion to permeablize the sample. The most popular enzymes are proteinase K, pepsin, pepsinogen, and trypsin, and each has its own optimal pH. Until cellular organization of DNA or RNA is fully understood, the best enzymes are those with broader substrate specificity. Optimal permeabilization is largely determined by the type of cell or tissue and the conditions of fixation. Consequently, it is very difficult to extrapolate these conditions for different samples and protocols and these should always be determined empirically. Unfortunately, there is no objective criteria for measuring this process. The simplest approach is to titrate enzyme activity varying enzyme concentration, time, or temperature and assess the results by IS-PCR using *in situ* hybridization for detection of the signal. However, this is contingent on having previously optimized the PCR *in situ* conditions, which in turn is contingent on how the tissue was prepared. This makes for a complicated and extensive set of experiments, even without considering the additional controls.

An alternative strategy is to isolate each of the three steps to IS-PCR and optimize them individually. One of the best methods for optimizing permeabilization is to exploit the 5′ to 3′ exonuclease activity of most thermostable DNA polymerases. Over a wide range of PCR conditions the enzyme will repair single-stranded nicks to the DNA in the absence of primers. In the presence of a labeled nucleotides, e.g., with biotin or digoxigenin, the nuclei will become labeled only if the PCR components have proper access to the template. Cross-linking fixatives and the tissue processing can induce extensive nicking to the DNA. This damage is usually evident in formalin-fixed, paraffin-embedded tissue sections rather than in ethanol-fixed cytospin preparations. Optimal permeabilization conditions should be obtained at the lowest titration of enzyme that labels virtually all the cells. Just how many cells can be labeled is unknown, but this approach will provide the researcher with an excellent starting point.

Such a strategy can be modified for a RNA target by incorporating the labeled nucleotide along with random hexamers during the reverse transcription step prior to PCR. Again, the best conditions will probably be those that need the least amount of proteolytic digestion to give a strong cytoplasmic signal. The only provisions are (i) the need to eliminate competing genomic DNA and (ii) that the window for optimal permeabilization is narrower, limited to the step of reverse transcription, whereas for DNA the window can be extended over a number of cycles because permeabilization by heating will continue during PCR.

Other methods for permeabilizing sample for PCR include ionic and nonionic detergents, acid and base hydrolysis, antigen retrieval, and microwaves (Lewis, 1996a; Mancuso *et al.*, 1997). Clearly, the process of thermal cycling will contribute to permeablization and must be considered when optimizing. Mild acid hydrolysis is known to be an affective technique for selectively removing histones from chromatin, thereby exposing the DNA (Moran *et al.*, 1985; Lewis, 1996a). Man *et al.* (1996) suggest an overnight incubation at 80°C to reverse the cross-links of formalin fixation and eliminate the need for enzymatic digestion. The improvements offered by antigen retrieval have been used successfully to improve the sensitivity of ISH (Teo and Griffin, 1990). They should be equally applicable to IS-PCR, reducing the crude proteolytic pretreatments and/or the number of thermal cycles, both of which should reduce diffusion.

To amplify from mRNA *in situ* requires the transcription of the message in the absence of competing genomic DNA. The DNA can be eliminated either by DNase pretreatment or by designing primers that preferentially amplify from the cDNA. Effective removal of the DNA from formalin-fixed tissue requires extensive DNase treatment (O'Leary, 1998a,b). For unknown reasons this treatment has a deleterious effect on the overall integrity of the specimen. The more elegant approach is to design primers that will only anneal across a splicing junction of a transcript or bracket an extremely long (>2 kbp) intron. Under optimal conditions, either strategy will only amplify the cDNA, thereby avoiding the need for DNase pretreatment. Initial attempts to perform IS-RTPCR required the use of two enzymes, a reverse transcriptase and a thermostable DNA polymerase. This cumbersome procedure has been superceded by one-step methods using either a single enzyme exhibiting both activities, such as r*Tth*, or a premixed blend of a reverse transcriptase and a "hot start" thermostable DNA polymerase such as AmplitaqGold (Nuovo and Forde, 1995; Peters *et al.*, 1997; Uhlmann *et al.*, 1998). AMV and MMLV reverse transcriptase give comparable results but neither are as efficient as Superscript II (GIBCO BRL Life Technologies, Gaithersburg, MD; *http://www.lifetech.com*) (Martínez *et al.*, 1995; Bagasra and Hansen, 1997; Szczepek *et al.*, 1997). As is readily apparent throughout this review, the optimal conditions for reverse transcription can only be determined empirically.

Some authors recommend a brief fixation following these preparative steps, presumably to stabilize the sample and restore rigidity, although this will reestablish cross-links that the very preparative steps were designed to remove (Chen and Fuggle, 1993).

PCR

It is important to first test the integrity of the fixed template before attempting IS-PCR, even if fixation and all subsequent treatments are known. There are a number of protocols and commercial kits designed specifically to isolate DNA or RNA from fixed, paraffin-embedded sections for regular PCR analysis (Jackson *et al.*, 1990; Heller *et al.*, 1992; Finke *et al.*, 1993; Koopmans *et al.*, 1993; Going and Lamb, 1996; Mancuso *et al.*, 1997). However, even if the nucleic acid proves amplifiable, the thermal and chemical parameters used may not be applicable to the PCR performed on a glass slide.

There are two basic strategies for labeling the amplified product. One method is to tag the amplicon during the PCR and is generally known as direct IS-PCR or just IS-PCR. The alternative uses a labeled probe after the PCR to hybridize to the amplicon and has been called indirect IS-PCR or PCR-ISH. The direct approach results in a simplified detection protocol and fewer preparative steps. Consequently, the technique has been regarded as easier, quicker, and cheaper, but these advantages come with serious reservations regarding specificity. The direct labeling of the amplicon during PCR can be accomplished in two ways. The reporter molecule (typically biotin, digoxigenin, or fluorescein) is either attached to a nucleotide (typically dUTP) and added to the PCR or incorporated during the synthesis of one or both primers, usually at the 5′ end. Although subtle, this difference in labeling is important because using a labeled nucleotide during amplification has been shown repeatedly to give false positive results (Komminoth *et al.*, 1994; Sällström *et al.*, 1993; Teo and Shaunuk, 1995a). Heniford *et al.* (1993) clearly illustrated the dangers when they showed that a false positive signal was evident after as few as five cycles. This phenomenon has been attributed principally to the native 3′ to 5′ exonuclease activity of DNA polymerase repairing single-strand nicks to the DNA or to endonuclease activity in dying or apoptotic cells. Strategies for inhibiting this nonspecific incorporation, including hot start, 3′ to 5′ exonuclease-deficient DNA polymerases, and capping, have proved unsuccessful (Sällström *et al.*, 1994) but there have been successful attempts to repair the degraded DNA prior the PCR (Gosden and Hanratty, 1993; Tokusashi *et al.*, 1995). Consequently, the use of labeled nucleotides should be avoided or at the very least be accompanied by a parallel reaction that lacks primers to check that the signal is primer driven. This form of direct labeling for DNA is best reserved for optimizing cellular permeablization as described previously or as a modification to reducing diffusion (see Diffusion). One intriguing idea is to take advantage of genomic DNA's slow rate of reanneal following a rapid denaturation. If the genomic DNA can be rendered single stranded for the duration of the PCR, there should be no primer-independent DNA synthesis (Lewis, 1996a). The use of the one-step methods for amplifying an mRNA target overcomes most of these reservations concerning use of labeled nucleotides because there is no DNA and the cDNA should only be transcribed by the antisense primer. However, the specificity of the signal must still be confirmed to eliminate the possibility of mispriming.

A better approach for direct IS-PCR that avoids the problem of primer-independent signal is to use labeled primer(s) as mentioned previously (Oates *et al.*, 1997; Lewis, 1996a). However, certain labels, such as fluorescein and digoxigenin, show a nonspecific affinity for cellular structures and can exhibit high background staining. Biotin can present similar problems because it is a naturally occurring vitamin and known to be expressed at detectable levels in certain tissues. A promising label that does not display these drawbacks is dinitrophenyl, which generates exceptionally low background when used in a labeled primer for IS-PCR and IS-RTPCR (F. Lewis, personal communication).

Both approaches to performing direct IS-PCR can generate a false positive signal because of mispriming and subsequent extension. This can be minimized by carefully optimizing the annealing temperature of the PCR and confirming the specificity of the signal by performing a parallel indirect IS-PCR. As a result of the problems discussed previously, it is recommended that anyone contemplating IS-PCR should start with the indirect strategy for detecting the amplicon. Apart from the inefficiency of amplification, there are no caveats associated with performing indirect detection during the thermal cycling.

It is generally accepted that the greater the size of the amplicon, the less likely it is to diffuse out of a cell and give rise to false positives. However, the maximum length of an amplicon will be limited by the fixation and the efficiency of PCR (Greer *et al.*, 1991a,b). To overcome the damaging effects of fixation and reduce diffusion, a number of authors have tried using multiple primer sets with complimentary tails or overlapping sequences to form products of every-increasing length (Haase *et al.*, 1990; Staskus *et al.*, 1991; Chiu *et al.*, 1992).

There should be no need to exceed 40 cycles and in theory as little as 10 should generate enough signal for detection if amplification is close to exponential on a slide (Heniford *et al.*, 1993; Kovalenko *et al.*, 1997). Unfortunately, *in situ* amplification is relatively linear and inefficient compared to conventional PCR. Quantitative estimates suggest as little as 10- to 300-fold increase following 30 cycles of IS-PCR (Teo and Shaunuk, 1995a; Embretson, *et al.*, 1993). Despite variations between cell types, Mee *et al.* (1997) never found more than a 2- to 7-fold increase in signal after 10 cycles when compared by performing *in situ* hybridization. This inefficiency is probably a combination of factors but the most important is that the PCR environment on a slide is quite different to that in a tube. There is a significant increase in the ratio of glass/silane surface area to reaction volume and there are relatively vast amounts of cellular material surrounding the template, which all contribute to affect the availability of reagents and compromise PCR efficiency. Other contributing factors include the presence of potential inhibitors in the tissue (An and Fleming, 1991), silane coating (Teo and Shaunuk, 1995a), variable accessibility of the target, loss of amplicon by diffusion, local concentration gradients of reagents or amplicons unfavorable to exponential amplification, and the slower cycling inherent to the dedicated *in situ* machines.

Consequently, recommendations for the thermal and biochemical parameters for performing PCR on slides can only be general and a brief inspection of the literature shows little uniformity in the published protocols (Teo and Shaunuk, 1995a; O'Leary *et al.*, 1996). It is imperative that the researcher design a matrix to analyze all the major variables of PCR using a purified template on blank glass slides which have undergone the same preparative steps as if there was tissue present. The most important variables are the concentrations of Mg^{2+}, thermostable DNA polymerase, and primers, the annealing temperature, and cycle number. Improvements in amplification efficiency will only come from understanding of how the unique *in situ* conditions affect the performance of PCR on a slide. Various additives to the PCR mixture have been suggested to overcome the relative inefficiency or possible inhibitory effect of glass slides on the PCR process. The results have been inconsistent, although a case can be made for including albumin and gelatin (Yap and McGee, 1991; Lewis, 1996a).

The integration of hot start into conventional PCR eliminated most of the problems of nonspecific amplification due to mispriming. However, it is not clear if it is beneficial for IS-PCR; until proved otherwise, it would be prudent to include this modification (Nuovo *et al.*, 1993; Bagasra and Hansen, 1997).

Signal Detection

Following PCR, the samples are washed to remove the PCR reagents and free amplicons. To prevent any subsequent loss of amplicon, a number of authors suggest a brief fixation in formalin (O'Leary *et al.*, 1994b; Bagasra and Hansen, 1997). Detection of sufficient amplified target (100 or more copies) does not present any novel problems for standard *in situ* hybridization and immunocytochemical techniques. Consult the following references for published protocols to perform *in situ* hybridization after PCR; Choo (1994), Lewis (1996b), Bagasra and Hansen (1997), and Herrington and O'Leary (1997). The only provision is that during PCR the thermal effects may further denature proteins, causing nonspecific binding of the detection reagents. Addition of a blocking step before detection should resolve this problem. At the same time, the cycling may offset the need to specifically block for the endogenous activity of substrates such as peroxidases or alkaline phosphatases because the high temperature may destroy the enzymes. The indirect detection protocol requires additional steps to hybridize a labeled probe to the amplicon. The probes can be single- or double-stranded DNA or riboprobes, and they are amenable to a variety of reporter molecules. The reporter is detected by conventional immunocytochemistry, preferably with the Fab fragment of the antibody conjugated to a chromagen. Figures 1 and 2 illustrate the detection of single-copy gene elastin in normal sheep lung following IS-PCR.

Controls

Each IS-PCR experiment must be accompanied by a set of control reactions. Ideally, these should include a known positive and negative example and controls to address potential sources of false positives and false negatives. The false positives can be generated through primer-independent DNA synthesis and mispriming if using direct IS-PCR, genomic DNA if studying mRNA, nonspecific binding of labeled primers, diffusion, and nonspecific binding of detection reagents. False negatives can be caused by too much or too little pretreatment resulting in the loss or unavailability of template, respectively, diffusion, and failure of the PCR or signal detection stages.

Controls required for every experiment

- Perform PCR using a purified template on a blank microscope slide that has been through the preparative stages. As described earlier, this approach will optimize the PCR conditions and should be included in each experiment to monitor successful and specific amplification. The reaction mixture should be analyzed on an agarose gel.

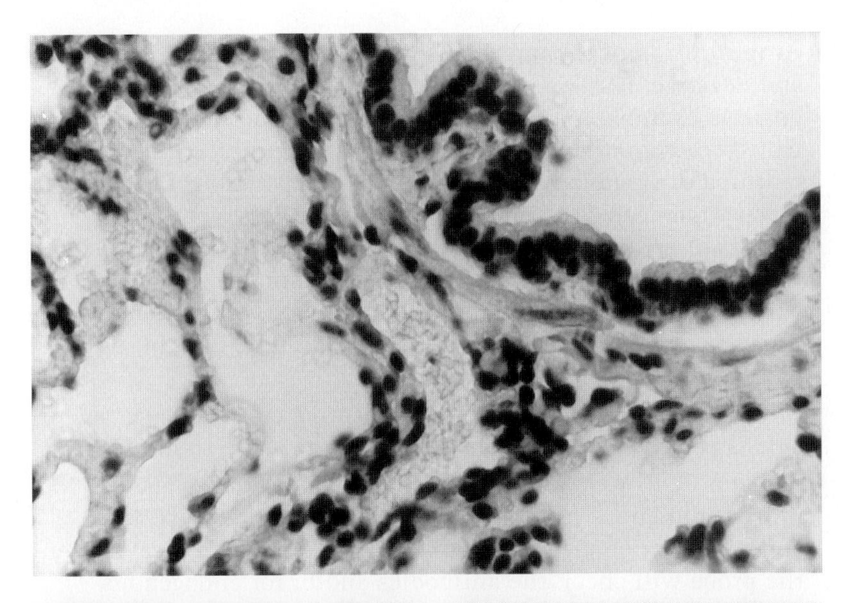

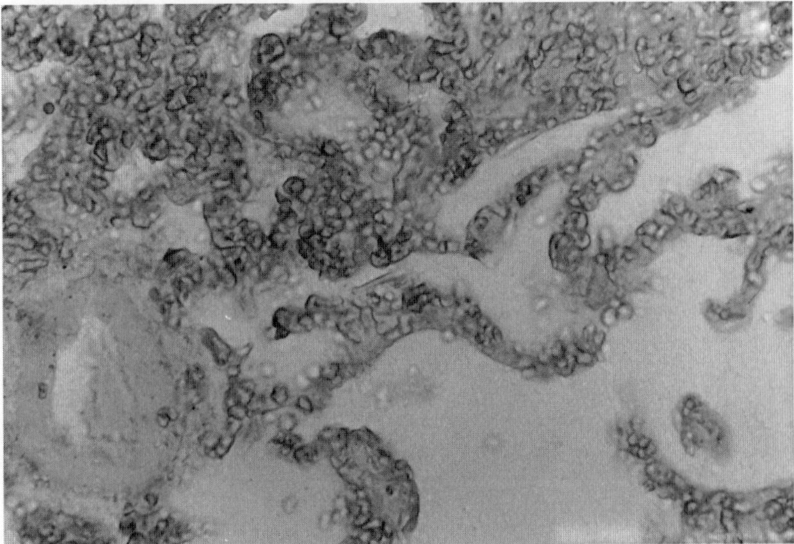

Figure 1 and 2 Micrographs of normal sheep lung following *in situ* PCR to detect the single sheep copy elastin gene using DNP-labeled primers. Figure 1 shows crisp nuclear staining of virtually all the cells. Note the complete absence of cytoplasmic or extracellular staining. Figure 2 is a negative control for the sheep elastin gene; *in situ* PCR was performed in the absence of dNTPs. Note the absence of nonspecific staining. Magnification, ×500.

- Amplification of the target in specimens known to be positive and negative: For tissues this may not be feasible because the positive example may simply not exist. Furthermore, to be a truly representative they should be histologically identical. For cultured or single cells, these could be preparations of wild-type versus transfected or gene knockout cells. Fortunately, for studies of mRNA, the transcripts tend to vary within a given tissue and therefore can be used as an internal control. A suitable alternative is to amplify a reference control gene within the same sample, e.g., pyruvate dehydrogenase, β-actin, or β-globin.
- Mix known ratios of positive and negative cells or place, side by side, a positive and negative sample of the same tissue.
- Perform PCR without the DNA polymerase or any labeled moieties to check for nonspecific binding of the probe or antibodies, or endogenous activity of the substrate. Leaving out the probe will check the antibodies and/or for endogenous substrate, whereas the absence of both the probe and the antibodies will identify any endogenous activity. Depending on the starting copy number, the probe may generate a faint positive signal.

Additional controls for IS-RTPCR

Pretreat samples with RNase or leave out reverse transcriptase. This will help confirm that the amplicon was specific to mRNA or cDNA. If using a single enzyme such as r*Tth,* leave out either the $Mn(OAc)_2$ salt or the 60°C reverse transcription incubation.

Additional controls if performing indirect IS-PCR or IS-RTPCR

Perform indirect IS-PCR but hybridize using an irrelevant/sense probe to check for specificity of probe.

Additional controls if performing direct incorporation of labeled primers

- Perform indirect IS-PCR to confirm specificity of amplification.
- Perform PCR in the absence of the DNA polymerase to check for nonspecific binding of labeled primers.

Additional controls if performing direct incorporation of labeled nucleotides

- Perform PCR in the absence of the primers to check for primer-independent DNA synthesis.
- Perform indirect IS-PCR to confirm specificity of amplification.

- Perform PCR in the absence of the DNA polymerase to check for nonspecific binding of labeled nucleotides.

Analyzing the reaction mixture that overlies the sample following PCR may be misleading. The absence of amplicon does not necessarily indicate PCR failure; it could also mean that there was little or no diffusion. Although its presence indicates specific amplification, it may also indicate excessive diffusion rather than any localized signal.

Diffusion

The most serious issue questioning the validity of IS-PCR is the phenomenon of diffusion. While most authors acknowledge that it is a significant source of false positives, most choose to ignore it and very few actually control for it. Diffusion artifacts are caused by amplicons or templates leaking out of positive cells and either diffusing into negative cells or serving as templates for extracellular amplification which then too goes back into cells. The degree of diffusion is very much dependent on cell phenotype, fixation, and the preparative steps prior to PCR. Clearly, making the cells sufficiently permeable to facilitate entry of the PCR reagents is the biggest contributing factor to diffusion. In addition, according to Charles' Law the changes in temperature and therefore in volume may accelerate the random movement of amplicons away from their site of synthesis.

The clearest evidence for or against diffusion has come from mixing experiments using known ratios of positive and negative cells (Bagasra *et al.*, 1992; Patterson *et al.*, 1993; Jin *et al.*, 1995; Teo and Shaunuk, 1995a). Alternatively, amplification of sequences in fixed metaphase spreads failed to show evidence of diffusion (Gosden and Hanratty, 1993; Troyer *et al.*, 1995). These authors independently confirmed the localization of the signal by *in situ* hybridization and linkage mapping.

In theory, generating longer amplicons would lessen diffusion, but Teo and Shaunak (1995a) noted that larger amplicons generated more false positives through diffusion. Incorporating nucleotides labeled with fluorescein or digoxigenin into the PCR should make the amplicon "stickier" or bulkier by increasing the effective size of the product and therefore less likely to diffuse (Komminoth *et al.*, 1992; Patterson *et al.*, 1993). Reducing the number of cycles or improving PCR efficiency are the simplest methods for minimizing the effect of diffusion (Komminoth *et al.*, 1992). The optimum number of cycles will vary according to the factors mentioned earlier and should to be determined empirically using some form of quantitation. Mee *et al.* (1997), using radioactivity, determined that within the first 10 cycles amplicon had been lost, possibly by diffusion. Other strategies for reducing diffusion include an agarose overlay (Chiu *et al.*, 1992) and antibodies to the cytokeratin AE1/AE3 to create a "sponge" to sequester newly formed cDNA (Man *et al.*, 1996).

Not until there is a clearer understanding of fixation, permeablization of cellular nucleic acids, and amplification inefficiency will there be successful solutions to diffusion. Novel approaches to solving these issues include using labeled probes or particles of known size, combined with scanning electron or confocal microscopy, and the use of highly purified proteolytic enzymes with specific substrates.

Tools for IS-PCR

There are three specialized thermal cyclers for performing PCR directly on the microscope slide. In addition, most thermal cyclers with interchangeable blocks include one designed for slides. The dedicated machines are the GeneAmp *In Situ* PCR System 1000 (Luehrsen *et al.*, 1995; Picton and Howells, 1997) from Perkin–Elmer (Foster City, CA; *http://www.perkin-elmer.com*), the OmniSlide Temperature Cycling System (Starling, 1997) from Hybaid, Ltd. (Holbrook, NY; *http://www.hybaid.co.uk*), and the PTC-100-16MS from MJ Research, Inc. (Watertown, MA; *http://www.mjr.com*). The GeneAmp System 1000 thermal cycler is supported by an extensive line of reagents and consumables, including a unique reaction containment system and the recent release of the first-ever control kit for IS-PCR. One of the major drawbacks, if not using the GeneAmp system from Perkin–Elmer, is the crude and often unsuccessful methods for sealing the edges of the glass coverslips to limit evaporation during thermal cycling. The development of special adhesive reaction chambers and particularly Self-Seal has rendered these methods obsolete (Sullivan *et al.*, 1997).

Alternative Amplification or Detection Strategies

Alternative methods exist for amplifying intracellular targets, with some demonstrating single-copy sensitivity. These include primed *in situ* DNA synthesis (PRINS), cycling PRINS, self-sustained sequence replication or 3SR, and the ligase chain reaction (Koch *et al.*, 1989; Guatelli *et al.*, 1990; Terkelsen *et al.*, 1993; Tyagi *et al.*, 1996; Gosden, 1997; Herrington and O'Leary, 1998; Mueller *et al.*, 1997).

A different strategy is to amplify the reporter molecule rather than the target. The most effective method is the catalyzed reporter deposition technique using peroxidase-mediated deposition of hapten-labeled tyramides (Adams, 1992; van Gijlwijk *et al.*, 1996; Zehbe *et al.*, 1997). This can increase the sensitivity of conventional *in situ* hybridization by up to 1000-fold. Commercial kits for signal amplification are GenePoint from Dako Corp. (Carpinteria, CA; *http://www.dakousa.com*) and Renaissance from NEN (Life Science Products, Boston, MA; *http://www.nenlifesci.com*). This improvement has enabled *in situ* hybridization to rival IS-PCR in terms of sensitivity but only on integrated low-copy viral sequences. The technique can be applied to IS-PCR to reduce cycle number and therefore reduce diffusion. The relative merits of amplifying the reporter versus the target have been the subject of a recent review (Komminoth and Werner, 1997).

One of the most exciting technological developments of IS-PCR has been the modification of the 5′ nuclease or TaqMan assay to generate an *in situ* signal (Patterson *et al.*, 1996). It is simpler and quicker than direct IS-PCR, as specific as indirect IS-PCR, immune to mispriming or primer-independent DNA synthesis, and quantifiable. Furthermore, the possibility exists for real-time analysis and multiplexing. The only problems regard localization of the fluorescent signal after cleavage and the quenching effects of the increased number of signal and quencher molecules within the confined space of a nucleus.

Applications

Many researchers have had considerable success in studying the latent or occult disease states of the major human viruses (Bagasra *et al.*, 1992; Long *et al.*, 1993; Bagasra and Pomerantz, 1994; Gressens and Martin, 1994; Komminoth *et al.*, 1994; Boshoff *et al.*, 1995; Levin *et al.*, 1996; Foreman *et al.*, 1997). Other microbial applications include the identification of normal or genetically altered bacteria at the single cell level (Tsongalis *et al.*, 1994; Porter *et al.*, 1995; Tolker-Nielsen *et al.*, 1997; Tani *et al.*, 1998).

Molecular biologists and geneticists are increasingly using IS-PCR to map chromosomes, identify the histological sites of integration, rearrangements, or translocations, as well as gene expression (Sukpanichnant *et al.*, 1993; Kaplitt and Pfaff, 1996; Arnold *et al.*, 1997; Larochelle *et al.*, 1997; Yin *et al.*, 1998).

Any technique that utilizes single cells is an obvious candidate for flow analysis. A number of groups have successfully combined IS-PCR with flow cytometry to analyze histocompatibility (Garcia-Morales *et al.*, 1997), HIV infection (Gibellini *et al.*, 1995), translocations (Testoni *et al.*, 1996), and CD expression (Szczepek *et al.*, 1997).

The technique is providing adjunct diagnostic and prognostic value for clinicians in the area of oncology (Pestaner *et al.*, 1994; Nuovo *et al.*, 1995; Nuovo, 1996; Szczepek *et al.*, 1997). Other applications include minimal residual disease and toxicology (Malarkey and Maronpot, 1996).

Conclusions

In situ PCR continues to generate much interest and debate. With solution PCR being rapidly incorporated into diagnostic assays there is an expectation that IS-PCR will make the transition. However, until a robust, reliable protocol is developed and the issues of diffusion and therefore quantitation are resolved, IS-PCR will be regarded as something of an art and consequently less of a science.

Appendix

The following is a detailed protocol that has been employed successfully in different laboratories throughout the world for the detection of a single-copy gene in normal mammalian tissue. All reagents are at least reagent grade or better and available from major vendors. All stages are performed at room temperature unless stated otherwise. Note that the protocol has been optimized with the GeneAmp *In Situ* PCR System 1000 using the GeneAmp *In Situ* PCR DNA Tissue Control Kit and GeneAmp reagents and consumables from Perkin–Elmer.

Preparation of Tissue Samples for IS-PCR

Trim tissue into 1 × 1 × 0.5-cm blocks, and remove blood if necessary by washing with ice-cold phosphate-buffered saline (PBS). Fix in freshly prepared 10% formalin, pH 7.0–7.4, for 18–24 hr and infiltrated into paraffin wax using normal embedding protocols. Cut 5-μm sections onto silanized microscope slides and air-dry. To reduce background and increase attachment of the sections, bake the slides on a horizontal hot plate at 60°C for 36 hr.

Pretreatment of Tissue Sections

All the reagents should be prepared fresh and used within 24 hr. The proteinase K and acetic anhydride solutions must be prepared immediately before use.

1. Dewax slides through at least two changes of xylene, and rehydrate through a graded series of ethanols with a final wash in deionized water.
2. Incubate the slides in 0.02 *M* HCl for 10 min. This helps to strip away histones.
3. Wash the slides in 1X PBS for 2 min.
4. Wash the slides in 0.01% Triton X-100 in 1X PBS for 3 min.
5. Wash the slides in 1X PBS for 2 min.
6. Prewarm slides in the proteinase K buffer (0.1 *M* Tris–HCl, pH 7.5, 5 m*M* EDTA) preheated to 37°C for 5 min. This helps to minimize slight variations in proteolytic digestion.
7. Transfer the slides to fresh proteinase K buffer containing 4 μg/mL of proteinase K (Perkin–Elmer) at 37°C for 10–60 min depending on the type of tissue. This must be determined empirically.
8. Incubate slides for 5 min in 2X SSC preheated to exactly

80°C; there is no need to maintain the temperature. Not only does this destroy the proteinase K activity but also it is a form of antigen retrieval.

9. Incubate the slides for 10 min in freshly prepared 0.1 *M* triethanolamine and 0.3% acetic anhydride. This acetylates the tissue, abolishing any electrostatic interactions with the labeled primers.
10. Wash the slides for 15 sec in 20% acetic acid on ice. This step effectively fixes the section and will destroy any endogenous alkaline phosphatase activity.
11. Wash the slides in deionized water and dehydrate through a graded series of ethanols to 100% ethanol. Store in 100% ethanol until ready for PCR. The slides can be kept in 100% ethanol overnight.

PCR

Optimum thermal and chemical parameters for the PCR need to be determined empirically. The most important variables are the concentrations of Mg^{2+} (1.5–3.5 m*M*), *Taq*, IS (Perkin–Elmer), and primers (0.2–1 μM), the number of cycles (25–35), and annealing temperature. To optimize, perform the reaction on blank pretreated slides using a purified template. Each primer is labeled at the 5′ end with three dinitrophenyl moieties. The final concentration of the DNP-labeled primers in the PCR is 1 μM. See appropriate section for list of controls when using labeled primers.

Signal Detection

1. Remove AmpliCover discs/clips assembly (or dissolve off the coverslips) and wash slides in 2X SSC for 5 min. Do not let the sections become dry.
2. Transfer to fresh 2X SSC for 5 min.
3. Drain and blot dry the edges of the slide, outline sections with PAP or hydrophobic pen, and transfer to 1X Blocking Solution for 30 min. The detection reagents (1X Blocking Solution, Anti-DNP antibody, and NBT/BCIP substrates) are available as a detection kit from Perkin–Elmer.
4. Drain slides and apply prediluted (~0.5 μg/ml anti-DNP anti-

body conjugated to alkaline phosphatase) to each tissue section. Incubate in a humidified chamber for 30 min.
5. Wash slides three times in 1X PBS for 5 min. each.
6. Drain and apply freshly prepared alkaline phosphatase substrate. Incubate for an initial 10 min in a humidified chamber in the dark before visually inspecting the slides using a microscope at 100–400× magnification. The nuclei should be purple/black with no staining of the cytoplasm. Check every 10 min until a positive signal is observed in the cell nuclei.
7. Inhibit further colorimetric development by thoroughly washing in 10 m*M* Tris–HCl and 1 m*M* EDTA, pH8.0, for 10 min.
8. Apply aqueous mounting medium and coverslip.

References

Adams, J. C. (1992). Biotin amplification of biotin and horseradish peroxidase signals in histochemical stains. *J. Histochem. Cytochem.* **40,** 1457–1463.

An, S. F., and Fleming, K. A. (1991). Removal of inhibitor(s) of the polymerase chain reaction from formalin fixed, paraffin wax embedded tissues. *J. Clin. Pathol.* **44,** 924–927.

Arnold, T. E., *et al.* (1997). In vivo gene transfer into rat arterial walls with novel adeno-associated virus vectors. *J. Vasc. Surg.* **25,** 347–355.

Bagasra, O., and Hansen, J. (1997). *In Situ PCR Techniques.* Wiley-Liss, New York.

Bagasra, O., and Pomerantz, R. J. (1994). In situ polymerase chain reaction and HIV-1. *Clin. Lab. Med.* **14,** 351–365.

Bagasra, O., *et al.* (1992). Detection of human immunodeficiency virus type 1 provirus in mononuclear cells by in situ polymerase chain reaction. *N. Engl. J. Med.* **326,** 1385–1391.

Bagasra, O., *et al.* (1994). Application of in situ PCR methods in molecular biology: I. Details of methodology for general use. *Cell Vision* **1,** 324–335.

Ben-Ezra, J., *et al.* (1991). Effect of fixation on the amplification of nucleic acids from paraffin-embedded material by the polymerase chain reaction. *J. Histochem. Cytochem.* **39,** 351–354.

Boshoff, C., *et al.* (1995). Kaposi's sarcoma-associated herpesvirus infects endothelial and spindle cells. *Nature Med.* **1,** 1274–1278.

Chen, R. H., and Fuggle, S. V. (1993). In situ cDNA polymerase chain reaction. A novel technique for detecting mRNA expression. *Am. J. Pathol.* **143,** 1527–1534.

Chiu, K.-P., *et al.* (1992). Intracellular amplification of proviral DNA in tissue sections using the polymerase chain reaction. *J. Histochem. Cytochem.* **40,** 333–341.

Choo, K. H. A. (Ed.) (1994). *In Situ Hybridization Protocols. Methods in Molecular Biology.* Humana Press, Totowa, NJ.

Dyanov, H. M., and Dzitoeva, S. G. (1995). Method for attachment of microscopic preparations on glass for *in situ* hybridization, PRINS and *in situ* PCR studies. *BioTechniques* **18,** 822–826.

Embretson, J., *et al.* (1993). Analysis of human immunodeficiency virus-infected tissues by amplification and in situ hybridization reveals latent and permissive infections at single-cell resolution. *Proc. Natl. Acad. Sci. U.S.A.* **90,** 357–361.

Ertsey, R., and Scavo, L. M. (1998). Coverslip mounted-immersion cycled in situ RT-PCR for the localization of mRNA in tissue sections. *BioTechniques* **24,** 93–100.

Finke, J., *et al.* (1993). An improved strategy and a useful housekeeping gene for RNA analysis from formalin-fixed, paraffin-embedded tissue by PCR. *BioTechniques* **14,** 448–453.

Foreman, K. E., *et al.* (1997). In situ polymerase chain reaction-based localization studies support role of human herpesvirus-8 as the cause of two AIDS-related neoplasms: Kaposi's sarcoma and body cavity lymphoma. *J. Clin. Invest.* **99,** 2971–2978.

Fox, C. H., *et al.* (1985). Formaldehyde fixation. *J. Histochem. Cytochem.* **33,** 845–853.

Garcia-Morales, R., *et al.* (1997). The effects of chimeric cells following donor bone marrow infusions as detected by PCR-flow assays in kidney transplant recipients. *J. Clin. Invest.* **99,** 1118–1129.

Gibellini, D., *et al.* (1995). *In situ* polymerase chain reaction technique revealed by flow cytometry as a tool for gene detection. *Anal. Biochem.* **228,** 252–258.

Goad, D. W., *et al.* (1991). Cytogenetic mapping of alfalfa. Proceedings of the 22nd Central Alfalfa Improvement Conference, Iowa State University, Ames.

Going, J. J., and Lamb, R. F. (1996). Practical histological microdissection for PCR analysis. *J. Pathol.* **179,** 121–124.

Gosden, J. R. (Ed.) (1997). *PRINS and in Situ PCR Protocols. Methods in Molecular Biology.* Humana Press, Totowa, NJ.

Gosden, J., and Hanratty, D. (1993). PCR in situ: A rapid alternative to in situ hybridization for mapping short, low copy number sequences without isotopes. *BioTechniques* **15,** 78–80.

Greer, C. E., *et al.* (1991a). PCR amplification from paraffin-embedded tissues: Recommendations on fixatives for long-term storage and prospective studies. *PCR Methods Appl.* **1,** 46–50.

Greer, C. E., *et al.* (1991b). PCR amplification from paraffin-embedded tissues. *Am. J. Clin. Pathol.* **95,** 117–124.

Gressens, P., and Martin, J. R. (1994). In situ polymerase chain reaction: Localization of HSV-2 DNA sequences in infections of the nervous system. *J. Virol. Methods* **46,** 61–83.

Gu, J. (1994). Principles and applications of *in situ* PCR. *Cell Vision* **1,** 8–19.

Guatelli, J. C., *et al.* (1990). Isothermal, *in vitro* amplification of nucleic acids by a multienzyme reaction modeled after retroviral replication. *Proc. Natl. Acad. Sci. U.S.A.* **87,** 1874–1878.

Haase, A. T., *et al.* (1990). Amplification and detection of lentiviral DNA inside cells. *Proc. Natl. Acad. Sci. U.S.A.* **87,** 4971–4975.

Heller, M. J., *et al.* (1992). DNA extraction by sonication: A comparison of fresh, frozen, and paraffin-embedded tissues extracted for use in polymerase chain reaction assays. *Mod. Pathol.* **5,** 203–206.

Heniford, B. W., *et al.* (1993). Variation in cellular EGF receptor mRNA expression demonstrated by in situ reverse transcriptase polymerase chain reaction. *Nucleic Acid Res.* **21,** 3159–3166.

Herrington, C. S., and O'Leary, J. J. (Eds.) (1998). *PCR 3: PCR in Situ Hybridization: A Practical Approach,* the Practical Approach Series. Oxford Univ. Press, Oxford.

Hodson, R. E., *et al.* (1995). In situ PCR for visualization of microscale distribution of specific genes and gene products in prokaryotic communities. *Appl. Environ. Microbiol.* **61,** 4074–4082.

Hybaid, Ltd. (1994). *A Guide to in Situ.* Hybaid, Ltd., Teddington, UK.

Jackson, D. P., *et al.* (1990). Tissue extraction of DNA and RNA and analysis by the polymerase chain reaction. *J. Clin. Pathol.* **43,** 499–503.

Jin, L., *et al.* (1995). Quantitative comparison of messenger RNA expression detected by in situ polymerase chain reaction and other methods. *Lab. Invest.* **72,** PA165.

Johansen, B. (1997). In situ PCR on plant material with sub-cellular resolution. *Ann. Bot.* **80,** 697–700.

Kaplitt, M. G., and Pfaff, D. W. (1996). Viral vectors for gene delivery and expression in the CNS. *Methods* **10,** 343–350.

Kareem, B. N., *et al.* (1997). A novel grid polymerase chain reaction (G-PCR) approach at ultrastructural level to detect target DNA in cell cultures and tissues. *J. Pathol.* **183,** 486–493.

Koch, J. E., *et al.* (1989). Oligonucleotide-priming methods for the chromosome-specific labelling of alpha satellite DNA in situ. *Chromosoma* **98,** 259–265.

Komminoth, P., and Long, A. A. (1993). In-situ polymerase chain reaction. An overview of methods, applications and limitations of a new molecular technique. *Virchows Arch. B* **64,** 67–73.

Komminoth, P., and Werner, M. (1997). Target and signal amplification: Approaches to increase the sensitivity of in situ hybridization. *Histochem. Cell Biol.* **108,** 325–333.

Komminoth, P., *et al.* (1992). In situ polymerase chain reaction detection of viral DNA, single-copy genes, and gene rearrangements in cell suspensions and cytospins. *Diagn. Mol. Pathol.* **1,** 85–97.

Komminoth, P., *et al.* (1994). Evaluation of methods for hepatitis C virus detection in archival liver biopsies. *Pathol. Res. Pract.* **190,** 1017–1025.

Koopmans, M., *et al.* (1993). Optimization of extraction and PCR amplification of RNA extracts from paraffin-embedded tissue in different fixatives. *J. Virol. Methods* **43,** 189–204.

Kovalenko, S. A., *et al.* (1997). Methods for in situ investigation of mitochondrial DNA deletions. *Hum. Mutat.* **10,** 489–495.

Larochelle, N., *et al.* (1997). Efficient muscle-specific transgene expression after adenovirus-mediated gene transfer in mice using a 1.35 kb muscle creatine kinase promoter/enhancer. *Gene Ther.* **4,** 465–472.

Levin, M. C., *et al.* (1996). PCR-in situ hybridization detection of human T-cell lymphotropic virus type 1 (HTLV-1) *tax* proviral DNA in peripheral blood lymphocytes of patients with HTLV-1-associated neurologic disease. *J. Virol.* **70,** 924–933.

Lewis, F. (1996a). *An Approach to in Situ PCR.* Perkin–Elmer, Foster City, CA.

Lewis, F. (1996b). A standard method for *in situ* hybridization on formalin fixed paraffin embedded tissue using the Perkin–Elmer GeneAmp *In situ* PCR System 1000. P. E. Research News No. 791703-001.

Lewis, F. A. (1996c). In situ PCR—Myth or magic. *J. Cell. Pathol.* **1,** 13–23.

Long, A. A., *et al.* (1993). Comparison of indirect and direct in-situ polymerase chain reaction in cell preparations and tissue sections. *Histochemistry* **99,** 151–162.

Long, A. A., and Komminoth, P. (1997). *In situ* PCR: An overview. *In PRINS and in Situ PCR Protocols* (J. R. Gosden, Ed.), Vol. 71, pp. 141–161. Humana Press, Totowa, NJ.

Luehrsen, K. R., *et al.* (1995). The GeneAmp *in situ* PCR system 1000: A new tool to amplify nucleic acid targets in cells and tissues. *Cell Vision* **2,** 348–350.

Malarkey, D. E., and Maronpot, R. R. (1996). Polymerase chain reaction and in situ hybridization: Applications in toxicological pathology. *Toxicol. Pathol.* **24,** 13–23.

Man, Y.-G., *et al.* (1996). Detailed RT-ISPCR protocol for preserving morphology and confining PCR products in routinely processed paraffin sections. *Cell Vision* **3,** 389–396.

Mancuso, T., *et al.* (1997). A fast and simple *in situ*-PCR method to detect human papilloma virus infection in archival paraffin-embedded tissues. *Int. J. Oncol.* **11,** 527–532.

Martínez, A., *et al.* (1995). Non-radioactive localization of nucleic acids by direct in situ PCR and in situ RT-PCR in paraffin-embedded sections. *J. Histochem. Cytochem.* **43,** 739–747.

Mee, A. P., *et al.* (1997). Quantification of vitamin D receptor mRNA in tissue sections demonstrates the relative limitations of *in situ*-reverse transcriptase-polymerase chain reaction. *J. Pathol.* **182,** 22–28.

Moran, R., *et al.* (1985). Detection of 5-bromodeoxyuridine (BrdUrd) incorporation by monoclonal antibodies: Role of the DNA denaturation step. *J. Histochem. Cytochem.* **33,** 821–827.

Mueller, J. D., *et al.* (1997). Self-sustained sequence replication (3SR): An alternative to PCR. *Histochem. Cell Biol.* **108,** 431–437.

Muro-Cacho, C. A. (1997). In situ PCR. Overview of procedures and applications. *Frontiers Biosci.* **2,** C15–C29.

Nuovo, G. J. (1994). In situ detection of PCR-amplified DNA and cDNA: A review. *J. Histotechnol.* **17,** 235–246.

Nuovo, G. J. (1996). Detection of viral infections by in situ PCR: Theoretical considerations and possible value in diagnostic pathology. *J. Clin. Lab. Anal.* **10,** 335–349.

Nuovo, G. J., and Forde, A. (1995). An improved system for reverse transcriptase in situ PCR. *J. Histotechnol.* **18,** 295–299.

Nuovo, G. J., *et al.* (1993). Importance of different variables for enhancing in situ detection of PCR-amplified DNA. *PCR Methods Appl.* **2,** 305–312.

Nuovo, G. J., *et al.* (1995). Correlation of the in situ detection of polymerase chain reaction-amplified metalloproteinase complementary DNAs and their inhibitors with prognosis in cervical carcinoma. *Cancer Res.* **55,** 267–275.

Oates, J. L., *et al.* (1997). Labelled primer in situ polymerase chain reaction (LPISPCR): A direct method for demonstration of t(14;18) in cell lines. *J. Pathol.* **182,** 20A.

O'Leary, J. (1998a). In situ amplification. In *Clinical Applications of PCR* (Y. M. D. Lo, Ed.), p. 16. Humana Press, Totowa, NJ.

O'Leary, J. J. (1998b). PCR in situ hybridization (PCR ISH). In *PCR 3: PCR in Situ Hybridization: A Practical Approach* (C. S. Herrington and J. J. O'Leary, Eds.), pp. 53–86. Oxford Univ. Press, Oxford.

O'Leary, J. J., *et al.* (1994a). PCR in situ hybridisation detection of HPV 16 in fixed CaSki and fixed SiHa cell lines. *J. Clin. Pathol.* **47,** 933–938.

O'Leary, J. J., *et al.* (1994b). The importance of fixation procedures on DNA template and its suitability for solution-phase polymerase chain reaction and PCR *in situ* hybridization. *Histochem. J.* **26,** 337–346.

O'Leary, J. J., *et al.* (1996). *In situ* PCR: Pathologist's dream or nightmare. *J. Pathol.* **178,** 11–20.

Patterson, B. K., *et al.* (1993). Detection of HIV-1 DNA and messenger RNA in individual cells by PCR-driven in situ hybridization and flow cytometry. *Science* **260,** 976–979.

Patterson, B. K., *et al.* (1996). Detection of HIV-1 DNA in cells and tissue by fluorescent *in situ* 5′-nuclease assay (FISNA). *Nucleic Acid Res.* **24,** 3656–3658.

Pestaner, J. P., *et al.* (1994). Potential of the in situ polymerase chain reaction in diagnostic cytology. *Acta Cytol.* **38,** 676–680.

Peters, J., *et al.* (1997). Detection of rare RNA sequences by single-enzyme in situ reverse transcription-polymerase chain reaction. *Am. J. Pathol.* **150,** 469–476.

Picton, S., and Howells, D. (1997). The GeneAmp In Situ PCR System 1000. In *PCR 3: PCR in Situ Hybridization: A Practical Approach* (C. S. Herrington and J. J. O'Leary, Eds.), pp. 164–174. Oxford Univ. Press, Oxford.

Porter, J., *et al.* (1995). Flow cytometric detection of specific genes in genetically modified bacteria using in situ polymerase chain reaction. *FEMS Microbiol. Lett.* **134,** 51–56.

Sällström, J., *et al.* (1993). Pitfalls of in situ polymerase chain reation (PCR) using direct incorporation of labelled nucleotides. *Anticancer Res.* **13,** 1153–1154.

Sällström, J. E., *et al.* (1994). Nonspecific amplification in *in situ* PCR by direct incorporation of reporter molecules. *Cell Vision* **1,** 243–251.

Spann, W., *et al.* (1991). In situ amplification of single copy gene segments in individual cells by polymerase chain reaction. *Infection* **19,** 242–244.

Starling, J. A. (1997). Automation and the Omnislide System. In *PCR 3: PCR in Situ Hybridization: A Practical Approach* (C. S. Herrington and J. J. O'Leary, Eds.), pp. 174–194. Oxford Univ. Press, Oxford.

Staskus, K. A., *et al.* (1991). In situ amplification of visna virus DNA in tissue sections reveals a reservoir of latently infected cells. *Microbiol. Pathogen.* **11,** 67–76.

Sukpanichnant, S., *et al.* (1993). Detection of clonal immunoglobulin heavy chain gene rearrangements by polymerase chain reaction in scrapings from archival hematoxylin and eosin-stained histologic sections: Implications for molecular genetic studies of focal pathologic lesions. *Diagn. Mol. Pathol.* **2,** 168–176.

Sullivan, D. E., *et al.* (1997). Self-Seal reagent: Evaporation control for molecular histology procedures without chambers, clips or fingernail polish. *BioTechniques* **23,** 320–325.

Szczepek, A. J., *et al.* (1997). CD34+ cells in the blood of patients with multiple myeloma express CD19 and IgH mRNA and have patient-specific IgH VDJ gene rearrangements. *Blood* **89,** 1824–1833.

Tani, K., *et al.* (1998). Development of a direct in situ PCR method for detection of specific bacteria in natural environments. *Appl. Environ. Microbiol.* **64,** 1536–1540.

Teo, C. G., and Griffin, B. E. (1990). Visualization of single copies of the Epstein–Barr virus genome by *in situ* hybridization. *Anal. Biochem.* **186,** 78–85.

Teo, I. A., and Shaunak, S. (1995a). PCR *in situ*: Aspects which reduce amplification and generate false-positive results. *Histochem. J.* **27,** 660–669.

Teo, I. A., and Shaunak, S. (1995b). Polymerase chain reaction *in situ*: An appraisal of an emerging technique. *Histochem. J.* **27,** 647–659.

Terkelsen, C., *et al.* (1993). Repeated primed in situ labeling: Formation and labeling of specific DNA sequences in chromosomes and nucleoli. *Cytogenet. Cell. Genet.* **63,** 235–237.

Testoni, N., *et al.* (1996). A new method of in-cell reverse transcriptase-polymerase chain reaction for the detection of BCR/ABL transcript in chronic myeloid leukemia patients. *Blood* **87,** 3822–3827.

Tokuda, Y., *et al.* (1990). Fundamental study on the mechanism of DNA degradation in tissue fixed in formaldehyde. *J. Clin. Pathol.* **43,** 748–751.

Tokusashi, Y., *et al.* (1995). Differentiation of the normal and mutant rat albumin genes on hepatic tissue sections by *in situ* PCR. *Nucleic Acid Res.* **23,** 3790–3791.
Tolker-Nielsen, T., *et al.* (1997). Visualization of specific gene expression in individual *Salmonella typhimurium* cells by in situ PCR. *Appl. Environ. Microbiol.* **63,** 4196–4203.
Troyer, D. L., *et al.* (1995). Use of DISC-PCR to map a porcine microsatellite. *Anim. Biotechnol.* **6,** 51–58.
Tsongalis, G. J., *et al.* (1994). Localized in situ amplificatin (LISA): A novel approach to in situ PCR. *Clin. Chem.* **40,** 381–384.
Tyagi, S., *et al.* (1996). Extremely sensitive, background-free gene detection using binary probes and Qb replicase. *Proc. Natl. Acad. Sci. U.S.A.* **93,** 5395–5400.
Uhlmann, V., *et al.* (1998). A novel, rapid in cell RNA amplification technique for the detection of low copy mRNA transcripts. *J. Clin. Pathol. Mol. Pathol.* accepted for publication.
van Gijlwijk, R. P. M., *et al.* (1996). Improved localization of fluorescent tyramides for fluorescence in situ hybridization using dextran sulfate and polyvinyl alcohol. *J. Histochem. Cytochem.* **44,** 389–392.
Yap, E. P. H., and McGee, J. O. D. (1991). Slide PCR: DNA amplification from cell samples on microscope glass slides. *Nucleic Acid Res.* **19,** 4294.
Yin, J., *et al.* (1998). In situ PCR for in vivo detection of foreign genes transferred into rat brain. *Brain Res.* **783,** 347–354.
Zehbe, I., *et al.* (1997). Sensitive in situ hybridization with catalyzed reported deposition, streptavidin-nanogold, and silver acetate autometallography. *Am. J. Pathol.* **150,** 1553–1561.
Zhang, H., and Wadler, S. (1997). Micropreparation of cultured cells for in situ reverse transcription PCR. *BioTechniques* **22,** 618–624.

Part Two

QUANTITATIVE PCR

13

STANDARDS FOR PCR ASSAYS

Dwight B. DuBois, Cindy R. WalkerPeach, Matthew M. Winkler, and Brittan L. Pasloske

The wide versatility of PCR has created numerous niches for this technology in research and clinical diagnostic laboratories. In both settings, well-defined DNA and RNA standards, compatible with the exquisite sensitivity and specificity of PCR, are needed. Because of their relatively advanced state of development, standards for infectious agents will be emphasized in this chapter.

PCR-based assays have been formatted as qualitative and quantitative tools. As the premiere example, the utility of quantitative reverse transcriptase (RT)-PCR for plasma HIV-1 RNA has been well established in clinical practice and in antiretroviral drug development (Mulder *et al.,* 1994). Standards developed for quantitative assays present a special challenge since they must be precisely calibrated. Additionally, independent verification of the amount of very low numbers of nucleic acid templates is restricted by the limited number of corresponding technologies with sensitivities comparable to PCR.

Special Problems with Standardizing PCR Assays

Multiple Steps and Labile Templates

In general, PCR assays involve the following steps: (i) template preparation (generally including initial specimen lysis with a chaotropic agent in the case of RNA); (ii) reverse transcription (in the case of RNA); (iii) enzymatic amplification via thermocycling; and (iv) detection of PCR product. Additionally, quantitative concentration of virus by centrifugation is the initial step in the ultrasensitive RT-PCR (Sun *et al.*, 1998) and bDNA assays for HIV-1. Ideally, quantitative standards should be prepared, amplified, and detected in a form and manner exactly like that of the test specimen. DNA standards are generally stable and can be directly added to biological specimens. In contrast, naked RNA standards are notoriously susceptible to ribonucleases present in biological specimens and are prone to inadvertent degradation in handling. To circumvent the difficulty of using RNA standards, workers in the field have used DNA controls for quantitative assays for RNA (Gilliland *et al.*, 1990) or have added naked RNA to chaotropic lysis solutions independently of sample extraction (Mulder *et al.*, 1994).

The Exponential Nature of PCR

Exponential production of amplicons during the amplification step creates inherent difficulties in using PCR as a precise and accurate quantitative tool. The following equation shows the quantity of PCR product as a function of cycling conditions:

$$N = N_0(1 + \text{eff})^n$$

where N is the final amount of amplified PCR products, N_0 is the initial amount of target template, eff is the amplification efficiency, and n the number of amplification cycles.

In theory, the efficiency of PCR ranges from a maximum of 1.0 (two amplicons made from each template in every cycle) to 0 (no amplification occurring). In practice, the average efficiency of PCR during the exponential phase of amplification is about 0.85 (Nicoletti and Sassy-Prigent, 1996). Figure 1 illustrates the effect of small differences in amplification efficiency on the final yield of PCR product (a 1% difference in efficiency results in about a 15% change in total quantity of amplicon production in a 25-cycle PCR) and stresses the necessity of internal controls to correct for tube-to-tube variations.

Types of Standards

The Ideal Standard

Table 1 lists characteristics of an ideal standard for amplification-based assays. While current standards have many of the listed qualities, none fulfills all the desired criteria. As more assay formats are developed, it will be increasingly important that standards closely mimic the biological entity of interest. For example, recently developed ultrasensitive RT-PCR and bDNA assays for HIV rely on con-

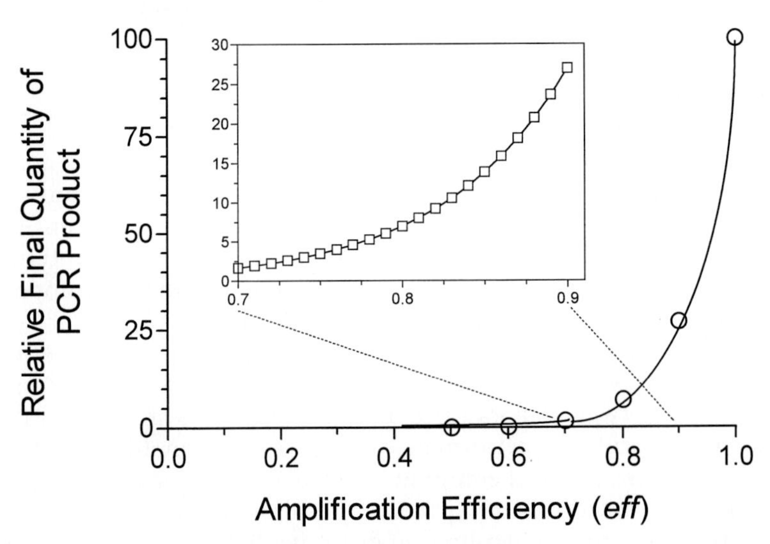

Figure 1 Plot of relative yield of DNA product as a function of amplification efficiency in a 25-cycle PCR. Relative final quantity at each amplification efficiency is a percentage of the final amount of PCR product amplified at a maximal efficiency (1.0). Inset shows expansion of the 0.7–0.9 range.

Table 1

Characteristics of An Ideal Standard for Amplification Based Assays

Characteristics
Chemical structure the same as the template of interest
Predefined sequence
Precisely quantified
Stable in clinical specimens (i.e., plasma, whole blood, and CSF).
Concentrating and partitioning properties (centrifugation and precipitation) which accurately mimic the native entity of interest
Controls for all steps in the assay
Easily reformatted to incorporate new mutations and quasispecies
Noninfectious and can be made inexpensively

centration of the virus by centrifugation prior to RNA isolation. However, without an adequate control for the centrifugation step, these assays are susceptible to inadvertent partial or complete aspiration of a nearly invisible pellet containing concentrated virus. Wide variability in the estimation of plasma viral RNA levels can result from subtle technical differences in handling pellets. Quantified standards directly spiked into clinical specimens with centrifugation properties similar to HIV would help alleviate this variation.

Internal and External Standards

Internal standards have two major purposes. First, the standard should act as a positive control for the nucleic acid extraction and amplification steps in qualitative assays to reduce the occurrence of sporadic false negative results (Cone *et al.,* 1992). Second, internal standards provide a known copy number of template in both competitive and noncompetitive quantitative PCR formats.

External standards are useful for monitoring the run-to-run performance of assays and for detecting systematic errors in PCR tests within and between laboratories. External standards are also invaluable in evaluating the performance of assays based on different technologies. For example, a common set of HIV-1 RNA standards, consisting of cultured virus spiked into seronegative plasma, was recently used to standardize quantitative RT-PCR, bDNA, and NASBA-based assays. In this study, use of common standards eliminated differences in the estimated absolute HIV-1 copy numbers (Brambilla *et al.,* 1998) between the three technologies.

Naked Nucleic Acid, Biological, and Engineered Standards

Naked RNA and DNA standards are widely used (Wang *et al.*, 1989; Ballagi-Pordany *et al.*, 1991; Pham *et al.*, 1998; Rosenstraus *et al.*, 1998). The major advantage of this approach is the relative ease of construction of defined sequences using molecular biology techniques, such as gene construction, cloning, site-directed mutagenesis, and *in vitro* transcription. These techniques also allow precision in the size and quantity of the final product. Figure 2 illustrates a variety of control template designs, which share primer binding sites

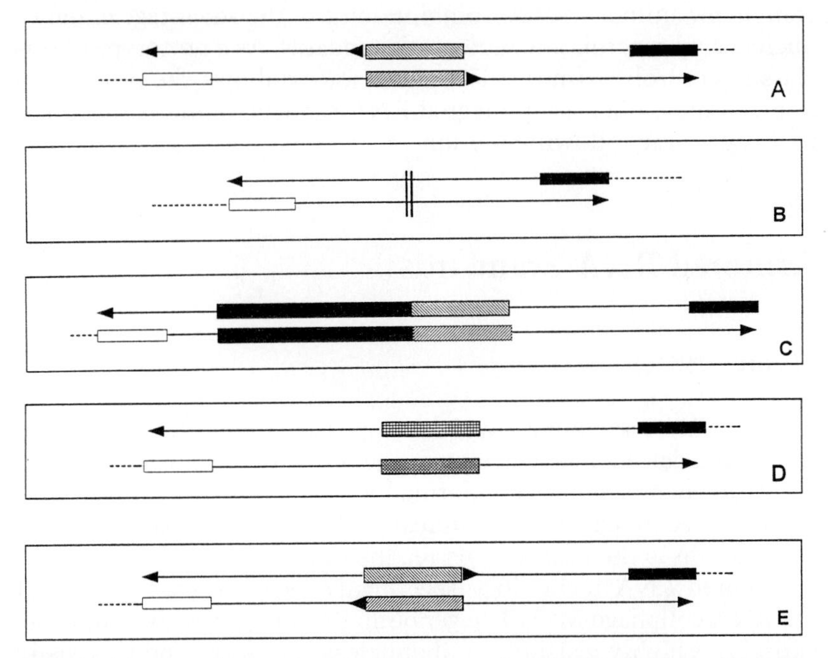

Figure 2 Diagram of internal control sequences for PCR-based assays. (A) Native template showing a internal probe binding site in the normal orientation. (B and C) Control sequences identical to sequence A but with either deletion of an internal sequence (B) or insertion of exogenous DNA (C). Size differences in the resulting PCR products allow differentiation from native template in gel-based assays. (D) Control sequence, the same size as native sequence, containing a substituted sequence. (E) Control sequence with an internal probe-binding sequence in the reversed orientation. Differential hybridization of capture or detection probes to the substituted or reversed sequences in these controls allows for independent detection in automated assay formats.

but vary either in size or in hybridization characteristics. Naked nucleic acid standards have the disadvantage of increased susceptibility to nucleases (particularly RNA standards) and the inability to control for the crucial steps of nucleic acid extraction.

Biological standards such as cultured virus, DNA bacteriophage, and viremic plasma have been used in both qualitative and quantitative assays. The main advantage of biological standards is control of specimen processing prior to and including template isolation. Significant drawbacks of this approach include the difficulty in supplying uniform batches of controls, the inability to culture some important pathogens (notably hepatitis C virus), and the genetic diversity created by serially passing virus strains.

Engineered standards such as recombinant bacteriophage and recombinant animal viruses, combine some of the advantage of naked nucleic acid controls and biological standards. As a prototype of this class of controls, Armored RNA technology allows the production of predefined, nuclease-protected RNA sequences and is discussed in detail in the following section.

Armored RNA Standards

Theory

Armored RNA technology was developed to protect RNA standards from degradation by inadvertant contamination by ribonucleases during assay procedures and during long-term storage. *In vivo* packaging of predefined RNA of interest within a shell of bacteriophage coat protein is the basis of this technology.

Armored RNA technology specifically uses the coat protein of the RNA coliphage MS2. The genomic RNA packaged within these particles is highly resistant to ribonuclease digestion and it is easily extracted from the bacteriophage coat protein by conventional methods of RNA isolation (Argetsinger and Gussin, 1966). We reasoned that a recombinant RNA (reRNA) could be packaged as pseudoviral particle, thereby conferring protection to the reRNA against ribonucleases.

MS2 has an icosahedral structure of 26 nm in diameter. It is composed of three different structural molecules: coat protein, maturase, and a 3.6-kb single-stranded RNA. There are 180 molecules of coat protein per phage particle. There is a single copy of coding strand RNA per phage particle encoding four different genes: maturase, coat protein, lysis protein, and replicase. Within the genomic RNA, there is a 19-nucleotide sequence having a stem-loop structure located at the start of the replicase gene called the operator sequence (Witherell, 1991). The operator binds with high affinity to coat protein, the K_D being 3×10^{-9} M (Lim, 1994). When a critical concentration of coat protein has been translated, it binds to the operator and initiates the formation of the bacteriophage particle.

Construction

Armored RNA technology is a plasmid-driven packaging system using the high-fidelity *Escherichia coli* RNA polymerase to transcribe the reRNA (DuBois *et al.*, 1997; Pasloske *et al.*, 1998). DNA encoding the MS2 coat protein, the RNA standard sequence, and the MS2 operator sequence are located downstream of an inducible promoter (Fig. 3). The recombinant packaging vector is transformed into *E. coli.* Addition of IPTG induces the transcription of the reRNA. The translation of the coat protein from the reRNA is tightly coupled to transcription. As the concentration of the coat protein reaches a critical concentration, it binds the operator sequence at the 3′ end of the reRNA, initiating the encapsidation of the reRNA (Fig. 3). Unlike MS2, in which the phage particles are released into the spent medium by lysing *E. coli,* the Armored RNA particles localize in the cytoplasmic fraction of *E. coli.* The particles are subjected to a stringent purification procedure involving several conventional protein purification steps. The final product is free of any detectable unpackaged nucleic acid and of any detectable nuclease contamination.

A wild-type HIV-1 Armored RNA (AR-HIV-1) standard was constructed which is compatible with the HIV Monitor assay (Kwok and Sninsky, 1993; Mulder *et al.*, 1994) and a wild-type HCV Armored RNA (AR-HCV-2b) standard (DuBois *et al.*, 1997) was constructed which is compatible with both the HCV Monitor assay and the Chiron Quantiplex HCV assay (Collins *et al.*, 1997). The AR-HIV-1 and AR-HCV-2b standards contain consensus RNA sequences generated by alignment of multiple sequences from clinical specimens. The AR-HIV-1 standard contains 172 nucleotides of HIV-1 sequence from the *gag* region, whereas AR-HCV-2b has 415 nucleotides representing the entire 5′ nontranslated region and a portion

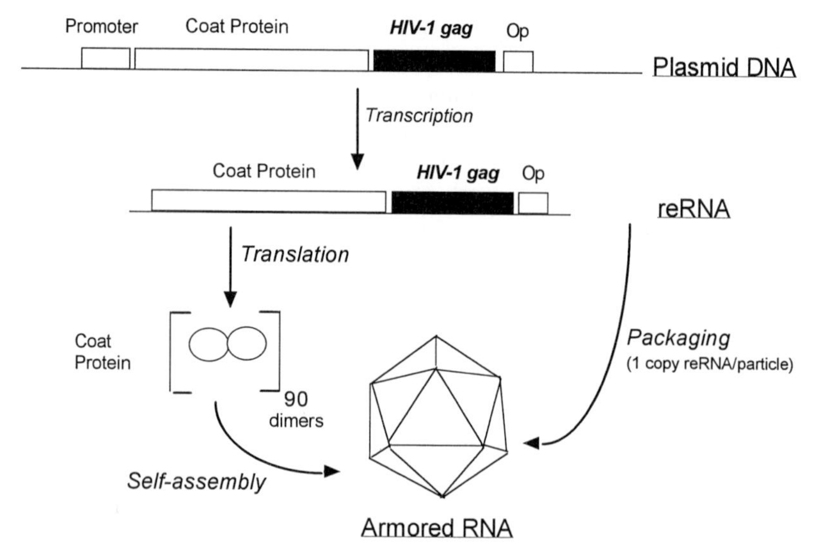

Figure 3 Armored RNA plasmid-driven packaging system. An expression/reRNA packaging plasmid was constructed with a *trc* promoter upstream from the MS2 coat protein gene and exogenous gene sequence. (HIV-1 *gag* is shown in this example.) Armored RNA is produced in the following sequence: (i) Following induction with IPTG, (ii) coat protein and *gag* genes are transcribed into RNA, and (iii) coat protein is translated; and (iv) in a precisely concerted process, encapsidation of the reRNA transcript, containing the HIV-1 *gag* RNA, occurs when coat protein reaches a critical concentration in the *E. coli* cytoplasm. Resulting HIV-1 Armored RNA particles are noninfectious particles containing exogenous RNA in place of the native MS2 replicase gene.

of the core region. The advantage of these Armored RNA controls is that it is possible to package any RNA sequence and, therefore, a panel of genotype-specific standards can be produced. With current technology, it is possible to package up to 1.5 kb of an exogenous RNA sequence within an Armored RNA particle.

Armored RNA standards are resistant to environmental nucleases which would destroy naked RNA rapidly. For example, Armored RNA is stable in human EDTA–plasma at 4°C for at least 11 months and in ACD–plasma at 37°C for at least 45 days. Armored RNA can also survive 45°C for at least 3 days in a stabilizing solution.

Utility as Internal, External, and Primary Standards

In addition to being used as an internal and external standard in accepted assay formats, Armored RNA can be used as a primary standard. In principle, *in vitro* transcripts function well as primary standards; however, the major shortcoming of naked RNA is the potential of contaminations, either during manufacture or at some later stage, with traces of ribonuclease, which can result in quantitative changes in the standard. A second problem is that *in vitro* transcripts often have a certain amount of length variability due to both premature termination during transcription and aberrant initiation of transcription. This heterogeneity mandates some kind of size selection, such as gel purification, which introduces another step in which RNase contamination can occur.

A large fraction of the RNA packaged within the coat protein is homogenous, full-length RNA. CsCl density gradient centrifugation further refines this quality by banding Armored RNA particles containing a specific ratio of RNA to protein. Calibration of the RNA copy number in a given Armored RNA preparation is accomplished by determining the absorbance at 260 nm and using an extinction coefficient derived from MS2 phage. This method yields RNA concentrations similar to those obtained using by limiting-dilution Poisson distribution approaches (D. DuBois *et al.*, unpublished data).

Future Needs for Standards

Standards for Technology Comparisons

In addition to PCR, there are a number of other ultrasensitive nucleic acid detection techniques, such as NASBA-TMA, bDNA, ligase chain reaction, and the cycling probe reaction. Common standards which will work with some or all of these techniques would be desirable. Due to the wide use of PCR, it can be assumed that this technology is the standard against which these other techniques will be compared. With the exception of bDNA, these other techniques detect a relatively short region of nucleic acid. Thus, a standard which will work well with PCR should work well with those other techniques, with the possible exception of bDNA. For example, the Quantiplex HIV assay (Chiron Corp., bDNA) targets a RNA sequence of approximately 3 kb in length.

Standards for Mutational Analysis

In addition to detection, PCR is frequently used as part of mutation detection assays for rare nucleic acid sequences. When PCR is used to detect mutations in genomic DNA, cloned DNA standards are adequate. PCR amplified standards may create a problem for some mutation detection procedures since the relatively high error rates of thermostable polymerases create a background level of random mutations. PCR is also frequently used to detect mutations in RNA viruses such as HIV and HCV. In this case, the standard ideally will be RNA to allow all steps in mutation detection assays to be controlled. Predefined sequences, such as *in vitro* transcripts or Armored RNA, would be well suited for this application. Such standards could also be mixed in predetermined quantities to objectively assess the ability of these mutation detection assays to detect minority populations of mutant sequences within heterogenous mixtures.

Blood Product Screening

In the near future, PCR and other nucleic acid amplification technologies will be used to screen individual whole blood donors for infectious agents, such as HIV1/2, HCV, and human T cell lymphotrophic viruses I and II. Screening pools of plasma donors with PCR has already become a standard practice in some European countries. Using a sensitive assay system such as PCR, prone to sporadic inhibition, to screen low-incidence populations of donors can be problematic. The vast majority of test results generated from screening asymptomatic donors will be negative. Thus, without the proper use of internal controls, the possibility of random PCR failures could confound any screening strategy. Addition of precisely calibrated controls to individual specimens before nucleic acid extraction would ensure the reliability of the testing system above a defined virus copy threshold.

Standards for Multiplexed Assays

Multiplexed PCR formats are assays in which a number of templates, each with unique primer binding regions, are coamplified. These reactions are prone to an overall decrease in sensitivity. In such assay formats, incorporation of internal controls for each template

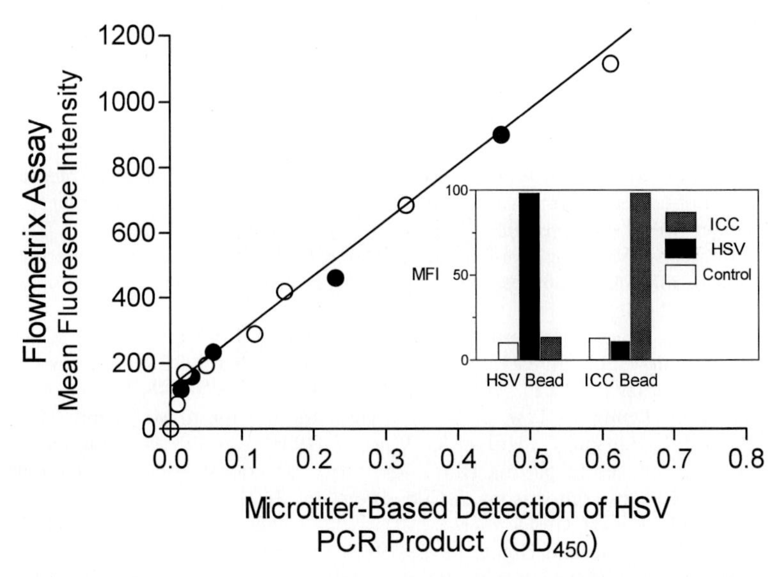

Figure 4 Multiplex detection of a 148-bp PCR product from the gB region of herpes simplex virus (HSV-1) (○) and an internal control sequence (ICC) that contains a reversed internal probe binding site (●). PCR products were detected using two sets of capture microspheres labeled with orange and red fluorescence dyes in uniquely identifiable ratios. Each dyed microsphere set is covalently coupled to oligonucleotide capture sequences specific for either HSV or ICC. The limits of detection of the FlowMetrix detection system and the microtiter-based system are approximately 10^7 and 5×10^7 amplicons/tube, respectively. Inset shows the specificity of binding of HSV and ICC to their respective beads.

will be important to avoid false negative results. The development of multiplexed assay formats has been hampered by the lack of corresponding multiplexed detection systems. We have recently developed a PCR-based assay for herpes simplex virus DNA (Walker-Peach *et al.*, 1997) using the FlowMetrix cytometric microsphere system (Luminex Corp.), which has true multiplexing capabilities and allows the simultaneous detection and quantitation of dozens of analytes (Fulton *et al.*, 1997). Figure 4 shows the tritration of HSV amplicon and an internal control (ICC) DNA using the cytometric microsphere system and a microtiter-based detection system. The ICC is the same size as the HSV amplicon and contains the same primer binding sites. Both formats produce linear signals and have comparable sensitivities. The FlowMetrix system also allows very specific detection of the wild-type HSV sequences and the ICC in the same hybridization and detection step.

References

Argetsinger, J. E., and Gussin, G. (1966). Intact ribonucleic acid from defective particles of bacteriophage R17. *J. Mol. Biol.* **21,** 421–434.

Bai, X., Hosler, G., Rogers, B. B., Dawson, D. B., and Scheuermann, R. H. (1997). Quantitative polymerase chain reaction for human herpes diagnosis and measurement of Epstein–Barr virus in posttransplant lymphoproliferative disorder. *Clin. Chem.* **43**(10), 1843–1849.

Ballagi-Pordany, A., Ballagi-Pordany, A., and Funa, K. (1991). Quantitative determination of mRNA phenotypes by the polymerase chain reaction. *Anal. Biochem.* **196**(1), 89–94.

Becker-Andre, M., and Hahlbrock, K. (1989). Absolute mRNA quantification using the polymerase chain reaction (PCR). A novel approach by a PCR aided transcript titration assay (PATTY). *Nucleic Acids Res.* **17**(22), 9437–9446.

Brambilla, D., Leung, S., Lew, J., Todd, J., Herman, S., Cronin, M., Shapiro, D. E., Bremer, J., Hanson, C., Hillyer, G. V., *et al.* (1998). Absolute copy number and relative changes in plasma HIV-1 RNA determinations: Effect of an internal standard on kit comparisons. *J. Clin. Microbiol.* **36**(1), 311–316.

Collins, M. L., Zayati, C., Detmer, J. J., Daly, B., Kolberg, J. A., Cha, T. A., Irvine, B. D., Tucker, J., and Urdea, M. S. (1995). Preparation and characterization of RNA standards for use in quantitative branched DNA hybridization assays. *Anal. Biochem.* **226,** 120–129.

Collins, M. L., Irvine, B., Tyner, D., Fine, F., Zayati, C., Chang, C., Horn, T., Ahle, D., Detmer, J., Shen, L. P., *et al.* (1997). A branched DNA signal amplification assay for quantification of nucleic acid targets below 100 molecules/ml. *Nucleic Acids Res.* **25**(15), 2979–2984.

Cone, R. W., Hobson, A. C., and Huang, M.-L. W. (1992). Coamplified positive control detects inhibition of polymerase chain reactions. *J. Clin. Microbiol.* **30**(12), 3185–3189.

DuBois, D. B., WalkerPeach, C. R., Pasloske, B. L., and Winkler, M. (1997). Universal ribonuclease resistant RNA standards (Armored RNA) for RT-PCR and bDNA-based hepatitis C virus RNA assys. *Clin. Chem.* **43**(11), 2218. [Abstract]

DuBois, D. B., Winkler, M. M., and Pasloske, B. L. (1997). Ribonuclease resistant viral RNA standards. USA Patent No. 5, 677, 124.

Forster, E. (1993). An improved general method to generate internal standards for competitive PCR. *BioTechniques* **16,** 18–20.

Fulton, J. R., McDade, R. L., Smith, P. L., Kienker, L. J., and Kettman, J. R. (1997). Advanced multiplexed analysis with the FlowMetrix system. *Clin. Chem.* **43**(9), 1749–1756.

Gilliland, G., Perrin, S., Blanchard, K., and Bunn, H. F. (1990). Analysis of cytokine mRNA and DNA: Detection and quantitation by competitive polymerase chain reaction. *Proc. Natl. Acad. Sci. U.S.A.* **87**(7), 2725–2729.

Kohler, T., Rost, A.-K., and Remke, H. (1997). Calibration and storage of DNA competitors use for contamination-protected competitive PCR. *BioTechniques* **23**(4), 722–726.

Kwok, S., and Sninsky, J. J. (1993). PCR detection of human immunodeficiency virus type 1 proviral sequence. In *Diagnostic Molecular Biology: Principles and Applications* (D. H. Persing, T. F. Smith, T. J. Tenover, and T. J. White, Eds. ASM Press, Washington, DC.

Lim, F., and Peabody, D. S. (1994). Mutations that increase the affinity of a translational repressor for RNA. *Nucleic Acids Res.* **22,** 3747–3752.

Mulder, J., McKinney, N., Christopherson, C., Sninsky, J., Greenfield, L., and Kwok, S. (1994). Rapid and simple PCR assay for quantitation of human immunodeficiency virus type 1 RNA in plasma: Application to acute retroviral infection. *J. Clin. Microbiol.* **32**(2), 292–300.

Nicoletti, A., and Sassy-Prigent, C. (1996). An alternative quantitative polymerase chain reaction method. *Anal. Biochem.* **236**(2), 229–241.

Pasloske, B. L., WalkerPeach, C. R., Obermoeller, R. D., Winkler, M. W., and DuBois, D. B. (198). Armored RNA technology for the production of ribonuclease resistant viral RNA controls and standards. *J. Clin. Micro.* **36**(12), 3590–3594.

Pham, D. G., Madico, G. E., Quinn, T. C., Enzler, M. J., Smith, T. F., and Gaydos, C. A. (1998). Use of lambda phage DNA as a hybrid internal control in a PCR-enzyme immunoassay to detect *Chlamydia pneumoniae. J. Clin. Microbiol.* **36**(7), 1919–1922.

Rosenstraus, M., Wang, Z., Chang, S. Y., DeBonville, D., and Spadoro, J. P. (1998). An internal control for routine diagnostic PCR: design, properties, and effect on clinical performance. *J. Clin. Microbiol.* **36**(1), 191–197.

Sun, R., Ku, J., Jayakar, H., Kuo, J-O., Brambilla, D., Herman, S., Rosenstraus, M., and Spadoro, J. (1998). Ultrasensitive reverse transcription-PCR assay for quantitation of human immunodeficiency virus type 1 RNA in plasma. *J. Clin. Micro.* **30**(10), 2964–2929.

Tang, Y.-W., Procop, G. W., and Persing, D. H. (1997). Molecular diagnostics of infectious diseases. *Clin. Chem.* **43**(11), 2021–2038.

WalkerPeach, C. R., Smith, P. L., DuBois, D. B., and Fulton, R. J. (1997). A novel, rapid multiplexed assay for herpes simplex virus DNA using the FlowMetrix cytometric microsphere technology. *Clin. Chem.* **43**(11), 2216. [Abstract 21].

Wang, A. M., Doyle, M. V., and Mark, D. F. (1989). Quantitation of mRNA by the polymerase chain reaction. *Proc. Natl. Acad. Sci. USA* **86**(24), 9717–9721.

Witherell, G. W., Gott, J. M., and Uhlenbeck, O. C. (1991). Specific interaction between RNA phage coat proteins and RNA. *Proc. Nuc. Acid Res. Molec. Biol.* **40,** 185–220.

14

RAPID THERMAL CYCLING AND PCR KINETICS

Carl T. Wittwer and Mark G. Herrmann

Rapid Thermal Cycling for PCR

DNA amplification by PCR can be performed rapidly. Rapid temperature cycling for PCR, last reviewed in 1994, is defined as completing 30 cycles of amplification in <30 min (Wittwer *et al.,* 1994). By using high surface area to volume sample containers and circulating air for heating and cooling (Wittwer *et al.,* 1989), amplification in 15 min or less is readily achieved (Wittwer *et al.,* 1990).

PCR Paradigms

Usually, denaturation, annealing, and extension are considered three reactions that occur separately at three temperatures for three time periods. This is a sequential, equilibrium paradigm of PCR (Fig. 1, left). This paradigm ignores temperature transitions and excludes the possibility of simultaneous reactions, e.g., annealing and extension occurring concurrently. With the equilibrium paradigm, instruments

PCR Applications

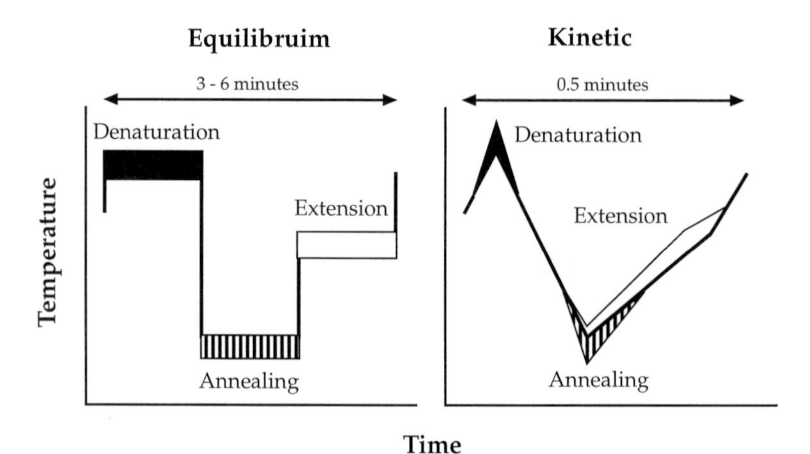

Figure 1 Equilibrium and kinetic paradigms of PCR.

are programmed using temperature and time steps, e.g., 94°C for 1 min, 55°C for 1 min, and 72°C for 2 min. Although easy to conceive, the equilibrium paradigm does not accurately reflect the temperature/time course of the sample. Sample temperatures do not change instantaneously. Indeed, during PCR, for most of the time the sample is usually in temperature transitions (Wittwer and Garling, 1991; Wittwer *et al.*, 1993). This is true with conventional instrumentation, and even more true during rapid cycling. Rapid cycling takes advantage of **momentary** denaturation and annealing. With precise temperature control, amplification yield and product specificity are optimal when denaturation and annealing times are <1 sec (Wittwer and Garling, 1991; Wittwer *et al.*, 1993). The denaturation and annealing temperatures need to be reached, but they **do not need to be held** (Wittwer *et al.*, 1994). Furthermore, for short products, extension is often completed during the transition to the "extension" temperature, and no extension hold is needed (Wittwer *et al.*, 1994). Under the equilibrium paradigm, such a rapid temperature cycle might be described as 94°C for 0 sec, 55°C for 0 sec, and 72°C for 0 sec. This is odd and not very helpful. It conveys only the temperature extremes, with no information about what happens between temperatures.

An alternate paradigm for rapid cycle PCR is a kinetic paradigm (Fig. 1, right). The temperature transitions are fully described to delineate the complete temperature history of the sample. Denaturation, annealing, and extension are considered to occur over a range of temperatures and may overlap temporally. If the kinetics of denaturation, annealing, and extension are known at all temperatures, a model of PCR could be devised that would predict amplification results for any given temperature cycle. Rapid and precise control of sample temperatures with temperature homogeneity within and between samples is necessary to test such a kinetic model.

Sample Vessel Geometry

Most rapid cycle instruments use capillary tubes as sample containers and control temperature with circulating air. It is easy to transfer solutions into and out of capillary tubes and they have a high surface area to volume ratio for rapid temperature control. The sample is spread out along a single axis, with heat transfer along the other two dimensions. An alternative is to use a planar sample compartment (e.g., a microfabricated silicon, glass, or plastic chip) that spreads the sample out over two dimensions, with heat transfer along the remaining axis. Sample introduction, recovery, and homogeneous temperature control are more difficult with the 2D design. The situation is analogous to sequencing on flat gels or capillaries; heat transfer is maximal with the capillary design, allowing higher voltages to speed separations. Current chips usually have conventional cycling times (Shoffner *et al.*, 1996). One rapid cycle implementation has been reported (Wooley *et al.*, 1996), although temperature control (e.g., overshooting temperature set points) appears to be a problem. The only drawback to capillaries is sample handling, particularly when large numbers of samples are needed. A composite plastic tube combined with a capillary has recently been used in rapid cycling (Wittwer *et al.*, 1997b).

Instruments and Protocols

Standard temperature cyclers for PCR complete 30 cycles in about 2–4 hr. These instruments are very slow compared to the times required for product denaturation, primer annealing, and polymerase action. Transition times between temperatures are long because of high thermal mass and globular sample geometry. These instruments are not well matched to the physical/enzymatic requirements of PCR that occur very quickly. With the proper instrumentation, amplification times can be reduced an order of magnitude from standard protocols.

Instruments

Rapid cycle PCR uses temperature cycles of 20–60 sec, allowing 30 cycles to be completed in 10–30 min. A commercial rapid cycler based on capillary tubes and circulating air is shown in Fig. 2. Ten-microliter samples are loaded by capillary action into 0.8-mm i.d. capillaries and sealed between air bubbles with a miniature butane torch (Blazer Corp., New York). The capillaries are held at microtiter spacing in holders so that multiple tubes can be loaded and sealed at once. The capillary holders fit into the top of the temperature-

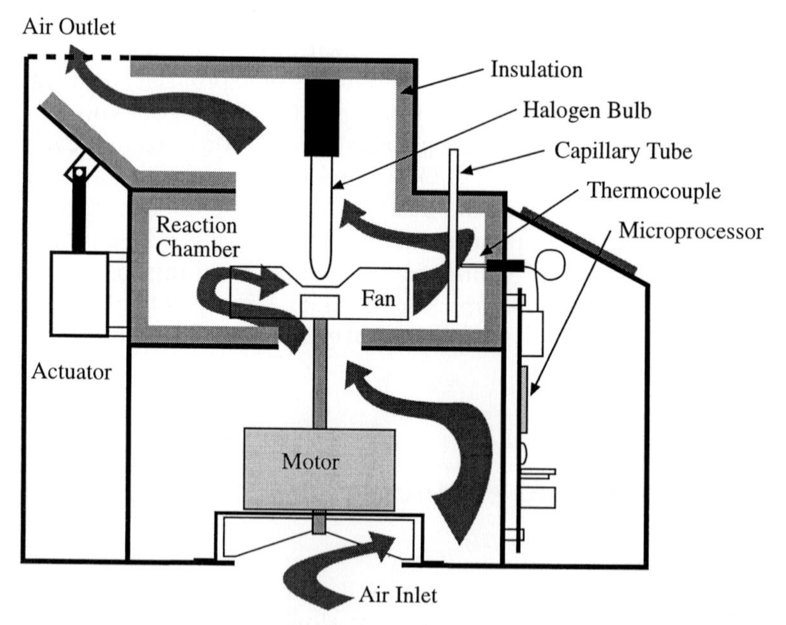

Figure 2 Commercial rapid air thermal cycler. A cross section of the RapidCycler (Idaho Technology) is shown. The air chamber is heated with a 500-W halogen bulb and cooled with ambient air. A high-speed fan ensures temperature homogeneity between the samples. See Wittwer *et al.* (1994) for further details.

controlled chamber and position the samples vertically inside the chamber. After amplification, the ends of the tubes are scored and removed, and the samples are directly loaded into gel wells with a microaspirator. Thirty-second cycles are easily obtained (Wittwer *et al.*, 1994).

We have further modified the capillary/air system to increase the heating and cooling rates and to incorporate transition rate control (Wittwer *et al.*, 1995). Temperature transition rates up to 20°C/sec have been obtained. Temperature cycling with controlled transitions of 10°C/sec is shown in Fig. 3. These 18-sec cycles can be reduced to only 9 sec if the hold times are removed, allowing 30 cycles to be completed in <5 min.

Reagents

In general, any amplification reagents can be used for rapid cycling. However, bovine serum albumin is usually included at 50 μg/ml to prevent surface inactivation of the polymerase on the glass capillary

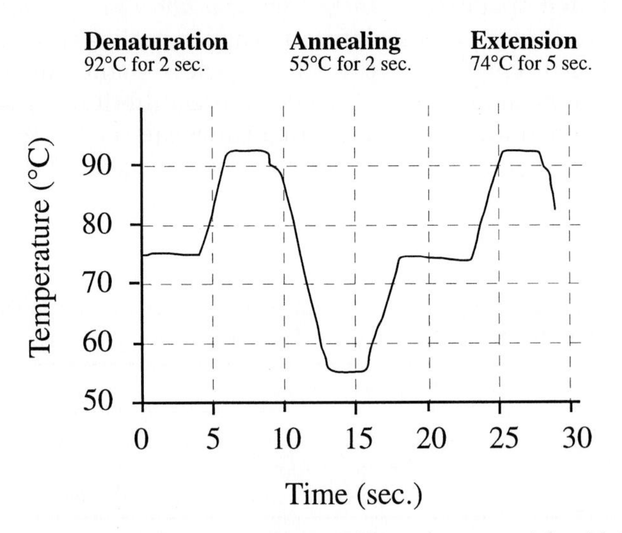

Figure 3 Controlled transition rate cycling at 10°C/sec. A commercial rapid air thermal cycler (see Fig. 2) was modified by using (i) four 500-W halogen bulbs for heating, (ii) larger air portals for more rapid cooling, and (iii) an analog slope control circuit (Wittwer *et al.*, 1995). The sample temperature within the capillaries was monitored with a microthermocouple (Wittwer *et al.*, 1994).

surface. For very rapid cycles (<25 sec), primer annealing rates can be increased by increasing the magnesium concentration and/or the primer concentration (Tan and Weis, 1992), although this is usually not necessary. The polymerase extension rate and the desired product size ultimately limit cycling speeds. Therefore, polymerases with rapid extension rates, like *Taq* polymerase, allow faster amplifications. If "hot starts" are needed, anti-*Taq* antibodies (Taq-Start, Clon-Tech, Palo Alto, CA) are preferred over heat-activated enzymes (AmpliTaq Gold, Perkin–Elmer, Foster City, CA) that usually take longer than 30 cycles of rapid cycle amplification to activate. If electrophoresis will be performed, gel loading dyes can be included in the amplification solution (Wittwer and Garling, 1991) so that the capillary, doubling as a sample container and a transfer pipette, can directly load the reaction products into gel wells.

Rapid Cycle Optimization

Optimization of temperature/time parameters for rapid-cycle amplification has been recently reviewed (Brown *et al.*, 1998). A "Rapid Cyclist User's Group" maintains a web site with helpful hints, protocols, and applications (*http://www.idahotec.com/rapideylist/Default.html*). In general, denaturation and annealing times are minimized so that temperature/time tracing show momentary denaturation and annealing "spikes." Under the equilibrium paradigm of PCR, the denaturation and annealing times are set to "0"; that is, the extreme temperatures are attained but not held. Table 1 lists

Table 1

Suggested Temperature and Time Parameters for Rapid-Cycle DNA Amplification under the Equilibrium Paradigm

	Temperature (°C)	Time (sec)
Denaturation	94[a]	0
Annealing	30 + 0.5 · (primer GC%)	0
Extension	74[b]	0.03 · (product length)[c]

Note. Reproduced with permission from Brown *et al.* (1998).

[a] For products with a high GC domain, consider adding DMSO or formamide and/or increasing the temperature.

[b] May need to be lower for products with a high AT domain.

[c] For products <100 bp use a 0-sec extension time.

suggested temperature/time parameters for rapid cycle PCR. Given the suggested parameters, a simple Mg^{2+} titration almost always results in a specific amplification product.

Instrument Variations

There are many ways to rapidly temperature cycle samples for PCR. The capillary/air design discussed previously is generally useful and commercially available. Here, we describe other specialized instruments for rapidly changing temperature and for continuous monitoring of PCR by fluorescence.

Water Bath Temperature Control

Very rapid changes to equilibrium temperatures can be made by transferring capillaries between water baths. More than 95% of any transition occurs within 1 sec, and equilibrium (±0.2°C) is achieved by 2 sec. A simple device for manually transferring a capillary between water baths is shown in Fig. 4. To change temperatures, the capillary is slid from one temperature-controlled compartment into

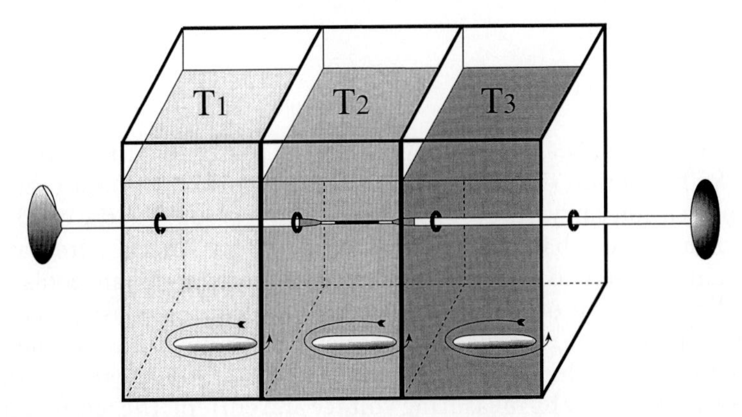

Figure 4 Water bath temperature control. A polycarbonate box, divided by insulating panels into three chambers, is filled with water. Each chamber is independently regulated at a constant temperature and stirred magnetically. The capillary sample is pulled between chambers for very rapid temperature equilibration.

another. Such a system can be used to study annealing rates and is the closest physical approximation to the equilibrium paradigm for PCR. In contrast, transition rates are poorly controlled in this system. In fact, transition rates are poorly controlled in all designs in which the sample is transferred between set temperature baths.

Adiabatic Compression

In order to control temperature transition rates, the samples are usually kept stationary and the temperature of the surrounding medium is changed continuously. With the capillary/air system, heat is provided by a light bulb (Wittwer *et al.*, 1994) or resistive heater (Wittwer and Garling, 1991; Wittwer *et al.*, 1997b) and rapidly mixed air transfers the heat to the capillaries. This process takes time and there is always a lag between the application of heat and sample temperature response.

One way to instantaneously change the temperature of air is to change its pressure by compression or expansion. Figure 5 diagrams an adiabatic temperature cycler based on air compression/expansion with a motor-driven piston. The samples are placed radially on a spinning platform to increase convective heat transfer. Compression ratios of 10 : 1 have produced sample temperature transition rates of 20°C/sec. However, transfer of heat through the capillary wall still takes time and the sample temperature response lags behind the change in air pressure.

Direct Electrolyte Heating

Direct solution heating can be obtained by passing a current through the PCR reaction (Fig. 6). The electrolytes normally present during amplification carry the current. The solution column in the capillary forms a "wire" that is evenly heated along its length. Alternating current is used to prevent bulk electrophoresis. A fan cools the capillary from the outside when necessary. Because viscosity changes with temperature, the electrical resistance through the capillary is related to the solution temperature. Since resistance can be calculated dynamically by measuring voltage and current, the temperature of the solution can be directly monitored through its electrical characteristics. Although the physical implementation shown in Fig. 6 is awkward, the ability to both control and monitor temperature

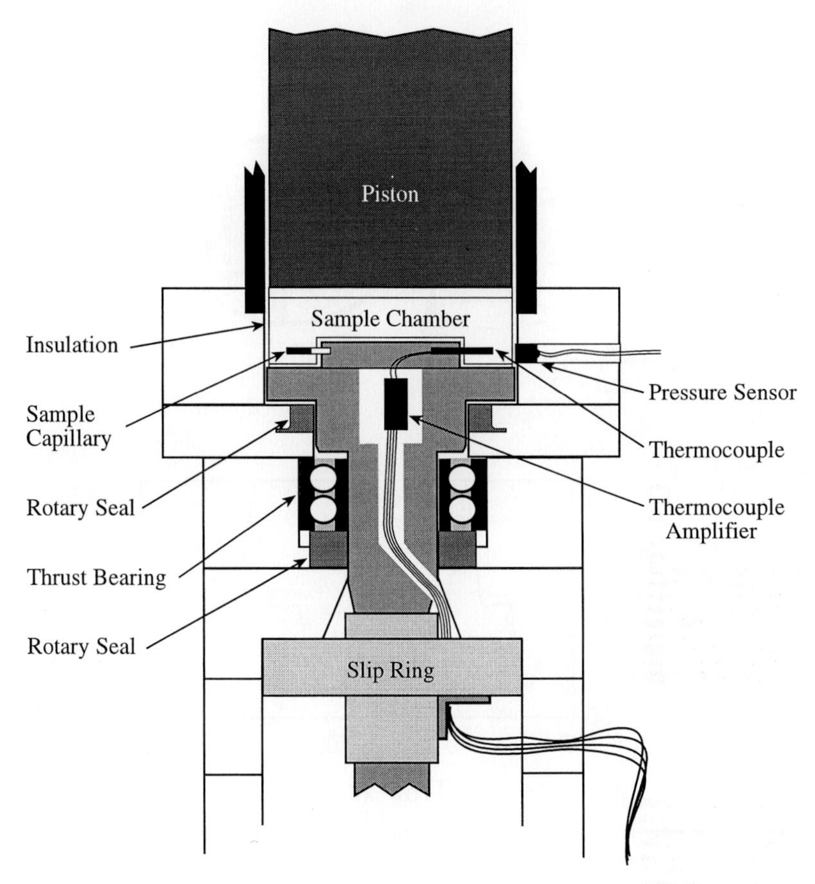

Figure 5 Temperature cycling by adiabatic compression. Heating/cooling is achieved by air compression/expansion. A piston is used to adjust the air volume within a pressurized sample chamber.

electrically has an inherent simplicity that may find eventual use in micromachined devices. Direct electrolyte heating is potentially the fastest thermal cycling method described here, although currently achieved cycle times are only modestly in the rapid cycling realm (Fig. 6).

Fluorescence Monitoring

Optically clear glass capillaries make natural cuvettes for fluorescence analysis. By integrating flow cytometry optics into a rapid

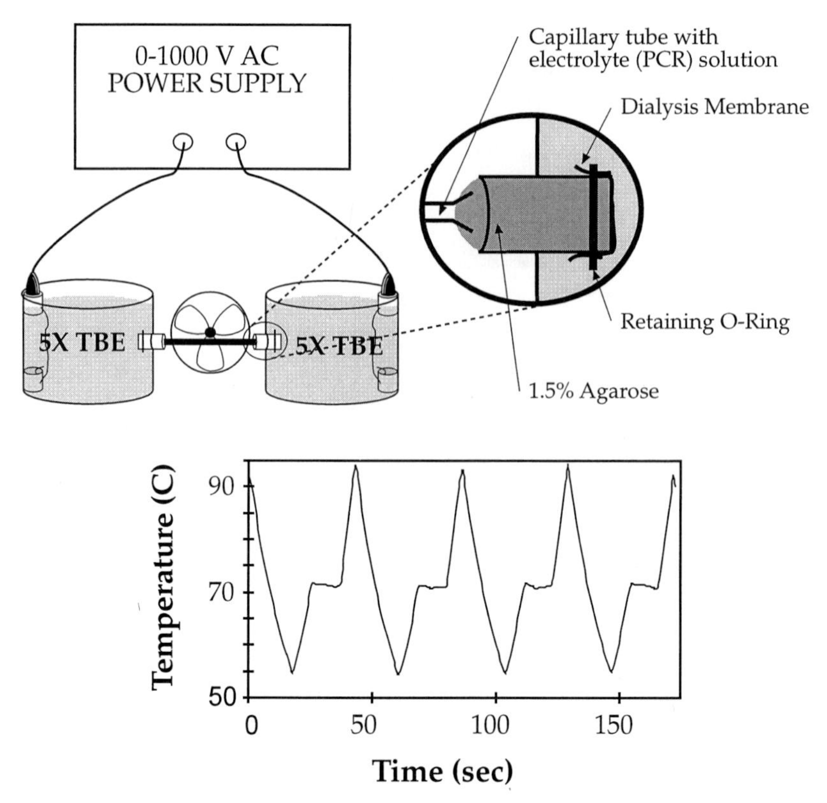

Figure 6 Heating by electrolyte conductance. A PCR sample within a capillary tube provides a conductive path between low-resistance buffer chambers. A variable AC source allows adjustable heating, and cooling is by ambient air. The sample temperature can be monitored by following the resistance of the sample.

capillary/air cycler (Wittwer *et al.*, 1997b), continuous monitoring of fluorescence during amplification is possible (Wittwer *et al.*, 1997a). The design of our prototype fluorescence thermal cycler is shown in Fig. 7. A xenon arc source is spectrally filtered and focused onto a capillary in a rapid cycling air chamber. Fluorescence is observed by photomultiplier tubes after spatial and spectral filtering. Multiple samples are monitored by rotating a sample carousel. Fluorescence, temperature, and time are acquired during temperature cycling as desired.

Commercial fluorescence temperature cyclers are available and one design is shown in Fig. 8 (Wittwer *et al.*, 1997b). The benchtop footprint is <1 ft^2. The optics are placed underneath the sample

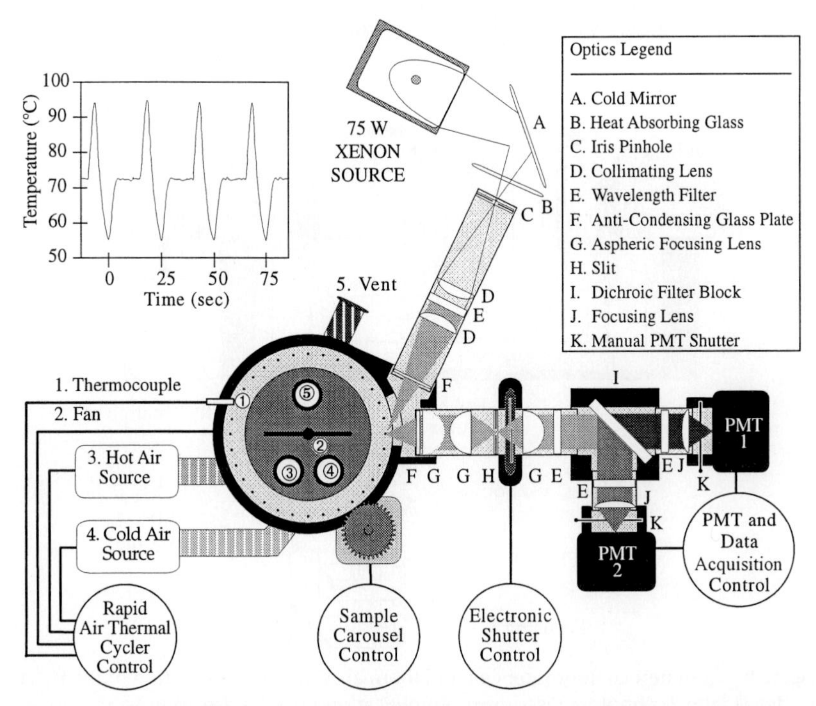

Figure 7 Prototype fluorescence rapid thermal cycler. Optical components are listed in the upper right legend and representative temperature cycles are diagrammed in the upper left graph. Samples are placed circumferentially around a cylindrical rapid air chamber. Continuous fluorescence monitoring is possible through optics adopted from flow cytometry (reprinted with permission from Wittwer *et al.,* 1997b).

chamber for epiillumination of the capillary tip. A blue light emitting diode provides excitation and the fluorescence is measured with photodiodes. The instrument is interfaced to a personal computer for acquisition of temperature, time, and three colors of fluorescence.

Applications

One advantage of rapid cycling is to minimize the amplification time, which is especially useful when multiple sequential amplifications are required (Brown *et al.,* 1998). When combined with con-

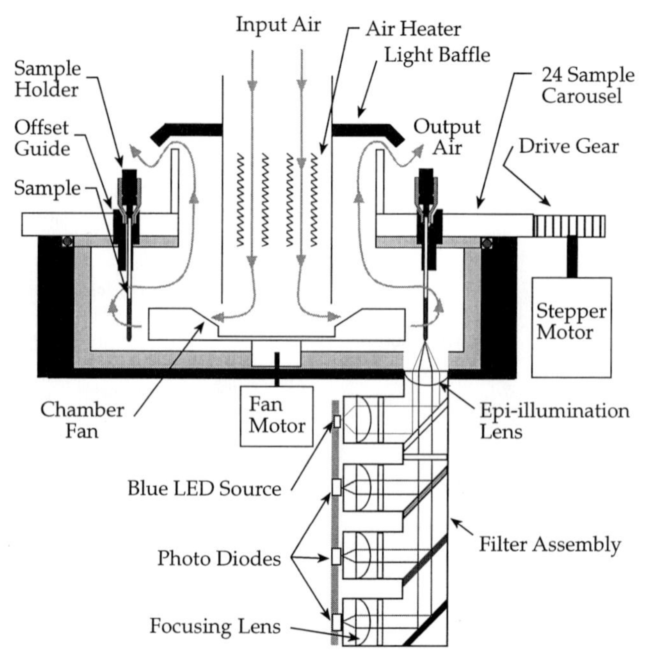

Figure 8 Commercial fluorescence rapid thermal cycler. A cross section of the LightCycler (Idaho Technology) is shown. Amplification samples are spun to the tips of composite plastic/glass sample cuvettes. Epiillumination optics focus on the capillary tip, increasing sensitivity by light piping at the glass/air surface (reprinted with permission from Wittwer *et al.*, 1997b).

tinuous fluorescence monitoring, simultaneous **amplification** and **analysis** for quantification (Wittwer *et al.*, 1997a, 1998) or mutation detection (Lay and Wittwer, 1997; Bernard *et al.*, 1998) are particularly attractive for rapid diagnosis. If fine product sizing is required, rapid cycling can be directly coupled to capillary electrophoresis (Swerdlow *et al.*, 1997; Woolley *et al.*, 1996).

Rapid and precise temperature control also improves the quality of amplification. Greater specificity is achieved by limiting the time for primer annealing (Wittwer and Garling, 1991). Products longer than the desired amplicon can be selected against by restricting the extension time. Degradation of target DNA is minimized by using short denaturation times (Gustafson *et al.*, 1993). Optimization is simplified with rapid cycle PCR because a greater range of annealing temperatures can be used, for example, in allele-specific amplification (Wittwer *et al.,* 1993).

PCR Kinetics

The rates of the reactions that occur during PCR are dependent on temperature and can be individually measured. These include both productive reactions (denaturation, primer anealing, and extension) and nonproductive reactions (e.g., reassociation of single-stranded product to double-stranded product).

For example, double-stranded PCR product can be denatured by boiling and then transferred to a 75°C water bath. The single strands will reassociate to double strands at a certain rate. At any time, the strand status can be "frozen" by rapid cooling in ice water. The device described under Water Bath Temperature Control (Fig. 4) is used with T_1 regulated by boiling, T_2 kept at 75°C with a circulator, and T_3 held at 0°C with crushed ice. After ice water quenching, gel electrophoresis can be used to separate reassociated double strands from the original single strands (Fig. 9). The product strand status

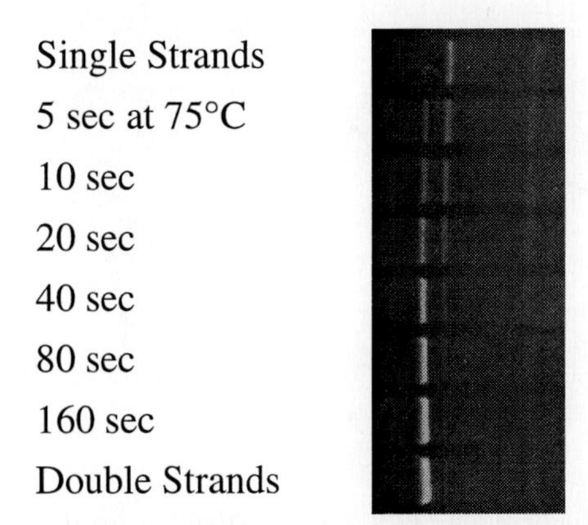

Figure 9 Product reassociation kinetics at 75°C. Fifty-nanogram aliquots of a 536-bp amplicon (Wittwer and Garling, 1991) were placed into glass capillaries and attached to the pull rods shown in Fig. 4). Only single-stranded product is observed (top lane) when the capillary is pulled from a boiling water bath directly to an ice water bath. If the capillary is transferred from boiling to a 75°C bath and finally to an ice water bath, more double-stranded product forms as the time at 75°C increases. Separation on a 1.5% agarose gel in 0.5 × TBE effectively separates single-stranded and double-stranded products.

during PCR can be followed by pulling tubes from a rapid air cycler at various parts of the cycle (Fig. 10). Similarly, primer annealing rates can be measured (Fig. 11).

Continuous monitoring of reaction kinetics is also possible with fluorescence. For example, product reassociation after denaturation can be followed with the double-strand-specific DNA dye, SYBR Green I (Wittwer *et al.*, 1998). Continuous monitoring greatly simplifies rate constant estimation. Another highly useful application of SYBR Green I, when present during PCR, is to identify products by their T_m (Ririe *et al.*, 1997). When fluorescence is continuously monitored during PCR and plotted against temperature, product melting is observed as a rapid drop in fluorescence as the product denatures (Fig. 12). During the transition from denaturation to anealing temperatures, product reassociation is very rapid, suggesting

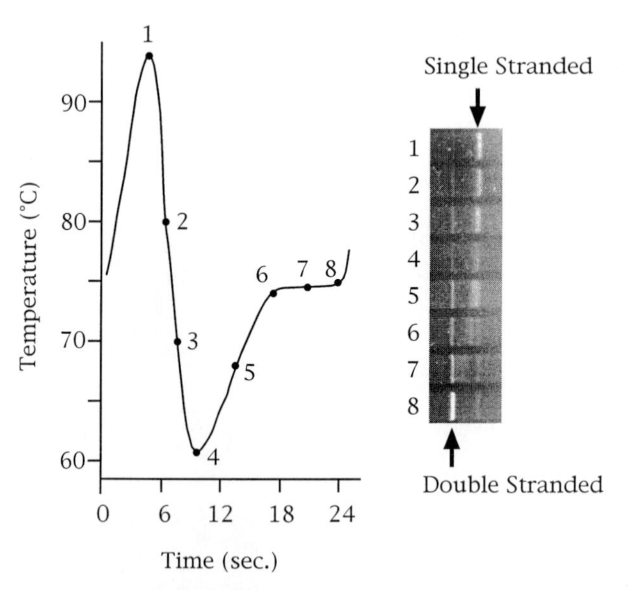

Figure 10 Strand status during PCR. Product strand status during cycle 30 of PCR was visualized by quenching multiple samples at different time points. The capillaries were manually pulled from a rapid air cycler (Fig. 2) and immediately immersed in ice water. At the denaturation temperature peak, all of the product is single stranded. As the temperature falls to primer annealing temperatures, double-stranded product reassociates from single-stranded product. At extension temperatures, a portion of the single-stranded band appears to migrate toward the double-stranded band, presumably reflecting intermediate DNA products during polymerase extension. The amplicon and electrophoresis conditions of Fig. 9 were used.

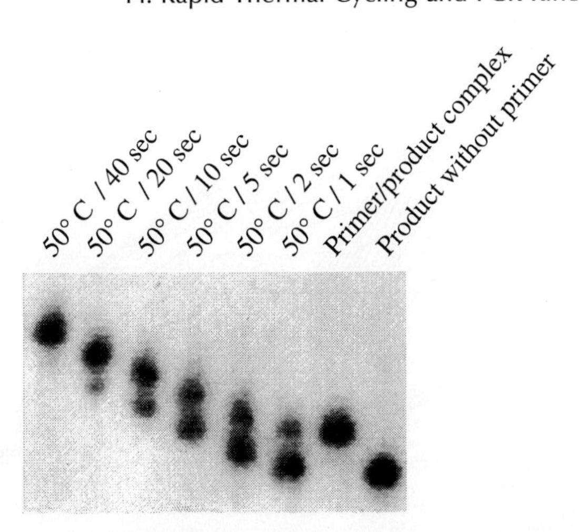

Figure 11 Primer annealing kinetics at 50°C. A 110-bp product (Wittwer *et al.,* 1989) was amplified with a 5′ biotin-labeled primer. Single-stranded product was isolated by streptavidin-coated magnetic beads and the single strands were kinased with ^{32}P. ^{32}P-labeled single strand (5 n*M*) and complementary primer (0.5 μM) were mixed in PCR buffer (Wittwer and Garling, 1991). Aliquots were added to capillary tubes and the apparatus shown in Fig. 4 was used to study primer annealing kinetics. The samples were passed through boiling water, a 50°C bath for a variable period of time, and finally quenched in ice water. Electrophoresis was performed immediately at 0–4°C on an 8% polyacrylamide gel in 1X TBE, transferred to filter paper, and imaged on a phosphorimager. To maintain conformational quenching of the single strands and prevent further primer annealing, the solution was kept cold, applied to the gel, and electrophoresed as rapidly as possible, i.e., immediately after quenching. Therefore, samples were applied to the different lanes at different times, resulting in apparent retarded migration at longer annealing times.

that it is a major cause of the well-known "plateau effect" (Wittwer *et al.,* 1997a).

Initial Template Quantification

Fluorescence monitoring once each cycle is very useful for quantitative PCR, i.e., estimating initial template copy number. The fluorescence of double-strand-specific DNA dyes (Higuchi *et al.,* 1993; Wittwer *et al.,* 1997a) or resonance energy-transfer hairpin primers (Nazarenko *et al.,* 1997) can be used to directly monitor product formation. However, signal specificity is entirely dependent on the PCR primers, and undesired amplification products (primer oligo-

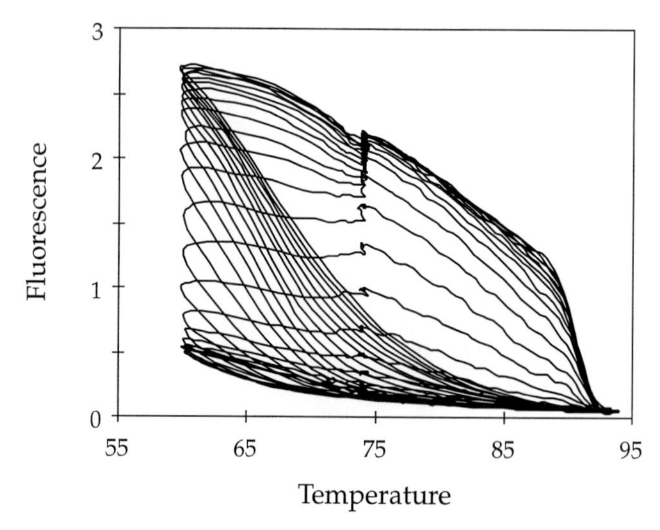

Figure 12 Continuous monitoring of PCR with SYBR Green I. A 536-bp product (Wittwer *et al.*, 1997a) was amplified from human genomic DNA in the presence of a 1:20,000 dilution of SYBR Green I (Molecular Probes). Temperature cycling conditions were 94°C maximum, 60°C minimum, and 74°C for 15 sec in a fluorescence-capable rapid temperature cycler (LightCycler, Idaho Technology). The fluorescence vs temperature plot eliminates time as a variable and displays the strand status over temperature. Both the high-temperature melting and low-temperature reassociation of product are observed for each cycle.

mers, etc.) may be a problem. Another alternative is to use fluorescence resonance energy-transfer probes dependent on hybridization to the PCR product. These include hairpin probes (Tyagi and Kramer, 1996), probes that are hydrolyzed by polymerase 5′-exonuclease activity (Kalinina *et al.*, 1997), and singly labeled adjacent hybridization probes (Wittwer *et al.*, 1997a, 1998).

Mutation Detection

Fluorescence monitoring of probe hybridization and its dependence on temperature is an elegant method for mutation detection (Lay and Wittwer, 1997; Bernard *et al.*, 1998). Genotyping by resonance energy transfer can produce a "dynamic dot blot" in <30 min after

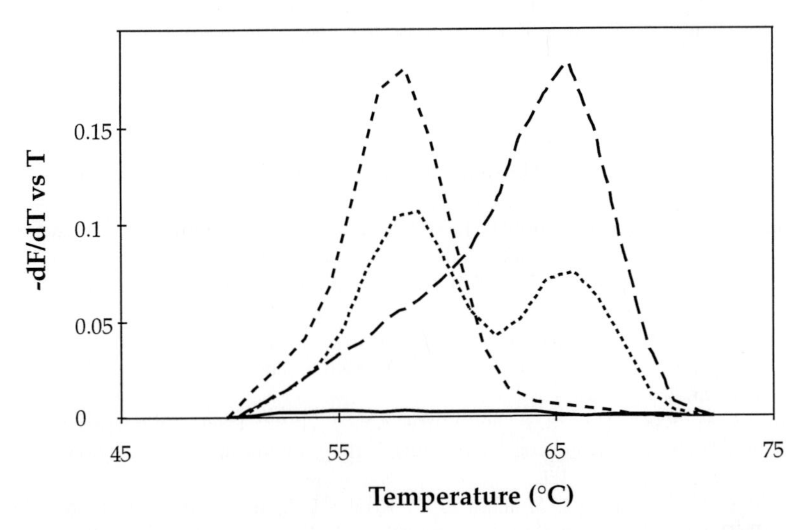

Figure 13 Detection of Factor V Leiden by probe melting curves. Derivative melting curve plots of the three possible genotypes at the factor V Leiden mutation locus are shown. Curves for homozygous wild type (long dashes), homozygous mutant (midlength dashes), heterozygous (short dashes), and a no template control (solid line) are shown. Amplification and analysis were completed in <30 min after the start of amplification without user intervention (reprinted with permission from Lay and Wittwer, 1997).

the start of amplification (Fig. 13). Fluorescein is used as the resonance energy-transfer donor, and Cy5 or another long-wavelength dye is used as the acceptor. One fluorophore is attached to the probe that crosses the mutation site, and the other is attached to either one primer or another probe. As the product is amplified, resonance energy transfer increases. A melting curve of the complex allows genotyping during amplification (Lay and Wittwer, 1997). It appears likely that all single base changes can be detected (Bernard *et al.*, 1998).

References

Bernard, P. S., Lay, M. J., and Wittwer, C. T. (1998). Integrated amplification and detection of the C677T point mutation in the methylenetetrahydrofolate reductase gene by fluorescence resonance energy transfer and probe melting curves. *Anal. Biochem.* **255,** 101–107.

Brown, R. A., Lay, M. J., and Wittwer, C. T. (1998). Rapid cycle amplification for construction of competitive templates. *In Genetic Engineering with PCR* (R. M. Horton and R. C. Tait, Eds.), pp. 57–70. Horizon Scientific, Norfolk, UK.

Gustafson, C. E., Alm, R. A., and Trust, T. J. (1993). Effect of heat denaturation of target DNA on the PCR amplification. *Gene* **123,** 241–244.

Higuchi, R., Fockler, C., Dollinger, G., and Watson, R. (1993). Kinetic PCR analysis: Real-time monitoring of DNA amplification reactions. *Bio/Technology* **11,** 1026–1030.

Kalinina, O., Lebedeva, I., Brown, J., and Silver, J. (1997). Nanoliter scale PCR with TaqMan detection. *Nucleic Acids Res.* **25,** 1999–2004.

Lay, M. J., and Wittwer, C. T. (1997). Real-time fluorescence genotyping of factor V Leiden during rapid cycle PCR. *Clin. Chem.* **43,** 2262–2267.

Nazarenko, I. A., Bhatnagar, S. K., and Hohman, R. J. (1997). A closed tube format for amplification and detection of DNA based on energy transfer. *Nucleic Acids Res.* **25,** 2516–2521.

Ririe, K. M., Rasmussen, R. P., and Wittwer, C. T. (1997). Product differentiation by analysis of DNA melting curves during the polymerase chain reaction. *Anal. Biochem.* **245,** 154–160.

Shoffner, M. A., Cheng, J., Hvichia, G. E., Kricka, L. J., and Wilding, P. (1996). Chip PCR. I. Surface passivation of microfabricated silicon-glass chips for PCR. *Nucleic Acids Res.* **24,** 375–379.

Swerdlow, H., Jones, B. J., and Wittwer, C. T. (1997). Fully automated DNA reaction and analysis in a fluidic capillary instrument. *Anal. Chem.* **69,** 848–855.

Tan, S. S., and Weis, J. H. (1992). Development of a sensitive reverse transcriptase PCR assay, RT-RPCR, utilizing rapid cycle times. *PCR Methods Appl.* **2,** 137–143.

Tyagi, S., and Kramer, F. R. (1996). Molecular beacons: Probes that fluoresce upon hybridization. *Nature Biotechnol.* **14,** 303–308.

Wittwer, C. T., and Garling, D. J. (1991). Rapid cycle DNA amplification: Time and temperature optimization. *BioTechniques* **10,** 76–83.

Wittwer, C. T., Fillmore, G. C., and Hillyard, D. R. (1989). Automated polymerase chain reaction in capillary tubes with hot air. *Nucleic Acids Res.* **17,** 4353–4357.

Wittwer, C. T., Fillmore, G. C., and Garling, D. J. (1990). Minimizing the time required for DNA amplification by efficient heat transfer to small samples. *Anal. Biochem.* **186,** 328–331.

Wittwer, C. T., Marshall, B. C., Reed, G. B., and Cherry, J. L. (1993). Rapid cycle allele-specific amplification: Studies with the cystic fibrosis delta F508 locus. *Clin. Chem.* **39,** 804–809.

Wittwer, C. T., Reed, G. B., and Ririe, K. M. (1994). Rapid cycle DNA amplification. *In The Polymerase Chain Reaction* (K. B. Mullis, F. Ferre, and R. A. Gibbs, Eds.), pp. 174–181. Birkhauser, Boston.

Wittwer, C. T., Hillyard, D. R., and Ririe, K. M. (1995). Rapid thermal cycling device. U.S. Patent No. 5,455,175.

Wittwer, C. T., Herrmann, M. G., Moss, A. A., and Rasmussen, R. P. (1997a). Continuous fluorescence monitoring of rapid cycle DNA amplification. *BioTechniques* **22,** 130–138.

Wittwer, C. T., Ririe, K. M., Andrew, R. V., David, D. A., Gundry, R. A., and Balis, U. J. (1997b). The LightCycler: A microvolume, multisample fluorimeter with rapid temperature control. *BioTechniques* **22,** 176–181.

Wittwer, C. T., Ririe, K. M., and Rasmussen, R. P. (1998). Fluorescence monitoring of rapid cycle PCR for quantification. *In Gene Quantification* (F. Ferre, Ed.), pp. 129–144. Birkhauser, Boston.

Woolley, A. T., Hadley, D., Landre, P., deMello, A. J., Mathies, R. A., and Northrup, M. A. (1996). Functional integration of PCR amplification and capillary electrophoresis in a microfabricated DNA analysis device. *Anal. Chem.* **68,** 4081–4086.

15

KINETICS OF COMPETITIVE REVERSE TRANSCRIPTASE-PCR

Amanda L. Hayward, Peter J. Oefner, Daniel B. Kainer, Cruz A. Hinojos, and Peter A. Doris

Competitive reverse transcriptase (RT)-PCR is a versatile tool for quantification of gene expression. It is a technique whose past has been shaded by doubts and questions concerning its reliability (Ferré, 1992; Wiesner *et al.*, 1993). Such suspicions arose in large part from a disbelief that the competitor and target templates would behave similarly in reverse transcription and PCR amplification. These suspicions emerged principally from intuitive beliefs about the use of a competitor which was obligatorily different from the native target. Initially, however, such intuitions lacked any stringent test. Of course, a similar lack of stringency was naturally frequent among early incarnations of the competitive RT-PCR technique. Thus, in the absence of information, well-formed conclusions were often held by both practitioners and skeptics of this technique without the experimental underpinnings necessary to support such conclusions. Fortunately, the passage of time and the entry of new analytical techniques has filled this void and replaced it with exciting and powerful new opportunities to explore biological function as manifested in variations in the level of gene expression.

Why Measure Gene Expression?

Approaches directed toward protein rather than nucleic acid quantitation may appear more relevant to understanding biological function. After all, protein is the molecular form through which most regulated cellular function is manifest. However, there are limitations in the utility of protein-directed approaches to quantitation. For example, it is clear that antibody approaches generally fail to distinguish between active and inactive pools in the case of the many cell proteins in which phosphorylation state can determine whether the specialized function of a protein is active or latent. Antibodies may also fail to recognize other posttranslational modifications (e.g., glycosylation and ADP-ribosylation) which influence protein function. Isoforms also pose a problem for protein-based approaches; if isoforms are products of different genes, then antibodies may or may not be selective between one isoform of a protein and another. The possibility of antibody selectivity is usually further reduced if proteins exist in isoforms resulting from alternative RNA splicing of transcripts from the same gene. Nucleic acid techniques can be absolutely selective across these same targets.

Quantitation of a specific isoform of a protein by Western blot may be possible if the goal is a relative comparison between one sample and another. Quantitation may even be achievable to some degree in very small tissue sample sizes (depending on protein abundance and antibody affinity). However, obtaining highly specific, well-calibrated reference standards for proteins can place constraints on the quantitative information which antibodies can provide.

The following is an example from our own research interest: Antibody approaches would not permit a comparison of the amount of sodium–proton exchanger isoform 1 (NHE1) with the amount of NHE isoform 3 (NHE3) in individual renal proximal tubules. If each antibody is specific for a single isoform, it can only be used to compare that isoform in different samples. Comparing the amount of NHE1 with NHE3 in the same sample is not possible. However, the amount of each isoform present per millimeter of tubule may be an important issue in the specialized functions of these isoforms. Nucleic acid-based approaches using competitive RT-PCR can accomplish such comparison at the level of gene expression because, when appropriately designed, they are capable of estimating the actual number of molecules of each specific message present in the same sample, permitting comparison on the basis of number of molecules not only across samples but also across genes encoding different proteins. When competitive RT-PCR methods are carefully designed and applied, both precision and accuracy are high, even when microscopic tissue samples are analyzed (Hayward-Lester *et al.*, 1995, 1996, 1997).

It is necessary to accept that all quantitative approaches, whether they are at the level of gene expression, protein abundance, or enzyme activity, suffer from limitations inherent to the tools used and the targets assayed. Thus, an important goal is to assemble as much reliable information as possible from each source as may be required.

Application of Quantitative Gene Expression

There are two general arenas in which quantitative approaches to gene expression or specific sequence abundance are useful. One is in the clinical laboratory, in which assessing viral titers has become an important tool in diagnosis and therapy. For example, it is possible to quantify the serum levels of HIV sequence (Piatak *et al.*, 1993a,b; Mulder *et al.*, 1994; Revets *et al.*, 1996). Using this information both individual and public health can be improved. For the individual, this information may be vital for assessing efficacy of therapy (O'Brien *et al.*, 1996), predicting outcome (Mellors *et al.*, 1995), adjusting dosage, and switching between therapeutic agents when resistance is suspected (Saag *et al.*, 1996). The importance to the individual patient of knowing that drugs which often have unpleasant side effects are nonetheless producing beneficial effects on the "level" of infection is also not to be underestimated. In public health, new drugs to treat chronic viral infection are only beneficial while the infecting organisms remain susceptible. Mutation rate is proportional to viral load. Therefore, the ability to determine that therapy has reduced the level of infection low enough for mutation rate to be insignificant provides a means to ensure a low likelihood of the emergence of new drug-resistance mutations (Schuurman *et al.*, 1995).

In the research laboratory there are many reasons for examining the level of gene expression. Many such methods (Northern blotting, solution hybridization, and others) have been in use for many years. Unfortunately, some are limited with respect to the detail contained in the quantitative information they provide and in their ability to be applied to small tissue samples containing genes whose expression is not abundant. It is this latter feature which has driven the development of PCR-based gene quantitation techniques.

In the clinical lab a relatively small number of assays need to be capable of high throughput, whereas in the research setting there is a vast range of gene targets (assays) on which individual laboratories might focus. Competitive RT-PCR is especially appealing in the research setting because the ability to design and construct the components of a reliable assay system are within the means of any molecular biology lab. In contrast, high-throughput assays have tended to center on techniques based on probes, reagents, and equipment which are not generally available in the research laboratory and require proprietary components which cannot be generated in the typical laboratory. A number of such assays have been developed and commercialized (Piatak *et al.*, 1993a,b; Urdea *et al.*, 1993; Dewar *et al.*, 1994; Mulder *et al.*, 1994).

Unraveling the Critical Elements of Performance in Competitive RT-PCR

Any assay system requires a thorough understanding of the components of the assay and the factors which impinge upon the behavior of these components. Such understanding has been slow to evolve in the field of competitive RT-PCR because of the relative complexity of the components and their comingling in a single reaction tube. Figure 1 illustrates the steps typical of competitive RT-PCR reactions.

At each step, variation can be added to the quantitation system. Furthermore, since estimates of gene expression are based on the ratio of reaction products, an inviolable principle is that the steps affect identically the native and competitor templates which are being reversed transcribed and amplified in the same reaction tube (Fig. 1). Thus, it is essential to identify and analyze each critical issue at each step in the pathway.

Preparation of Reagent Templates

Competitive reactions are performed by preparation of native RNA (usually total RNA from a tissue source). Obviously, all general

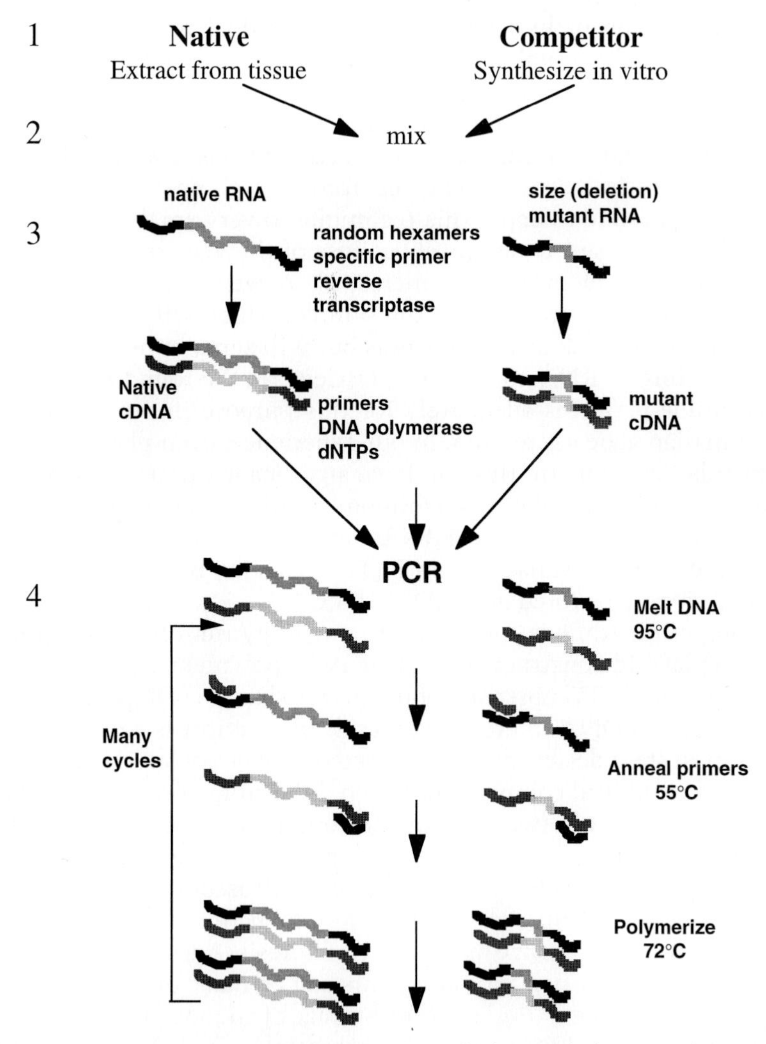

Figure 1 Major steps in competitive RT-PCR at which critical elements of accuracy, precision, and sensitivity can be influenced.

precautions must be taken to ensure that RNA is recovered intact and is not degraded by RNAses present in the tissue sample or introduced by contaminated reagents or utensils used during extraction. RNA is extracted promptly from tissue that is freshly collected and immediately placed into chaotropic solutions (RNAzol B, Tel-Test, Friendswood, TX). In some of our work we have performed microdissection of kidney to recover individual renal tubule segments. Since this involves more than brief tissue handling, samples are dissected in the presence of the RNAse inhibitor, vanadyl ribonucleoside complex (New England Biolabs, Beverly, MA). Meticulous attention to maintaining RNAse-free glass and plasticware and pipetting supplies is essential. The sensitivity of competitive RT-PCR makes it possible to apply this technique to very small tissue samples. In such cases, concerns about loss of the very small amounts of RNA being recovered are increased. A wise precaution is the addition of carrier material to the sample, which will not interfere with the assay or detection systems but will serve to block nonspecific binding of RNA to glass or plasticware. We found that linear acrylamide performs adequately for this purpose in our studies of mammalian gene expression. In our experience, even plastic microfuge tubes and pipette tips can have significant variation in performance between supplies from different vendors. We use exclusively plasticware which is certified RNAse and DNAse free (manufactured by Sorensen and available from DOT Scientific, Burton, MI). RNA preparations are stored at −70 to −86°C.

Competitor RNA is synthesized *in vitro* by run-off transcription from a plasmid construct containing RNA polymerase promoter sequence (usually T7) upstream of the mutated target cDNA. Mutation of the target cDNA to create a suitable competitor is a critical element in system design, as will be discussed in detail later. In general, we have employed competitors which differ from corresponding native sequences by between 14 and 28 contiguous bases. Purity of the synthetic competitor RNA is critical in permitting accurate calibration of this standard by UV absorbance spectroscopy. Extensive precipitation and washing should therefore be performed, followed by analysis of transcript size by gel electrophoresis. Adequate and reproducible UV absorption measurements can be made on *in vitro* transcribed RNA using a 40-μl cuvette. Storage of aliquots of the competitor standard at −70°C to −86°C is essential; stored aliquots should include yeast transfer RNA added after UV calibration.

Mixing of Competitor with Native RNA

Incorporation of internal standards into competitive RT-PCR has permitted important advances in the accuracy and reliability of the approach (Becker-André and Hahlbrock, 1989; Wang *et al.*, 1989; Gilliland *et al.*, 1990). It should be noted, however, that some early incarnations of the technique used competitor DNA templates added after completion of the RT step (Gilliland *et al.*, 1990). While the use of an RNA standard to control for the RT step is an advantage, it poses new questions. The first concerns when the competitor should be added to the sample of native RNA. A suitable occasion for addition of the competitor to the tissue sample appears to be during RNA extraction. However, we have chosen not to add it until the RNA from the sample has been purified. Our concern is that the true control for extraction is a tissue sample in which both the competitor and native RNA begin extraction as intracellular molecules. However, this is not possible using an exogenous competitor. Thus, the competitor is initially exposed to the cell surface, the extracellular materials, and the broken cell material which may already be present in a sample during tissue harvesting. In this medium, the competitor could become preferentially degraded. In contrast, the tissue RNA which will be recovered remains inside intact, freshly harvested cells. Obviously, recovery of RNA from tissue is $<100\%$. Thus, some loss of native RNA occurs during extraction which is not incurred by the competitor RNA. This loss represents a problem only when it is highly variable between samples or when it is necessary to make comparisons across samples of different types in which the relative recovery of native RNA varies substantially between types. In practice, we have not found either of these cases to be limiting.

Another issue involved in mixing native and competitor RNAs is that RT reactions are typically established using 1 μl of RNA. This volume must contain RNA from two different sources. Obviously, RNA extracted from large tissue samples can be mixed in relatively large volumes with competitor RNA so that pipetting accuracy is not a significant limiting factor to precision. However, in samples in which the amplification power of PCR is to be fully exploited so as to quantitate gene expression in RNA from very small tissue samples, the RNA sample may be contained in a very small volume. Nonetheless, we have found it feasible to combine 1 μl each of native and competitor RNA, mix by repeated pipetting, remove 1 μl to the RT reaction tube, and still obtain precise measurements in repeated observations on the same starting materials. However, skill acquired through practice, well-maintained instruments, and care are all prerequisite.

Reverse Transcription

Perhaps because the reverse transcription reaction involves no cycle-dependent amplification, less attention has been placed on performance of the native and competitor templates in this step than in the PCR step (Bouaboula *et al.*, 1992; Horikoshi *et al.*, 1992; Volkenandt *et al.*, 1992; Santagati *et al.*, 1993; Wiesner *et al.*, 1993; Chelly and Kahn, 1994). Our work shows that this is an oversight. Within the parameters we have examined, this oversight may result in loss of accuracy (the ability to estimate the actual number of template molecules in the starting RNA preparation) from the quantitation. It is clear from our observations that identical RT efficiency of native and competitor templates cannot be assumed. The consequence that emerges directly from dissimilar RT efficiency between native and competitor is loss of accuracy as narrowly defined previously. This limits the ability to compare results obtained between labs using different reagents or techniques.

We have performed competitive RT-PCR reactions in which both the native RNA templates and the competitor templates were produced by *in vitro* transcription and were added to reactions in known amounts. This allows the competitive reaction to be used to estimate the amount of native template present, and this amount can then be compared with the known input. Such reactions were always performed against a background of yeast RNA. Table 1 shows the

Table 1

Relationship between Actual Known Inpute RNA Amount and Amount Estimated by Competitive RT-PCR[a]

	Mean	SEM
Estimation discrepancy	3.79	0.2
Slope	1.01	0.013
R^2	0.99	0.005

[a] Estimation discrepancy is the difference between input amount and estimated amount. Slope and R^2 refer to the titrations (log competitor amount versus log ratio of native and competitor reaction product amounts) used to estimate the unknown. According to Raeymaekers' (1993) model, amplification of native and competitor templates which have identical amplification efficiencies throughout PCR should result in straight-line titrations with a slope of 1. SEM, standard error of the mean (originally published in *Nucleic Acids Research*).

relationship between the known input RNA amount and the estimated amount in a competitive system for the α_1 subunit of rat sodium, potassium-ATPase. In this system, the competitor differed from native by the presence of a 145-bp insert of contiguous, nonhomologous sequence. There is clearly an estimation error (3.79-fold overestimation) which is constant regardless of input amount. Two questions are relevant to this error: Is it derived from RT or PCR or both? Can it be removed?

We investigated the first question by repeating this experiment using known starting quantities of DNA (native and copetitor plasmid cDNA) and limiting the comparison to PCR. The results are summarized in Table 2 and indicate that no significant estimation discrepancy results when known quantities of DNA inputs are employed. This test has been repeated for a number of templates and no significant estimation discrepancy has been a consistent finding for all. We conclude from these experiments that RT efficiency differences can occur in RNA templates which are over 60% homologous. Such differences, however, are remarkably constant (as reflected in the SEM for estimation discrepancy in Table 1). Because the efficiency difference is constant over a wide concentration range (at least 4 log) (Hayward-Lester *et al.*, 1995), once a correction factor is determined, accurate quantification (accurate means, in this context, an estimate which accurately reflects the actual number of target template copies in the initial tissue RNA sample) can be achieved.

Over time, several laboratories might generate quantitative gene expression methods for a single target and the goal of comparing actual copy number between laboratories might be sought. Indeed, in the clinical setting in which recommendations concerning treatment regimens may be established at certain copy number cut-off levels,

Table 2

Estimation Error When Known DNA Inputs Are Used Instead of Known RNA Inputs

	Mean	SEM
Estimation discrepancy	1.07	0.064
Slope	1.02	0.023
R^2	0.99	0.026

it is essential that some means of determining the accurate copy number be obtained. For this reason, we have investigated variables that might be responsible for the correction factor and whether alterations in RT reaction conditions might influence these correction factors.

The first variable examined was the effect of increasing homology between the native and competitor templates. A new competitor was synthesized which contained a 14-base deletion compared to the native sequence, and the examination of known RNA inputs of estimated native quantity was repeated (Table 3). In this case, the estimation discrepancy was removed and the system generated accurate estimates of starting copy quantity.

The versatility of competitive RT-PCR allows similar quantitative approaches to be devised for any target sequence. This flexibility has led us to generate gene quantitation systems for numerous genes of interest in our work. We have applied the same analysis to several of these genes. For example, a system for quantitating expression of rat cyclophilin-like protein (Iwai and Inagami, 1990) employed a competitor constructed by ligating a 24-bp oligonucleotide into the native cDNA. The resulting RNA showed identical RT efficiency compared to the native. A system for quantitating rat calcineurin A subunit expression employed a 28-bp deletion mutation to construct a competitor. This system revealed a 0.45 ± 0.031-fold underestimation of actual copy number present in the sample which was due to an RT efficiency difference between the templates. Although persistent, the RT efficiency difference is again constant.

This indicated that simply reducing the internal mutation from 145 to 14–28 bases would not always eliminate the RT efficiency difference. Since others have reported that the presence of stem-loop structures predicted by RNA folding programs was correlated with

Table 3

RT-PCR Quantification of Known Inputs of RNA Using a Highly Homologous Competitor RNA (14-Base Deletion)

	Mean	SEM
Estimation discrepancy	0.945	0.064
Slope	0.978	0.032
R^2	0.989	0.005

reduced reverse transcription (Pallansch *et al.*, 1990), we analyzed whether any pattern of secondary structure formation determined by RNA secondary structure modeling was correlated with the RT efficiency difference (Zuker, 1994). Our results indicate no simple relationship between secondary structure or its relative stability and RT efficiency of the homologous templates, but they suggest that the creation or loss of stem-loop structures at the location of the mutation may underlie the RT efficiency difference. Removal of a predicted stem-loop structure by mutation of the competitor correlated with increased RT efficiency relative to the corresponding native template, whereas creation of new stem-loop structure by the mutation correlated to reduced RT efficiency. These observations are limited to a small number of correlations and may prove to be reasonable explanations. However, the design of mutations intended to manifest predicted secondary structures from the native template for which they control and the verification of the expected effect on RT efficiency is necessary in order to confirm these interesting but inconclusive early observations on the role of stem-loop structure formation in RT efficiency. Furthermore, it is not possible to ascertain what effect, if any, intermolecular RNA interactions occurring between the heterogenous RNAs expressed by differentiated cells might have on relative RT efficiency.

Additional experiments were performed to modify RT reaction conditions in order to determine whether efficiency differences between homologous templates (calcineurin was used) could be reduced or eliminated. Broad-ranging modifications were made in magnesium, dNTP, and random hexamer concentrations, the use of specific primers versus random priming, RNA preheating, RT reaction temperature, and RT enzyme used (MMLV RTase, AMV RTase, and rTTh polymerase). In some cases, small, significant improvements were achieved; however, the persistent RT efficiency differences were never eliminated. Efforts to combine each of the optimized conditions for maximum effects did not show any additive benefit on final reaction product ratio (Table 4).

PCR Coamplification of Homologous Templates

The results in Table 2 answer the question which has most troubled skeptics of competitive PCR: Can two dissimilar templates be amplified with identical efficiency? For two DNA inputs to be amplified simultaneously in the same reaction tube using a single pair of

Table 4

Effect of Varying Several RT Reaction Components on Relative RT Efficiency Difference Observed between Calcineurin A and Its 28-bp Deletion Competitor Template[a]

Reaction Component	Modification Tested	RT Efficiency Ratio
Reverse transcriptase		
Standard conditions	MMLV	0.33 ± 0.01
Test conditions	rTth	0.44 ± 0.03; $p < 0.01$
	AMV	0.31 ± 0.05; NS
RNA preheating		
Standard conditions	No RNA preheating	0.38 ± 0.004
Test conditions	95°C preheat	95° = 0.21 ± 0.02; $p < 0.001$
RNA preheating + DMSO (90%)		
Standard conditions	No preheating, no DMSO	0.39 ± 0.005
Test conditions	50°C preheat ± DMSO	0.21 ± 0.03; $p < 0.001$
Reverse transcription reaction temperature		
Standard conditions	42°C	
Test conditions	37°C–77°C	Positive linear relationship between RT temperature and product ratio; R^2 = 0.46; $p < 0.05$; range of ratios = 0.31–0.49
$MgCl_2$		
Standard conditions	5 m*M*	0.38
Test conditions	1–7.5 m*M*	No consistent effect observed
dNTPs		
Standard conditions	1 m*M*	
Test conditions	0.25–10 m*M*	Inverse linear relationship between dNTP at constant Mg^{2+} (5 m*M*); $R^2 = 0.76$; $p < 0.01$; range of ratios = 0.17–0.53
Random hexamers		
Standard conditions	2.5 μ*M*	
Test conditions	1.25–5 μ*M*	Inverse linear relationship between hexamer concentration and product ratio; $R^2 = 0.9$; $p = 0.004$; at 1.25 μ*M*, ratio = 0.62

(*continues*)

Table 4 (*Continued*)

Reaction Component	Modification Tested	RT Efficiency Ratio
RT priming		
Standard conditions	Random hexamers	0.331 ± 0.019
Test conditions	18-bp CaN-specific primer	0.459 ± 0.016

[a] The columns indicate the individual reaction component or procedure which was varied, the range of conditions or type of modification tested, and the effect on the RT efficiency difference indicated as a ratio of reaction products. The goal was to identify reaction conditions which eliminate the difference in RT efficiency between the two templates, resulting in a ratio of 1 (originally published in *Nucleic Acids Research*).

primers and for one, by virtue of knowledge of its initial amount, to be used to accurately and precisely quantify the initial amount of the other by analyzing the ratio of reaction products at the end of the reaction requires identical rates of expansion of reaction product amounts for each template throughout the reaction. Dissimilarity between native and competitor templates is necessary so that they can be separately quantified at the conclusion of the reaction. However, even for amplicons that share only 63% homology (as is the case in Table 2), this dissimilarity does not prevent identical amplification.

Because many of our applications of competitive RT-PCR involve very small RNA samples from microdissected tissue, it was important to determine if it was critical that all reactions remain in the linear phase. With small initial inputs, reactions which comprise 35–45 amplification cycles are sometimes necessary to yield sufficient product for analysis. This results in few cycles in the linear phase in which products are detectable. The implications of entry into the plateau phase are not insignificant. A theoretical analysis is helpful in understanding the characteristics imposed by amplification efficiency variation on competitive PCR. Performance characteristics should be assessed and accuracy validated for each new assay system as it is developed.

Raeymaekers (1993, 1994) has provided an insightful mathematical description of competitive PCR which will be followed here. Biologists examine the rate of accumulation of PCR products during a reaction and discern two distinct phases: linear amplification (the exponential phase which is linear on a log plot) and plateau. A mathematical discipline, however, requires the recognition that these two points are constituent parts of the same continuum and the factors which create them contribute, albeit to different degrees, throughout the entire PCR reaction. The two principal parameters which are difficult to conceptualize simultaneously are initial amplification efficiency (i.e., efficiency in the first cycle of the reaction) and rate of decline of amplification efficiency. The biologist examines PCR product accumulation and concludes that initial amplification efficiency is preserved throughout the linear phase and then at some point an undefined factor begins to exert, resulting eventually in the termination of new product accumulation. The mathematical approach acknowledges that this factor is present from the completion of the first amplification cycle onwards. However, its growth in magnitude is exponential. Therefore, to describe the coamplification of two related templates in a single PCR reaction it is necessary not only to describe amplification mathematically but also to include terms which describe the initial amplification efficiency and those which describe the decline in amplification efficiency that produces the plateau phenomenon. In order for a competitor to be a valid reference, both initial amplification efficiency and decline in efficiency must be identical for both competitor and native templates. If this is the case, a simplified mathematical model can be derived which makes testable predictions about the relationship between accumulated reaction products in relation to the initial known amount of competitor present.

Equation 1 describes the amplification of a single template in PCR in simplified terms, where N is the final amount of reaction product, N_0 is the initial amount of DNA in the reaction, n is the number of cycles, and E is the efficiency of the reaction:

$$N = N_0(1 + E)^n \tag{1}$$

If the initial unknown amount of a gene U in a competitive RT-PCR reaction is U_0 and that of its specific (mutant) competitor RNA is C_0 and these are passed through n reaction cycles in which the efficiency of amplification is E_u and E_c for the unknown and competitor, respectively, then from Eq. (1) we can describe the amount of reaction products at the end of n cycles by

$$U_n = U_0 \cdot (1 + E_u)^n \tag{2}$$

$$C_n = C_0 \cdot (1 + E_c)^n \tag{3}$$

Taking the logarithm of the ratio of Eq. (2) and Eq. (3) gives

$$\log(U_n/C_n) = \log U_o - \log C_o + n \cdot \log[(1 + E_u)/(1 + E_c)] \quad (4)$$

An assumption of competitive RT-PCR is that E_u and E_c remain equal throughout the reaction. However, this is a critical assumption involving two distinct components: identical initial amplification efficiencies and identical rate of decline in amplification efficiencies through every cycle of the reaction. If these are valid assumptions, Eq. (4) can be simplified to

$$\log(U_n/C_n) = \log U_o - \log C_o \quad (5)$$

In calculating competitive RT-PCR reactions to obtain the unknown amount of a gene (U_0) present in a sample, a plot can be made between $\log(U_n/C_n)$ and $\log C_0$, the known amount of starting competitor RNA. This allows Eq. (5) to be solved and U_0 to be estimated.

Because it is of the form $y = bx \pm a$, Eq. (5) indicates that such a plot will form a straight line having a slope of -1 [or 1 if $\log(C_n/U_n)$ is plotted]. The value of b must be 1 or -1 (depending on whether U_n or C_n is chosen as the denominator of the reaction product ratio) in order to correspond to the form of Eq. (5). Thus, the assumption of identical amplification efficiencies throughout the reaction may be tested by examining the linearity and slope properties of this relationship.

Most work reported using competitive PCR has paid attention to the property of linearity in this relationship, but less attention has been applied to the slope. There have been numerous reports containing data in which the slope requirement fails (Gilliland *et al.*, 1990; Siebert and Larrick, 1992; Zachar *et al.*, 1993). For this reason, it is neither surprising nor improper that suspicion about the reliability of competitive RT-PCR to generate accurate estimates has arisen.

There has been a general assumption that competitive PCR reactions must be analyzed while still in a linear phase of the reaction to prevent loss of accuracy due to different rates of decline by the native and competitor templates from their initial amplification efficiencies. However, the conclusion that a reaction has remained in linear phase has generally been reached without test and certainly without test of every individual sample. Such tests are also cumbersome and reveal that for many reactions the number of cycles in which detectable product is present and amplification remains linear can be few. New methods of cycle by cycle PCR product accumulation analysis indicate that numerous factors may influence the rate at which plateau is reached and some may be beyond the control of the investigator (Higuchi *et al.*, 1993; Heid *et al.*, 1996).

Therefore, if there is uncertainty about whether some or all reactions have left the linear phase, the question of whether competitive RT-PCR can be successfully performed in reactions which extend into the plateau phase gains importance. An affirmative answer to this question offers the important appeal of increased sensitivity. Samples with low abundance can make use of the full amplification power of more extensive repetitive cycling. It would also preclude the need to demonstrate that each individual reaction has not proceeded to plateau.

Regarding the mathematical approach to explore the effect of initial amplification efficiency and the rate of decline of amplification efficiency on competitive PCR reactions, Rayemaekers (1993) constructed equations to express the stepwise alteration in amplification efficiency which occurs as reactions progress. For example,

$$E_{u_k} = E_{U_i} - x \cdot (U_k + C_k)/P_{\max} \quad (6)$$

is an equation in which the PCR amplification efficiency of the unknown template at the kth reaction cycle (E_{u_k}) is related to the amplification efficiency in the first cycle (E_{U_i}). The efficiencies are related by subtraction of the unknown and competitor products ($U_k + C_k$) accumulated at the beginning of the kth cycle, divided by the maximum amount of both products which can be produced by this reaction (i.e., the plateau amount), and multiplied by a constant, x. This constant has a positive value of <1 and reflects the effect of increasing amplification cycle number on the efficiency of amplification of U. For example, if k is the second cycle, the value of $U_k + C_k$ will be very small compared to $P_{\max}$, so even if increasing cycle number rapidly reduces amplification efficiency of U (i.e., even if x is a large value closer to 1 than 1), E_{u_k} will still be very close to the E_{U_i} in the second cycle. However, as $U_k + C_k$ becomes close to $P_{\max}$, the importance of x in determining how much new reaction product is made in each cycle increases.

A similar equation can be constructed to describe changes in amplification efficiency of the competitor during the reaction:

$$E_{c_k} = E_{c_i} - y \cdot (U_k + C_k)/P_{\max} \quad (7)$$

The following critical question can be derived from Eqs. (6) and (7): Does $y = x$ for any particular combination of native and competitor amplicons?

The problem created by the effects which initial amplification efficiency and rate of decay of initial efficiency have on competitive PCR is not a simple one to address in practical terms. One method that can be used is computer modeling of competitive PCR based on the previous equations. Such models will make predictions about the performance of these reactions when individual variables are altered singly or in combination. The consequences of such changes on data analysis can then be observed. We have used Microsoft Excel software to create a competitive PCR model which uses the previous equations and the variables set at values indicated in Table 5 to estimate cycle by cycle product accumulation. This model is available for download to persons who wish to further explore the impact of changing these parameters on the outcome of competitive RT-PCR reactions at *http://www.grad.ttuhsc.edu/archive/index.html.* Table 5 summarizes the tests performed and compares the actual amount of target entered into the model with the estimated target amount. The titrations which result from the parameters are plotted in Fig. 2.

Several important observations are apparent from Table 5. First, when initial amplification efficiencies are identical and rates of decline of amplification efficiency of each template are also identical, titrations are generated which are straight lines with a slope of unity. When this is the case, the starting amount of the unknown native template is accurately estimated by the model. These results indicate that the model has accurately incorporated the equations defining ideal template behavior in competitive PCR reactions. The model also suggests that the property of linearity is not one which reliably reflects whether or not templates are demonstrating ideal behavior in competitive PCR. In no instance was an R^2 value of <0.99 obtained. Slope is perhaps better able to reflect whether reaction components are diverging from ideal behavior. However, the utility of slope in this regard is strongly influenced, under the conditions modeled here, by the starting copy number present in the reaction. Our modeling also reflects the fact that deviation from ideal behavior can have effects on accuracy which range from negligible (e.g., Table 5, line J) to extreme (Table 5, line P), at least within the range of the parameter values we have tested.

These modeling observations are instructive in cautioning against placing an excessive reliance during evaluation of assay performance on conformation with theoretical ideals of linearity and slope of titrations. For example, line H in Table 5 is almost ideal in these properties but still manages a nearly 70-fold inaccurate estimation.

Table 5

Computer Modeling of Competitive RT-PCR Kinetics

Line	E_{u_1}	E_{c_1}	x	y	R^2	Slope	Actual Starting Copies	Estimated Starting Copies	Estimated/ Actual
A	1	1	1	1	1	1	5	5	1
B	1	1	1	1	1	1	50	50	1
C	1	0.8	1	1	1	0.998	5	107.9	21.6
D	1	0.8	1	1	1	0.998	50	1,156.45	23.1
E	1	1	1	1	1	1	5,000	5,000	1
F	1	1	1	1	1	1	50,000	50,000	1
G	1	0.8	1	1	1	0.996	5,000	237,411	47.5
H	1	0.8	1	1	1	1.002	50,000	3,411,412	68.2
I	1	1	1	1	1	1	5	5	1
J	1	1	0.5	0.9	0.997	0.904	5	5.6	1.12
K	1	1	1	1	1	1	50	50	1
L	1	1	0.5	0.9	0.998	0.910	50	92.5	1.85
M	1	1	1	1	1	1	5,000	5,000	1
N	1	1	0.5	0.9	0.999	1.538	5,000	447,554	89.5
O	1	1	1	1	1	1	50,000	50,000	1
P	1	1	0.5	0.9	0.991	1.400	50,000	31,158,133	623.2

Note. Line labels refer to the plots in Fig. 2. E_{u_1} and E_{c_1} initial amplification efficiencies of the unknown and competitor templates, respectively; x and y, constants which determine the rate of decline in their amplification efficiency through the course of the reaction. Both of these pairs of parameters have been set at identical or differing values. The correlation coefficient and slope of the titration lines generated in simple linear regression as in Fig. 2 have been shown (R^2 and slope). Actual starting copies is the number of units of native template entered into the model at the "beginning" of the reaction. Estimated starting copies is calculated from the equation of the titration line which is generated from the model by setting the value of the log product ratio to 0 (i.e., log 1/1 = 0, the point where competitor and native are estimated to be present in identical quantity). The ratio of the estimated to actual copy number is also shown for each simulation. Each simulation was for 30 cycles and the maximum total product accumulation in each reaction. (P_{max}) was set at 2×10^{10} copies for all simulations. Titration lines are generated from five simulated reactions which usually include individual reactions that reach P_{max} and others that do not. Plateau of any particular reaction is strongly influenced by template initial amplification efficiency rates, rates of decline of amplification efficiency, and starting copy number of both templates.

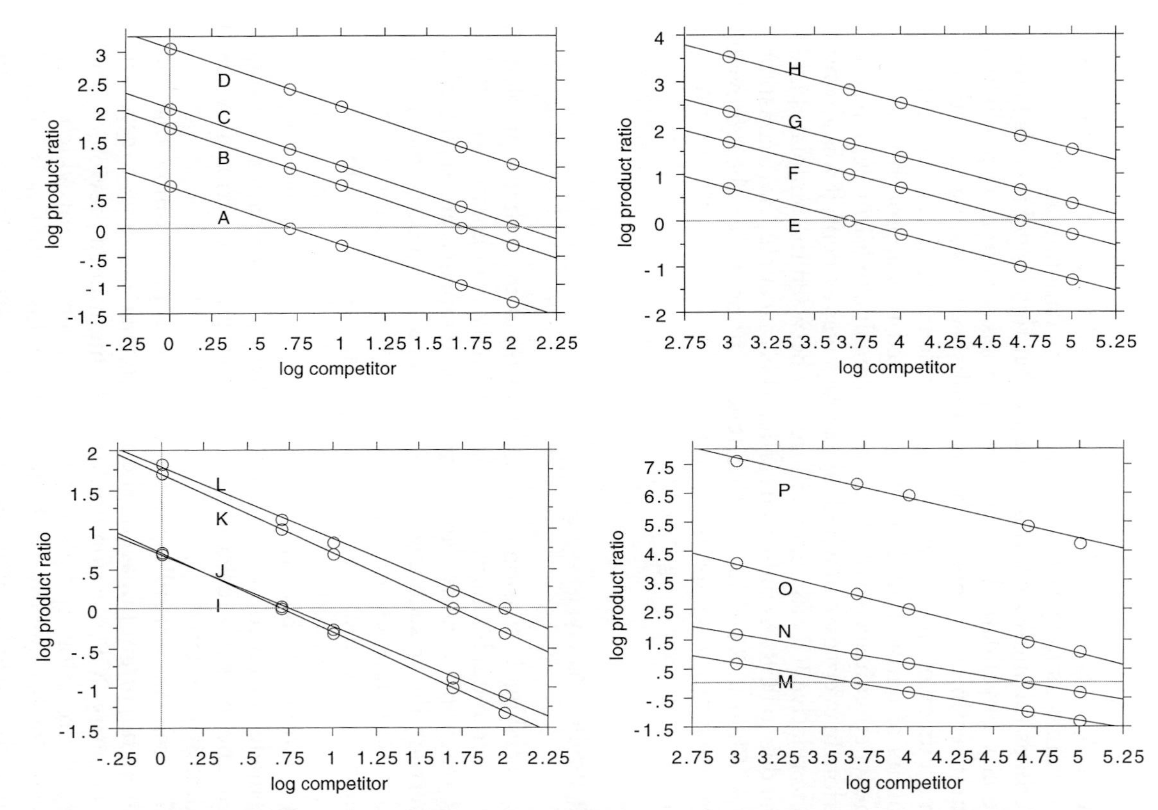

Figure 2 Simulated competitive RT-PCR titrations. Simulated reaction conditions for each titration line are given in Table 5. Each plot represents the effect of changing a single amplification efficiency parameter on simulations performed at each of two initial native input levels.

They also indicate that entry into plateau in no way reduces the accuracy of estimation as long as initial amplification efficiency and rate of decay of amplification efficiency are identical. (It is not necessary for these parameters to be set at a value of 1, as in Table 5, lines A, B, E, F, I, K, M, and O: Identical results are obtained at other values as long as $E_{n_1} = E_{c_1}$ and $x = y$; data not shown).

Perhaps the most important insight that this model reveals is the necessity to validate that any particular assay system is producing accurate estimates. This can be accomplished by testing the estimation accuracy of the system against a wide range of known starting amounts of native template. We employed this approach to determine whether inaccurate estimations were due to RT or PCR amplification. In every case in which inaccuracy was evident, it was always due to relative RT efficiency differences between the templates and never due to PCR. As indicated earlier, this inaccuracy has a constant value. When an RT efficiency difference is present it can be removed by redesigning the competitor template or by determining the correction factor and applying it to the estimates.

PCR Product Analysis

The preceding discussion omitted mention of the difficult problem which heteroduplex formation plays in analysis of competitive RT-PCR reactions. If native and competitor templates are selected to be very similar in sequence except for a small engineered mutation, then heteroduplexes can readily form between the similar complementary native and competitor DNA strands. Heteroduplexes are reaction products and therefore accumulated heteroduplexes represent information about the amount of native and competitor products made in a reaction. It is essential that this information is recognized and properly used if the reaction product ratio is to be accurately estimated.

The problem of handling heteroduplexes is not a simple one and it has an important impact on the selection of techniques to separate and quantify PCR reaction products. As mentioned previously, the entry of a reaction into plateau is an occurrence about which little information is generally available unless sophisticated tools are used to monitor cycle by cycle reaction product accumulation without opening reaction tubes (Higuchi *et al.*, 1993; Heid *et al.*, 1996). Using these tools, it has become clear that increasing the primer concentration can delay the cycle in which a reaction "enters" plateau. The simple conclusion to be drawn is that primer depletion causes plateau. However, it should also be considered that primers are not the only templates in PCR reactions which have the ability to anneal to primer binding sites. Reaction products are also completely complementary at these sites and, as their concentration grows through the course of a reaction, products will begin to compete with primers for binding. Product strands which bind at the primer binding sites continue their complementary binding throughout their regions of shared sequence and form a heteroduplex. Since these cannot be extended into new product, heteroduplex formation reduces new product accumulation and amplification rates subsequently decline through progressive cycles while heteroduplexes become increasingly abundant.

The practical problems caused by heteroduplexes are complex and numerous. First, a competitor may be constructed which differs only by one or two bases from the native amplicon, so as to create (or eliminate) a restriction site unique to the competitor. This will allow the competitor product to be distinguished by restriction digestion and permit it to be separately quantified. However, competitor strands participating in heteroduplex formation will create an incomplete restriction site which will not be digested. These competitor strands will be erroneously included in the products counted as native products (or vice versa) and the reaction product ratio will be incorrectly calculated.

Heteroduplexes often do not resolve on nondenaturing agarose gel electrophoresis (Hayward-Lester *et al.*, 1995, 1996). Our experience is that generally heteroduplexes migrate with or very close to the band containing the homoduplexes made of the longest strand present in the heteroduplex; however, this is not always the case and migration may be influenced by the agarose and running buffer used. Furthermore, even when heteroduplexes can be completely resolved on agarose, their ability to intercalate dyes used to identify and quantify products may be sufficiently different so as to introduce significant error into determining the reaction product quantities formed. Furthermore, an additional step is required after gel separation of the reaction products to count the amount of each product made.

An important advance in the successful application of quantitative RT-PCR is the use of a newly developed chromatographic technique for analysis of nucleic acids. Denaturing high-performance liquid chromatography (dHPLC) is a technique performed on a stationary phase composed of reversed-phase, nonpermeable polystyrene-divinylbenzene beads (DNAsep, Sarasep, Inc., San Jose, CA). The eluting solvents contain an ion-pairing agent, usually triethylammonium acetate, which mediates interaction between nucleic acids and the C18 nonpolar coating of the beads. This interaction is modified by increasing organic solvent strength (acetonitrile). As reported by Oefner and colleagues (Huber *et al.,* 1995), the elution profile of double-stranded DNA from these columns is tightly related to the number of paired bases in the DNA. Heteroduplexes present an interesting profile to these columns. They comprise two different strands, one of which (the competitor) has been designed to share large sequence similarity with the other (the native), differing only by a small insertion or deletion. The effective length of paired bases in the heteroduplex can never exceed the number of bases paired in the shorter of the two strands from which the heteroduplex is formed. Indeed, partial denaturation in the region adjacent to the unmatched pairs suggests a product with slightly fewer paired bases than the shorter homoduplex. Competitive reactions containing heteroduplexes behave as if three reaction products have been formed in them: two products of different sizes reflecting the native and competitor homoduplexes, and heteroduplexes. The heteroduplexes, while existing in two forms (forward strand of native paired with reverse strand of competitor and vice versa), will both present to the column a similar number of paired bases. Figure 3 illustrates the analysis of competitive reactions by agarose gel electrophoresis and dHPLC.

Several other important advantages necessary for accurate analysis of competitive PCR reactions are provided by dHPLC. First, the eluting DNA can be detected and quantified by on-line UV absorbance, eliminating the need for intercalating dyes. The separation is rapid, usually 5 min per sample, and easily automated. Column reequilibration time at the conclusion of one run and prior to the beginning of the subsequent run is limited only by dead space volume in the HPLC system: The phase regenerates very rapidly. Separation of DNA differing by only 5% in size is readily accomplished, allowing competitors to be constructed which are 95% similar in shared base sequence to the native amplicon. Elution times between runs are highly reproducible and the system is easily optimized to accommodate additional assay systems in which PCR products of various lengths are generated. The range of product sizes resolved is approximately 150–800 bp. Furthermore, HPLC is a well- and long-established technique with a very wide installed equipment base, so access to equipment is often simple. Finally, dHPLC column life and perfor-

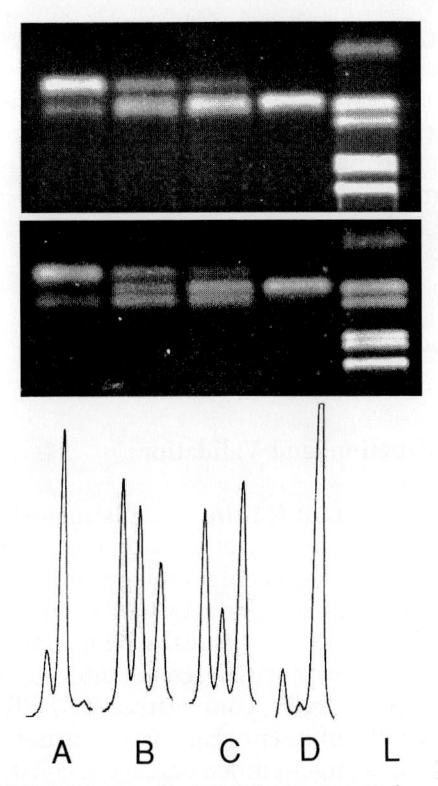

Figure 3 (Top) RT-PCR reaction products A–D were separated on a 4% high-resolution agarose gel (Metaphor, FMC, Rockland, ME). Heteroduplexes were unresolved. Lane L is a pUC18/*Hae*III ladder. The native product is 246 bp and the competitor product is 218 bp. (Middle) Analysis of the same reaction products on a 4% NuSieve gel reveals much better separation of heteroduplexes. (Bottom) HPLC analysis of products A–D indicates that heteroduplexes can be readily resolved. The elution sequence (from left to right of each chromatogram) is heteroduplex, 218-bp competitor product, 246-bp native product (originally published in *Nucleic Acids Research*).

mance are remarkable. We have run thousands of reactions over a single column without noticeable reduction in separation efficiency or an increase in back pressure. A program which runs in Excel (Q-RT-PCR.xla) has been developed and released into the public domain to assist in the analysis of these reactions (also available at *http://www.grad.ttuhsc.edu/archive/index.html*). It will calculate reaction product ratios, correcting the size differences between products, reallocate heteroduplex strands to the appropriate reaction product to which it belongs, perform regression analysis of titrations, provide statistical analysis (titration linearity, slope, and p value of the regression), and estimate unknowns. Eventually, we hope to interface this data analysis software directly with chromatography software so that the hands-on involvement required to estimate unknowns is no more than identifying samples and providing elution time ranges of the products of interest.

Performance Characteristics

Versatility, Construction, and Validation

Perhaps the most important feature that distinguishes competitive RT-PCR from other techniques for gene quantitation is that it is a method which almost any molecular biology laboratory can tailor to accommodate its own particular targets. Unlike commercial systems which have been devised for the clinical setting in which proprietary reagents and hardware have been integrated into simplified solutions for a limited number of targets, competitive RT-PCR can be applied to any target for which sufficient sequence information is available. The technique places a larger burden of reagent construction, validation, and quality control on the user. However, as the important issues in these areas have begun to emerge and solutions have been devised to address these issues, there is no reason why high-quality assays cannot be developed quickly and simply by adapting methods shown to be successful for other templates. During the past 3 years, we and our colleagues have generated assay systems for 10 targets and the process has been simplified and made faster with each additional piece of knowledge about the impact of design features on performance.

Construction of reagent templates from cloned cDNA sequence can often be achieved in about 1 week. This may begin with subcloning of the cDNA into a vector containing an RNA polymerase promoter site to permit runoff transcription. Identification of a strategy to generate a mutation in the cDNA to provide a suitable template for transcription of the competitor RNA will depend on individual sequence. Occasionally, simple restriction/ligation strategies can be successful in producing a sequence variation of about 3% of the contiguous bases in the native cDNA. We use 3% as a target because amplicons will always be smaller than the RNA generated by run-off transcription and because dHPLC can readily separate 5% sequence variations, including heteroduplexes. We have also employed a mutation strategy which involves ligation of a blunt-ended, double-stranded DNA oligomer. This was constructed by annealing two complementary primers to create a 24-bp oligomer which contains a rare restriction site helpful in subsequently diagnosing subclones after ligation. PCR-mediated mutagenesis can also be a simple path to mutant construction (Frotschl *et al.*, 1996).

Thorough validation of any new assay system is critical to producing reliable data and can generally be accomplished in an additional 2–4 weeks. Critical steps involve

1. Analysis of titrations for slope and linearity properties over the range of native template abundance expected to be analyzed in samples
2. Validation that the assay system can accurately estimate known input quantities over a wide range of inputs and PCR cycles
3. Determination of any RT efficiency difference which may exist and its consistency within defined conditions (this difference might be avoided if preliminary information concerning the role of stem-loop structures in generating these differences is substantiated and reagents are designed to avoid secondary structure differences caused by mutation creation)
4. Assessment of precision of the assay system, i.e., the variation observed when the assay system is used to analyze the same sample repeatedly.

Sensitivity

An original scientific motive in developing competitive RT-PCR was to provide quantitative information which may reflect altered function in very small tissue samples (e.g., microdissected nephron segments containing approximately 100 cells). Therefore, a reasonable sensitivity limit would be the ability to quantify expressed genes, in replicate, from a single sample with sufficient resolution to detect differences of 40% between two populations using 6–10 representative individuals per population. We have achieved this goal, as illustrated in Fig. 4.

In order to prove further the relationship between sensitivity and precision in the lower range of competitive RT-PCR, we performed stepwise dilutions of similar RNA preparations to reduce the initial target copy number. Our early attempts to quantify rat cyclophilin gene expression in RNA extracted from renal nephron segments often failed to consistently yield detectable products. However, we knew from quantitations of whole kidney that the abundance of this template in whole kidney was high. Amplification of a similarly abundant signal (α isoform of sodium, potassium-ATPase) was easily achieved in nephron samples. We investigated the possible influence of primers on sensitivity (Fig. 5). We noted that low sensitivity was associated with reactions that appeared to produce only small amounts of reaction products which were difficult to accurately quantify. Resynthesizing the primers improved sensitivity and accuracy (Fig. 5, bottom line) but only partly, suggesting a limited role for between-batch differences in primer quality in determining sensitivity. However, replacement of a single primer with one directed

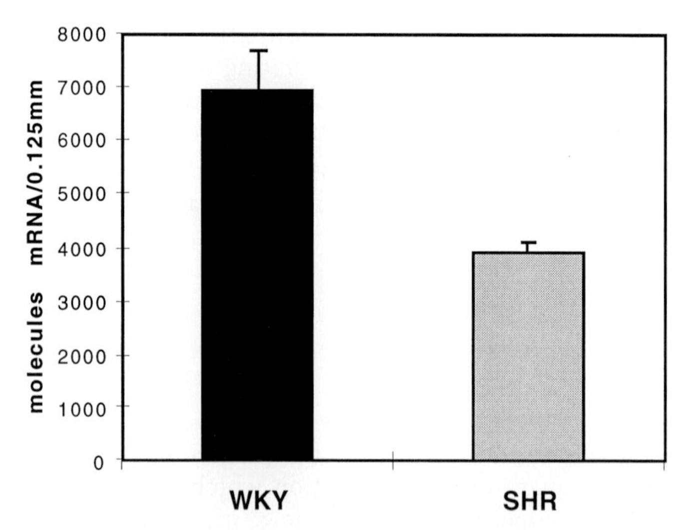

Figure 4 Expression of α_1 isoform of sodium, potassium-ATPase in total RNA from cortical collecting duct segments of 16-week-old spontaneously hypertensive rats (SHR) is reduced compared with matched Wistar–Kyoto (WKY) rats ($p = 0.004$). Expression is reported as the number of molecules of specific mRNA detected per 0.125 mm of nephron.

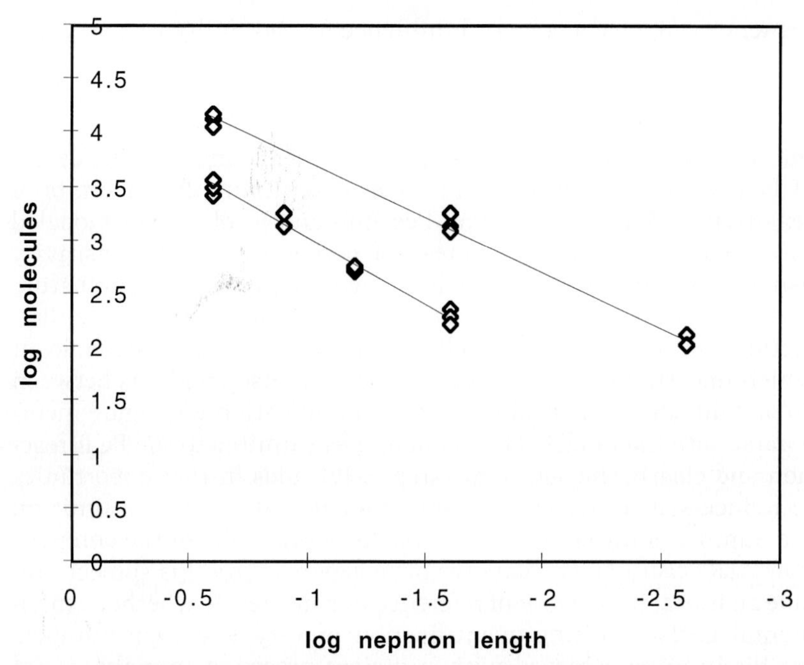

Figure 5 The bottom regression line shows the results of quantification by competitive RT-PCR of rat cyclophilin gene expression in rat nephron samples. Serial dilution of starting RNA resulted in progressive reduction in the amount of gene expression estimated. Although the relationship between dilution and estimation was linear (R^2 = 0.99), the slope (1.22) differed from the expected value of 1. Furthermore, the primers employed in this reaction produced reaction products of low abundance which were difficult to completely resolve by HPLC, resulting in low accuracy in the quantitations. The top regression line was obtained when the same RNA preparations were reanalyzed after one of the two primers had been redesigned. In this case, much more abundant reaction products were obtained in assays which continued for the same number of cycles (indicating higher overall reaction efficiency with this primer pair). The regression which resulted from this dilution indicates that the expected reduction in abudance due to dilution was obtained (R^2 = 1.00, slope = 1.02). Accuracy was very well preserved even with initial copy numbers estimated to be as low as 100 per sample.

to a different region of the template resulted in a high sensitivity able to precisely determine initial template abundance as low as 100 copies (Fig. 5 top line). Below 100 copies, reactions were frequently encountered which failed to make detectable products or which made nonspecific products. It is clear that both primer quality and sequence can have a powerful influence on the ability to quantitate low-abundance inputs.

The ability to quantify at lower levels of abundance with accuracy and precision is influenced by many factors. Stochastic events in the various reactions which take place during RT and the early cycles of PCR have an important impact in determining lower limits of sensitivity (Melo *et al.*, 1996). Peccoud and Jacob (1996) modeled these phenomena in the context of competitive PCR sensitivity using a computer simulation which incorporates the random nature of chemical reactions involving very low abundance reagents. Predictions from this model indicate that the critical threshold at which uncertainty in quantitation begins to rise rapidly is between 20 and 40 copies, depending on the amplification efficiency incorporated into the model. This modeling was limited to the PCR reaction and clearly the additional step of RT adds further opportunity to reduce sensitivity because RT efficiency also places a limit on the number of initial template molecules available for the competitive PCR reaction. The data we have generated on this subject provide an initial framework of reference in considering whether experimental goals requiring quantification of very low copy numbers are likely to be achieved with sufficient precision to make useful comparisons.

Acknowledgments

This work was supported by Grants DDK RO1 45538 (PAD), T32 DDK07556-20 (AHL), and PO1-HG00205 (PJO) from NIH. CAH was the recipient of a HHMI Fellowship. We are grateful to Dr. Sandra Sabatini for assistance in obtaining nephron specimens and to Betty Lonis for excellent technical contributions.

References

Becker-André, M., and Hahlbrock, K. (1989). Absolute mRNA quantification using the polymerase chain reaction (PCR). *Nucleic Acids Res.* **17,** 9437–9446.

Bouaboula, M., Legoux, P., Pessegue, B., Delpech, B., Dumont, X., Piechaczyk, M., Casellas, P., and Shire, D. (1992). Standardization of mRNA titration using a polymearse chain reaction method involving co-amplification with a multispecific internal control. *J. Biol. Chem.* **267,** 21830–21838.

Chelly, J., and Kahn, A. (1994). RT-PCR and mRNA quantification. In *The Polymerase Chain Reaction* (K. B. Mullis, F. Ferré, and R. A. Gibbs, Eds.), pp. 97–109. Birkhauser, Boston.

Dewar, R. L., Highbarger, H. C., Sarmiento, M. D., Todd, J. A., Vasudevachari,

M. B., Davey, R. T., Jr., Kovacs, J. A., Salzman, N. P., Lane, H. C., and Urdea, M. S. (1994). Application of branched DNA signal amplification to monitor human immunodeficiency virus type 1 burden in human plasma. *J. Infect. Dis.* **170,** 1172–1179.

Ferré, F. (1992). Quantitative or semi-quantitative PCR: Reality versus myth. *PCR Methods Appl.* **2,** 1–9.

Frotschl, R., Kleeberg, U., Hildebrandt, A. G., Roots, I., and Brockmoller, J. (1996). Polymerase chain reaction (PCR)-based construction of a competitor target for quantitative reverse transcriptase PCR measurement of cytochrome P450 1A1 mRNA. *Anal. Biochem.* **242,** 280–282.

Gilliland, G., Perrin, S., Blanchard, K., and Bunn, H. F. (1990). Analysis of cytokine mRNA and DNA: Detection and qunatitation by competitive polymerase chain reaciton. *Proc. Natl. Acad. Sci. U.S.A.* **87,** 2725–2729.

Hayward-Lester, A., Oefner, P. J., Sabatini, S., and Doris, P. A. (1995). Accurate and absolute quantitative measurement of gene expression by single tube RT-PCR and HPLC. *Genome Res.* **5,** 494–499.

Hayward-Lester, A., Oefner, P. J., and Doris, P. A. (1996). Rapid quantification of gene expression by competitive RT-PCR and ion-pair reversed-phase HPLC. *BioTechniques* **20,** 250–257.

Hayward-Lester, A., Chilton, B. S., Underhill, P. A., Oefner, P. J., and Doris, P. A. (1997). Quantification of specific nucleic acids, regulated RNA processing and genomic polymorphisms using reversed-phase HPLC. In *Gene Quantification* (F. Ferré, Ed.). Birkhauser, Boston.

Heid, C. A., Stevens, J., Livak, K. J., and Williams, P. M. (1996). Real time qunatitative PCR. *Genome Res.* **6,** 986–994.

Higuchi, R., Fockler, C., Dollinger, G., and Watson, R. (1993). Kinetic PCR analysis: Real-time monitoring of DNA amplification reactions. *Bio/Technology* **11,** 1026–1030.

Horikoshi, T., Danenberg, K. D., Stadlbauer, T. H., Volkenandt, M., Shea, L. C., Aigner, K., Gustavsson, B., Leichman, L., Frosing, R., Ray, M., *et al.* (1992). Quantitation of thymidylate synthase, dihydrofolate reductase, and DT-diaphorase gene expression in human tumors using the polymearse chain reaction. *Cancer Res.* **52,** 108–116.

Huber, C. G., Oefner, P. J., and Bonn, G. K. (1995). Rapid and accurate sizing of DNA fragments on alkylated nonporous poly(styrene-divinylbenzene) particles. *Anal. Chem.* **67**(3), 578–585.

Iwai, N., and Inagami, T. (1990). Molecular cloning of a complementary DNA to rat cyclophilin-like protein mRNA. *Kidney Int.* **37,** 1460–1465.

Mellors, J. W., Kingsley, L. A., Rinaldo, C. R., Jr., Todd, J. A., Hoo, B. S., Kokka, R. P., and Gupta, P. (1995). Quantitation of HIV-1 RNA in plasma predicts outcome after seroconversion. *Ann. Int. Med.* **122,** 573–579.

Melo, J. V., Yan, X. H., Diamond, J., Lin, F., Cross, N. C., and Goldman, J. M. (1996). Reverse transcription/polymerase chain reaction (RT/PCR) amplification of very small numbers of transcripts: The risk is misinterpreting negative results. *Leukemia* **10,** 1217–1221.

Mulder, J., McKinney, N., Christopherson, C., Sninsky, J., Greenfield, L., and Kwok, S. (1994). Rapid and simple PCR assay for qunatitation of human immunodeficiency virus type 1 RNA in plasma: Application to acute retroviral infection. *J. Clin. Microbiol.* **32,** 292–300.

O'Brien, W. A., Hartigan, P. M., Martin, D., Esinhart, J., Hill, A., Benoit, S., Rubin,

M., Simberkofff, M. S., and Hamilton, J. D. (1996). Changes in plasma HIV-1 RNA and CD4+lymphocyte counts and the risk of progression to AIDS. *N. Engl. J. Med.* **334,** 426–431.

Pallansch, L., Beswick, H., Talian, J., and Zelenka, P. (1990). Use of an RNA folding algorithm to choose regions for amplification by the polymerase chain reaction. *Anal. Biochem.* **185,** 57–62.

Peccoud, J., and Jacob, C. (1996). Theoretical uncertainty of measurements using quantitative polymerase chain reaction. *Biophys. J.* **71,** 101–108.

Piatak, M., Jr., Saag, M. S., Yang, L. C., Clark, S. J., Kappes, J. C., Luk, K. C., Hahn, B. H., Shaw, G. M., and Lifson, J. D. (1993a). High levels of HIV-1 in plasma during all stages of infection determined by competitive PCR. *Science* **259,** 1749–1754.

Piatak, M., Luke, K.-C., Williams, B., and Lifson, J. D. (1993b). Quantitative competitive polymerase chain reaction for accurate quantification of HIV DNA and RNA species. *BioTechniques* **14,** 70–80.

Raeymaekers, L. (1993). Quantitative PCR: Theorectical considerations with practical implications. *Anal. Biochem.* **214,** 582–585.

Raeymaekers, L. (1994). Comments on quantitative PCR. *Eur. Cytokine Network* **5,** 57.

Revets, H., Marissens, D., de Wit, S., Lacor, P., Clumeck, N., Lauwers, S., and Zissis, G. (1996). Comparative evaluation of NASBA HIV-1 RNA QT, AMPLICOR-HIV monitor, and QUANTIPLEX HIV RNA assay, three methods for quantification of human immunodeficiency virus type 1 RNA in plasma. *J. Clin. Microbiol.* **34,** 1058–1064.

Saag, M. S., Holodniy, M., Kuritzkes, D. R., O'Brien, W. A., Coombs, R., Poscher, M. E., Jacobsen, D. M., Shaw, G. M., Richman, D. D., and Volberding, P. A. (1996). HIV viral load markers in clinical practice. *Nature Med.* **2,** 625–629.

Santagati, S., Bettini, E., Asdente, M., Muramatsu, M., and Maggi, A. (1993). Theoretical considerations for the application of competitive polymerase chain reaction to the quantitation of a low abundance mRNA: Estrogen receptor. *Biochem. Pharmacol.* **46,** 1797–1803.

Schuurman, R., Nijhuis, M., van Leeuwen, R., Schipper, P., de Jong, D., Collis, P., Danner, S. A., Mulder, J., Loveday, C., Christopherson, C. *et al.* (1995). Rapid changes in human immunodeficiency virus type 1 RNA load and appearance of drug-resistant virus populations in persons treated with lamivudine (3TC). *J. Infect. Dis.* **171,** 1411–1419.

Siebert, P. D., and Larrick, J. W. (1992). Competitive PCR. *Nature* **359,** 557–558.

Urdea, M. S., Wilber, J. C., Yeghiazarian, T., Todd, J. A., Kern, D. G., Fong, S. J., Besemer, D., Hoo, B., Sheridan, P. J., Kokka, R., *et al.* (1993). Direct and quantitative detection of HIV-1 RNA in human plasma with a branched DNA signal amplification assay. *Aids* 7(Suppl. 2), S11–S14.

Volkenandt, M., Dicker, A. P., Banerjee, D., Fanin, R., Schweitzer, B., Horikoshi, T., Danenberg, K., Danenberg, P., and Bertino, J. R. (1992). Quantitation of gene copy number and mRNA using the polymerase chain reaction. *Proc. Soc. Exp. Biol. Med.* **200,** 1–6.

Wang, A. M., Doyle, M. V., and Mark, D. F. (1989). Quantification of mRNA by the polymerase chain reaction. *Biochemistry* **86,** 9717–9721.

Wiesner, R. J., Beinbrech, B., and Ruegg, J. C. (1993). Quantitative PCR. *Nature* **366,** 416.

Zachar, V., Thomas, R. A., and Goustin, A. S. (1993). Absolute quantification of target

DNA: A simple competitive PCR for efficient analysis of multiple samples. *Nucleic Acids Res.* **21,** 2017–2018.

Zuker, M. (1994). Prediction of RNA secondary structure by energy minimization. In *Computer Analysis of Sequence Data* (A. M. Griffina nd H. G. Griffin, Eds.), pp. 267–294. CRC Press, Totowa, NJ.

16

KINETIC PCR ANALYSIS USING A CCD CAMERA AND WITHOUT USING OLIGONUCLEOTIDE PROBES

Russell Higuchi and Robert Watson

The real-time monitoring of PCR amplifications, or kinetic PCR analysis, allows one to follow PCR DNA replication on a cycle-by-cycle basis. This was first accomplished by including in the PCR at less than inhibitory levels a dye whose fluorescence increases upon binding dsDNA. Such a dye is the commonly used ethidium bromide (*EtBr*), although many other dyes may be used. As an increasing amount of double-stranded DNA (dsDNA) is generated by the PCR, so is an increasing amount of fluorescence, which can be monitored from outside the reaction vessel. The initial use of this dye was in the qualitative detection of target sequences by the presence or absence of a fluorescence increase (Higuchi *et al.,* 1992). Subsequently, the cycle-by-cycle monitoring of fluorescence increase was shown to facilitate quantitation of the target sequence (Higuchi *et al.,* 1993). Compared with other methods of quantitation based on PCR, kinetic PCR has the advantages of being a "closed-tube" assay, including mechanical simplicity, high throughput, and containment of potential carryover contamination. Kinetic PCR also has the ability to quantify over a range of many logs of target concentration without the sample dilution needed by other methods.

Cycle-by-cycle monitoring allows one to trace a "growth curve" or profile for a PCR that is perfectly analogous to that of a bacterial culture's growth. In PCR, as in bacterial replication, there is an initial, exponential growth phase followed by a slackening of growth that leads ultimately, upon exhaustion of available resources, to a stationary phase in which there is no additional replication. Quantitation is based on the fact that the more target sequence there is to begin with, the fewer cycles it will take for a PCR to go through these stages. If a threshold level of fluorescence is set in the (detectable) growth phase, one can ask, How many cycles does it take to surpass this threshold? This number of cycles, which can be interpolated to fractions of cycles, is inversely and linearly related to the logarithm of the initial number of target sequences (Higuchi *et al.*, 1993).

A concern with the "dye-alone" method of PCR product detection is that nonspecific, unintended amplification products will also fluoresce, with the potential of creating falsely positive results. Sometimes, these unintended amplification products arise from partial homologies between the primers and nontarget sequences contained elsewhere in the nucleic acid sample. This can usually be overcome by using alternative primer sequences and/or more stringent primer annealing conditions. Often, however, the unintended amplification product is the result of a primer extension that uses the other PCR primer as its template (Chou *et al.*, 1992). This inefficient process creates a dsDNA that has been termed "primer dimer." It is important to note that primer dimer, when monitored in real time, usually arises only after many cycles and does not interfere with the quantification of target sequences unless the target is at relatively low amounts (<10–1000 copies). Also, additional means can be taken to significantly delay the appearance of primer dimer (see Enhancing Specificity and Sensitivity).

Alternatively, the sensitivity limit imposed by primer dimer can be overcome by using a specific, fluorescent oligonucleotide probe whose fluorescence increases only upon its binding to a complementary sequence in the target amplification product (Holland *et al.*, 1991; Lee *et al.*, 1993; Heid *et al.*, 1996; Tyagi *et al.*, 1998; Wittwer *et al.*, 1997a). The disadvantage of using probes is their synthesis and optimization, which can be expensive and time-consuming. In addition, because the number of fluor molecules per molecule of PCR product is one or a few for probe methods versus potentially hundreds for dye-alone methods, the fluorescent signal is inherently weaker, necessitating more sensitive instrumentation. An advantage of the probe methods is the ability to use fluorescent moieties with different emission spectra for different probes, such that the detection in one amplification of multiple PCR products (different targets or internal amplification controls) is possible by spectral analysis. However, for most researchers requiring simple, specific nucleic acid quantification, a single channel of fluorescence detection and a sensitivity to a few hundred copies per PCR suffices. Additionally, as described later, there are ways to verify that the correct amplification has taken place and there can be internal amplification controls without using oligonucleotide probes.

In order to monitor fluorescence emanating from a PCR, a bifurcated fiber-optic cable was first used both to deliver excitation light from a spectrofluorometer to the PCR and to send emmitted light back to the spectrofluorometer (Higuchi *et al.*, 1992). A fluorescence-monitoring, thermal cycling instrument capable of spectral analysis and based on fiber-optic transmission and a laser light source is commercially available (Applied Biosystems Model 7700; Perkin–Elmer, Norwalk, CT). In order to monitor 96 PCRs simultaneously, this instrument uses 96 fiber-optic cables and an optical multiplexer to sample rapidly the output of each, one at a time. Mechanically, this becomes increasingly complex and expensive as throughput is further increased.

Another approach to monitoring is the use of both photodiode emitters and detectors. PCRs in small-volume glass capillary tubes are rapidly heated and cooled in an airstream and are interrogated one at a time each cycle during a run (Wittwer *et al.*, 1997b). A carousel is used to rotate 24 PCRs past the emitter/detector. Again, obtaining higher throughputs becomes increasingly complex. This commercially available system (the LightCycler, Idaho Technology, Idaho Falls, ID) is also sensitive enough for the detection of fluorescent probes.

A simpler, less expensive of method of monitoring multiple PCRs is to use a charged-coupled device (CCD) camera to do quantitative, digital imaging of the PCRs as they sit in the thermocyler block (Higuchi *et al.*, 1993). With this method, the number of samples monitored is mechanically less relevant; a 384-well sample block can be imaged as easily as a 96-well block. On the other hand, mainly because of the requirement of some distance between the light source and the block and between the block and the camera, the sensitivity of camera-based detection is generally less than that of fiber-optic cable-based detection, which suffers little transmission loss over distance. This lessened sensitivity is generally not a problem for PCR product detection with dye alone, however.

CCD Camera-Based Kinetic Thermal Cycler

The camera system originally described by Higuchi *et al.* (1993) has been modified to increase its utility. A compact, light-tight enclosure, fitted over a thermal cycler, avoids the need for a darkroom, requires less space, and decreases the distance between the camera and the samples, making the collection of fluorescence more efficient. The use and modification of a commercially available, digital imaging system—based on a low-cost, eight-bit video camera and originally intended for recording gel electrophoresis images—significantly reduces the cost and complexity of the system and provides simple to use software that automates much of the collection and processing of the data.

A schematic diagram of the system is shown in Fig. 1A and a photograph of the system is shown Fig. 1B. The light-tight enclosure, constructed of opaque plastic, is fitted over the sample compartment of the thermal cycler (GeneAmp PCR System 9600, Perkin–Elmer). The heated lid assembly, supplied with the thermal cycler, is not used in this design and is lifted off its track, pulled to the limit of its cable, and set on top of the thermal cycler. A midrange UV lamp, modified to hold two fluorescent tubes, is mounted on top of the optical enclosure to provide 300-nm excitation of the sample block (Model UVM-57, UV PRoducts, San Gabriel, CA). A solenoid-actuated shutter is mounted under the UV lamp to limit the exposure of the samples to UV light. At the desired times during a PCR cycle, a signal is generated by the controlling software which opens the shutter. The shutter is then held open with reduced current to limit heating of the solenoid. A large, 6 × 5 in. 3-mm dichroic mirror, made of silica, is mounted at 45° between the UV light and the sample block. It is designed to transmit the 300-nm excitation light and to reflect the 600-nm emission, characteristic of ethidum bromide, to the rear of the enclosure (540DCSP, Omega Optical Corp., Brattle Boro, VT). A plastic Fresnel lens with a focal length of 10 in. is mounted between the dichroic mirror and the back of the enclosure and serves to limit the parallax in imaging the samples (Part No. 32685;6.7 × 6.7 in.; Edmond Scientific Co., Barrington, NJ).

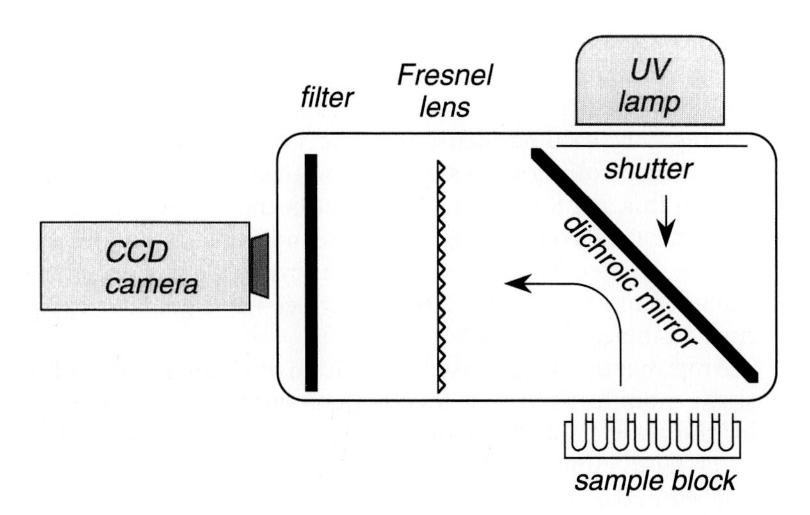

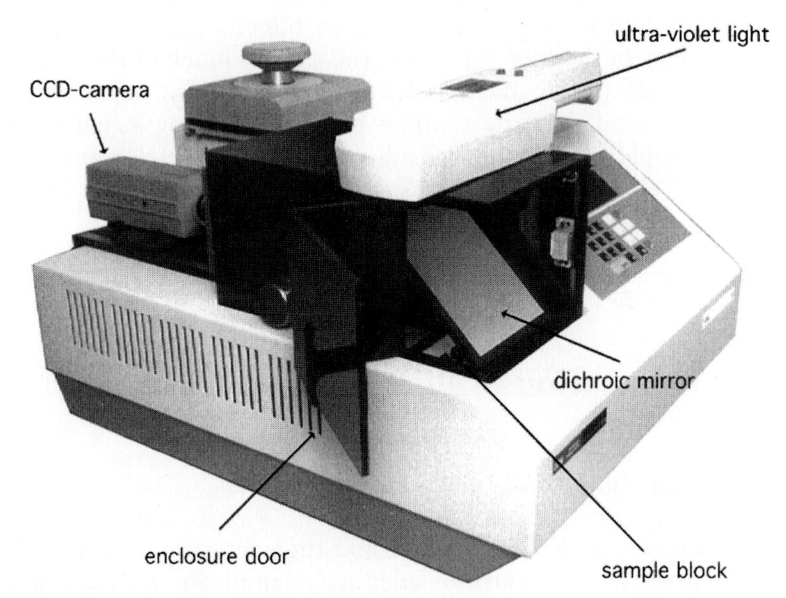

Figure 1 (A) Schematic diagram of light-tight enclosure that fits over the sample block of a thermal cycler. The light path flows from the UV lamp, through the dichroic mirror to illuminate the sample block; from the sample block, it is reflected from the dichroic mirror, through the Fresnel lens and interference filter, to the CCD camera. (B) The light-tight enclosure mounted on a Perkin–Elmer thermal cycler. The heated lid assembly has been moved aside to accommodate the enclosure.

The video camera is mounted at the back of the enclosure and pointed at the dichroic mirror. An interference filter with a peak bandpass of 600 nm is fitted on a 12.5- to 75-mm zoom lens. The camera output is sent to a frame-grabber board in a personal computer (PC) in which the images of the samples are evaluated for sample fluorescence as described later. The video camera, frame grabber, PC, and controlling software are a modification of the Alpha Innotech Digital Imaging System, IS-1000 (San Leandro, CA). While the thermal cycler and controlling PC operate independently, their activities are coordinated by using the thermal cycler's output to a printer—intended to log the completion of cycles—to signal the appropriate time to image the samples.

Amplification reactions for kinetic analysis are set up with several slight modifications that together compensate as much as possible for the absence of the heated lid. Each reaction is overlaid with 50 μl of light mineral oil to prevent evaporation. Small air bubbles trapped under the oil overlay will expand and contract during thermal cycling and can cause signal variations. The set of reactions is therefore centrifuged in a swinging bucket rotor, at about 1500 rpm, for 1 min to ensure that all air bubbles are driven out. After the array of samples is set into the sample block of the thermal cycler, a black plastic mask with holes corresponding to the 96-well format is placed over the samples to block a strong signal from any one sample interfering with adjacent samples. A UV transparent plastic cover is placed over the samples that, together with the mask, provides an insulating, convective heat barrier.

Generation of PCR Growth Curves

Setup for Digital Imaging

Once samples are in place and the enclosure is closed, the shutter to the UV light can be opened and a sample image taken and displayed on the PC monitor at varying lengths of exposure. The basal fluorescence from EtBr in the samples can be seen in the images. Using the imaging software, the individual samples are indicated by overlaying an array of 96 circular "areas of interest" (AOIs) over the image (Fig. 2). The array of AOIs may be moved, and if necessary distorted, to more exactly match the sample positions in the captured image. The circle size can be varied but is kept constant within any one array. Analog-to-digital conversion of CCD pixel voltages is accomplished in the frame-grabber board described previously. The eight bits of digital resolution result in 256 levels of light measurement at each pixel. During a run the software sums all the pixel measurements within each AOI and records the values.

The data stream from a simple kinetic PCR experiment is a series of such sample-specific fluorescence measurements over the course of thermal cycling. Although such measurements can be taken at all stages of a PCR cycle (Wittwer *et al.*, 1997a), most frequently the measurements focus on the extend phase (or the anneal/extend phase if monitoring a "two-temperature" PCR). It is during these phases that double-stranded DNA is made and indicator dyes can stably bind. Typically, for each PCR there is a single fluorescence measurement toward the end of this phase for each cycle of PCR. With 96 PCRs the total number of fluorescence measurements is 96 times the number of PCR cycles (typically 30–60). A straightforward way of handling such a data matrix is with a spreadsheet program for PCs, such as Microsoft's Excel, which can tabulate and graph for each PCR a fluorescence growth curve (Fig. 3).

Normalization of Fluorescence Measurements

In order to generate growth curves for which meaningful comparisons can be made, two levels of normalization are needed: (i) the normalization of differences in fluorescence measurement which occur from cycle to cycle and (ii) the normalization of measurement differences which occur from sample to sample. Cycle-to-cycle variation is caused by drift and flicker of the UV lamp and by photobleaching of the fluorescent dye. Sample-to-sample variation has many sources. In particular, there is nonuniformity of illumination which arises from the dichroic mirror and it's incident angle-dependent transmission of the UV light. This results in stronger illumination in the center of the sample block. In addition, there are differences in the transmissiveness of caps and oil overlays over each sample.

To correct for cycle-to-cycle variation, an invariant fluorescence standard or standards are used in the image field. Deviations in the measured fluorescence of these standards are corrected (to equal the average of the measurements for all cycles) by using cycle-specific

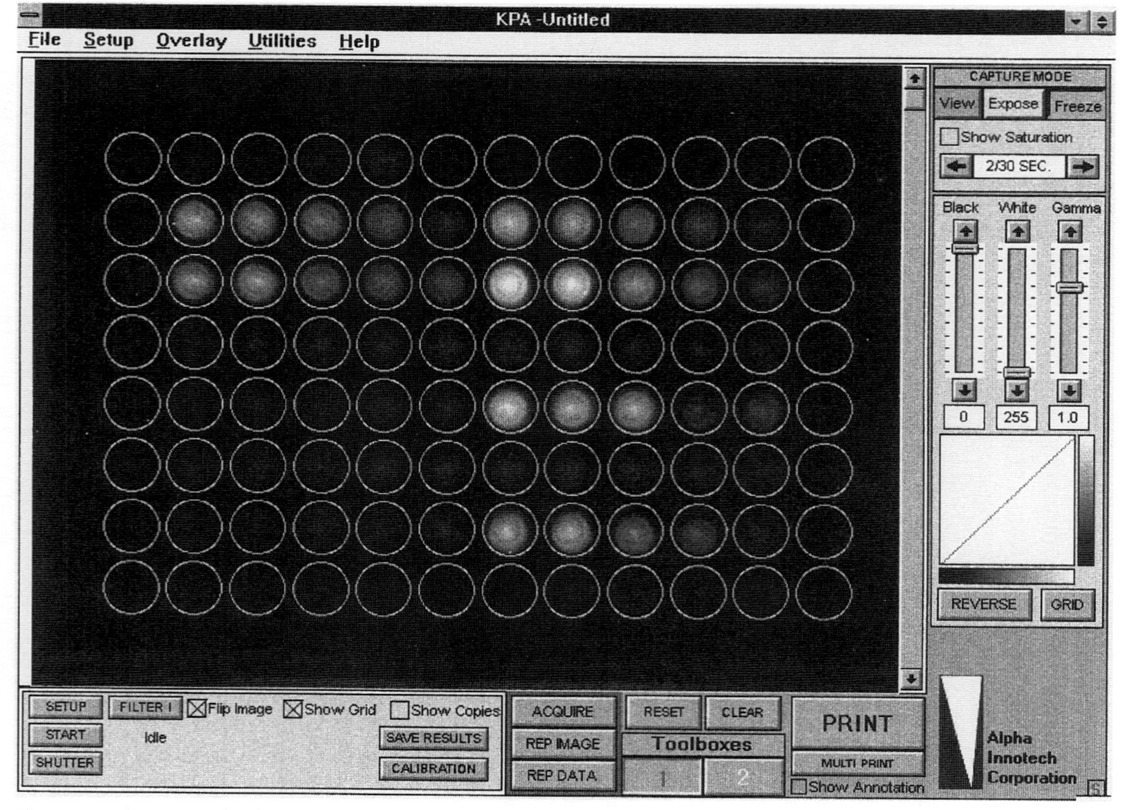

Figure 2 The user interface for the collection and analysis of video images of an array of samples. A grid of 96 Areas of Interest is shown on the screen as are many user-selectable menus and control buttons.

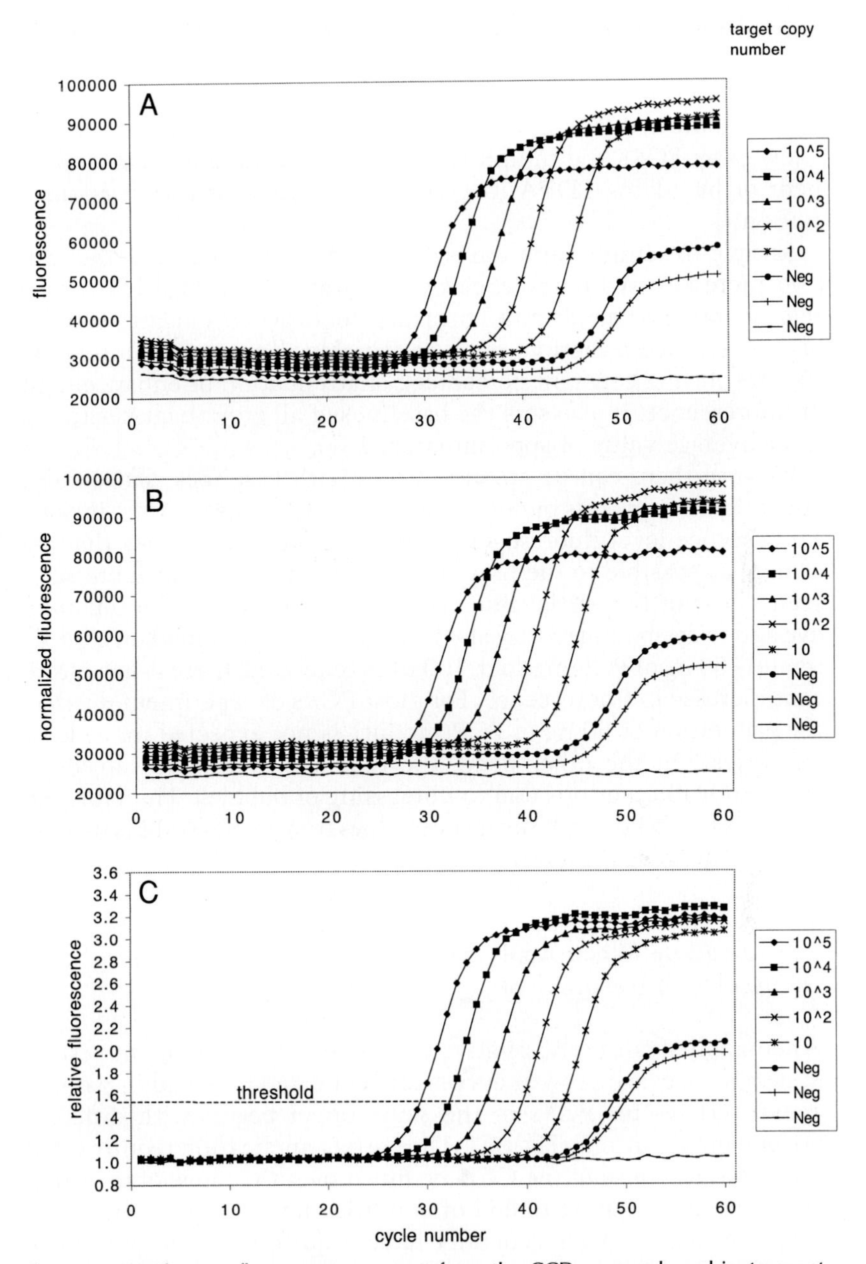

Figure 3 (A) The raw fluorescence output from the CCD camera-based instrument showing growth curves from eight RT-PCRs from the indicated starting copy numbers of HIV RNA template. (B) The data from A, cycle-to-cycle normalized as described in the text. (C) The data from B, sample-to-sample normalized as described in the text. The threshold level of fluorescence from which C_t's (see text) are derived is indicated.

normalization factors. These factors are then applied to each cycle's measurements of all samples (Figs. 3A and 3b). Such standards work best if their fluorescence is about the same as the initial fluorescence of the monitored PCRs. A simple way to accomplish this is ti make a few extra PCRs and inhibit them by leaving out a critical component or by adding EDTA in excess of the divalent cation concentration.

Sample-to-sample variation is corrected for by calculating the relative increases in fluorescence over initial values (Fig. 3C). Using relative rather than absolute measurements accommodates sources of fluorescence variation, such as variable illumination intensity, that result in a constant percentage of attenuation or enhancement of fluorescence. It also sets the baselines of all growth curves to the same average value of approximately 1.

Because there can be growth curves with baselines that exhibit slight "drift" upwards or downwards, it is best to use as the "initial" fluorescence level fluorescence measurements from cycles that are as near as possible to the first cycle at which significant increase is seen. Most of the initial fluorescence level is due to the unbound dye and to dye bound to the ssDNA primers. Until more than approximately 50 ng of PCR product/100 μl is generated, there is no significant increase in fluorescence. For most PCRs of experimental value, the generation of 50 ng of PCR product is not expected for at least 12 cycles. On the other hand, the first few cycles are subject to significant fluctuations due to outgassing of bubbles. Therefore, an average of each PCR's fluorescence values from cycles 6–12 is usually used as the initial fluorescence level.

Considerations When Applying Relative Fluorescence Measurement

When applied correctly, relative fluorescence measurement results in nearly identical growth curves for replicate PCRs and in parallel growth curves for PCRs of the same target begun with different target concentrations (Fig. 3C). The correct application requires that the signal response of the CCD be linear over the range of measurements from beginning to end of amplification. If kept within their dynamic range, CCDs generally have a quite linear response. To keep within this dynamic range, the camera gain and exposure time used in a run must not be set so low at the beginning of the run that some pixels do not register nor so high that at the end of the run some pixels are saturated. The image analysis software allows the false coloration of pixels registering either the lowest (green) or highest (red) levels. Using these false colors, a general calibration of gain and exposure can be done against fluorescence standards of known dye or dye/DNA concentrations. Once camera gain is set, it is usually not changed. When setting up a given experiment, the exposure is set at twice that at which green pixels are seen in any of the wells. This allows enough "headroom" to accomodate fluorescence increases seen using EtBr without risk of saturating pixels.

The correct application of relative fluorescence must also take into account nonsample contributions to baseline fluorescence. These include dark current (the accumulation of charge in a CCD in the absence of light), stray light entering the system, and the occasional reaction tube that has a marked endogenous fluorescence (see Protocol for method of prescreening reaction tubes). All these can contribute to fluorescence measurement to different extents for different samples. If these contributions are significant compared to the actual fluorescence from the sample, significant error in relative fluorescence measurement can result. In the systems described previously, dark current can be easily measured before each run with the excitation lamp off and shutter closed using the same exposure that will be used for the experiment. These dark current values for each well can then be subtracted from subsequent, in-run values. Stray light should be minimized as much as possible by using nonreflective surfaces and baffling. If still significant, it can be measured for each run and subtracted, just as can dark current, by putting a nonreflective, black field over the thermal cycler block and taking an image—again at the same exposure used during the run. Note that this measurement will also include dark current contribution.

Incorrect results may also arise when different PCRs in the same run have markedly different compositions. For example, PCRs begun with more than approximately 200 ng of genomic DNA have a higher initial fluorescence and a reduced relative fluorescence gain compared to the same amplification of lesser amounts of genomic DNA. This is despite the denaturation of DNA after the initial thermal cycle and is probably due to a proportion of quickly reassociating "snapback" DNA sequences. There can also be differences between PCRs in the ionic strength, PH, and cosolvents that result in differential quenching of fluorescence. These differences can result from design or from carryover from sample preparation. There is also the possibility of carryover from sample preparation of fluorescent or fluorescence quenching substances. Sample preparation procedures which ensure high purity of nucleic acids are therefore recommended.

Quantitative Analysis

Once normalized growth curves have been generated, a single threshold level of fluorescence can be set for quantitative comparisons among the PCRs (Fig. 3C). The threshold is set high enough so that it is not inadverdently crossed by fluctuation or drift in the baseline but not so high that the linear phase of any growth curve is exceeded. The number of cycles needed for each growth curve to surpass this threshold can be calculated by first fitting a regression line to the fluorescence measurements contiguous to the threshold. We most frequently use the two contiguous measurements below and the three above for this line. From the equation of this line, a fractional cycle value (C_t) is calculated that intercepts the threshold.

C_t values from PCRs containing target template of known concentration are used to establish a standard quantitation curve. Such a curve is shown in Fig. 4, based upon the C_t's derived from Fig. 3C, which demonstrates a log-linear relationship between target copy number and C_t. The occurrence of template-independent PCR product (primer dimer) is delayed enough so that the log-linear relationship persists down to 10 target copies. The equation for such a standard curve can be obtained simply by regression and the target copy number of "unknowns" determined by entering into the equation the unknowns' C_t. Under optimal conditions, replicate PCRs at more than a few hundred copies per PCR show a reproducibility in C_t of about ± 0.2 cycle, which translates to a copy number reproducibility of about $\pm 15\%$. Replicate PCRs with less than about 100 copies begin to show additional variation in C_t due to stochastic fluctuation.

The slope of the log-linear range of the standard curve also reflects the initial DNA replication efficiency of the PCR. Given two otherwise identical PCRs starting at different target concentrations, the slope derives from the number of cycles it takes for the PCR with less target to catch up, in product copy number, to the other PCR. For example, if there is 100% replication efficiency (a doubling in copy number per cycle), it takes exactly one cycle for a PCR begun with a twofold dilution of starting target to catch up. Lower efficien-

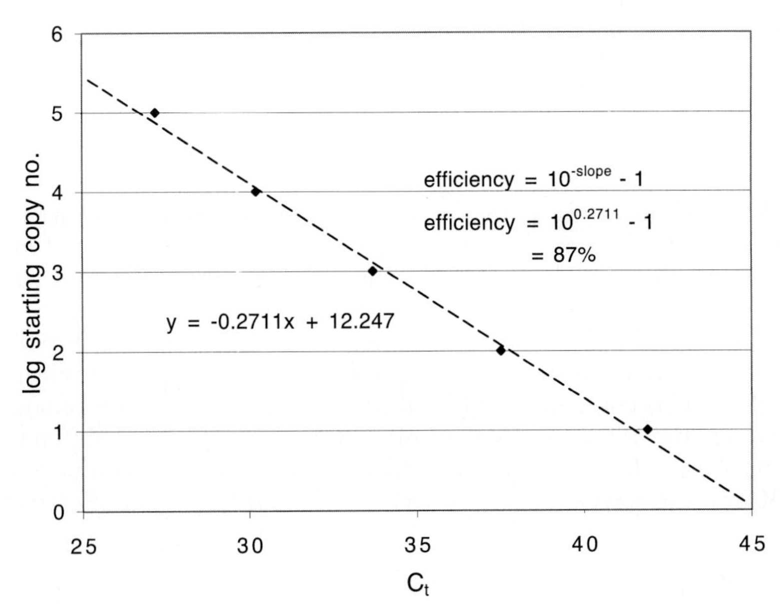

Figure 4 A standard quantitative curve regressed from the C_t's of the growth curves shown in Fig. 3C. A calculation of the initial amplification efficiency from the slope of the standard curve is shown (see text for explanation).

cies result in larger cycle offsets. This offset, once established, persists over the whole growth curve even as the per-cycle replication efficiency diminishes as the PCRs progress through different phases, from exponential to linear (where the threshold is) to plateau. Note that there is a presumption that the threshold level represents the same amount of product for each growth curve. This will be true if relative fluorescence measurement is applied correctly as described previously.

This efficiency can be calculated from the slope of the standard curve (Fig. 4). In the first cycle of PCR, the resultant copy number of product, c_1, is related to the initial target copy number, c_0 by

$$c_1 = c_0(1 + \text{efficiency})$$

By converting to logarithms and rearranging, it can be seen that

$$\log c_1 = \log c_0 + \log(1 + \text{efficiency})$$

$$\log c_1 - \log c_0 = \log(1 + \text{efficiency})$$

Since $\log c_1 - \log c_0$ is equivalent to the negative slope of the standard curve, this can be restated as

$$-\text{slope} = \log(1 + \text{efficiency}), \text{ or}$$

$$10^{-\text{slope}} - 1 = \text{efficiency}$$

When quantitating a large number of different targets simultaneously, it is inconvenient to run standard curves for each target. Kang *et al.* (1998) have shown that a single quantitative standard can suffice for the reverse transcriptase (RT)-PCR-based quantitation of large numbers of different RNA transcripts. Compared to an outside reference, the measured concentrations were comparable to within twofold. This could only be true if there were no large discrepencies between the RT and DNA amplification efficiencies of different transcripts. This is accomplished by keeping the amplicon size to <300 bp and screening primer sets for efficient DNA amplification, using the above algorithm.

Enhancing Specificity and Sensitivity

As discussed previously, the primary limit to the sensitivity of dye-alone kinetic PCR is the generation of a template-independent, dsDNA artifact commonly referred to as primer dimer (Chou *et al.*, 1992). This artifact is the result of one primer serving as the template for chain extension of the other, even though there is no site to which the other primer can anneal. This is a relatively rare event, but once such an extended primer is generated, it contains both primer-annealing sequences and thus can serve as a legitimate template for PCR amplification. The generation of such an extended primer is more likely at lower temperatures, perhaps as a result of metastable base pairing of the 3′ end of one primer to the other primer.

A significant reduction in primer dimer was seen if an essential reagent, such as the polymerase, was left out when setting up the PCRs and was only added when the temperature was as high as possible, just before the first denaturation step of thermocycling (Mullis, 1991). Such a "hot start" has been made most convenient through the use of a chemically modified DNA polymerase (AmpliTaq Gold, Perkin–Elmer; Birch, 1996) that requires an incubation at >90°C to be activated. This allows the enzyme to be present during PCR setup. For DNA amplifications, the use of the thermally activated polymerase commonly delays the C_t for primer dimer to >45 cycles—more cycles than an amplification from a single DNA template copy would require.

An additional hot start occurs as an ancillary effect of the use of the enzyme uracil-*n*-glycosylase (UNG) and the incorporation of dUTP in place of dTTP that is intended to help control carryover contamination by PCR product (Longo *et al.*, 1990; Persing and Cimino, 1993). Illegitimate primer extensions that incorporate dU are cleaved by UNG until the temperature is raised to >50°C, which inactivates the UNG. For DNA amplifications, we have found that thermally activated polymerase and UNG hot starts, when combined, can delay primer dimer more than either hot start alone. The UNG hot start is of significant benefit for single-tube, single-enzyme (e.g., *Tth* polymerase; Myers and Sigua, 1995) RNA amplifications, for which a thermally activated polymerase is unavailable.

For DNA and RT-PCR we have found that both primer sequence and primer quality can affect the occurrence of primer dimer. R. Saiki (personal communication) has shown that primer dimer is usually reduced when both primers have 3′ ends that terminate in a series of at least two dA residues. For this reason, it is worthwhile to choose amplification primers which have such termini and at the same time yield as small an amplicon as possible to maintain as high an RT and DNA amplification efficiency as possible. Additionally, different syntheses of the same primer sequences can yield markedly different levels of primer dimer in PCR. For this reason, the resynthesis of primers that give high levels of primer dimer can sometimes help. The growth curves shown in Fig. 3, in which the C_t for primer dimer is >45 cycles, were generated by using these suggestions for primer design in single-tube RT-PCRs using *Tth* polymerase and dU incorporation with UNG.

Verification of PCR Product Identity without Probes

Using dye detection, amplification of the desired target sequence can be validated in two ways. First, under the right conditions, the level of fluorescence reached at the plateau phase of a PCR growth curve is proportional to the amplicon size. Thus, amplification of anything other than amplicon of the expected size for the specific target results in an unexpectedly high or low plateau fluorescence. Second, it is possible to perform DNA melting analysis in a thermocycling instrument based on the ability of these dyes to distinguish between ss and dsDNA (Ririe *et al.*, 1997). This allows one to distinguish specific from nonspecific amplification products based on the melting temperature of the duplexes.

The relationship between fluorescence and amplicon size was first noted in "primer-limited" PCRs. The amount of PCR product that can be made is limited by the starting primer concentration when it is less than about 0.4 μM each (Higuchi *et al.*, 1993). One result of such primer limitation is a growth curve with a very flat plateau phase like those in the growth curves of Fig. 3. The lack of additional growth probably results when the concentration of product exceeds that of remaining primer by enough such that product reassociation occurs before primer annealing and extension can occur. Since the concentration of product at which this occurs is mostly independent of the size of the amplicon, amplicons of different molecular weight "plateau" at the same concentration. This means that the mass of dsDNA made at plateau is proportional only to the amplicon size. With EtBr at 4 μg/ml, the fluorescence of dsDNA is proportional to amplicon mass up to at least 4 μg dsDNA/100 μl PCR (Higuchi *et al.*, 1993). Taken together, these linear relationships explain why the plateau fluorescence should be a direct measure of amplicon size.

This can be seen in the growth curves shown in Fig. 3C in which the product present in the no template reactions—low-molecular-weight primer dimer—plateaus at a much lower relative fluorescence that does the specific target product that is more than twice as large. Mixtures of primer dimer and target product show intermediate fluorescence levels. Detecting primer dimer in this way has so far been the most useful role for this mode of analysis.

More generally, Fig. 5 shows the directly proportional relationship between amplicon size and relative fluorescence at plateau for six different amplicons ranging in size from about 100 to 550 bp. When relative fluorescence measurement is correctly performed as described previously, it is possible to distinguish amplicons that differ by more than about 20% in size.

Ririe *et al.* (1997) have shown in their kinetic PCR instrument that by slowly denaturing or annealing PCR product after kinetic PCR with SYBR green detection, they can follow the melting or

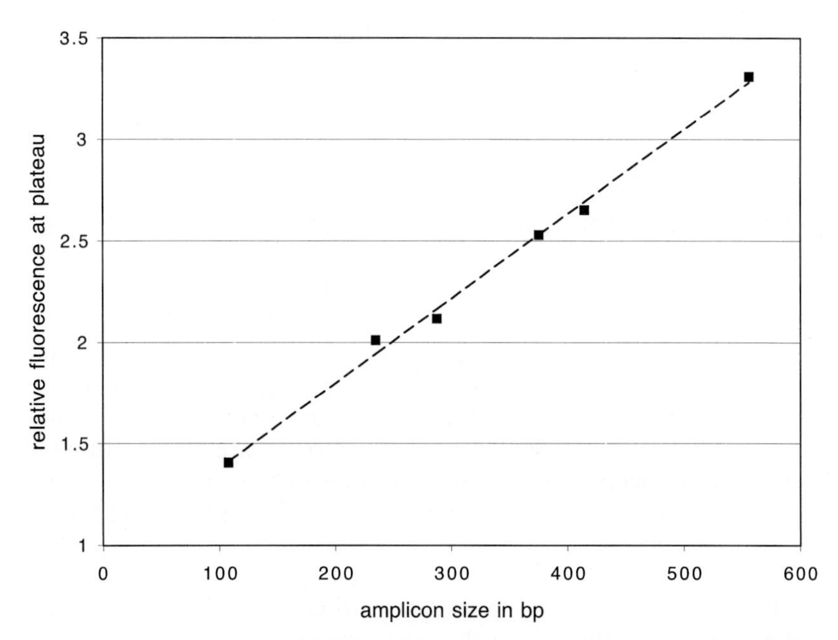

Figure 5 The relationship between amplicon size and relative fluorescence yield is demonstrated for six different amplicons. In separate PCRs, three forward primers increasingly distant from one common back primer were used on two M13 ssDNA templates (with and without a 115-base cloned insert) to generate the six different-sized products. The common back primer was always at limiting concentration.

reannealing of the duplex DNA. The temperature at which this occurs is a function of the sequence of the duplex. Primer dimer, because of its short length, generally has a melting temperature (T_m) well below that of the correct target sequence. It is usually possible to resolve small amounts of target amplicon in the presence of large amounts of primer dimer. The closer the T_m's, the more difficult is the resolution; it is possible to resolve equal amounts of amplicon that differ by about 2°C in T_m. Such melting temperature analysis has been performed using EtBr on the CCD-based instrument.

Detecting Inhibition

Inhibition of the PCR can cause falsely negative results or underestimation of the target sequence concentration. However, the shape of kinetic PCR growth curves can indicate whether or not this is occurring. Partial inhibition usually results in a growth curve that is shallower than expected (Higuchi *et al.*, 1993), resulting in a delayed C_t. Total inhibition results in a flat growth curve profile, which is an obvious indicator of inhibition when at least primer dimer is expected; if no primer dimer is expected, this does not distinguish a totally inhibited PCR from a truly negative one. Another "gray area" is a partially inhibited PCR displaying a growth curve that is delayed to where primer dimer would be expected. Usually, however, the exact shape of the inhibited growth curve is still distinctive.

Since these methods of detecting inhibition are subjective and qualitative, it is preferable to develop a method that is analogous to the use of an internal quantitative standard or "mimic" (Ferré, 1992) that would allow the calculation of the correct target copy number even with the occurrence of partial inhibition. An approach to this method is shown in Fig. 6. An internal standard template in a known, high amount is added to the PCR. Primers to this template are used at a low, limiting concentration. If target sequence is present, the amplification results in a biphasic growth curve as shown, with the first phase being the primer-limited amplification of the control template and the second being target amplification. The presence of an inhibitor would delay or eliminate the first phase. The target

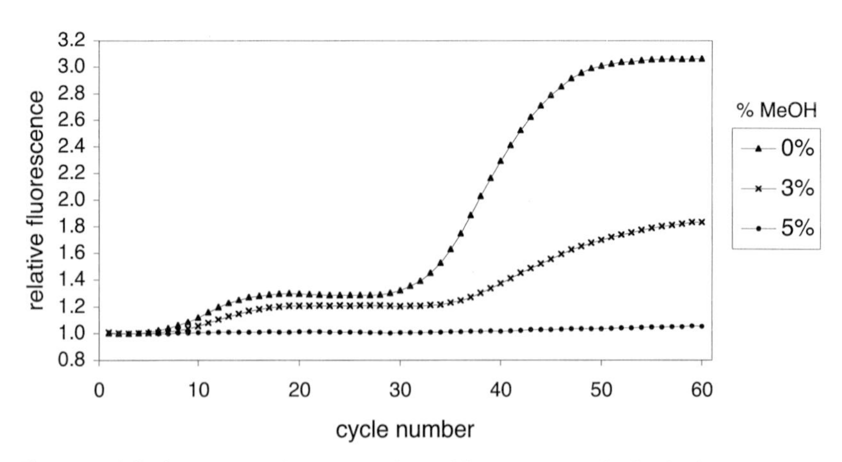

Figure 6 A high-copy number, internal amplification control. The biphasic growth curve results from the inclusion, at limiting concentration, of a high-copy number template (M13 ssDNA) and two additional (M13) primers into a PCR directed to an unrelated (human β-globin), low-copy template. Adding methanol (MeOH) inhibits both phases of the growth curve.

phase would also be delayed in a way that could be related to the delay of the control phase. Knowing this relationship, a correct quantitation could still be performed. Figure 6 shows the effect of 3 and 5% methanol. A total inhibition with 5% methanol eliminates both phases of the growth curve; a partial inhibition with 3% delays the two phases.

It has been problematic to obtain an internal control template of this type whose response to inhibitor can mimic that of the target well enough for accurate quantification. It may be that the best use of this type of control will be to design it such that it is always more sensitive to inhibition than is the target sequence. This would provide a "red flag" indicating that the inhibitors are present and that repurification of the sample is necessary to obtain valid quantification.

Protocol

This protocol is for a RT-PCR using r*Tth* enzyme (Myers *et al.,* 1995), UNG with dUTP incorporation, and a 50-μl sample addition.

Reagents

5X buffer	0.25 *M* Tricine–HCl, pH 8.3, 0.6 *M* potassium acetate, 25% glycerol
dNTP mix	20 m*M* each dG-, dA-, and dCTP; 2 m*M* dTTP; 20 m*M* dUTP
UNG	1 U/μl (Perkin–Elmer)
EtBr	100 μg/ml in H_2O
Tth DNA polymerase	2.5 U/μl (Perkin–Elmer)

1. Prescreen all the MicroAmp tubes (Perkin–Elmer) and caps, if used, by spreading the tubes out on a UV transilluminator covered with plastic wrap. Remove any tubes that show strong fluorescence.
2. Program the thermal cycler as follows and save for further use as a series of linked files

50°C, 2 min	UNG incubation step
60°C, 30 min	Reverse transcription step

95°C, 30 sec	Denaturation step
55°C, 30 sec	Anneal step
72°C, 30 sec	Extend step (set controller to take image 25 sec into this extend step)

Repeat last three steps for desired number of PCR cycles
4°C hold or 72°C hold

3. Assemble the TC9600 tray and rack. Place the required number of screened PCR tubes in the rack, including tubes for positive and negative controls and for fluorescence standards. The fluorescence standard is made by adding 50 μl of a 40 m*M* EDTA solution to 50 μl of the master mix (step 4).
4. Prepare the master mix as follows: listed is the volume of each component to add to the master mix per single 100-μl reaction:

Distilled H_2O	14.5 μl (DEPC treated to inactivate RNase)
5X buffer	20 μl
dNTP mix	2.5 μl
UNG	2 μl
25 μ*M*, forward primer	1 μl
25 μ*M* reverse primer	1 μl
EtBr	1 μl
Tth DNA polymerase	4 μl
0.1 *M* $Mn(OAc)_2$	4 μl

   The master mix can be prepared in advance by adding all the components except $Mn(OAc)_2$.
5. Thoroughly mix the master mix immediately after adding $Mn(OAc)_2$ and pipette 50 μl of the master mix into each reaction tube. Add 50 μl of sample to each PCR tube. Avoid introducing air bubbles.
6. There are two options for preventing refluxing of the samples. First, they can be covered with one drop of mineral oil, then the rack of samples can be spun in a centrifuge in a swinging bucket rotor (1500 rpm for 1 min) to remove tiny air bubbles that otherwise congeal into large air bubbles and interfere with detection as the samples heat. Second, one can add one AmpliWax Gem 100 (Perkin–Elmer) to each tube, being careful to add only one bead. Tube caps may be used, but we find it convenient to not use them at all, being careful to avoid jostling of the rack. Once melted and hardened, the AmpliWax barrier seals the reaction tubes.
7. Transfer the rack to the kinetic thermal cycler and ensure that the tubes are seated in the block. Set the exposure (see section

IVc, above) and position the grid over the sample image and start the analyzing software. Make sure that the kinetic PCR software is "looking" for the right set point No. in the run setup dialog box. For a three-temperature thermal cycle, the software should be set for "Setpt#3." Start the programmed series of thermal cycles on the thermal cycler.

Acknowledgments

We acknowledge the contributions of Carita Elfstrom, Mary Fisher, and Stanley Wang, who performed much of the experimental work, and of Rob Watson and Tom Vess, who helped with data analysis. We also thank Daryl Ray and Haseeb Chaudry for their help with the imaging system. We thank John Sninsky for his long-standing support of this technology.

References

Birch, D. E. (1996). Simplified hot start PCR. *Nature* **381,** 445–446.

Chou, Q., Russell, M., Birch, D. E., Raymond, J., and Bloch, W. (1992). Prevention of PCR mis-priming and primer dimerization improves low copy-number amplifications. *Nucleic Acids Res.* **20,** 1717–1723.

Ferré, F. (1992). Quantitative or semi-quantitative PCR: Reality versus myth. *PCR Methods Appl.* **2,** 1–9.

Heid, C. A., Stevens, J., Livak, K. J., and Williams, P. M. (1996). Real time quantitative PCR. *Genome Res.* **6,** 986–994.

Higuchi, R., Dollinger, G., Walsh, P. S., and Griffith, R. (1992). Simultaneous amplification and detection of specific DNA sequences. *Bio/Technology* **10,** 413–417.

Higuchi, R., Fockler, C., Dollinger, G., and Watson, R. (1993). Kinetic PCR analysis: Real-time monitoring of DNA amplification reactions. *Bio/Technology* **11,** 1026–1030.

Holland, P. M., Abramson, R. D., Watson, R., and Gelfand, D. H. (1991). Detection of specific polymerase chain reaction product by utilizing the 5′ to 3′ exonuclease activity of *Thermus aquaticus* DNA polymerase. *Proc. Natl. Acad. Sci. U.S.A.* **88,** 7276–7280.

Kang, J. J., Watson, R. M., Fisher, M. E., Higuchi, R., Gelfand, D. H., and Holland, M. J. (1998). Transcript quantitation in total yeast cellular RNA using kinetic PCR. Submitted for publication.

Lee, L. G., Connell, C. R., and Bloch, W. (1993). Allelic discrimination by nick-translation PCR with fluorogenic probes. *Nucleic Acids Res.* **21,** 3761–3766.

Longo, M. C., Berninger, M. S., and Hartley, J. L. (1990). Use of uracil DNA glycosylase to control carry-over contamination in polymerase chain reactions. *Gene* **93,** 125–128.

Mullis, K. B. (1991). The polymerase chain reaction in an anemic mode: How to avoid cold oligodeoxyribonuclear fusion. *PCR Methods Appl.* **1,** 1–4.

Myers, T. W., and Sigua, C. L. (1995). Amplification of RNA: High-temperature reverse transcription and DNA amplification with Thermus thermophilus DNA polymerase. *In PCR Strategies* (J. Sninsky *et al.*, Eds.), pp. 58–68. Academic Press, San Diego.

Persing, D. H., and Cimino, G. D. (1993). Amplification product inactivation methods. In *Diagnostic Molecular Microbiology: Principles and Applications* (D. H. Persing *et al.*, Eds.), pp. 105–121. American Society for Microbiology, Washington, DC.

Ririe, K. M., Rasmussen, R. P., and Wittwer, C. T. (1997). Product differentiation by analysis of DNA melting curves during the polymerase chain reaction. *Anal. Biochem.* **245,** 154–160.

Tyagi, S., Bratu, D. P., and Kramer, F. R. (1998). Multicolor molecular beacons for allele discrimination. *Nat. Biotechnol.* **16,** 49–53.

Wittwer, C. T., Herrmann, M. G., Moss, A. A., and Rasmussen, R. P. (1997a). Continuous fluorescence monitoring of rapid cycle DNA amplification. *Biotechniques* **22,** 130–131.

Wittwer, C. T., Ririe, K. M., Andrew, R. V., David, D. A., Gundry, R. A., and Balis, U. J. (1997b). The LightCycler: A microvolume multisample fluorimeter with rapid temperature control. *BioTechniques* **22,** 176–181.

17

QUANTIFICATION OF TELOMERASE ACTIVITY USING KINETIC TELOMERIC REPEAT AMPLIFICATION PROTOCOL

Sheng-Yung P. Chang

Telomeres at the ends of eukaryotic chromosomes consist of tandem repeat sequences, and telomeres are essential for maintaining the stability of the chromosomes (Blackburn, 1991). Telomerase is a ribonucleoprotein complex which maintains the telomeric length by extending telomeric DNA using a small region of its RNA component as a template. In most normal tissues, telomeres become shorter with each cell division due to the "end-replication" problem. When telomeres reach a certain critical length, cells are withdrawn from the cell cycle and become senescent. A rare cell may escape the control of senescence, continue to proliferate, and become immortal. Using a PCR-based telomeric repeat amplification protocol (TRAP), Kim *et al.* (1994) determined the telomerase activity in many human immortal cell lines, normal germline and somatic tissues, and biopsy samples from various types of tumors. They concluded that telomerase was repressed in most normal somatic cells but reactivated in immortal tumor cells. However, normal germlike tissues are telomerase positive. Subsequently, many studies confirmed that telomerase activity was detected in ~85% of various types of cancer (Shay and Wright, 1996; Bacchetti and Counter, 1995). Therefore, the presence of telomerase activity in tumors may be a useful marker for diagnosis and prognosis of cancer and follow-up of recurrence of the disease. However, in addition to germline tissues and some hematopoietic sperm cells (Counter *et al.*, 1995), a low level of telomerase activity was also detected in some normal self-renewing tissues (Ramirez *et al.*, 1997; Taylor *et al.*, 1996) and some benign tumors (Hiyama *et al.*, 1996; Brien *et al.*, 1997). Consequently, quantification of the level of telomerase activity becomes essential to study the clinical correlation between telomerase activity and tumor progression. In the past 3 years, several modified quantitative TRAP methods were described and used for small clinical studies (Wright *et al.*, 1995; Kim and Wu, 1997; Aldous and Grabill, 1997).

Kinetic PCR monitors the increase of fluorescence of ethidium bromide bound to double-stranded PCR product during amplification (Higuchi *et al.*, 1992). A digital camera captures an image of all the PCR reactions at the annealing step of each cycle. At the end of amplification, images of the entire amplification are analyzed by the computer and quantitative results are generated. This chapter describes the conversion of gel-based TRAP to the kinetic format of TRAP. This simple, fast, nonradioactive, and quantitative format requires very little hands-on time and is suitable for analyzing a large number of samples.

Protocols

Preparation of Cell Extract

The method described in Piatyszek *et al.* (1995) was used to prepare cell extract from cells grown in culture and clinical specimens. The protein concentration of cell extract from each sample was determined using Coomassie Protein Assay (Pierce) and then diluted to an appropriate concentration with CHAPS lysis buffer and stored at −70°C in aliquots.

Kinetic Thermal Cycler

The instrument kinetic thermal cycler (KTC) monitoring the fluorescence increase during amplification was first described by Higuchi *et al.* (1993). Subsequent modifications of KTC are discussed in Chapter 27.

Primer Design and Quantitation Standard

In order to delay the formation of primer dimer, sequences of telomerase-specific (TS) primer (Kim *et al.*, 1994) and anchored return (ACX) primer (Kim and Wu, 1997) were modified. The two nucleotides, dT, at the 3′ terminus in TS primer were changed to dA, and the modified TS prime, 5′-AATCCGTCGAGCAGAGAA-3′, was designated as SYC555. The two nucleotides, dC, at the 3′ terminus in ACX were eliminated, and the truncated ACX primer, SYC′556 (5′-GCGCGGCTTACCCTTACCCTTACCCTAA-3′), was designated as SYC556. The sensitivity of TRAP assay was not affected using modified primers SYC555 and SYC556.

A synthetic 81-mer oligonucleotide, QS (5′-GCGCGGCTTACCCTTACCCTTACCCTAAATTGGACTCAGTCCCTCGGTCATCTCACCTTCTTTCTCTGCTCGACGGATTCC*p*-3′), containing both SYC555 and SYC556 sequences was used as DNA template in reactions to generate a quantitation standard curve. The copy number of QS was calculated based on the absorbance at 260 n*M* of the oligonucleotide. All aliquots containing $<10^6$ copies/μl of QS were diluted using TE buffer containing 20 ng/μl poly rA (Pharmacia) as the carrier.

Kinetic TRAP Reactions

All reagents were prepared using DEPC-treated H_2O. MicroAmp reaction tubes (Perkin–Elmer) used for kinetic TRAP were selected by examining on an UV transilluminator; a very small number of reaction tubes with high fluorescence background were eliminated.

A typical 100-μl reaction contained the following components:

1X modified TRAP buffer (20 m*M* Tris–HCl, pH 8.3, 2 m*M* $MgCl_2$, 63 m*M* KCl, 0.005% Tween 20, 1 m*M* EGTA, 0.1 mg/ml BSA)

100 μM dATP

100 μM dGTP

100 μM dCTP

200 μM dUTP

1 μg/ml ethidium bromide (Calbiochem) [the concentration (50 μg/ml) of ethidium bromide stock solution was determined by measuring the absorbance (0.73) at 480 n*M*]

0.2 μM SYC555

0.2 μM SYC556
5 units *Taq* DNA polymerase
5-μl cell extract prepared using CHAPS lysis buffer

For positive control reactions, cell extract from an immortalized cell line (e.g., kidney 293 or HeLa cells) were used. For negative control reactions, CHAPS lysis buffer was added to the reactions mix. Duplicate reactions using 10-fold serially diluted oligo QS (5×10^1 copies to 5×10^4 copies per reaction) were always included in each experiment. Four to six reactions containing 1X modified TRAP buffer and 1 μg/ml ethidium bromide were included in each experiment as "external standards" to control cycle-to-cycle variations in fluorescence level.

One drop of mineral oil (Perkin–Elmer) was added to each reaction, and reaction tubes were then loosely covered with MicroAmp 8- or 12-cap strips. The MicroAmp base with the assembled reaction tubes was centrifuged briefly in a tabletop centrifuge to eliminate air bubbles generated by the addition of cell extract. The cap strips were removed before amplification was initiated.

The kinetic TRAP was started at 25°C for 10 min for the telomerase reaction and followed by amplification of 50–60 cycles of 95°C for 20 sec and 60°C for 20 sec. A digital image of all reactions was taken at the annealing temperature of each cycle. Raw data collected at each cycle was imported into a Microsoft Excel spreadsheet. The cycle-to-cycle variation of fluorescence level due to temperature fluctuation was first normalized with the average of external standards. The relative fluorescence increase throughout the entire amplification of each reaction was then calculated. Furthermore, an arbitrary fluorescence level (AFL), at least 10 standard deviations above the baseline fluorescence level, was used to determine the cycle threshold value (C_t) of each sample. Therefore, C_t is the cycle number at which the fluoresence level reaches the AFL value. A quantitation standard curve was first generated by plotting C_t values versus the log of copy numbers of a serial dilution of oligo QS. The telomerase activity in the sample was then calculated using the quantitation standard curve.

Results and Discussion

Figure 1 demonstrates the relative fluorescence increases of reactions using serially diluted cell extract from 293 cells. The sensitivity of

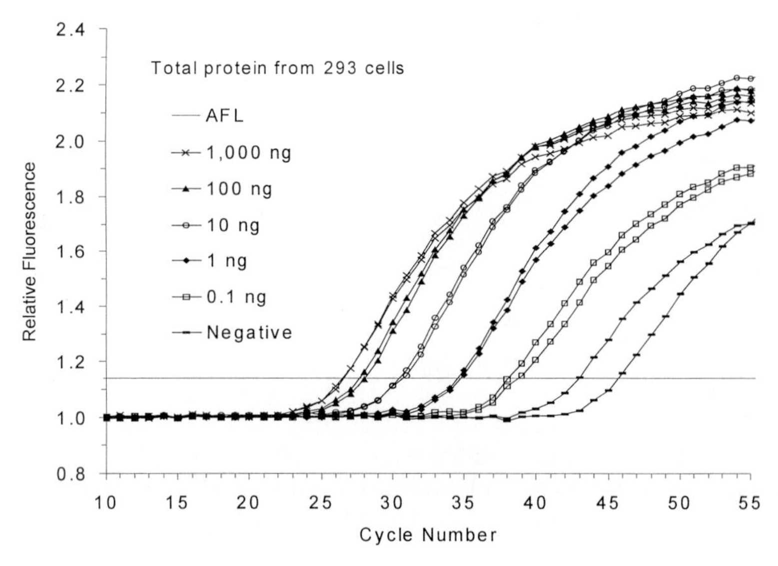

Figure 1 Relative fluorescence increase of reactions using serially diluted cell extract from kidney cell line 293. The preparation of cell extract, the determination of total protein concentration, and amplification conditions were described under Protocols.

the kinetic TRAP assay is ~0.1 ng total protein, which represents cell extract from approximately a single 293 cell. Primer dimer formation was observed after 40 cycles using the improved primer pair. Figure 2 illustrates that the dynamic range of the assay is ~0.1 to 10^3 ng of total protein in cell extract from 293 cells. The upper limit of the dynamic range was due to the presence of inhibitors in the concentrated cell extract. It was also observed that the level of telomerase activity varied slightly in different preparations of cell extract from the same cell line. The difference was likely due to the growth conditions of cells, the efficiency of lysis, the presence of PCR inhibitors, and the instability of telomerase in cell extract. Therefore, reactions with serially diluted oligo QS as the quantitation standards were included in each experiment. The remaining technical challenges of developing of a reliable TRAP assay include the development of a simpler and more robust procedure for cell lysis.

The precision of kinetic TRAP is demonstrated in Table 1. The C_t values of duplicate reactions of four experiments using serially diluted oligo QS and cell extract from 293 cells are listed. The coefficient of variation (CV) was ~1 or 2% for all the reactions with oligo QS and 1–100 ng cell lysate from 293 cells. The higher CV for

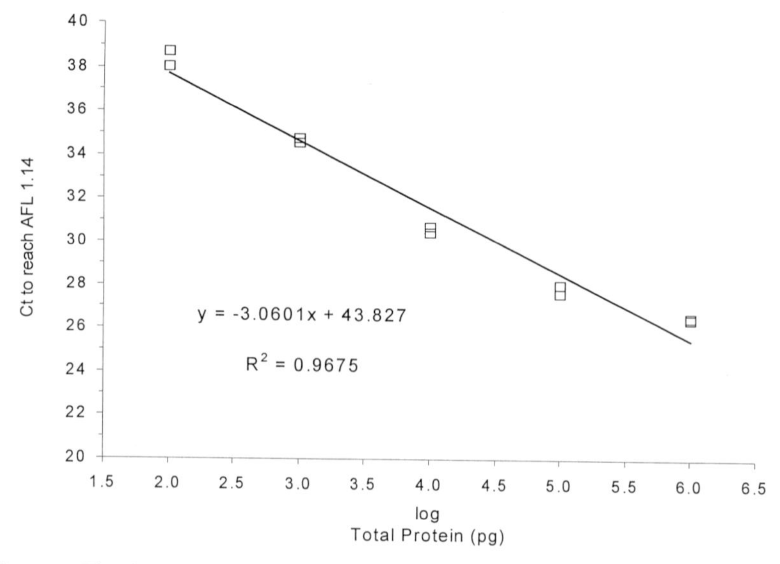

Figure 2 The dynamic range of kinetic TRAP assay. The arbitrary fluorescence level (AFL) is 1.2. The cycle numbers intercepting with AFL1.14 of each reaction in Fig. 1 are plotted with the log of total protein concentration of each reaction.

reactions with 0.1 ng cell extract from 293 cells was due to the dropout reactions in Experiment 2. A quantitation standard curve was generated using the average C_t values and log of copy numbers of oligo OS in Table 1. Using the equation $y = 3.351x + 42.35$, from the trendline of the standard curve, the copy number in each sample can be calculated. For example, the copy numbers for the two reactions with 10 ng cell extract from 293 cells in experiment 1 in Table 1 were ~ 920 and ~770 copies for C_t of 32.41 and 32.68, respectively.

In order to test whether a small number of tumor cells can be detected in the presence of a large number of normal cells, a mixing experiment was carried out. One microgram of total protein from telomerase-negative (or a very low undetectable level) WIN cells was mixed with various amounts of total protein from telomerase-positive 293 cells. The telomerase activity was then determined by kinetic TRAP. Figure 3 demonstrates that kinetic TRAP is able to detect telomerase activity in the mixture of 1 ng total protein (10 telomerase-positive 293 cells) in the presence of 1 μg of WIN total protein (10,000 telomerase-negative WIN cells).

Table 1

The Precision of Kinetic TRAP

Sample	C_t[a] at AFL = 1.14				Average	SD	CV (%)
	Experiment 1	Experiment 2	Experiment 3	Experiment 4			
Oligo QS (copy No.)							
50,000	26.73	26.19	27.00	26.27	26.57	0.27	1.02
	26.73	26.68	26.42	26.51			
5,000	30.19	29.83	29.90	29.64	29.94	0.22	0.75
	30.19	29.76	30.22	29.83			
500	33.65	32.90	33.61	33.16	33.39	0.27	0.80
	33.65	33.30	33.33	33.51			
50	37.11	35.67	36.87	36.09	36.59	0.61	1.66
	37.11	35.87	37.00	37.00			
293 cell extract (ng)							
100	29.08	29.09	29.29	28.49	28.83	0.32	1.11
	28.95	28.70	28.70	28.38			
10	32.41	31.79	32.22	31.86	32.14	0.34	1.05
	32.68	32.32	32.10	31.71			
1	36.73	36.93	36.54	35.23	36.25	0.58	1.60
	36.62	36.28	36.00	35.66			
0.1	39.75	40.88	38.56	38.88	40.12	2.38	5.94
	38.96	45.74	39.24	38.92			
Negative	41.32	40.44	42.70	43.09	41.01	1.27	3.10
	39.91	39.65	40.80	40.19			

Note. Abbreviations used; C_t, cycle threshold; CV, coefficient of variation.

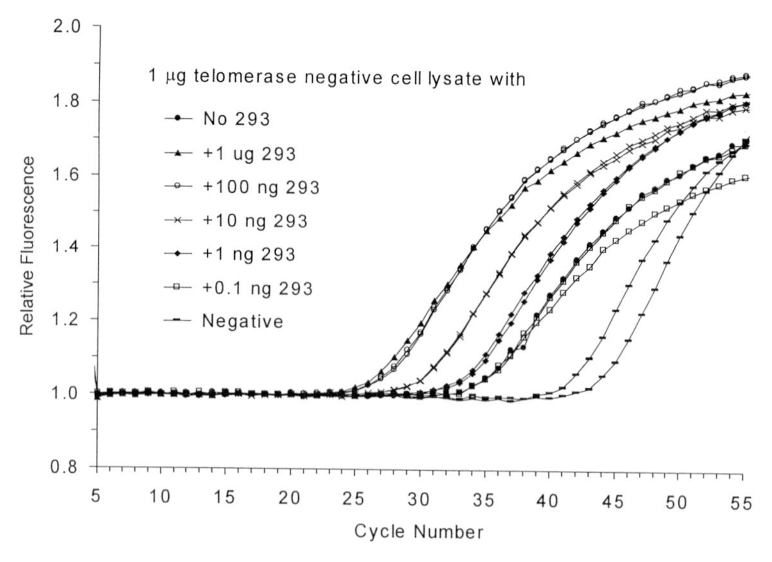

Figure 3 Detection of telomerase-positive 293 cells in the presence of telomerase-negative WIN cells.

The kinetic TRAP is simple, sensitive, nonradioactive, accurate, and requires little hands-on time after amplification. This format is suitable for screening a large number of clinical specimens to demonstrate the clinical utility of using telomerase as a diagnostic and prognostic tumor marker. For example, readily available urine samples were used to correlate the presence of telomerase activity and recurrence of bladder cancer (Kinoshita *et al.,* 1997).

Kinetic reverse transcriptase (RT)-PCR has been used to quantify the level of infected human immunodeficiency virus 1, hepatitis virus C, and mRNA levels of various genes in yeast and humans (Kang and Holland, 1997, unpublished results). The gene encoding the catalytic subunit of human telomerase (hTERT/hEST2) was cloned recently (Nakamura *et al.,* 1997; Meyerson *et al.* (1997), and the steady-state level of hTRT mRNA also correlated with the level of telomerase activity. Therefore, kinetic RT-PCR may be used to quantify hTRT mRNA levels and circumvent the issues of efficiency of cell lysis and the instability of telomerase activity during storage.

References

Aldous, W. K., and Grabill, N. R. (1997). A fluorescent method for detection of telomerase activity. *Diagn. Mol. Pathol.* **6,** 102–110.

Bacchetti, S., and Counter, C. M. (1995). Telomeres and telomerase in human cancer. *Int. J. Oncol.* **7,** 423–432.

Blackburn, E. H. (1991). Structure and function of telomeres. *Nature* **350,** 569–573.

Brien, T. P., Kallakury, B. V., Lowry, C. V., Ambros, R. A., Muraca, P. J., Malfetano, J. H., and Ross, J. S. (1997). Telomerase activity in benign endometrium and endometrial carcinoma. *Cancer Res.* **57,** 2760–2764.

Counter, C. M., Gupta, J., Harley, C. B., Leber, B., and Bacchetti, S. (1995). Telomerase activity in normal leukocytes and in hematologic malignancies. *Blood* **85,** 2315–2320.

Higuchi, R., Dollinger, G., Walsh, P. S., and Griffith, R. (1992). *Biotechnology* **10,** 413–417.

Higuchi, R., Fockler, C., Dollinger, G., and Watson, R. (1993). Kinetic PCR analysis: Real-time monitoring of DNA amplification reactions. *Biotechnology* **11,** 1026–1030.

Hiyama, E., Gollahon, L., Kataoka, T., Kuroi, K., Yokoyama, T., Gazdar, A. F., Hiyama, K., Pistyszek, M. A., and Shay, J. W. (1996). Telomerase activity in human breast tumors. *J. Natl. Cancer Inst.* **88,** 116–122.

Kim, N. W., and Wu, F. (1997). Advances in quantification and characterization of telomerase activity by the telomeric repeat amplification protocol (TRAP). *Nucleic Acids Res.* **25,** 2595–2597.

Kim, N. W., Piatyszek, M. A., Prowse, K. R., Harley, C. B., West, M. D., Ho, P. L. C., Coviello, G. M., Wright, W. E., Weinrich, S. L., and Shay, J. W. (1994). Specific association of human telomerase activity with immortal cells and cancer. *Science* **266,** 2011–2015.

Kinoshita, H., Ogawa, O., Kakehi, Y., Mishina, M., Mitsumori, K., Itoh, N., Yamada, H., Terachi, T., and Yoshida, O. (1997). Detection of telomerase activity in exfoliated cells in urine from patients with bladder cancer. *J. Natl. Cancer Inst.* **89,** 724–730.

Meyerson, M., Counter, C. M., Eaton, E. N., Ellisen, L. W., Steiner, P., Caddle, S. D., Ziaugra, L., Beijersbergen, R. L., Davidoff, M. J., Liu, Q., Bacchetti, S., Haber, D. A., and Weinberg, R. A. (1997). hEST2, the putative human telomerase catalytic subunit gene, is upregulated in tumor cells and during immortalization. *Cell* **90,** 785–795.

Nakamura, T. M., Morin, G. B., Chapman, K. B., Weinrich, S. L., Andrews, W. H., Lingner, J., Harley, C. B., and Cech, T. R. (1997). Telomerase catalytic subunit homologs from fission yeast and human. *Science* **277,** 955–959.

Piatyszek, M. A., Kim, N. W., Weinrich, S. L., Hiyama, K., Hiyama, E., Wright, W. E., and Shay, J. W. (1995). Detection of telomerase activity in human cells and tumors by a telomeric repeat amplification protol (TRAP). *Methods Cell Sci.* **17,** 1–15.

Ramirez, R. D., Wright, W. E., Shay, J. W., and Taylor, R. S. (1997). Telomerase activity concentrates in the mitotically active segments of human hair follicles. *J. Invest. Dermatol.* **108,** 113–117.

Shay, J. W., and Wright, W. E. (1996). Telomerase activity in human cancer. *Curr. Opin. Oncol.* **8,** 66–71.

Sugino, T., Yoshida, K., Bolodeoku, J., Tahara, H., Buley, I., Manek, S., Wells, C., Goodison, S., Ide, T., Suzuki, T., Tahara, E., and Tarin, D. (1996). Telomerase activity in human breast cancer and benign breast lesions: Diagnostic applications in clinical specimens, including fine needle aspirates. *Int. J. Cancer* **69,** 301–306.

Talyor, R. S., Ramirez, R. D., Ogoshi, M., Chaffins, M., Piatyszek, M. A., and Shay, J. W. (1996). Detection of telomerase activity in malignant and nonmalignant skin conditions. *J. Invest. Dermatol.* **106,** 759–765.

Wright, W. E., Shay, J. W., and Piatyszek, M. A. (1995). Modifications of a telomeric repeat amplification protocol (TRAP) result in increased reliability, linearity and sensitivity. *Nucleic Acids Res.* **23,** 3794–3795.

Part Three

GENE DISCOVERY

18

DIFFERENTIAL DISPLAY

Klaus Giese, Hong Xin, James C. Stephans, and Xiaozhu Duan

Differential display (DD) is a powerful technique to compare patterns of gene expression in RNA samples of different types or under different biological conditions (Liang and Pardee, 1992; Welsh *et al.*, 1992). The technique produces partial cDNA fragments by a combination of reverse transcription (RT) and PCR of randomly primed RNA. Changes in the expression level of genes are identified after separation of the cDNAs on sequencing-type gels.

Compared to traditional differential and subtractive gene cloning methods (Scott *et al.*, 1983; Lau and Nathans, 1985; Sive and St, 1988; Rothstein *et al.*, 1993), the DD technique offers several advantages. The most significant of these are (i) the simplicity of the DD technique in detecting both upregulated and downregulated genes in the same experiment, (ii) the analysis of several samples in parallel, and (iii) the identification of both rare and abundant messages (Liang and Pardee, 1992; Wan *et al.*, 1996). Moreover, the DD technique requires only small amounts of RNA and is very rapid.

In our laboratory we are using the DD technique to identify candidate genes that encode positive and negative factors that are associated with tumor cell invasion and metastasis. The process by which a single malignant cell colonizes a distant site is still an unsolved mystery but this process represents the major contribution to cancer-related mortality (Kerbel, 1989; Fidler and Radinsky, 1990; Liotta *et al.*, 1991). Knowledge about dysregulated genes may help in early detection of the metastatic state and in the development of therapeutic products.

The most significant drawbacks of the original described DD technique are the reliability and reproducibility of the differential expression pattern which together with the generation of relatively short, uninformative cDNA fragments (<400 nucleotides) account for many false positives (Debouck, 1995). Recently, the use of longer primers in the amplification steps has resulted both in increased reproducibility of the differential expression pattern and in the ability to display larger cDNA fragments of up to 2 kb in length which provides more coding information for gene identification. In addition, improvements in the gel electrophoresis equipment facilitate the separation of these large cDNA fragments (Zhao *et al.*, 1995; Liang *et al.*, 1995; Averboukh *et al.*, 1996; Diachenko *et al.*, 1996). Other factors also influence the rate of false positives, including poor-quality RNA, poorly matched cell types, and differences in cell growth conditions. In this chapter, we will highlight aspects of sample preparation and the use of several control steps in the DD technique which together markedly decrease the number of false positives. We will also discuss aspects of the time-consuming post-differential display work to validate putative candidate genes.

RNA Source for the DD Approach

RNA for the DD was isolated from human cancer cell lines grown in tissue culture or in immunodeficient animals. These model systems are more likely to produce a greater quantity of RNA and a higher quality of RNA compared to RNA isolated from biopsies which generally consists of a small amount of tumor material with uncertain quality due to the admixture of malignant and normal cell types. The quality of the RNA samples is of critical importance because even small differences can generate a large number of false positives. Our system consists of human cancer cell lines displaying a low- and high-metastatic phenotype in which the high-metastatic variant is derived from a low-metastatic parent. This model system should, in principle, focus only on those genes which are involved in tissue invasion and metastasis, decreasing the total number of differentially expressed cDNAs. Gene expression in model systems, however, can differ from what is observed in cancer tissue by many criteria. Therefore, it will require a combination of different gene discovery approaches to validate the importance of a particular candidate gene in the metastatic process.

Experimental Protocol

RNA was isolated essentially as described by Chirgwin *et al.* (1979). To ensure high-quality material for the DD analysis we purify our RNA preparation by CsCl centrifugation (Sambrook *et al.*, 1989) followed by DNase I treatment (DNA cleaning kit, GenHunter, Nashville, TN). Only the combination of these steps provided in our hands RNA of high quality. In particular, the CsCl gradient step, which removes DNA and other contaminating materials, is very important. The quality of the RNA is analyzed by agarose gel electrophoresis, divided into aliquots, and stored at −80°C.

DD-PCR

Figure 1 outlines schematically the basic steps in the DD-PCR. Each reaction is performed in duplicate. First-strand cDNA synthesis (RT) reactions is performed at 42°C in a final volume of 20 μl in a thermal cycler with a heated lid (MJ Research PTC200, Watertown, MA) using 200 ng of total RNA and one of the 10 double-anchored oligo(dT) primers supplied in the HIEROGLYPH mRNA profile system (Genomyx Corp., Foster City, CA). [Anchored primers contain two base-specfic nucleotides at the 3′ end of the oligo(dT) stretch which determine the specificity of the RT reaction]. DD-PCR is performed in 20-μl reactions as follows: 1 cycle, 95°C for 2 min; 4 cycles, 92°C for 15 sec, 46°C for 30 sec, and 72°C for 2 min; 25 cycles, 92°C for 15 sec, 60°C for 30 sec, and 72°C for 2 min; 72°C for 10 min followed by hold at 4°C. Note the reduced number of cycles and higher annealing temperature of 60°C compared to the original method in which 40 cycles are performed at an annealing temperature of 40°C (Liang and Pardee, 1992). The shorter cycle number ensures a better representa-

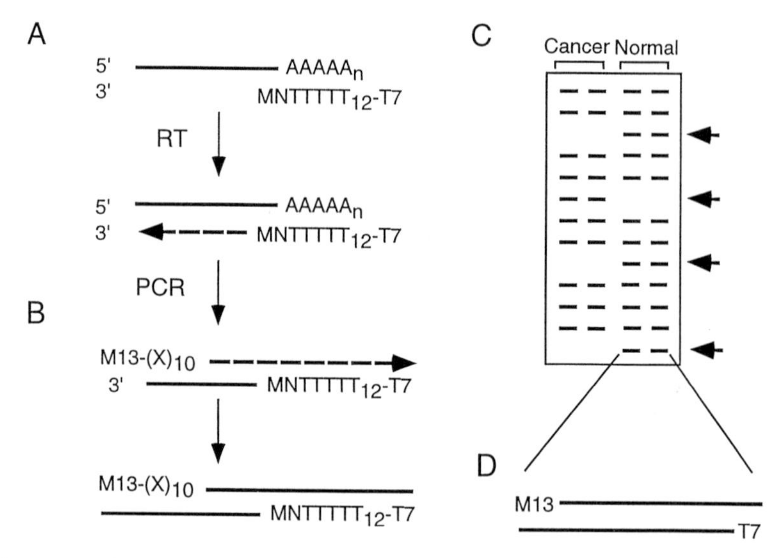

Figure 1 Schematic outline of the differential display technique including high-resolution gel electrophoresis and reamplification. (A) Reverse-transcription of total RNA using specific double-anchored oligo(dT) primers. (B) PCR amplification using an arbitrary upstream primer and the same oligo(dT) primer as used in A. (C) A differential display gel; shown are arbitrary cDNA fragments. The arrows indicate differentially expressed (repressed/induced) cDNAs. (D) Reamplification of a differentially expressed cDNA fragment using M_{13} reverse and T_7 primers. $T_{12}MN-T_7$: (M is A, G, C, N is A, G, C, T, and T_7 represents the recognition sequence for T_7 RNA polymerase. $M_{13}-X_{10}$; where M_{13} represents the M_{13} reverse primer site and X = A, G, C, T.

tion of the prevalence of each message in the mRNA population and the higher annealing temperature increases the specificity of the PCR. We include in the PCR step an RNA sample in which the RT reaction was omitted to control for low-level DNA contamination. To visualize the cDNA fragments [α-^{33}P]dATP (0.25 μl per reaction; 1000–3000 Ci/mmol) is added to the PCR reaction. Labeled cDNA products are electrophoresed using the extended-format programmable GenomyxLR electrophoresis apparatus which resolves DNA fragments in the range of 400–2000 bp in length (Genomyx Corp).

Reamplification and Characterization of the Differentially Expressed cDNA Fragments

After the run, the gel is washed to remove urea, dried, and exposed to X-ray films (Bio-Max MR, Kodak, Rochester, NY). Differentially expressed cDNA fragments are isolated from the dried gel by aligning the autoradiogram on top of the glass plate. The band of interest is cut with a sharp scalpel, and the gel slice is wetted using 3 μl of distilled water and transferred into a PCR tube for reamplification. The presence of a T7 RNA polymerase promoter sequence at the 5′ end of each anchored oligo(dT) primer and the M13 reverse sequence at the 5′ end of each arbitrary primer facilitates the amplification of each band with this unique set of primers. The use of only two unique primers for reamplification is a major advantage over the previous method which required the same primer pairs as those used in the DD (Liang and Pardee, 1992). Moreover, direct sequencing of the cDNA fragments is possible using both primers, increasing the process of cDNA identification (Linskens *et al.*, 1995; Wang and Feuerstein, 1995; Martin *et al.*, 1997). Direct sequencing of the amplified DD fragments results, in most cases, in a readable sequence. A readable sequence, however, does not necessarily indicate the presence of a single cDNA fragment. In practice, a gel slice isolated from the GenomyxLR electrophoresis system (Genomyx Corp.) contains one to three different cDNA fragments. Several strategies exist to isolate the correct DNA fragments. A common approach is to clone the reamplified cDNA fragments and to analyze several colonies (5–10) by restriction digest and sequencing. An alternative approach is to use the single-strand conformation polymorphism technique which separates mixtures of DNA fragments by nondenaturing polyacrylamide gels (Mathieu-Daude *et al.*, 1996). Another approach is to design specific nested primers based on the readable nucleotide sequence and to reamplify a specific DNA fragment (Martin *et al.*, 1997).

Results and Conclusions

We found it very important to use RNAs for the DD which were isolated under identical conditions because small differences in the samples will generate false positives. An example for a growth condition-related false positive is shown in Fig. 2. In this experiment, we compare gene expression in a low- and a high- metastatic human cancer cell line. The RNA for the display shown in panel I in Fig. 2 is

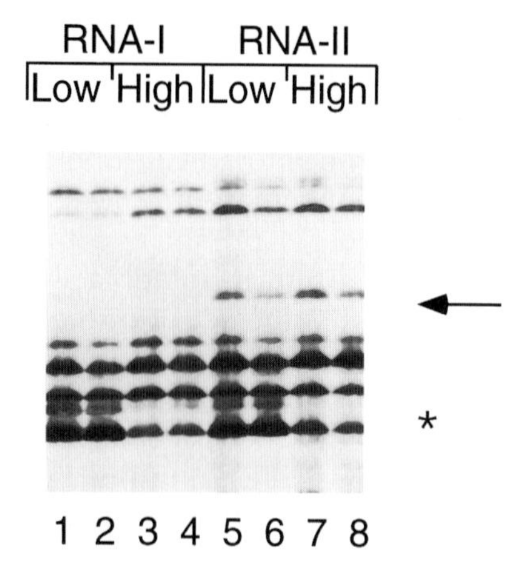

Figure 2 Differential display analysis using RNA isolated under different growth conditions. RNA was isolated from cell lines displaying a low- or high-metastatic phenotype at 60–70% density (lanes 1–4, RNA-I) or at complete confluency (lanes 5–8, RNA-II). The arrow indicates a gene upregulated only in cells grown at high density. The asterisk indicates a differentially expressed cDNA.

isolated from cells at 60–70% density to ensure logarithmic growth, whereas the RNA used in panel II is isolated from cells which were 100% confluent. Comparison of the cDNA expression pattern from the low- and high-metastatic cell lines showed several differentially (Fig. 2, asterisk) displayed cDNAs. These differences were detected irrespective of the growth conditions. In contrast, the cDNA indicated by the arrow in Fig. 2 was highly upregulated in the RNA samples isolated from cells which were grown at high density. Therefore, it is important to perform the DD reaction with RNA samples obtained from cells which are as closely matched as possible. This aspect of matched samples is of particular interest when DD is performed using RNA isolated from fast-growing tumor tissue and compared to RNA isolated from adjacent normal tissue.

Figure 3 represents another common result in the differential display approach. The expression of the indicated cDNA is markedly upregulated in RNA samples from a high (H)-metastatic cancer cell line irrespective of the tissue culture growth conditions (Fig. 3A). A Northern blot analysis, however, showed only a minor (less than two-fold) upregulation of this message in the high-metastatic cell line (Fig. 3B). This result could be interpreted to suggest a false positive DD result. Next, we performed a RT-PCR analysis using

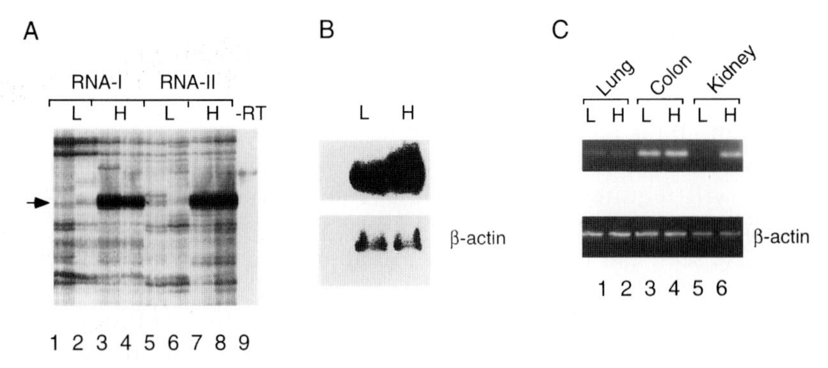

Figure 3 Differential display analysis and validation of the differential expression pattern by Northern blot analysis and RT-PCR. (A) RNA was isolated from cell lines displaying a low- or high-metastatic phenotype at 60–70% density (lanes 1–4, RNA-I) or at complete confluency (lanes 5–8, RNA-II). Shown is part of a differential display analysis indicating an upregulated cDNA in a high-metastatic kidney cell line (arrow). -RT, differential display reaction without reverse transcription. (B) Northern blot analysis using the differentially expressed cDNA fragment identified in A. (C) RT-PCR using RNA isolated from different tissue culture cell lines displaying a low- and high-metastatic phenotype. RT-PCR primers were designed according to the nucleotide sequence of the DD cDNA fragment shown in A.

specific primers designed from the nucleotide sequence of the differentially expressed cDNA fragment. We included in this experiment RNA isolated from two other human cancer cell lines also displaying a low- and high-metastatic phenotype. We generally include these additional cancer lines in our validation studies to determine whether the gene is misregulated in a cell type-specific manner or generally associated with the metastatic phenotype. Figure 3C shows the result of a quantitative RT-PCR reaction using 25 cycles. Consistent with the DD result, the cDNA was highly overexpressed in the high-metastatic kidney cell line (compare lanes 5 and 6) but not in the other two cancer cell lines irrespective of the metastatic phenotype (lanes 1–4). Further experiments indicated that the upregulated cDNA represents a splice variant which is missing in the low-metastatic kidney cell line (data not shown). Whether the presence of this unique splice variant is of importance for the metastatic phenotype has yet to be determined. Based on these results, differentially expressed cDNAs which have been isolated with only one primer pair should be validated using more than one technique.

Future Problems

Although DD is a flexible and simple method to scan the approximately 10,000–15,000 different messages in a given cell, a complete profile will require up to 80 different arbitrary upstream primers with downstream primers (Liang *et al.*, 1995; Bauer *et al.*, 1993). A complete scan can therefore result in approximately 1000 different reactions which does not include the post-DD work required for validation of the expression pattern and the isolation of full-length clones. Therefore, further improvements in the technique, in particular in the post-DD work, are required to perform a genomewide gene expression scan. Recently, increases in throughput of the DD analysis have been reported based on fluorescence-labeling techniques (Jones *et al.*, 1997; Kito *et al.*, 1997). This technique facilitates the analysis of multiple DD reactions in one lane using different fluoroescently labeled primer combinations.

Several methods exist to validate the differential expression pattern, including Northern blot analysis and RNA protection assays. Although both methods will give valid results, their use in a large-scale analysis is rather limited. Recently, post-DD work has been facilitated by array technologies which can validate large numbers of differential-displayed cDNA fragments in one experiment (Schena *et al.*, 1995; Poirier *et al.*, 1997; Corton and Gustafsson, 1997). The principle of these validation methods is to spot the differentially expressed cDNA fragments onto a membrane or glass slide and hybridize with a radiolabeled cDNA prepared by reverse transcription of total RNA. Although these methods validate hundreds or thousands of differentially expressed cDNAs at the same time, labeling rare mRNAs can be very challenging. In addition, other gene discovery methods, such as the serial analysis of gene expression (Velculescu *et al.*, 1995) and the representational difference analysis (Lisitsyn *et al.*, 1993), have been reported to facilitate the identification of novel and disease-relevant genes. The identification of differentially expressed genes is only the first step in genomics. A much bigger problem will be to quickly address specific functions to individual genes and to find ways for their regulation.

Acknowledgment

The authors thank Joan Zakel for expert technical assistance and Robert Tressler and Patrice Pitot for providing cell lines and expertise in animal model systems. We thank our oligonucleotide synthesis and DNA sequencing departments for their excellent assistance in this project. The authors are grateful to Peter Curtin, David Duhl, Michael Innis, and Jörg Kaufmann for helpful comments and discussions. We thank Katy Buckley-Smith for help in manuscript preparation.

References

Averboukh, L., Douglas, S. A., Zhao, S., Lowe, K., Maher, J., and Pardee, A. B. (1996). Better gel resolution and longer cDNAs increase the precision of differential display. *BioTechniques* **20,** 918–921.

Bauer, D., Mueller, H., Reich, J., Riedel, H., Ahrenkiel, V., Warthoe, P., and Strauss, M. (1993). Identification of differentially expressed mRNA species by an improved display technique (DDRT-PCR). *Nucleic Acids Res.* **21,** 4272–4280.

Chirgwin, J. M., Przybyla, A. E., MacDonald, R. J., and Rutter, W. J. (1979). Isolation of biologically active ribonucleic acid from sources enriched in ribonuclease. *Biochemistry* **27**(24), 5294–5299.

Corton, J. C., and Gustafsson, J. A. (1997). Increased efficiency in screening large numbers of cDNA fragments generated by differential display. *BioTechniques* **5,** 802–810.

Debouck, C. (1995). Differential display or differential dismay? *Curr. Opin. Biotechnol.* **6,** 597–599.

Diachenko, L. B., Ledesma, J., Chenchik, A. A., and Siebert, P. D. (1996). Combining the technique of RNA fingeprinting and differential display to obtain differentially expressed mRNA. *Biochem. Biophys. Res. Commun.* **219**(3), 824–828.

Fidler, I. J., and Radinsky, R. (1990). Genetic control of cancer metastasis. *J. Natl Cancer Inst.* **3,** 166–168.

Jones, S. W., Cai, D., Weislow, O. S., and Esmaeli-Azad, B. (1997). Generation of multiple mRNA fragments using fluorescence-based differential display and an automated DNA sequenzer. *BioTechniques* **3,** 536–543.

Kerbel, R. S. (1989). Towards an understanding of the molecular basis of the metastatic phenotype. *Invasion Metastasis* **9,** 329–337.

Kito, K., Ito, T., and Sakaki, Y. (1997). Fluorescent differential display analysis of gene expression in differentiating neuroblastoma cells. *Gene* **184,** 73–81.

Lau, L., and Nathans, D. (1985). Activation of mouse genes in transformed cells. *EMBO J.* **4,** 3145–3151.

Liang, P., and Pardee, A. B. (1992). Differential display of eukaryotic messenger RNA by means of the polymerase chain reaction. *Science* **257,** 967–971.

Liang, P., Bauer, D., Averboukh, L., Warthoe, P., Rohrwild, M., Muller, H., Strauss, M., and Pardee, A. B. (1995). Analysis of altered gene expression by differential display. *Methods Enzymol.* **254,** 304–321.

Linskens, M. H. K., Feng, J., Andrews, W. H., Enlow, B. E., Saati, S. M., Tonkin, L. A., Funk, W. D., and Villeponteau, B. (1995). Cataloging altered gene expression in young and senescent cells using enhanced differential display. *Nucleic Acids Res.* **23,** 3244–3251.

Liotta, L. A., Steeg, P. S., and Stetler-Stevenson, W. G. (1991). Cancer metastasis and angiogenesis: An imbalance of positive and negative regulation. *Cell* **64**(2), 327–336.

Lisitsyn, N., Lisitsyn, N., and Wigler, M. (1993). Cloning the differences between two complex genomes. *Science* **259,** 946–951.

Martin, K., Kwan, C.-P., and Sager, R. (1997). A direct-sequencing-based strategy for identifying and cloning cDNAs from differential display gels. *In Methods in Molecular Biology. Differential Display: Methods and Applications* (A. B. Pardee and P. Liang, Eds.). Humana Press, Totowa, NJ.

Mathieu-Daude, F., Cheng, R., Welsh, J., and McClelland, M. (1996). Screening of differentially amplified cDNA products from RNA arbitrary primed PCR fingerprints using single strand conformation polymorphism (SSCP) gels. *Nucleic Acids Res.* **24,** 1504–1507.

Poirier, G. W., Pyati, J., Wan, J. S., and Erlander, M. G. (1997). Screening differentially expressed cDNA clones obtained by differential display using amplified RNA. *Nucleic Acids Res.* **4,** 913–914.

Rothstein, J. L., Johnson, D., Jessee, J., Skowronski, J., DeLoia, J. A., Solter, D., and Knowles, B. B. (1993). Construction of primary and subtracted cDNA libraries from early embryos. *Methods Enzymol.* **225,** 587–610.

Sambrook, J., Fritsch, E. F., and Maniatis, T. (1989). *Molecular cloning—A Laboratory Manual. Cold Spring Habor Laboratory Press,* Cold Spring Harbor, NY.

Schena, M., Shalon, D., Davis, R. W., and Brown, P. O. (1995). Quantitative monitoring of gene expression patterns with a complementary DNA microarray. *Science* **270,** 467–470.

Scott, M. D., Westphal, H.-H., and Rigby, R. W. J. (1983). Identification of a set of genes expressed during the G0/G1 transition of cultured mouse cells. *Cell* **34,** 557–567.

Sive, H. L., and St, J. T. (1988). A simple subtractive hybridization technique employing photoactivatable biotin and phenol extraction. *Nucleic Acids Res.* **16,** 10937.

Velculescu, V. E., Zhang, L., Vogelstein, B., and Kinzler, K. W. (1995). Serial analysis of gene expression. *Science* **270,** 484–487.

Wan, J. S., Sharp, S. J., Poirier, G. M.-C., Wagaman, P. C., Chambers, J., Pyati, J., Hom, Y.-L., Galindo, J. E., Huvar, A., Peterson, P. A., Jackson, M. R., and Erlander, M. G. (1996). Cloning differentially expressed mRNAs. *BioTechniques* **14,** 1685–1691.

Wang, X., and Feuerstein, G. Z. (1995). Direct sequencing of DNA isolated from mRNA differential display. *BioTechniques* **18**(3), 448–453.

Welsh, J., Chada, K., Dalal, S. S., Cheng, R., Ralph, D., and McClelland, M. (1992). Arbitrarily primed PCR fingerprinting of RNA. *Nucleic Acids Res.* **20,** 4965–4970.

Zhao, S., Ooi, S. L., and Pardee, A. B. (1995). New primer strategy improves precision of differential display. *BioTechniques* **18,**(5), 842–846.

19

SINGLE-CELL cDNA LIBRARIES

Peter S. Nelson

The set of genes expressed by a given organism, tissue, or cell confers developmental and functional specificity and determines the physical characteristics reflected as phenotypes. Methods to identify, isolate, and characterize these expressed genes are fundamental for the study of a wide range of biological processes from development to carcinogenesis. Most methods for expression analysis at the transcript level are designed for evaluating one or at most a few genes per experiment. Alternatively, capturing the information represented by all of the cellular transcripts as a library of complementary DNA molecules (cDNAs) offers a means to simultaneously assess the entire repertoire of genes expressed in a given biological sample. Such a library of expressed genes conveys the identity and level of expression for each gene transcript in a given population of cells and thus reflects the "transcriptome" (Velculescu *et al.*, 1997), the dynamic link between the genome and the proteome.

cDNA libraries have proven useful for a variety of applications, including the identification of tissue-specific genes, differential gene expression studies, and the discovery of homologous genes between species. Conventional methods for cDNA libary construction involve purifying messenger RNA (mRNA) from cell lines or from large heterogenous tissue samples composed of several different cell types (Gubler and Hoffman, 1983). However, cDNA libraries constructed from whole tissues constitute a composite of the transciptomes of several different cell types that may each have distinctly different expression profiles and phenotypes. Abundant transcripts in an abundant cell type may dominate a tissue expression profile and reduce the odds of recognizing expressed genes from a less abundant cell type. The gene expression profiles of individual cells would provide a further level of detail in the analysis of cellular identity, diversity, and function.

The amount of mRNA within a single cell is estimated to be between 0.1 and 1 pg representing between 10,000 and 30,000 different transcripts; each is present at a different frequency ranging from highly abundant to rare (Bishop *et al.*, 1974). This minute quantity of a highly complex mixture of mRNAs is difficult to manipulate experimentally and thus requires some type of amplification. The PCR, a method originally developed to multiply copies of a known sequence of DNA, can be adapted to amplify a complex mixture of nucleic acids such as the entire cellular transcriptome. A limiting factor in PCR-based amplification of complex DNA or cDNA populations is the requirement for known sequence flanking the target in order to specify oligomer-directed amplification. This limitation can be overcome by using the poly-A^+ tail present on the 3′ end of most mRNA molecules as one common amplification point, and the 5′ end (3′ end of the newly synthesized cDNA) of each molecule can be modified to carry a common sequence by either "tailing" or adapter ligation (Dumas Milne Edwards *et al.*, 1997; Jena *et al.*, 1996).

Aside from the amplification step, the basic concepts and caveats for constructing cDNA libraries from plentiful amounts of RNA also pertain to single-cell library construction. The overall procedure involves the reverse transcription (RT) of mRNA to generate first-strand cDNA, the synthesis of second-strand cDNA molecules, size-selection of cDNA to eliminate truncated cDNA molecules, ligation into a plasmid or phage vector, transformation into suitable host cells, and quality analysis of the resultant library. A detailed review of the molecular biology underlying each of these steps is beyond the scope of this discussion. Several cDNA library protocols have been published and are worthy of review prior to initiating a library-construction project (Ausubel *et al.*, 1995; Gubler and Hoffman, 1983). The high quality of commercially available enzymes and reagents has markedly improved the quality, ease, and speed with which cDNA libraries can be constructed. Nevertheless, attention to detail and a working knowledge of basic molecular techniques is critical for successful library construction.

Few cDNA libraries are sufficiently versatile for all potential uses. Thus, the desired end use should dictate the specifics of library construction. For example, if expressing the cDNA is desirable, then the choice of cloning vector will dictate whether a cDNA can be expressed in prokaryotic or eukaryotic cells. If random sequencing to produce expressed sequence tag data sets is desired, then a directionally cloned library will allow the investigator to know the cDNA orientation and obtain sequence selectively from the 5′ end in order to avoid the poly-A tract and increase the likelihood of obtaining coding sequence.

Potential limitations of the amplification step may dictate specific modifications of the protocol for constructing cDNA libraries from single cells, depending on the projected uses for the library. The kinetics of the PCR reaction are such that short DNA sequences are amplified with a greater efficiency than longer sequences. Amplifying a complex population of short and long cDNAs may result in a skewed population favoring the short cDNAs. This bias can be addressed by limiting the length of the original cDNA strand to ~500–700 bases regardless of the size of the original RNA template (Brady and Iscove, 1993) but at the expense of reducing the number of full-length cDNA clones in the library. Alternatively, a combination of thermostable polymerases capable of "long" PCR may serve to reduce this amplification bias (Barnes, 1994).

Generally, a high-quality cDNA library will faithfully represent the sequence, size, and complexity of the original mRNA population. The primary consideration in designing the following protocol was to ensure that at least one copy of each expressed gene is represented. Other design considerations were for the cDNAs to represent full-length mRNA transcripts, for the library to have a minimum number of "contaminating" clones such as ribosomal RNAs and empty vectors, and for the library to maintain the fidelity of transcript ratios such that abundant transcripts remain abundant cDNAs and rare transcripts remain rare cDNAs.

Protocol

The tremendous utility of cDNA libraries has spawned the development of many construction protocols. Variations exist for essentially every step of the construction process. This protocol was adapted from a variety of sources, including published protocols by Gubler and Hoffman (1983), Brady and Iscove (1993), and Schweinfest *et al.* (1995) and from several commercial cDNA construction methods developed by Clontech (*http://www.clontech. com*), Life Technologies (*http://www.lifetech.com*), and Stratagene (*http://www.stratagene. com*). The protocol has been used successfully to construct cDNA libraries from single prostate epithelial cells. However, each step has not been rigorously compared with all available alternatives. Recognizing that different cell types and applications may require different strategies, alternative protocols for several steps are provided, as are references for others (see *Method Notes* and *Alternative Methods*).

In general, because such small quantities of starting materials are used, potential losses are minimized by limiting tube transfers, purification steps, etc. It is important to emphasize the establishment of an RNAse-free working environment to avoid contamination and degradation of RNA that will directly affect the ultimate quality of the library (see *method note 1*).

The basic protocol is divided into five steps:

Step A: Obtaining single cells using microdissection from tissue sections

Step B: cDNA synthesis

Step C: cDNA amplification by PCR

Step D: Directional cDNA cloning and transformation

Step E: Assessment of library quality

A flow schema for the entire cDNA construction process is depicted in Fig. 1.

Step A: Obtaining Single Cells

Single cells are obtained from frozen or fixed 7- to 10-μm tissue sections using glass capillary micropipettes drawn to a 10- to 20-μm tip lumen. Several detailed protocols for micropipette production and microdissection have been published (Gordon, 1993). Alternatives to microdissection for obtaining single cells are described under *Alternative Methods* (No. 1).

General Reagents

Phenol: chloroform:IAA (24:23:1) (Ambion, Austin, TX, No. 9732), buffer saturated to pH 7.5

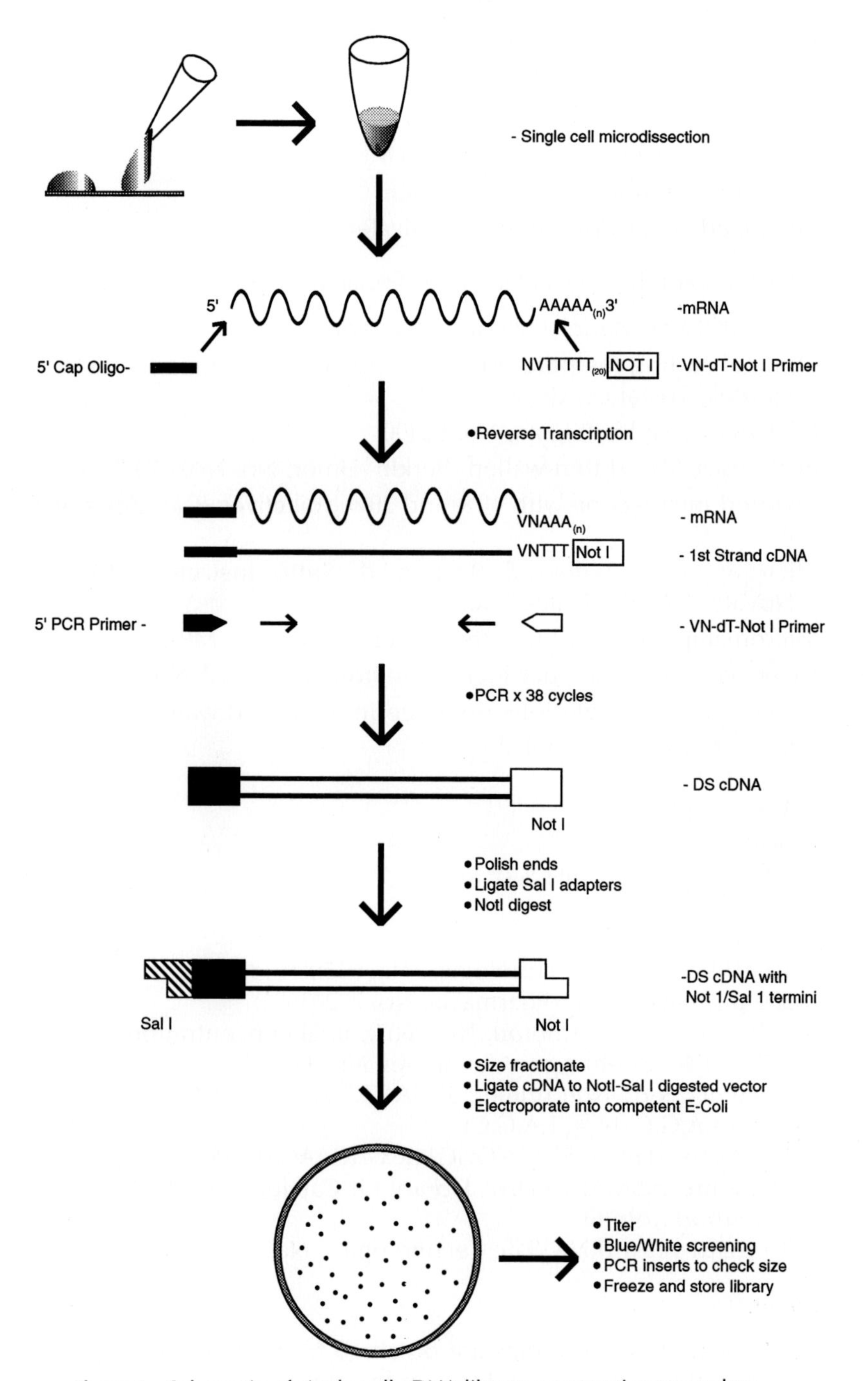

Figure 1 Schematic of single-cell cDNA library construction procedure.

Chloroform (Fisher Scientific)
Ethanol 95%
Ethanol 75%
3 *M* sodium acetate (NaOAc), pH 4.8
7.5 *M* ammonium acetate (NH_4OAc)
deionized water (Millipore or equivalent)

Equipment and Reagents for Step A (See method note 2)

Cryostat (Leitz Model 1720 or equivalent)
Optimal cutting temperature (OCT) tissue freezing media (Sakura Finntek, Torrance, CA)
Microscope slides (Sigma, No. S8400)
PCR tubes, 500-μl thin-walled (Perkin–Elmer, No. N801-737)
Inverted microscope with 10× and 20× objectives (Carl Zeiss, Inc., Aalen, Germany)
Micropipettes, 1.0-mm o.d., 0.5-mm i.d. (Sutter Instrument Co., Novato, CA, No. B100-50-10)
Micromanipulator and pipette holder (Sutter, No. MP-85)
Glass syringe, 25-μl leuer lock (Hamilton, Reno, NV, No. 80222)
Tissue buffer (all solutions prepared in deionized water)
  50 m*M* Tris–HCl (pH 8.3)
  75 m*M* KCl
  3 m*M* $MgCl_2$
Lysis/RT buffer
  50 m*M* Tris–HCl (pH 8.3)
  75 m*M* KCl
  3 m*M* $MgCl_2$
  10 m*M* DTT (Life Technologies, No. 15508-013)
  500 μ*M* each dNTP (Pharmacia, No. 27203501)
  RNAse inhibitor (Ambion, No. 2682), final concentration 1 U/μl
  0.5% NP40 (Boehringer-Mannheim, No. 1332473)
  50 ng/μl VN-dT-*Not*I primer: 5′-GACTCTAGAGCGGCCGCC$(T)_{20}$VN, (V = A,G,C, N = T,A,G,C)
  1 μ*M* cap adapter: 5′-TACGGCTGCGAGAAGACGACAGAAGGG-3′ (Clontech.SMART cDNA Library Kit Catalog No. PT3000-1) (see *method note* 6)
  0.5 μg/μl yeast tRNA (Life Technologies, No. 15401-011)

Method

A1. Prostate tissue samples are obtained from the operating room, embedded in OCT, and frozen by immersion in isopentane/

liquid nitrogen. Sections are cut to 7 μm on a cryostat and mounted onto untreated microscope slides. Sections are either used immediately or fixed in 4% paraformaldehyde and frozen at −80°C for future use (Eberwine *et al.*, 1992).

A2. The tissue section is placed on the stage of a disssecting microscope and covered with ~100 μl of tissue buffer.

A3. Single epithelial cells are drawn into a glass micropipette of 10- to 20-μm tip diameter with the aid of a micromanipulator by applying a gentle suction through a connected plastic polyethylene tubing filled with mineral oil connected to a 25-μl glass syringe.

A4. After withdrawing the pipette containing a single cell and a small amount (1 or 2 μl) of tissue buffer from the slide surface, the cell and the tissue buffer solution are ejected into a 500-μl Eppendorf tube containing 5 μl of lysis/RT buffer, and the tubes are placed on ice. The cell manipulation can be documented with a video camera attached to the microscope.

Note: It is useful to prepare control samples using the same procedure, but without the aspiration of cells, to use as media controls in the subsequent RT and PCR reactions.

Step B: First-Strand cDNA Synthesis

REAGENTS FOR STEP B (SEE METHOD NOTE 3)

Superscript II RNase H$^-$ reverse transcriptase (Life Technologies, No. 18064-014): 200 U/μl

METHOD

B1. Heat the samples at 65°C for 1 min to denature the RNA, allow to cool to room temperature for 3 min to anneal the VN-dT-*Not*I primer, and place on ice.

B2. Spin briefly in a microcentrifuge to collect the sample, and add 100 U (0.5 μl of 200 U/μl) of superscript reverse transcriptase. Pipette gently to mix and incubate at 42°C for 60 min.

B3. Stop the reaction by heating at 65°C for 10 min and place on ice (see *method note 4*). (Note: The sample may be stored at −20°C for this stage.)

Note: It is adviseable to set up an identical reaction except for the exclusion of the reverse transcriptase [RT (−) control]. This reaction will serve as a control for products resulting from DNA contamination (see *method note 5*).

Step C: Second-Strand cDNA Synthesis and Amplification by PCR

PCR amplification of the cDNA is carried out using oligonucleotides homologous to the primer used for RT and the cap primer (see *method note 5*). A mix of thermostable DNA polymerases is employed for high-fidelity, long-distance PCR (see *method note 6*). Alternatively, a protocol of homopolymer tailing the cDNA using terminal deoxytransferase (TdT) can be used to provide a known sequence for PCR (see *alternate method 2*).

REAGENTS FOR STEP C

10× Advantage polymerase buffer (Clontech, No. K1905-1): 400 m*M* tricine–KOH (pH 9.2), 150 m*M*KOAc, 35 m*M* $Mg(OAc)_2$, 750 μg/ml bovine serum albumin

50× Advantage polymerase mix (Clontech, No. 8417-1: 50% glycerol, 40 m*M* Tris–HCl (pH 7.5), 50 m*M* KCl, 25 m*M* $(NH4)_2SO4$,
0.1 m*M* EDTA, 5.0 m*M* β-mercaptoethanol, 0.25% thesit, KlenTaq-1 DNA polymerase, Deep Vent DNA polymerase, TaqStart antibody, 1.1 μg/μl.

10 m*M* dNTP mix: (see reagents for step A) 10 m*M* each dATP, dCTP, dGTP, and dTTP

10 μ*M* VN-dT-*Not*I primer: see materials for step A

10 μ*M* cap primer: 5′-TACGGCTGCGAGAAGACGACAGAA-3′

Mineral oil (Sigma, No. M-3516)

Proteinase K (Boerhinger-Mannheim, No. 1413783): 20 mg/ml

T4 DNA polymerase (Life Technologies, No. 18005-017): 5 U/μl

5× ligation buffer (Life Technologies, No. 15224-017: supplied with T4 DNA ligase): 250 m*M* Tris–HCl (pH 7.6), 50 m*M* $MgCl_2$, 5 m*M* ATP, 5 m*M* DTT, 25% PEG-8000

*Sal*I–*Xho*I adapters (Strategene, No. 901125) (see Method note 8): 1 μg/μl

T4 DNA ligase (Life Technologies, No. 15224-017): 1 U/μl

10× *Not*I digestion buffer (Life Technologies, React 3 buffer): 500 m*M* Tris–HCl (pH 8.0), 100 m*M* $MgCl_2$, 1000 m*M* NaCl

1-kb ladder DNA size markers (Life Technologies, No. 15615-016)

Thermal cycler (Perkin-Elmer, Model 480)

Method

PCR amplification

C1. Add the following reagents to the 6-μl first-strand reaction product. (Note: Set up identical reactions with control tubes: RT-free control and cell-free control.)

6 μl first-strand cDNA
44 μl deionized water
6 μl 10× Advantage PCR buffer (or equivalent)
1 μl 10 m*M* dNTPs
1 μl VN-dT-*Not*I primer
1 μl cap primer
1 μl 50× Advantage DNA polymerase mix (1 U/μl) (or equivalent)
60 μl total volume

C2. Gently mix and centrifuge in a microfuge for 2 sec to collect the reaction components.

C3. Preheat the thermal cycler to 94°C.

C4. Overlay the reaction mixture with two drops of mineral oil.

C5. Proceed with the PCR using the following parameters:

(95°C for 1 min) × 1 cycle
(95°C for 15 sec, 68°C for 5 min) × 30 cycles
4°C hold

C6. Analyze 8 μl of the reaction product adjacent to the 1-kb size markers on a 1% agarose Tris acetate EDTA (TAE) gel containing ethidium bromide. The cDNA should exhibit a smear ranging from ~0.5 to 5 kb (Fig. 2) (see *method note 8*).

C7. Remove the mineral oil by adding an equal volume of chloroform, vortexing thoroughly, and spinning at 14,000*g* in a microfuge for 5 min. Place the DNA solution in a fresh tube.

C8. Add 2 μl of proteinase K (20 μg/μl), mix gently, and spin briefly.

C9. Incubate at 45°C for 1 hr.

C10. Heat at 90°C for 10 min to inactivate the proteinase K, and place on ice.

Polishing the cDNA ends

C11. Transfer 50 μl of the reaction product to a fresh 500-μl tube

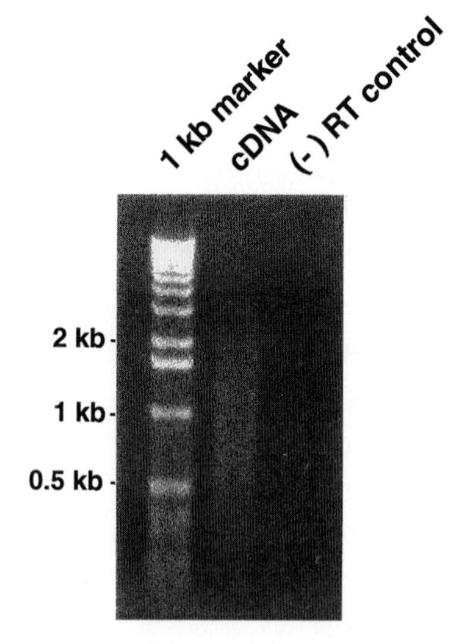

Figure 2 Typical results of a total single-cell cDNA amplification. The RT reaction from a single cell was subjected to 38 cycles of PCR. Five microliters of the 60-μl reaction was run on a 1% TAE gel. A cDNA smear from ~300 bp to >2 kb is seen. No product is present in control reaction, which did not include reverse transcriptase.

and add 3 μl (5 U/μl) of T4 DNA polymerase (store remainder of the reaction at −20°C).

C12. Incubate at 16°C for 30 min.

C13. Heat at 72°C for 10 min.

C14. Precipitate with 27.5 μl of 7.5 *M* NH_4OAc and 200 μl of room temperature 95% EtOH. Mix thoroughly. Spin at 14,000*g* for 30 min at room temperature.

C15. Remove the supernatant and overlay the pellet with 200 μl 75% ethanol (−20°C). Spin briefly and remove supernatant.

C16. Air-dry the pellet for ~10 min to evaporate EtOH.

C17. Add 14 μl deionized water, and resuspend the pellet by gentle pipetting.

Adapter ligation

C18. Ligate *Sal*I adapters to cDNA by adding the following to the resuspended cDNA on ice:

6 μl 5X ligation buffer

7 μl *Sal*I adapters (1 μg/μl)

3 μl T4 DNA ligase (1 U/μl)

30 μl total volume

Mix gently and spin briefly in a microfuge to collect reagents.

C19. Incubate at 16°C overnight.

C20. Add 3 μl of 0.2 *M* EDTA to terminate the reaction.

C21. Phenol:chloroform extract by adding 70 μl of water and 100 μl of phenol:chloroform, and mix by gentle inversion for 2 min.

C22. Centrifuge at 14,000*g* for 10 min at room temperature. Remove the top layer and place in a new 500-μl tube.

C23. Add 100 μl of chloroform and mix by gentle inversion for 2 min.

C24. Centrifuge at 14,000*g* for 10 min at room temperature. Remove the top layer and place in a 500-μl tube.

C25. Precipitate the cDNA by adding 1/10 vol. 3 *M* NaOAc and 2.5 vol. of cold (−20°C) 95% EtOH. Mix by gentle inversion, and place at −20°C for 1 hr.

C26. Centrifuge in a microfuge at 14,000*g* for 20 min, remove the supernatant, spin briefly, and remove residual liquid. Air-dry the pellet for ~10 min.

*Not*I *digestion*

C27. Add the following reagents on ice, in the order shown, to the cDNA from step C26:

41 μl deionized water

5 μl 10× *Not*I digestion buffer

4 μl *Not*I (15 U/μl)

50 μl final volume

C28. Mix gently and incubate the reaction for 2 hr. at 37°C.

C29. Add 50 μl phenol:chloroform:isoamyl alcohol (25:24:1), vortex thoroughly, and centrifuge at room temperature for 5 min at 14,000*g*.

C30. Remove 45 μl of the upper layer and place in a new 1.5-ml microfuge tube.

C31. Add 25 μl of 7.5 *M* NH_4OAc and 150 μl 95% EtOH (−20°C). Vortex thoroughly and immediately centrifuge at room temperature for 30 min at 14,000*g*.

C32. Carefully remove the supernatant and overlay the pellet with 500 μl of 75% EtOH (−20°C), centrifuge for 2 min as described previously, and carefully remove the supernatant.

C33. Evaporate residual EtOH for 10 min at room temperature. Place on ice or at 4°C until ready for size selection.

Step D: Size Selection and cDNA Cloning

It is important to remove small (<400 bp) cDNAs, which will clone more efficiently than larger fragments, and to remove the *Not*I restriction fragment and excess adapters. Size selection is performed using size-exclusion chromatography. Several vendors supply appropriate columns.

REAGENTS FOR STEP D

Sephacryl S-500 size-selection column (Life Technologies, No. 18092-015)
1× column buffer: Tris–HCl (pH 7.5), 0.1 m*M* EDTA, 25 m*M* NaCl
Glycogen (Boehringer-Mannheim, No. 901393)
pSPORT 1 *Not*I/*Sal*I cut (Life Technologies, No. 15383-011): 50 ng/μl
10× ligation buffer (see reagents for step C)
T4 DNA ligase (see reagents for step C)
LB agar plates with 100 μg/ml ampicillin
SOC media (Life Technologies)

Method

Size selection

D1. Preparation of size-selection columns (~1.5 hr):
Turn the column upside down and remove the bottom cap.
Aspirate any air bubbles with a pipette tip or syringe.
Replace the bottom cap, turn upright, remove top cap slowly, and then remove the bottom cap.
Attach the column to a stand and allow the column to drain completely.
Add 800 μl of column buffer to the top and allow to drain completely.
Repeat the 800-μl column buffer wash three times for a total of four washes.

D2. Label 18 sterile 1.5-ml microfuge tubes sequentially 1–18 and place in order on a rack beneath the column.

D3. Resuspend the pellet from step C33 in 50-μl column buffer.

D4. Add the sample to the top of the resin bed and allow to completely drain into the bed.

D5. Rinse the sample tube with 50 μl of column buffer and apply to the top of the resin bed.

D6. Collect the effluent into tube No. 1.

D7. Add 100 μl of column buffer to the column and allow to drain into collection tube No. 2.

D8. Add 100 μl of column buffer to the column and begin collecting single-drop fractions (~30 μl/drop) into individual microfuge tubes (Nos. 3–18). Continue adding 100-μl aliquots of column buffer until 18 fractions are collected.

D9. Run 3 μl of each fraction in a separate lane on a 1.2% agarose TAE minigel with 1-kb size markers for 10 min at 150 V. Determine the fractions that contain the greatest amount of high-molecular-weight product by visual inspection of the gel (Fig. 3). Typically, the cDNA elutes in fractions 6–11, with fractions 8 and 9 having the greatest size and intensity. Collect the two fractions of greatest size/intensity as well as the two fractions preceeding and the two fractions succeeding them (total of six fractions) (Fig. 3). Avoid fractions with low-molecular-weight material because these products will efficiently ligate and produce a library with small cDNA inserts.

D10. Pool the six fractions and measure the volume (approximately 180 μl).

D11. Add 5 μl yeast tRNA (1 μg/μl), 0.1 vol 3 *M* sodium acetate, and 2 vol 95% EtOH (−20°C).

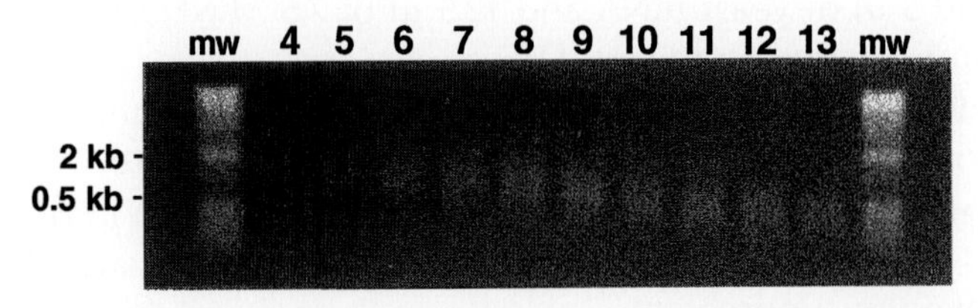

Figure 3 Size selection of cDNA. Sequential 30-μl fractions of size-selected cDNA were examined by running 3 μl of each fraction on a 1% TAE gel at 150 V for 10 min. Fractions 4–13 are shown adjacent to 1-kb ladder molecular-weight markers. The greatest amount of high-molecular-weight product is seen in fractions 7 and 8. Fractions 5–10 were pooled for construction of the library. In this example, fractions >10 contain short cDNAs and unligated adapters and are not used for the ligation reaction.

D12. Vortex briefly and immediately centrifuge at room temperature for 20 min at 14,000*g*.

D13. Carefully remove supernatant with a pipette, and overlay with 300 μl 75% EtOH (−20°C).

D14. Centrifuge for 2 min at 14,000*g*, and remove the supernatant. Allow the pellet to air-dry for ~10 min.

D15. Resuspend the pellet in 6 μl deionized water.

Ligation to plasmid vector (see method note 9)

The optimal ratio for cDNA insert to vector is an important determinant of transformation efficiency and must be determined empirically for the cDNA/vector combination. It is suggested to set up three parallel ligation reactions with three different cDNA/vector ratios.

D16. On ice, set up the ligations with the resuspended cDNA from step D15 in three 0.5-ml tubes:

	Ligation No.		
Component	**1 (μl)**	**2 (μl)**	**3 (μl)**
cDNA	0.5	1.0	1.5
pSPORT 1 (50 ng/μl)	1.0	1.0	1.0
5× ligation buffer	4.0	4.0	4.0
T4 DNA ligase	1.0	1.0	1.0
Water	13.5	13.0	12.5
Total	20	20	20

D17. Mix gently and incubate at 4°C for a minimum of 16 hr.

Electroporation into electrocompetent Escherichia coli (see alternative method 3)

D18. Add 5 μl of yeast tRNA and 12.5 μl of 7.5 *M* NH_4OAc to the ligation reaction from step D15.

D19. Add 70 μl of 95% EtOH (−20°C). Vortex and immediately centrifuge for 20 min at 14,000*g* at room temperature.

D20. Remove supernatant. Overlay the pellet with 500 μl of 75% EtOH (−20°C). Centrifuge for 2 min at 14,000*g* and carefully remove the supernatant.

D21. Allow pellet to air-dry for 10 min to remove residual EtOH.

D22. Resuspend the pellet in 5 μl of deionized water.

D23. Add 1 μl of the resuspended cDNA to 40 μl of Electromax DH10B cells and electroporate in a 0.1-cm gap electroporation cuvette in a BioRad Gene Pulser with the following settings: 1.8 kV, 200 ohm, 25 μF (see instructions included with the competent cells). Store the remaining ligation reaction at −20°C.

D24. Immediately add 1 ml of SOC media to the electroporated cells and incubate at 37°C at 300 rpm for 1 hour.

D25. Plate dilutions of 1.0, 0.1, and 0.01 μl on LB plates containing 100 μg/ml ampicillin, and incubate the plates overnight at 37°C. Store the remaining transformed cells at 4°C.

D26. Count the colonies on each plate and calculate (i) the total number of colonies in the 1-ml electroporation and (ii) the number of colonies per microgram of cDNA (see *alternative method 3)*.

D27. If additional transformants are desired, electroporate additional aliquots of the ligation reaction.

D28. Plate out the library for screening and/or store aliquots of the library by adding 1 ml of a sterile solution of 40% glycerol/60% LB (v/v) (final glycerol concentration is 20%) and aliquoting into cryovials for storage at −80°C or under liquid nitrogen. If desired, the library may be amplified to provide a large clone resource (Ausubel *et al.*, 1995).

Step E: Assessment of Library Quality

The quality of the cDNA library can be measured by several parameters, including the diversity and total number of cDNA clones, the percentage of clones with inserts, the length of the clone inserts, and the absence of contamination (Fig. 4). The total number of clones

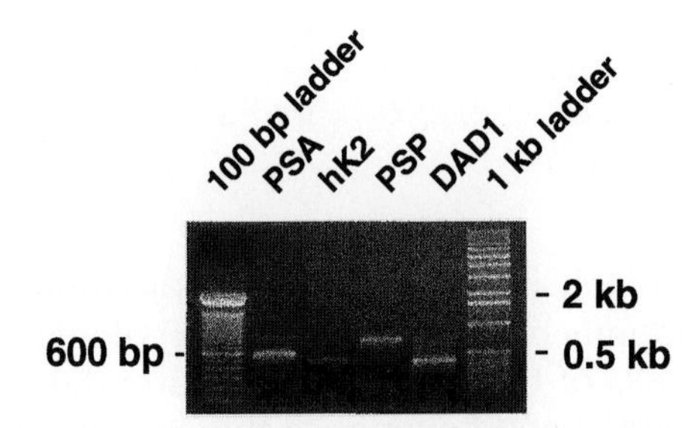

Figure 4 Analysis of library complexity. One-microliter aliquots of a 1:20 dilution of the adapter-ligated cDNA were subjected to PCR with intron-spanning PCR primers for the amplification of the prostate-specific genes, prostate-specific antigen (PSA), human glandular kallikrein 2 (hK2), and prostate secretory protein (PSP) and the ubiquitously expressed gene defender-against-death-1 (DAD1).

is determined by titering the library (see D25 and *alternative method 3*). The number of clones with inserts can be determined by blue/white screening with IPTG and X-gal (Maniatis *et al.*, 1989). The size of the cDNA inserts can rapidly be determined using PCR and insert-flanking vector primers of randomly selected clones, and the complexity can be determined by hybridization (Ausubel *et al.*, 1995) or by single-pass cDNA sequencing (Adams *et al.*, 1992).

Reagents

10 μM VN26 oligonucleotide: 5′-TTTCCCAGTCACGACGTTGTA-3′

10 μM VN27 oligonucleotide: 5′-GTGAGCGGATAACAATTTCAC-3′

Taq DNA polymerase, 5 U/μl (Life Technologies)

10 mM dNTPs (Pharmacia, Alameda, CA: 100 mM each dNTP, No. 27-2035-01)

10× polymerase buffer (Life Technologies)

50 mM $MgCl_2$

Method

Checking Recombinant Clones and Insert Size

Prepare a PCR reaction master mix by adding the following reagents on ice, and dispense 50 μl into each well of a 96-well thermocycler plate on ice.

Reagent	Volume (μl) × 1	Volume (μl) × 20
10× PCR buffer	5	100
Deionized water	41.5	830
$MgCl_2$ (50 mM)	1.5	30
dNTPs (10 mM)	0.6	12
VN26 primer (10 μM)	0.6	12
VN27 primer (10 μM)	0.6	12
Taq polymerase (5 U/μl)	0.2	4
Total	50	1000

Scale up appropriately if desired. Using sterile toothpicks or plastic pipette tips, randomly pick individual white colonies and place into the individual wells of the PCR plate. Prewarm the thermocycler to 96°C, insert the plate, and commence thermocycling:

(95°C for 30 sec) × 1 cycle

(95°C for 15 sec., 60°C for 30 sec., 72°C for 120 sec. × 35 cycles

(72°C for 10 min) × 1 cycle

Analyze 3 μl on a 1% agarose TAE gel with 0.1 μg/ml ethidium bromide and 1-kb DNA size markers. The PCR products can be rapidly purified and sequenced directly if desired (Wang *et al.*, 1995).

Note: the absence of a PCR product does not necessarily mean the clone lacks an insert. cDNAs >3 kb are difficult to amplify routinely using this method. Plasmid preps should be done on clones that do not amplify before concluding that a clone does not contain an insert.

Method Notes

Method note 1: An RNAse-free environment is essential during the initial stages of the protocol, prior to the conversion of mRNA to cDNA. Suggestions: wear gloves and use freshly deionized water directly or dH_2O treated with diethyl pyrocarbonate. Unopened sterile single-use plasticware is generally safe. For more detail, see Blumberg (1987).

Method note 2: RNAse inhibitors require DTT in the reaction mixture to prevent irreversible inactivation. "Lock docking" primers are used to prime cDNA synthesis at the junction of the gene-specific sequence and the poly-A tail (Borson *et al.*, 1992). The primer also incorporates a *Not*I restriction site to facilitate directional cloning of the cDNA.

Method note 3: Larger products have been shown to be produced with Superscript reverse transcriptase (SSRT) than with AMV RT, possibly due to lack of RNAse H activity in SSRT. This feature is particularly important when synthesizing cDNA from small amounts of mRNA. In contrast to nonrecombinant RTs, excess SSRT does not appear to be deleterious to the cDNA reaction.

Method note 4: The length of the incubation time for the RT reaction will influence the length of cDNAs synthesized. If it is essential to maintain the transcript ratios after amplification, one option is to limit the RT reaction time to 15 min to ensure that all cDNAs are of approximately equal length and thus eliminate a potential bias due to amplification (Brady and Iscove, 1993).

Method note 5: The initial isolation of RNA is not routinely performed in this protocol due to the losses incurred when attempting to purify RNA from single cells. Reaction specificity is provided by the oligo-(dT) primer and the cap primer. Single-cell expression studies have been reported using cDNA synthesized without RNA isolation steps (Berardi *et al.*, 1995; Klebe *et al.*, 1996; Trumper *et al.*, 1993). DNase treatment should be considered if the RT(−) reaction gives a high-molecular-weight product. [Note: A nonspecific product of <400 bp is often seen in both the cell (−) and RT (−) controls and does not imply DNA contamination.] To the cell lysis reaction in step A4 add 0.5 μl of 0.5 U/μl DNase I (GenHunter, Nashville, TN; 1 μl of 10 U/μl and 19 μl of 1X DNase buffer). Incubate for 15 min at 37°C followed by 75°C for 5 min to inactivate the DNase (Huang *et al.*, 1996). Proceed to step C1. Alternatively, a method for the direct isolation of RNA from single cells has been described (Ziegler *et al.*, 1995).

Method note 6: The method for providing a defined nucleotide sequence to the 3′ end of each newly synthesized cDNA molecule involves the use of a cap oligonucleotide that base pairs with deoxycytidine residues added by RT's terminal transferase activity. This activity appears to occur at the 7-methylguanosine cap structure present on the 5′ end of most mRNAs (Fromont-Racine *et al.*, 1993) and serves to extend the template for SSRT. Each new cDNA will thus contain the cap oligo sequence that can be used for PCR amplification. This technology is called SMART (patent pending, Clontech) for switch mechanism at the 5′ end of RNA templates (Zhu *et al.*, 1996). The 5′-end oligonucleotide is modified in a proprietary fashion, and simply synthesizing an oligo of appropriate sequence may not provide satisfactory results. Kits incorporating the SMART protocol and a long-distance thermostable polymerase mixture (*method note 7*) are available (Clontech, No. K1051-1).

Method note 7: Modifications to DNA polymerases and combinations of DNA polymerases have been shown to significantly increase the base pair fidelity and amplifiable target length of the PCR. A mixture of KlenTaq1 (a 5′-exo(−), N-terminal deletion variant of *Taq* DNA polymerase) and a proofreading polymerase, such as *Pfu*, Vent, or Deep Vent, is able to amplify targets up to 35 kb (Barnes, 1994). These mixtures are extremely useful for the amplifications of cDNA populations. Several long-distance polymerase mixtures are commercially available. The polymerase mix suggested in this protocol (Advantage cDNA polymerase mix, Clontech, No. 8417-1) is a combination of KlenTaq1 polymerase, Deep Vent polymerase, and TaqStart antibody to provide automatic hot start PCR (Kellogg *et al.*, 1994). Hot start refers to any method for assembling PCR reactions that keeps one or more of the reaction components physically or functionally separate from the rest of the components prior to the onset to the thermal cycling. This technique prevents background due to low-level DNA synthesis from nonspecifically primed sites.

Method note 8: The PCR parameters described in the protocol have been optimized to use thin-wall PCR tubes in a Perkin–Elmer System 480 thermal cycler. Other tubes and thermal cyclers may require modifications. The optimal cycling parameters are somewhat empirical and may need to be worked out for each cell type. In general, use the least number of cycles because fewer cycles produce fewer nonspecific PCR products. If a faint smear (or no smear) is seen following the PCR, place the PCR reaction back in the thermal cycler and cycle for 2–6 more cycles. If no product is seen, set up a fresh PCR reaction with 5 μl of the PCR product as a template and perform 10–15 cycles. For unknown reasons, a smear of low-molecular-weight material (<400 bp) is often seen in the (−) cell and (−) RT control reactions. A low-molecular-weight smear of <800 bp in the (+) RT tubes, or a total lack of PCR product in the amplification, may indicate degraded RNA template caused either by improper sample handling (RNA not preserved) or by a contaminant in a reagent, test tubes, etc.

Method note 9: Adapters are used to increase the efficiency of ligation by incorporating a sticky 5′ extension. Other investigator-prepared or commercially available adapters can be used with the appropriate manipulation of the cloning vector. The adapter structure is:

5′ TCGACCTCGAG 3′
3′ GGAGCTC$_p$ 5′

To eliminate self-ligation of the adapters during cDNA ligation, only one of the oligomers of the adapter is phosphorylated.

Alternative Methods

Alternative method 1—Obtaining single cells: Individual cells can be obtained from cells in suspension using fluorescent monoclonal antibody staining by aspiration through micropipettes (Trumper *et al.*, 1993) or using limiting dilution assays (Sharrock *et al.*, 1990). Additional methods include flow cytometric sorting using single-cell deposition devices (Bertram *et al.*, 1995) and laser capture microdissection techniques (Emmert-Buck *et al.*, 1996).

Alternative method 2—Homopolymer tailing: The method of homopolymer tailing of cDNAs to provide a known sequence suitable for subsequent amplification is derived from the rapid amplification of cDNA ends (RACE) strategy to clone rare transcripts (Frohman *et al.*, 1988). This method has been further adapted for cDNA library construction (Belyavsky *et al.*, 1989; Brady and Iscove, 1993). Briefly, TdT is used to add a homopolymer stretch of any nucleotide to the ends of the first-strand cDNAs (step B3 of the protocol). Primers are designed complementary to the homopolymer tail, and the cDNAs are amplified by the PCR. An additional method for adding a common sequence to the cDNA ends involves attaching a linker of defined sequence using RNA ligase (Dumas Milne Edwards *et al.*, 1997).

Alternative method 3—Heat shock transformation: Electroporation of the ligated cDNA will generally yield a greater number of transformants from the same amount of ligated cDNA compared to the heat shock method. However, for many purposes heat shock transformation into high-efficiency competent cells (MAX Efficiency DH10B, Life Technologies) will produce a sufficiently complex library. Transform a 100-μl sample of competent cells with 5 μl (12.5-ng vector) of the ligation reaction from protocol step D17. Follow the instructions accompanying the competent cells, but do not dilute the ligation reaction. Following the transformation, plate aliquots as described in protocol step D25 to determine the transformation efficiency. Approximately 2.5–5.0 $\times$ 10^7 transformants/μg cDNA (0.5–1.0 $\times$ 10^6 transformants/μg of vector or 1 $\times$ 10^5 transformants per 100-μl transformation reaction) are obtained using the heat shock protocol, and electroporation typically yields $\sim$ 1–5 $\times$ 10^6 transformants per 40-μl transformation reaction. A library containing $\sim$1 $\times$ 10^6 independent clones is representative of the mRNA complexity in most cases.

Acknowledgments

I thank Dr. James Eberwine, Dr. Victor Ng, Dr. Barbara Trask, and Oanh Nguyen for helpful discussions and review of the manuscript. I thank Vilaska Nguyen for assistance with figure preparation. This work was supported in part by a grant from the CaPCURE Foundation.

References

Adams, M. D., Dubnick, M., Kerlavage, A. R., Moreno, R., Kelley, J. M., Utterback, T. R., Nagle, J. W., Fields, C., and Venter, J. C. (1992). Sequence identification of 2,375 human brain genes. *Nature* **355,** 632–634.

Ausubel, F. M., Brent, R., Kingston, R. E., Moore, D. D., Seidman, J. G., Smith, J. A., and Struhl, K. (1995). In *Current Protocols in Molecular Biology* (K. Janssen, Ed.). Wiley, New York.

Barnes, W. M. (1994). PCR amplifications of up to 35-kb DNA with high fidelity and high yield from lambda bacteriophage templates. *Proc. Natl. Acad. Sci. U.S.A.* **91,** 2216–2220.

Belyavsky, A., Vinogradova, T., and Rajewsky, K. (1989). PCR-based cDNA library construction: General cDNA libraries at the level of a few cells [published erratum appears in *Nucleic Acids Res.* 1989 July 25; **17**(14), 5883]. *Nucleic Acids Res.* **17,** 2919–2932.

Berardi, A. C., Wang, A., Levine, J. D., Lopez, P., and Scadden, D. T. (1995). Functional isolation and characterization of human hematopoietic stem cells. *Science* **267,** 104–108.

Bertram, S., Hufert, F. T., Neumann-Haefelin, D., and von Laer, D. (1995). Detection of DNA in single cells using an automated cell deposition unit and PCR. *BioTechniques* **19,** 616–620.

Bishop, J. O., Morton, J. G., Rosbash, M., and Richardson, M. (1974). Three abundance classes in Hela cell messenger RNA. *Nature* **250,** 199–204.

Blumberg, D. D. (1987). Creating a ribonuclease-free environment. *Methods Enzymol.* **152,** 20–24.

Borson, N. D., Salo, W. L., and Drewes, L. R. (1992). A lock-docking oligo(dT) primer for 5′ and 3′ RACE PCR. *PCR Methods Appl.* **2,** 144–148.

Brady, G., and Iscove, N. N. (1993). Construction of cDNA libraries from single cells. *Methods Enzymol.* **225,** 611–623.

Dumas Milne Edwards, J. B., Valdenaire, O., and Mallet, J. (1997). Anchoring a defined sequence to the 5′ ends of mRNAs. The bolt to clone rare full-length mRNAs. *Methods Mol. Biol.* **67,** 261–278.

Eberwine, J., Spencer, C., Miyashiro, K., Mackler, S., and Finnell, R. (1992). Complementary DNA synthesis in situ: Methods and applications. *Methods Enzymol.* **216,** 80–100.

Emmert-Buck, M. R., Bonner, R. F., Smith, P. D., Chuaqui, R. F., Zhuang, Z., Goldstein, S. R., Weiss, R. A., and Liotta, L. A. (1996). Laser capture microdissection [see Comments]. *Science* **274,** 998–1001.

Frohman, M. A., Dush, M. K., and Martin, G. R. (1988). Rapid production of full-length cDNAs from rare transcripts: Amplification using a single gene-specific oligonucleotide primer. *Proc. Natl. Acad. Sci. U.S.A.* **85,** 8998–9002.

Fromont-Racine, M., Bertrand, E., Pictet, R., and Grange, T. (1993). A highly sensitive method for mapping the 5′ termini of mRNAs. *Nucleic Acids Res.* **21,** 1683–1684.

Gordon, J. W. (1993). Micromanipulation of gametes and embryos. *Methods Enzymol.* **225,** 207–238.

Gubler, U., and Hoffman, B. J. (1983). A simple and very efficient method of generating cDNA libraries. *Gene* **25,** 263–269.

Huang, Z., Fasco, M. J., and Kaminsky, L. S. (1996). Optimization of Dnase I removal of contaminating DNA from RNA for use in quantitative RNA-PCR. *BioTechniques* **20,** 1012–1014, 1016, 1018–1020.

Jena, P. K., Liu, A. H., Smith, D. S., and Wysocki, L. J. (1996). Amplification of genes, single transcripts and cDNA libraries from one cell and direct sequence analysis of amplified products derived from one molecule. *J. Immunol. Methods* **190,** 199–213.

Kellogg, D. E., Rybalkin, I., Chen, S., Mukhamedova, N., Vlasik, T., Siebert, P. D., and Chenchik, A. (1994). TaqStart antibody: "Hot start" PCR facilitated by a neutralizing monoclonal antibody directed against Taq DNA polymerase. *BioTechniques* **16,** 1134–1137.

Klebe, R. J., Grant, G. M., Grant, A. M., Garcia, M. A., Giambernardi, T. A., and Taylor, G. P. (1996). RT-PCR without RNA isolation. *Biotechniques* **21,** 1094–1100.

Maniatis, T., Fritsch, E. F., and Sambrook, J. (1989). *Molecular Cloning: A Laboratory Manual* (2nd ed.). Cold Spring Harbor Laboratory Press, Cold Spring Harbor, NY.

Schweinfest, C. W., Nelson, P. S., Graber, M. W., Demopoulos, R. I., and Papas, T. S. (1995). Subtraction cDNA libraries. *Methods Mol. Biol.*

Sharrock, C. E., Kaminski, E., and Man, S. (1990). Limiting dilution analysis of human T cells: A useful clinical tool [see Comments]. *Immunol. Today* **11,** 281–286.

Trumper, L. H., Brady, G., Bagg, A., Gray, D., Loke, S. L., Griesser, H., Wagman, R., Braziel, R., Gascoyne, R. D., Vicini, S., *et al.* (1993). Single-cell analysis of Hodgkin and Reed–Sternberg cells: Molecular heterogeneity of gene expression and p53 mutations. *Blood* **81,** 3097–3115.

Velculescu, V. E., Zhang, L., Zhou, W., Vogelstein, J., Basrai, M. A., Bassett, D. E., Hieter, P., Vogelstein, B., and Kinzler, K. W. (1997). Characterization of the yeast transcriptome. *Cell* **88,** 243–251.

Wang, K., Gan, L., Boysen, C., and Hood, L. (1995). A microtiter plate-based high-throughput DNA purification method. *Anal. Biochem.* **226,** 85–90.

Zhu, Y., Chenchik, A., and Siebert, P. D. (1996). Synthesis of high-quality cDNA from nanograms of total or poly A+ RNA with the CapFinder PCR cDNA library construction kit. *CLONTECHniques* **XI.**

Ziegler, B. L., Lamping, C. P., Thoma, S. J., Thomas, C. A., and Fliedner, T. M. (1995). Single-cell cDNA-PCR. In *Methods in Neurosciences,* Vol. 26, pp. 62–73. Academic Press, San Diego.

20

WHOLE CELL ASSAYS

James Snider

The fork-like structure-dependent synthesis-associated nuclease activity of *Taq* DNA polymerase, first described by Holland *et al.*, has emerged as a very powerful analytic technique for nucleic acids analysis. In 1993 (Lee, 1993) this 5′ nuclease assay evolved to incorporate a dually labeled fluorescent, sequence-specific oligonucleotide probe and the TaqMan assay was born.

This truly homogenous assay is very simple, requiring only a labeled probe selected for a sequence between the primers in addition to normal PCR reagents. Figure 1 illustrates the TaqMan concept. As the PCR primers are extended during the extension phase of the amplification cycle one of the strand extension complexes will run into the 5′ end of the labeled probe. *Taq* DNA polymerase will begin displacement of the opposing probe sequence and once the fork-like structure is created the probe will be nucleolytically cleaved. This process continues until the entire probe is displaced and the amplicon is completed. Measurable fluorescent signal is created when the probe is cleaved. When the probe is intact, the excited state energy of the reporter fluorescein dye, located at the 5′ end of the probe, is absorbed by the quencher dye, located at the 3′ end of the probe.

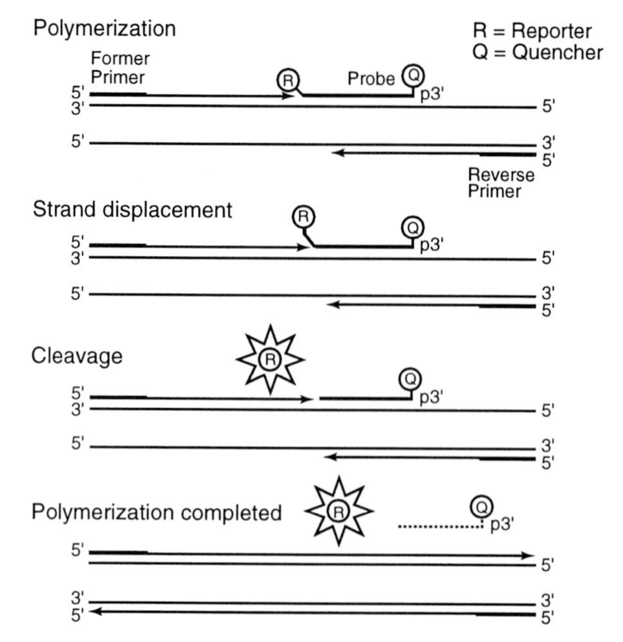

Figure 1 Schematic representation of the homogenous nature of the TaqMan reaction.

When the probe is cleaved by the enzyme the reporter dye can fluoresce freely; this increase in reporter dye signal represents the generation of an amplicon. Fluorescent signal accumulates with accumulation of amplicon and accurately represents the amount of amplicon created.

It is possible to measure TaqMan reaction fluorescent signal with instruments that provide either real-time data, reading signal each cycle, or endpoint data, reading signal only at the end of the reaction. Both real-time and endpoint measurements are valuable but are applicable for different techniques. For quantitation and optimization purposes real-time measurements have proved invaluable. A representative real-time TaqMan amplification plot generated by an ABI Prism 7700 Sequence Detector is illustrated in Fig. 2. A typical PCR amplification is characterized by two distinct phases, a log phase and a plateau phase. During the log phase the amplification is increasing exponentially, effectively doubling the amount of amplicon with each subsequent cycle. At some point in the amplification process

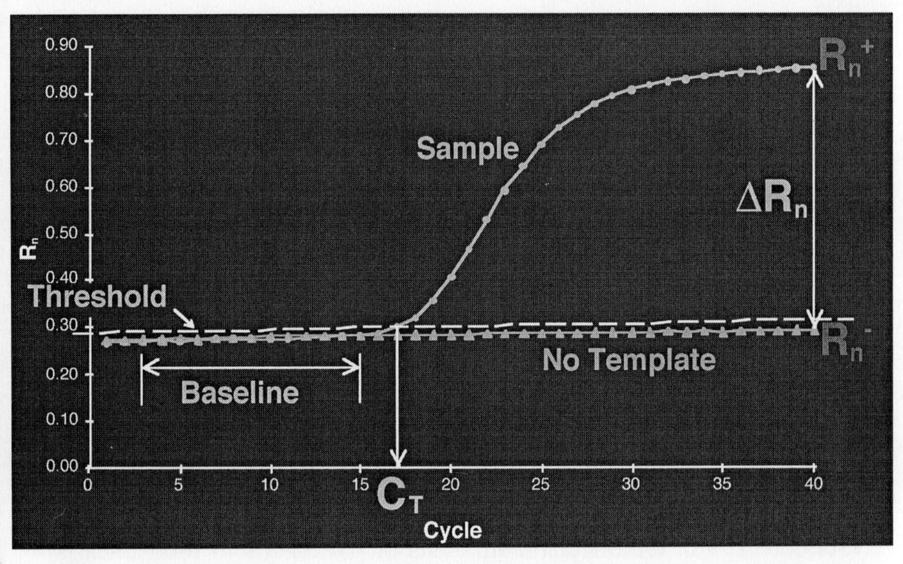

Figure 2 Representation of a real-time amplification plot with designations of the distinct phases.

a critical reaction component becomes rate limiting and the reaction transitions into the plateau, effectively creating an environment in which no net amplicon accumulation is occurring and the signal remains constant. The full value of real-time analysis for quantitation purposes has been discussed in detail by Livak *et al.* To summarize briefly, during log phase, reactions with the same primer/probe set behave almost identically and very consistently. Endpoint measurements, however, have been shown to be quite variable—much more variable than real-time measurements.

The amplification plot in Fig. 2 shows three phases: a baseline phase, a log phase, and a plateau phase. The nature of the 7700 detection instrument is such that in early log phase cycles there is not enough cleaved probe signal generated for the instrument to measure. At a point referred to as the threshold cycle (C_t) the log phase signal increases above the baseline. For the rest of the amplification process the signal accumulates as expected. More concentrated templates generate lower C_t values than do less concentrated templates. This phenomenon has created in new paradigm for PCR quantitation: when signal appears rather than how much signal is generated at a specific time.

There are several advantages to using real-time measurements; one of the most significant advantage is reaction robustness. The factors that drive a reaction into plateau and consequently create variable endpoint measurements are not as influential in the log phase of the reaction, resulting in very reproducible measurements. The inherent homogeneity and sensitivity of the TaqMan reaction, coupled with the power of real-time measurements, creates a technology well suited to generic analysis of complex samples or whole cell assays.

Whole cell assay has been interpreted to mean several things: (i) analysis using total cellular RNA or genomic DNA or (ii) analysis of viral or bacterial sequences from various sources or subsets thereof. Many laboratories are applying TaqMan technology to analysis of gene expression systems using total cellular RNA (Livak, 1996; Heid, 1996; Gibson, 1996; Gerard, 1996). Some laboratories have also published reports of analysis of viral sequences from typical sources (Schoen, 1996). Quantitative analysis of pathogens in foodstuffs is a prime application (Bassler, 1995; Witham, 1996). An application of emerging significance is biallelic genotyping using TaqMan probes (Livak, 1995). The rest of this discussion will focus on a specific example of a generic assay for *Escherichia coli* genomic sequence. The concepts for this assay are universally applicable to any sequence of interest.

The fundamental requirement to design a successful TaqMan assay is to select primers compatible with a probe designed to detect the sequence of interest. Because this is a sequence-specific assay, it is imperative that the sequence be known. TaqMan is not well suited to analysis of *de novo* sequences. The Appendix itemizes the steps for selecting the oligos and optimization of the reaction. If this protocol is adhered to rigorously, a new assay can be designed with a very high level of confidence. As we developed these protocols we recognized that the most influential variable in the design was the primer selection relative to the probe. An early assumption was that it was necessary to select a primer set, confirm by traditional techniques that this primer set worked appropriately, and then select a probe compatible with this primer set. With this design paradigm we experienced a significant assay optimization barrier. Through experience we discovered that if the probe is selected first and the primer sequences are selected to be compatible with this probe, then the optimization barrier is much lower and the overall success rate is improved. Figure 3 illustrates such an example.

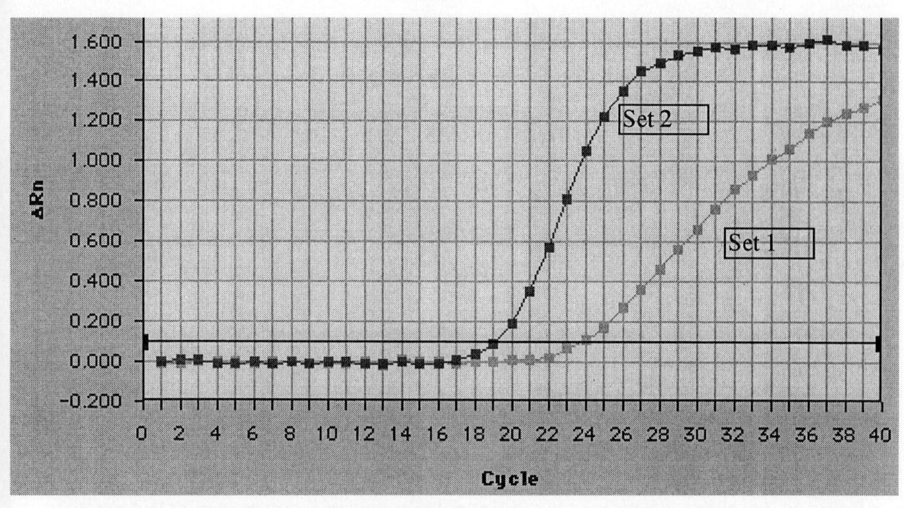

Figure 3 Comparative amplification plots illustrating the performance difference derived from primer sequence selection.

In these amplification plots of an assay for a typical DNA virus, the probe sequence is constant and the reverse primer sequence is constant; the only variable is the forward primer sequence. Primer set 1 had been shown to work well in traditional gel analysis of the amplicon. The best choice of probe for these primers was selected. For primer set 2 the same probe and reverse primer were used, but the forward primer was selected to be more compatible with the other two oligos. The amplification plots in Fig. 3 illustrate the dramatic performance difference achieved by selecting a new primer sequence. This phenomenon is not unique to this target sequence and has been observed many times. By following the protocol in the Appendix, the amplification plots typically resemble the example for primer set 2 in Fig. 3.

As in the previous examples, we designed a TaqMan oligo set to be used in the measurement of residual *E. coli* DNA levels in a quality control assay. The objective in this situation is to provide an accurate, consistent assay to measure residual DNA levels in a recombinant product. The principles employed are identical to those described earlier and can be applied to virtually any required assay. Figure 4 illustrates the amplification profiles generated for this assay and the resolution between template dilutions. As expected, fivefold dilutions are resolved after ~2.5 cycles.

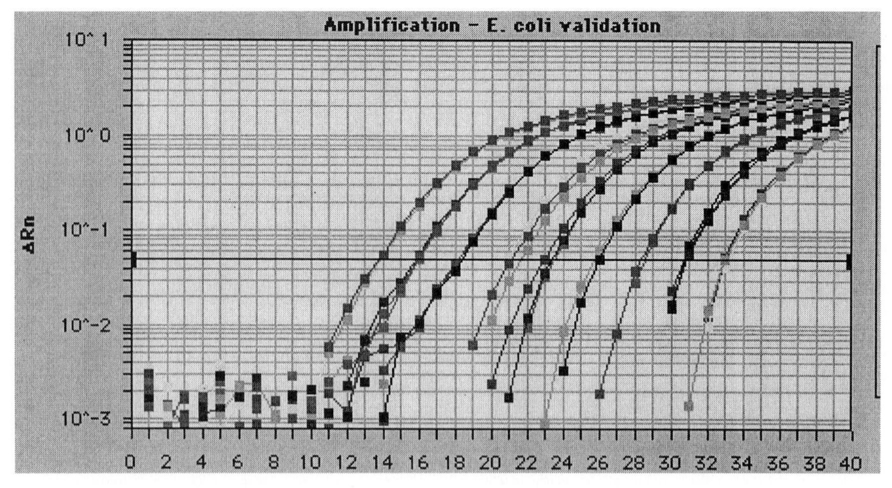

Figure 4 Amplification plots of fivefold serial dilutions of *E. coli* genomic DNA shown in semi-log view. The highest concentrations, plots on the extreme left, are 1–4 pgs. Each group represents three replicates. The black horizontal line represents the R_n value at which C_t was determined. These C_t values were plotted against the template amount to calculate a standard curve for which a series of unknowns were measured against.

Although this result is quite impressive, it does not address the consistency requirement. To assess the consistency of this assay, refer to Table 1, which summarizes an experiment in which an external standard curve was employed to quantitatively measure the template concentration in replicates of dilutions of three different unknown samples. The objective of this experiment is to evaluate the reproducibility of replicate measurements and the predictability of dilution measurements. Each group is composed of four replicates and shows a very high level of reproducibility. Furthermore, when dilutions are compared the compensated results are also very consistent. Certainly this does not completely validate this assay, but it does provide strong evidence that this assay and assays based on these principles should be generally applicable.

All the measurements shown in Table 1, with the exception of the 1:10 dilution from sample A, are within the bounds of the standard curve. This sample was not considered in subsequent analysis. By analyzing replicates of dilutions, we hoped to illustrate that (i) the technology would very reproducibly analyze replicates and (ii) dilution factors would be maintained and confirmed by the analy-

Table 1

Unknown Samples Applied to the Standard Curve Generated from the Example in Fig. 4[a]

Sample	Quantity (pg)	Normalized (pg)	Amount (ng)	Total Amount	% of Total
A 1:10	3.51 E +04	350,580	350.6	2600	23.8
A 1:100	5.91 E +03	590,870	590.9		
A 1:500	1.29 E +03	646,850	646.9		
B 1:10	1.08 E +03	10,814	10.8	1600	0.75
B 1:100	1.32 E +02	13,199	13.2		
B 1:500	2.36 E +01	11,802	11.8		
C 1:10	1.08 E +04	107,710	107.7	1800	7.1
C 1:100	1.32 E +03	131,600	131.6		
C 1:500	2.51 E +02	125,370	125.4		

[a] In this experiment, four replicates of three dilutions of three different unknowns were analyzed.

sis. Table 1 does not list the variation of the replicates but these samples were virtually identical to the amplifications shown in Fig. 4; there was almost no measurable variation. Table 1 clearly shows that the dilution factors were maintained during this analysis. This is most easily seen in the Amount column. With the noted exception of A 1:10, the dilutions compare very well.

An additional analysis was performed on these samples to determine the percentage of the total amount of DNA attributable to residual *E. coli.* Total amount column in Table 1 shows the DNA concentration as determined by an OD_{260} measurement. In these samples the *E. coli* DNA was spiked at a known level and this assay very accurately measured this concentration (shown in the last column).

In this thesis I have only considered the factors influencing amplification and detection. The quality of sample and the presence of reaction inhibitors can have a profound influence on the overall process. Sample preparation is too complex to fully develop here, but it is important to recognize that some compounds, which may copurify with the nucleic acids, may completely inhibit the reaction resulting in no amplification. TaqMan does not have any special sample purity requirements; our experience has shown that any purification method that yields PCR-quality nucleic acids is acceptable. In fact, TaqMan, due to its very sensitive characteristics, is quite powerful in assessing the quality of purification techniques.

Appendix

Primer and Probe Design Guidelines

1. Amplicon length: Robust TaqMan assay development is at its best with the shortest possible amplicons. Consistent and predictable results can be obtained for amplicons as short as 50 bp and as long as 200 bp. TaqMan technology is ideally suited for sequence detection within the domain of TaqMan probes. Forward and reverse primers should thus be designed as close as possible to each TaqMan probe.
2. Sequence guidelines for either probes or primers
   %GC range of 30–80%
   Less than 4 contiguous G′s
3. Design the probe
   No G on 5′ end
   Primer Express T_m of 68–70°C for single-probe applications
   Choose the strand with a probe having more C′s than G′s.
4. Design the primers
   Design the forward and reverse primers as close as possible to the probe without overlapping the probe.
   Primer Express T_m of 58–60°C
   Five nucleotides at the 3′ end should have only 1 or 2 G + C′s.
5. Annealing temperature
   If the probe T_m is determined empirically in the Sequence Detection Systems 7700, then annealing temperature = T_m (probe) −10°C.
   If the probe T_m is estimated only from Primer Express, then annealing temperature = 60°C.

Generic Master Mix Formulation

Reagents in the TaqMan PCR Core Reagent Kit (Product No. N8080228) and 20% glycerol (Product No. 402929) should be used to prepare a TaqMan generic master mix. The following is the composition of this master mix:

Reagent	Final concentration in TaqMan PCR reaction	Volume per 115 (25-μl) Wells (μl)
20% glycerol	6%	863
10× TaqMan buffer A	1X	288
25 m*M* $MgCl_2$	7.5 mM	863
dATP	200 μ*M*	58
dCTP	200 μ*M*	58
dGTP	200 μ*M*	58
dUTP	400 μ*M*	58
AmpliTaq Gold, 5 U/μl	0.05 U/μl	29
UNG, 1 U/μl	0.01 U/μl	29
Total		2304

Target Amplification

The initial conditions for the TaqMan PCR reaction should be as follows:

Reaction component	Final concentration in TaqMan PCR reaction	Volume (25 μl per Well)
Generic master mix	1X	20.00
Forward + reverse primers	200 n*M*	1.25
TaqMan probes	100 n*M*	1.25
DNA sample	0.4 ng/μl	2.50
Total		25.00

Thermal cycling	Temperature (°C)	Time	No. of cycles
Stage 1	50	2 min	1
Stage 2	95	10 min	1
Stage 3	95	15 sec	40
	60	1 min	

Optimization of Primer Concentrations

Primer concentrations are optimized at the 60°C elongation temperature defined previously. The forward and reverse primers are cooptimized by running the wells defined by the 3 × 3 matrix shown below at a 10 n*M* probe concentration. For allelic discrimination, this optimization may be carried out with only one of the two probes designed per mismatch site.

A minimum of four replicate wells are run for each of the nine conditions defined by this matrix. This matrix covers an effective T_m range of ±2°C around the nominal T_m for these primers.

Forward primer (n*M*) **Reverse primer (n*M*)**	**50**	**350**	**900**
50	50/50	350/50	900/50
350	50/350	350/350	900/350
900	50/900	350/900	900/900

At the end of these runs, tabulate the results for the R_n and C_t in a similar matrix, and choose the minimum forward and reverse primer concentrations which yield the maximum R_n and minimum C_t.

Optimization of Probe Concentrations

Probe concentration is optimized at the 60°C elongation temperature and optimum forward and reverse primer concentrations defined previously. The concentration is optimized by running wells at 50 nM intervals between 50 and 400 n*M*. The purpose of this optimization is to choose the minimum probe concentration yielding the maximum R_n and minimum C_t.

References

Bassler, H. (1995). Use of a fluorogenic probe in a PCR based assay for the detection of Listeria monocytogenes. *Appl. Environ. Microbiol.* **61**(10), 3724–3728.

Gerard, C. (1996). A rapid and quantitative assay to estimate gene transfer into retrovirally transduced hematopoietic stem/progenitor cells using a 96 well format PCR and fluorescent detection system universal for MMLV-based proviruses. *Human Gene Therapy* **7**, 343–354.

Gibson, U. (1996). A novel method for real time quantitative RT-PCR. *Genome Res.* **6**(6), 995–1001.

Heid, C. (1996). Real time quantitative PCR. *Genome Res.* **6**(6), 986–994.

Holland, P. (1991). Detection of specific polymerase chain reaction product by utilizing the 5′ → 3′ exonuclease activity of Thermus aquaticus DNA polymerase. *Proc. Natl. Acad. Sci. U.S.A.* **88,** 7276–7280.

Lee, L. (1993). Allelic discrimination by nick-translation PCR with fluorogenic probes. *Nucleic Acids Res.* **21**(6), 3761–3766.

Livak, K. (1995). Towards fully automated genome wide polymorphism screening. *Nature Genet.* **9,** 341–342.

Livak, K. (1996). Quantitation of DNA/RNA using real-time PCR detection. *PE-ABD Response Piece,* 1–8.

Schoen, C. D. (1996). Detection of potato leafroll virus in dormant potato tubers by immunocapture and a fluorogenic 5′ nuclease RT-PCR assay. *Pytopathology* **86**(9), 993–999.

Witham, P. (1996). A PCR based assay for the detection of Escherichia coli shiga-like toxin genes in ground beef. *Appl. Environ. Microbiol.* **62**(4), 1347–1353.

21

SCREENING DIFFERENTIALLY DISPLAYED PCR PRODUCTS BY SINGLE-STRAND CONFORMATION POLYMORPHISM GELS

Françoise Mathieu-Daudé, Nick Benson, Frank Kullmann, Rhonda Honeycutt, Michael McClelland, and John Welsh

Since first described in 1992, differential display (Liang and Pardee, 1992) and RNA abitrarily primed PCR (RAP-PCR; Welsh *et al.*, 1992) have been extensively used to investigate differential gene expression and coordinate regulation in a wide variety of situations, including cell responses to different treatments and growth conditions (Ralph *et al.*, 1993; McClelland *et al.*, 1994; Blanchard and Cousins, 1996; Holmes *et al.*, 1997) and comparisons of different developmental stages (Que *et al.*, 1996; Paterno *et al.*, 1997). The RNA fingerprinting approach has also found many applications in cancer research (Chen *et al.*, 1996; Chang *et al.*, 1997; Vogt *et al.*, 1997). Several modifications to the original protocols have been reported (Liang and Pardee, 1997), and the development of new technologies, such as automated sequencing or the use of fluorescent-tagged primers, is facilitating the use of this approach. The different variations of RNA fingerprinting have three common steps: (i) cDNA synthesis by reverse transcription (RT) of total or messenger RNAs followed by PCR amplification using arbitrary primers and gel display, (ii) isolation and characterization of differentially amplified transcripts, and (iii) confirmation of differential expression. The step of isolation and characterization of the PCR fragments representing differentially amplified products is often a bottleneck in this approach because it is laborious or gives products that are not differentially expressed because they are not the ones being targeted. These problems are associated with the comigration of other PCR products in the fingerprint gel. In this chapter we focus on the use of single-strand polymorphism gels in order to facilitate the isolation and purification of differentially amplified cDNAs from differential display and RAP-PCR fingerprinting experiments.[1] We provide the two protocols that are currently in use in our laboratory.

Principles

In our original protocol, the bands representing interesting differences in abundances of some products were excised from the RNA fingerprinting gels, reamplified using the primers originally employed, cloned, and sequenced. However, because of comigration of products in the original fingerprint, the reamplification of even a tiny portion of the gel usually generates multiple products of almost the same size. Thus, in the next steps, reamplified material contains a mixture of desirable and undesirable products. After cloning of this mixture, many clones must be sequenced and Southern blotted to determine which one of the sequences corresponds to the gene of interest. Assuming that the band of interest vastly predominates after PCR, the statistically predominant sequence should be the targeted sequence. However, we and others found that during PCR,

[1] The RAP-PCR protocol differs from the differential display protocol originally described by Liang and Pardee (1992) in that an arbitrary primer is used in both steps of the reaction, whereas differential display only uses arbitrary priming in the amplification step and anchored oligo(dT) for RT. The use of arbitrary primers in both directions allows the sampling of internal RNA fragments, rather than the noncoding 3′ end, increasing the probability of sampling protein coding regions. By using two different primers, one for the RT reaction and another added for the PCR reaction, fragments that are generated do not have the same primer at each extremity, allowing the orientation of the sequenced product to be inferred. Products with the same primer at each end are at a disadvantage due to the formation of "panhandles" (Welsh and McClelland, 1991).

the "Cot effect" tends to reduce the amplification rate of the most abundant products comparing to the less abundant ones, resulting in a partial normalization.[2]

A considerable amount of wasted effort can be avoided by the use of a native acrylamide gel to separate the cDNA product of interest away from other products of different sequence but of similar size based on single-strand conformation polymorphisms (SSCP). This application is less challenging than most applications of the method. Instead of needing to distinguish molecules that differ by only one or a few bases, as must be done in the typical application of the method, we need only distinguish between molecules that share a similar size but have entirely different sequences.[3] We have described this approach (Mathieu-Daudé *et al.*, 1996a) in which SSCP gels are used on the reamplified mixture. The gel allows the classification of the PCR products into (i) those that can be directly cloned because they are relatively pure, (ii) those that will be excised from the SSCP gel and reamplified, and (iii) those that consist of a too complex mixture and can be rejected for further study. This approach is the method of choice to isolate and characterize products from a set of fingerprints that have been generated some time ago. However, we found that problems associated with reamplification of the product isolated from the gel could be avoided by purifying the product of interest on the SSCP gel prior to reamplification. This latter approach is easier and faster; thus, it should be preferentially used, but the SSCP gel has to be run immediately, while the products from the original fingerprint are still radioactive.

[2] In later cycles of PCR, rehybridization of abundant products inhibits their amplification efficiency. Thus, the rate of amplification of abundant PCR products declines faster than that of less abundant products in the same tube, an effect referred to as the Cot effect (Suzuki and Giovannoni, 1996; Mathieu-Daudé *et al.*, 1996b). Differences in product abundances will decrease as the number of PCR cycles increases, and small differences in mRNA abundances will gradually be erased. Thus, the Cot effect causes a partial normalization of cDNAs. One way to minimize this effect is to reduce the number of PCR cycles. Reducing the number of cycles is particularly important for quantitative RT-PCR experiments used for confirmation of differential expression.

[3] The mobility of single-stranded DNAs in a native acrylamide gel is affected by their secondary structure, and the migration will rely on both the molecular weight and the conformation of the product. PCR-SSCP (Hayashi, 1991) was first described to detect mutations by alterations in mobility of separated single strands (band shifts). We use SSCP electrophoresis to separate targeted sequences from contaminating products which comigrate with differentially expressed fragments. Since resolution is based on conformational analysis, products that comigrate with the product of interest on the original fingerprint, but which have a totally different sequence, will have a different mobility on the SSCP gel.

Equipment and Reagents

Tris–EDTA (TE) buffer: 10 m*M* Tris–HCl, 1 m*M* EDTA, pH 7.5
20 m*M* dNTPs mix (5 m*M* each dNTP)
Arbitrary primers (100 μM), 10- to 20-mers (Operon Technologies, Alameda, CA, or Genosys Biotechnologies, The Woodlands, TX)
Taq polymerase (Ampli*Taq* DNA Polymerase, Perkin–Elmer–Cetus, Norwalk, CT)
10× PCR buffer: 100 m*M* Tris–HCl, 500 m*M* KCl, pH 8.3
25–100 m*M* $MgCl_2$ stock solution
[α-^{32}P]dCTP (3000 Ci/mmol)
Thermal cycler (e.g., GeneAmp PCR System 9600, Perkin–Elmer–Cetus)
10× Tris-borate–EDTA (TBE) buffer: 0.89 *M* Tris–borate, 20 m*M* EDTA, pH 8.3
4- or 5% polyacrylamide, 45% urea gel prepared in 1× TBE buffer
2× HydroLink MDE gel solution (FMC BioProducts, Rockland, ME)
10% (w/v) ammonium persulfate and TEMED (Sigma)
Sequencing gel electrophoresis apparatus with plates and combs
Denaturing loading buffers: 96% formamide, 0.1% bromophenol blue, 0.1% xylene cyanol, and 10 m*M* EDTA for polyacrylamide gels; and 96% formamide, 0.1% bromophenol blue, 0.1% xylene cyanol, and 10 m*M* NaOH for MDE gels
Power supply and gel dryer with vacuum pump
X-ray films (Kodak X-Omat and BioMax) and cassettes
QIAEX II gel extraction kit (Qiagen, Chatsworth, CA)

Method and Protocols

Isolation and Purification of Differentially Amplified Products

Here we describe two protocols currently in use for the isolation of the PCR products from the RNA fingerprinting experiments. The first protocol is preferred but requires RAP-PCR or differential display reactions that are still sufficiently radiolabeled. Alternatively, the second protocol can be used for old RAP-PCR fingerprints that are no longer radioactive. Both protocols begin with the gel display of the PCR reactions (differential display or RAP-PCR fingerprint). If interesting differences in abundances of some products are noticed on these gels, they are selected for the isolation of the differentially amplified bands of interest.

Protocol 1: Direct Purification on SSCP Gel

1. An aliquot of the amplification products of the RAP-PCR reaction (4 μl) is mixed with 15 μl of formamide dye solution, denaturated at 94°C for 3 min and chilled on ice; 2 μl is loaded onto a 4–6% polyacrylamide–urea gel. Electrophoresis is performed at 1500 V for approximately 4 hr. The gel is dried under vacuum and placed on a Kodak BioMax X-ray film for 16–48 hr. Several luminescent labels should be taped on the dried gel before exposure to allow the subsequent alignment of the autoradiogram with the gel when excising interesting bands out of the gel.
2. (Optional) The RAP-PCR reactions containing the bands of interest are reloaded onto a preparative polyacrylamide–urea gel in multiple adjacent lanes (three or four lanes, 3 μl/lane) and electrophoresed as described previously. This gel allows the resolution of a large quantity of the PCR product of interest. This step is not necessary if the original gel was performed with a wide well comb because the band to be excised will have a sufficient mass.
3. The autoradiogram and the gel are aligned and the bands of interest are excised from the gel using a needle and a razor blade. The excision should be precise and the size of the fragment cut should be kept to a minimum. This gel is reexposed to X-ray film to check for the accuracy of the excision.
4. Each band is eluted from the tiny piece of gel in 100 μl of TE at 65°C for 2 or 3 hr and left at room temperature a few more hours to allow more diffusion. Alternatively, the band can be eluted using the QIAEX II gel extraction kit (Qiagen).
5. After ethanol precipitation of the eluate, the radioactive pellet is directly resuspended in 4 μl of the MDE loading dye. The

sample is denatured at 94°C for 3 min, cooled on ice, and loaded onto an HydroLink MDE gel prepared in 0.6× TBE buffer with 5% glycerol according to the manufacturer's instructions. Electrophoresis is performed overnight (for approximately 16 hr) at 8 W.

6. The dried gel, with luminescent labels, is autoradiographed using an intensifying screen. If pure, each product is expected to produce two SSCP bands of the same intensity, one for each of the two DNA strands. Frequently, however, only one band is observed because only one strand was excised from the polyacrylamide denaturing gel. This is usually observed for low-molecular-weight bands in which the two strands are more easily separated in a denaturing gel. When more than two bands are observed, because of the carryover of contaminating products that comigrated in the initial polyacrylamide gel, the product of interest is the band of highest intensity (Fig. 1).
7. The product resolved on the SSCP gel is excised and eluted as in step 4 in 50 μl of TE; 8 μl of the eluted solution is reamplified using the same primers as in the original fingerprint. The template is mixed with 32 μl of PCR mixture for a 40-μl final reaction containing 10 m*M* Tris (pH 8.3), 50 m*M* KCl, 3 m*M* $MgCl_2$, 0.2 m*M* of each dNTP, 1 μM of each primer, and 4 U of Ampli*Taq* polymerase.
8. Thermocycling is performed using 25–30 cycles of 94°C for 30 sec, 37°C for 30 sec, and 72°C for 1 min, followed by a final 72°C extension for 5 min.
9. PCR products are run on a low-melting-point agarose gel (NuSieve GTG, FMC BioProducts), cleaned with the QIAEX II

Figure 1 Electrophoresis on a SSCP gel of differentially amplified products isolated from a RAP-PCR fingerprint gel. Bands representing interesting products were excised from the dried denaturing gel, eluted, precipitated, and run on the SSCP gel as described under Protocol 1. Arrowheads on both sides of the gel indicate each single strand of the products: two strands for products 1 (320 bases) and 2 (201 bases) and one strand only for product 3 (146 bases). Empty arrows indicate contaminant bands of lower intensity that were comigrating with the product of interest in the original fingerprint gel. Each single-stranded product (one strand only for products 2 and 3) is isolated from this gel, reamplified, and sequenced as described under Protocol 1. These products correspond to differentially expressed genes in *Salmonella typhimurium* in response to mitomycin C treatment.

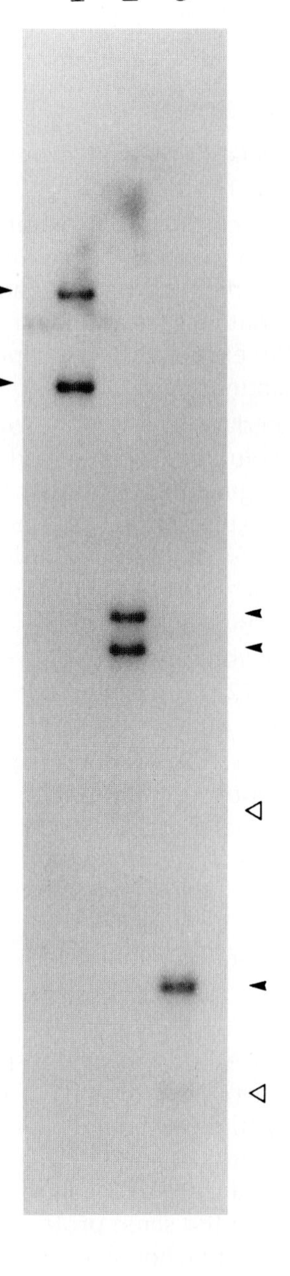
1
2
3

gel extraction kit (Qiagen), and used for direct sequencing or cloning followed by sequencing.

PROTOCOL 2: SSCP GEL PURIFICATION WITH PRIOR REAMPLIFICATION

1. RAP-PCR or differential display fingerprints have been performed as in step 1 of protocol 1.
2. Bands of interest are cut out from the original fingerprint as described in step 3 of protocol 1. In addition, for each of the bands, the corresponding region in an adjacent lane where the product is not present, or less abundant, is cut out as a control. The precision of the excision is very important to minimize the carry over of contaminants.
3. The gel slice is eluted as in step 4 of protocol 1 in 50 μl of TE; 5 μl of the eluted solution is reamplified using the same primers as in the original fingerprint. The template is mixed with 15 μl of PCR mixture for a 20-μl final reaction containing 10 mM Tris (pH 8.3), 50 mM KCl, 3 mM $MgCl_2$ 0.2 mM of each dNTP, 0.75 μM of each primer, 0.5 μCi [α-^{32}P]dCTP, and 2 U of Ampli*Taq* polymerase.
4. Thermocycling is performed using 25 cycles of 94°C for 30 sec, 37°C for 30 sec, and 72°C for 1 min, followed by a final 72°C extension for 5 min.
5. Four microliters of the PCR products is mixed with 18 μl of formamide dye solution containing 10 mM NaOH, denatured at 94°C for 3 min and cooled on ice; 2 μl of each sample is loaded onto an HydroLink MDE gel prepared in 0.6× TBE buffer according to the manufacturer's instructions. Electrophoresis is performed overnight (for approximately 16 hr) at 8 W. The dried gel is then autoradiographed with several luminescent labels.
6. Different profiles can be obtained.
    a. In some rare cases, the product displays only two bands of the same intensity, one for each of the DNA strands, or one band only (the two strands having the same mobility) but free of any contaminant. In that case, the product is pure, and the template of the PCR can be directly reamplified using the same protocol as in steps 3 and 4, but in 40–μl final reaction and omitting the radiolabeled dCTP.

b. In most cases, the profiles show one to three bands of low intensity in addition to the one or two bands of high intensity. The product of interest is most likely to be the band of highest intensity. This can be confirmed by the control lanes. In the control (from the adjacent lane in which the product was absent or less abundant), this band should be absent or of lower intensity since the PCR at a low number of cycles has preserved the difference in the abundances (Fig. 2). Once the correct product is identified, it is excised from the SSCP gel and eluted as previously described in 50 μl of TE. Eight microliters of the eluted solution is mixed with 32 μl of PCR mixture and reamplified in the same conditions as in steps 3 and 4 using 30 cycles and omitting the radiolabeled dCTP. Figure 3 shows examples of products of this second turn of reamplification, where [α-^{32}P]dCTP was deliberately added in the PCR and the products were run on a second SSCP gel to check for the purity of the products after reamplification. This figure demonstrates the usefulness of the purification step on the SSCP gel since now the products are free of any contaminants.

c. Occasionally, the profile has several or many bands of the same intensity that may or may not differ from the control. Products giving such a complex pattern can be eliminated and not considered for further studies. Even in these cases, the SSCP gel was useful since time was not wasted on cloning and sequencing of mixtures of products.

7. The nonradioactive PCR products are run on a low-melting-point agarose gel and cleaned as in step 9 of protocol 1 and then used for direct sequencing or cloning followed by sequencing.

Cloning, Sequencing, and Southern Blots

Fragments can be cloned into a PCR cloning vector containing 3′-T overhangs, such as the pCR 2.1 vector (Original- or TOPO- TA Cloning Kit, Invitrogen, Carlsbad, CA). Plasmids or purified PCR products can be sequenced using sequencing kits and automatic sequencers [i.e., ABI PRISM Dye Terminator Cycle Sequencing Kit (Perkin–Elmer) or Thermo Sequenase Fluorescent Sequencing Kit (Amersham, Arlington Heights, IL)]. Direct sequencing of the PCR products, using one of the original 10-mer primers, is usually possible except when the

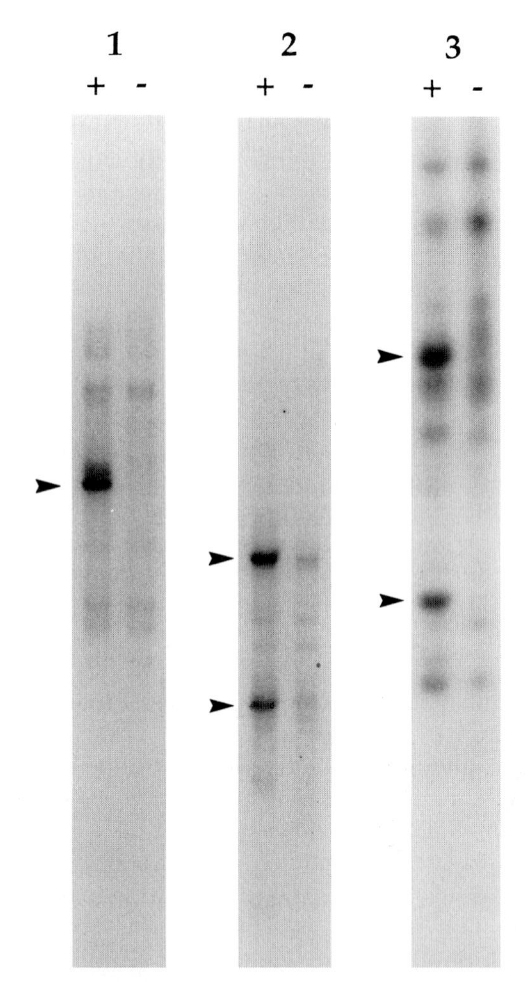

Figure 2 Electrophoresis on a SSCP gel of the reamplification products of differentially amplified fragments excised from a RAP-PCR fingerprint gel. Isolation, reamplification, and electrophoresis were performed as described under Protocol 2. For each of the three products, two reamplification experiments are run in parallel. Lane (+) corresponds to the reamplification of the product cut from the RAP-PCR gel in the lanes in which it was more abundant, and lane (−) corresponds to the reamplification of the portion of gel in which the product was of low abundance or apparently absent. Arrowheads indicate the single strands of the product of interest in + lanes, which are absent or of lower intensity in the corresponding control lanes (−). The single-stranded products identified as being the products of interest are then isolated from this gel, reamplified, and sequenced as described under Protocol 2.

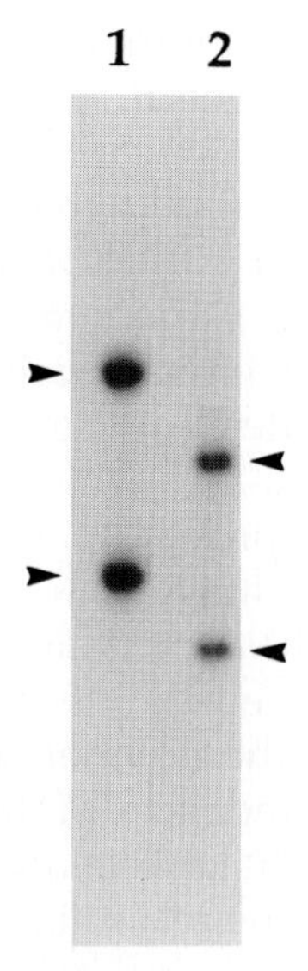

Figure 3 Electrophoresis on a SSCP gel of two products that were isolated from the SSCP gel and reamplified a second time in the presence of radiolabeled dCTP to check for the efficiency of the SSCP purification. Arrowheads indicate the two strands of the product free of any contaminants.

PCR product of interest has the same primers at both ends. If short primers are to be used for sequencing, then the cycling conditions should be adjusted by lowering the annealing temperature to 35°C and adding a ramp to reach the elongation temperature. Cloning of the correct fragment can be confirmed by Southern blot after capillary transfer of the RAP fingerprint polyacrylamide gel onto a nylon membrane (Hybond N+, Amersham, Buckinghamshire, UK) using standard conditions. Clones that serve as probes are labeled and hybridized to the membranes using the nonradioactive ECL direct nucleic acid labeling and detection system (Amersham) according to the manufacturer's instructions (Mathieu-Daudé *et al.,* 1998).

Confirmation of Differential Gene Expression

Once the differentially amplified transcripts isolated from RAP-PCR fingerprints have been sequenced, confirmation of the differential expression of the genes is an obligatory step. Relative abundances of RNAs between samples can be assessed by conventional methods, such as Northern blotting using standard protocols (Sambrook *et al.*, 1989) or RT-PCR using an internal control (Gilliland *et al.*, 1990; Bouaboula *et al.*, 1992; Ferré *et al.*, 1994). However, when a large amount of template RNA or a decent control are not available, an alternative approach can be used. In this method, two specific primers of the transcripts are used in RT-PCR under low-stringency conditions similar to those used to generate RAP-PCR fingerprints. In addition to the product of interest, arbitrary products which are largely invariant are generated and behave as internal controls for RNA quality and quantity and for reverse transcription efficiency. This simple approach has proved to be effective in determining relative abundances of specific RNA transcripts in microorganisms for which low quantities of RNA are available and an invariant internal standard is not known (Mathieu-Daudé *et al.*, 1998, 1999). In all quantitative RT-PCR approaches, the number of PCR cycles must be low (14–24 according to the abundances of the product) to preserve the differences in starting template mRNA abundances. End labeling one of the primers can increase the resolution by rendering the products highly radioactive.

Acknowledgments

This work was supported in part by the Tobacco-Related Disease Research Program of the University of California, grant number 6KT-0272, to F.M-D. and by a generous gift from Sidney Kimmel.

References

Blanchard, R. K., and Cousins, R. J. (1996). Differential display of intestinal mRNAs regulated by dietary zinc. *Proc. Natl. Acad. Sci. U.S.A.* **93,** 6863–6868.

Bouaboula, M., Legoux, P., Pességué, B., Delpech, B., Dumont, X., Piechaczyk, M., Casellas, P., and Shire, D. (1992). Standardization of mRNA titration using a polymerase chain reaction method involving co-amplification with a multispecific internal control. *J. Biol. Chem.* **267,** 21830–21838.

Chang, G. T., Blok, L. J., Steenbeek, M., Veldscholte, J., van Weerden, W. M., van Steenbrugge, G. J., and Brinkmann, A. O. (1997). Differentially expressed genes in androgen-dependent and -independent prostate carcinomas. *Cancer Res.* **57,** 4075–4081.

Chen, S. L., Maroulakou, I. G., Green, J. E., Romano-Spica, V., Modi, W., Lautenberger, J., and Bhat, N. K. (1996). Isolation and characterization of a novel gene expressed in multiple cancers. *Oncogene* **12,** 741–751.

Ferré, F., Marchese, A., Pezzoli, P. Griffin, S., Buxton, E., and Boyer, V. (1994). Quantitative PCR: An overview. *In The Polymerase Chain Reaction* (K. B. Mullis, F. Ferré, and R. A. Gibbs, Eds.), pp. 67–88. Birkhauser, Boston.

Gilliland, G., Perrin, S., Blanchard, K., and Bunn, H. F. (1990). Analysis of cytokine mRNA and DNA: Detection and quantitation by competitive polymerase chain reaction. *Proc. Natl. Acad. Sci. U.S.A.* **87,** 2725–2729.

Hayashi, K. (1991). PCR-SSCP: A simple and sensitive method for detection of mutations. *PCR Methods Appl.* **1,** 34–38.

Holmes, D. I., Abdel Wahab, N., and Mason, R. M. (1997). Identification of glucose-regulated genes in human mesangial cells by mRNA differential display. *Biochem. Biophys. Res. Commun.* **238,** 179–184.

Liang, P., and Pardee, A. (1992). Differential display of eukaryotic messenger RNA by means of the polymerase chain reaction. *Science* **257,** 967–971.

Liang, P., and Pardee, A. (1997). *Methods in Molecular Biology: Differential Display Methods and Protocols* (P. Liang and A. B. Pardee, Eds.). Humana Press Totowa, NJ.

Mathieu-Daudé, F., Cheng, R., Welsh, J., and McClelland, M. (1996a). Screening of differentially amplified cDNA products from RNA arbitrarily primed PCR fingerprints using single strand conformation polymorphism (SSCP) gels. *Nucleic Acids Res.* **24,** 1504–1507.

Mathieu-Daudé, F., Welsh, J., Vogt, T., and McClelland, M. (1996b). DNA rehybridization during PCR: The "Cot effect" and its consequences. *Nucleic Acids Res.* **24,** 2080–2086.

Mathieu-Daudé, F., Welsh, J., Davis, C., and McClelland, M. (1998). Differentially expressed genes in the *Trypanosoma brucei* life cycle identified by RNA fingerprinting. *Mol. Biochem. Parasitol.* **92,** 15–28.

Mathieu-Daudé, F., Trenkle, T., Welsh, J., Jung, B., Vogt, T., and McClelland, M. (1999). Identification of differentially expressed genes using RNA fingerprinting by arbitrarily primed PCR. *In Methods in Enzymology: cDNA Preparation and Display* (J. N. Abelson and M. I. Simon, Eds.). Academic Press, Orlando.

McClelland, M., Ralph, D., Cheng, R., and Welsh, J. (1994). Interactions among regulators of RNA abundance characterized using RNA fingerprinting by arbitrarily primed PCR. *Nucleic Acids Res.* **22,** 4419–4431.

Paterno, G. D., Li, Y., Luchman, H. A., Ryan, P. J., and Gillespie, L. L. (1997). cDNA cloning of a novel, developmentally regulated immediate early gene activated by fibroblast growth factor and encoding a nuclear protein. *J. Biol. Chem.* **272,** 25591–25595.

Que, X., Svard, S. G., Meng, T. C., Hetsko, M. L., Aley, S. B., and Gillin, F. D. (1996). Developmentally regulated transcripts and evidence of differential mRNA processing in *Giardia lamblia. Mol. Biochem. Parasitol.* **81,** 101–110.

Ralph, D., McClelland, M., and Welsh, J. (1993). RNA fingerprinting using arbitrarily primed PCR identifies differentially regulated RNAs in Mink lung (Mv1Lu) cells growth arrested by TGF-β. *Proc. Natl. Acad. Sci. U.S.A.* **90,** 10710–10714.

Sambrook J., Fritsch, E. F., and Maniatis T. (1989). *Molecular Cloning: A Laboratory Manual.* Cold Spring Harbor Laboratory Press, Cold Spring Harbor, NY.

Suzuki, M. T., and Giovannoni, S. (1996). Bias caused by template annealing in the amplification of mixtures of 16 S rRNA genes by PCR. *Appl. Environ. Microbiol.* **62,** 625–630.

Vogt, T., Welsh, J., Stolz, W., Kullmann, F., Jung, B., Landthaler, M., and McClelland, M. (1997). RNA fingerprinting displays UVB-specific disruption of transcriptional control in human melanocytes. *Cancer Res.* **57,** 3554–3561.

Welsh, J., and McClelland, M. (1991). Genomic fingerprinting using arbitrarily primed PCR and a matrix of pairwise combinations of primers. *Nucleic Acids. Res.* **19,** 5275–5279.
Welsh, J., Chada, K., Dalal, S. S., Ralph, D., Cheng, R., and McClelland, M. (1992). Arbitrarily primed PCR fingerprinting of RNA. *Nucleic Acids Res.* **20,** 4965–4970.

22

MICROSATELLITE PROTOCOLS

Yun Oh and Li Mao

New techniques are permitting molecular biologists and geneticists to systematically evaluate and compare large areas of the human genome. Much of this analysis is based on the PCR amplification of microsatellites. Mammalian genomes contain large amounts of repetitive DNA sequences, an increasing number of which are being identified as stretches of tandem repeat units. The repeat units ranging in size from 8 to 50 bp form stretches of DNA referred to as variable number tandem repeats or minisatellites. The repeat units ranging in size from 2 to 6 bp form stretches of DNA referred to as short tandem repeats or microsatellites.

Microsatellites have been particularly useful for comparative genetics and genomic mapping since their first description because of several fortuitous characteristics. First, microsatellite sequences show a high degree of length polymorphism. Human subjects average >70% heterozygosity at individual microsatellites (Anonymous, 1992). Second, microsatellite sequences are abundant and evenly distributed throughout the genome. In one survey microsatellite tandem repeats of >20 bp in length can be found at every 6 kb in a 745-kb sequence of human DNA (Beckman and Weber, 1992). Mark-

PCR Applications

ers flanking microsatellite sequences more than 100 bp long can be used as primers to PCR amplify tandem repeat sequences in specific genomic loci (Weber and May, 1989). These amplified microsatellite sequences can be resolved on polyacrylamide and high-resolution agarose gels, allowing rapid analysis of large areas of the genome.

Amplification of an individual's DNA with a panel of markers yields a characteristic pattern of allelic sizes. The larger the panel used, the more specific is the resulting microsatellite "profile" for the individual. This application of microsatellites has been utilized extensively in forensics to establish individual identity (Linquist *et al.*, 1996) and also in paternity testing to demonstrate allelic inheritance.

Microsatellite analysis is becoming instrumental in gene discovery. The detailed allelotyping provided by microsatellites has enabled geneticists to localize familial disease genes by linkage analysis, for example, the familial Mediterranean fever gene (Touitou *et al.*, 1996). Once localized to an area of a chromosome using a screening panel of microsatellites, the disease gene may then be isolated by a more closely focused panel.

Microsatellite analysis has been instrumental in the discovery of tumor suppressor genes deleted in malignant tumor cells. Tumor suppressor genes are usually inactivated by deletion of one allele and point mutation of the other allele. Microsatellites linked to the deleted alleles will not be amplified by PCR, resulting in loss of heterozygosity (LOH). Compared to restriction fragment-length polymorphisms (RFLP), the traditional method of LOH detection, microsatellites offer a higher rate of heterozygosity resulting in fewer noninformative loci. Microsatellite analysis also requires far less DNA sample than does RFLP analysis. Furthermore, finer resolution of genetic loci and ease of performance favor microsatellite analysis, which has practically replaced RFLP analysis for the detection of LOH (Lindquist *et al.*, 1996).

With microsatellite markers, paraffin-embedded archival tissue can now be screened for LOH at chromosomal loci; rare and scant biopsy specimens from malignancies and premalignancies can be microdissected and analyzed for deleted tumor suppressor loci (Boige *et al.*, 1997; Mao *et al.*, 1996). A typical allelotype screen analyzes 100–400 microsatellites per sample. Using multiplex PCR, a panel of 10–15 microsatellites can be analyzed simultaneously, further reducing the amount of sample needed (Lindquist *et al.*, 1996). Microsatellite analysis is also needed to determine replication error phenotype (RER$^+$) in tumors with defects of DNA mismatch repair genes (Eshleman and Markowitz, 1995) or other microsatellite instabilities in different types of tumors (Mao *et al.*, 1994).

Progressive discovery of new microsatellite sequences and their markers has enabled the development of a comprehensive map of the human genome (Dib *et al.*, 1996). Over 5000 microsatellites have been utilized in this endeavor as of 1996, and the number of microsatellite markers identified has increased 10-fold during the past 5 years, permitting finer resolution of the genetic mapping (Anonymous, 1992; Dib *et al.*, 1996). Currently, approximately 20,000 mapped microsatellite markers are available commercially (Research Genetics, Huntsville, AL).

Protocols

DNA Extraction from Fresh Tissue or Cells

Routine DNA Extraction Protocol

Up to 200 mg of fresh tissue or tissue stored at −80°C is homogenized mechanically in 2.7 ml TE-9 buffer (0.5 *M*, Tris, pH 8.8, 20 m*M* EDTA, 10 m*M* NaCl). Add 300 μl 10× digestion buffer [10% sodium dodecyl sulfate (SDS) and 10 mg/ml proteinase K in TE-9 buffer).

Similarly, cultured cells up to 10^7 cells can be resuspended directly into 200 μl of 1× digestion buffer (1% SDS and 1 mg/ml proteinase K in TE-9).

Incubate the digest suspension at 60°C for 4–6 hr.

Extract twice with 1 vol of phenol : chloroform in a 3 : 1 volume ratio, buffered in TE.

Transfer the aqueous layer to a fresh tube containing one-third volume of 10 *M* ammonium acetate, 3 vol of 100% ethanol, and 2 μl of glycogen (20 mg/ml). Incubate at room temperature for 5 min and centrifuge at 14,000 rpm for 20 min.

Remove supernatant. Wash DNA pellet in 70% ethanol once, dry lightly, and resuspend in a suitable volume of distilled H_2O.

DNA concentration can be measured by spectrophotometer. High-quality DNA samples have a 260/280 nm absorption ratio of 1.7–1.9.

DNA Extraction from Parafin-Embedded Tissue

The following discussion applies to microdissection of tissue sections from routine formalin-fixed, paraffin-embedded tissue archived in most pathology departments. Fixatives containing heavy metals, such as mercury, are strong inhibitors of PCR (Fiallo *et al.*, 1992). Tissue fixed in formalin for longer than 24 hr prior to paraffin embedding or in poorly buffered formalin will yield less DNA; however, archival tissue stored in formalin for more than 20 years can still yield DNA, albeit one or two orders of magnitude less (Kosel and Graeber, 1994).

Microdissection from Tissue Section

Obtain 4- to 12-μm sections of paraffin-embedded tissue for microdissection.

Stain an adjacent 4-μm section with hematoxalyn or other suitable dye to assess the quantity and homogeneity of desired cells in the specimen and to guide microdissection of the 12-μm section.

While observing the tissue section under a stereomicroscope, scrape desired tissue from the slide using a fresh scalpel blade and sterile steel needles. If the slides have not yet been deparaffinized, scrape the tissue into a 1.5-ml screw-top tube containing 1 ml xylenes. If sections were previously deparaffinized manually or by an automated slide processor, scrape the tissue directly into digestion buffer, skipping the next two steps.

Shake the tube for 15 sec. Add 250 μl of 70% ethanol and shake again. Pellet cell debris DNA by high-speed microcentrifugation for 5 min.

Discard supernatant and air-dry the pellet for 1 hr or vacuum centrifuge the pellet for 15 min.

Resuspend tissue pellet in 300 μl of 1× digestion buffer and incubate at 42°C for 24–48 hr. After the first 12 hr, another 30 μl of 10× digestion buffer is added to augment digestion.

Extract DNA using the same protocol as that for fresh tissue and cultured cells.

Tissue sections can be microdissected with even greater precision using a laser capture technique. Individual cells or clusters of cells can be selectively adhered onto a plastic membrane overlying the slide by photocoagulation; the cells can then be processed in a similar way for DNA extraction (Emmert-Buck *et al.*, 1996).

Labeling PCR Products for Microsatellite Analysis

Unlabeled microsatellite PCR products can be visualized on a high-resolution agarose gel by including ethidium bromide in the gel or by staining with other dyes, such as SYBR green (FMC, Rockland, ME). However, for reasons discussed later, most workers prefer to use polyacrylamide gels for analyzing microsatellites, which requires either radioisotope or fluorescence labeling of the PCR products.

Radioisotopic Labeling of PCR Products—Two Methods

1. 5′ end-label one of the primers of a microsatellite marker pair with γ-^{32}P and include this primer in the PCR reaction. To end-label a primer, add the following reagents together in a microfuge tube starting with the largest volume (usually H_2O) first and finishing with the smallest volume last:

   Distilled H_2O to a total volume of 50 μl
   40 pmol of one primer (~1 μl)
   1 μl bovine serum albumin
   5 μl 10× T4 kinase buffer
   40 μCi [γ-^{32}P]ATP (4.5 Ci/mmol; ICN Pharmaceuticals, Costa Mesa, CA)
   10 U (1 μl) T4 DNA polynucleotide kinase (Life Technologies, Gaithersburg, MD)

   Mix thoroughly. Incubate for 10 min at 37°C, and then heat inactive enzyme at 95°C for 5 min. One microliter of this labeling reaction, or ~0.5–1 μCi of labeled primer, is then included in the 25-μl PCR reaction outlined previously. Unused labeled primer can be stored at −20°C.
2. Incorporate radioisotope during PCR amplification by including 10 μCi of [α-^{32}P]ATP or [α-^{32}P]CTP (25 Ci/mmol; ICN Pharmaceuticals) in the 25-μl PCR reaction mixture outlined previously.

Nonisotopic Labeling of the PCR Products

Use fluorescent dye-labeled primers in PCR which are available from most manufacturers of custom oligos and microsatellite markers.

Fluorescent products can then be detected and analyzed by a dedicated gel scanning apparatus.

PCR Amplification

PCR conditions, especially the annealing temperature (T_M), must be optimized for each pair of primers used. Optimal annealing temperature is usually the primer sequence's $T_M \pm 4°C$ but must be determined empirically. The reaction's final $MgCl_2$ concentration may also need to be optimized between 1 and 5 m*M* for most primer pairs.

General PCR Amplification Protocol

Add the following into a microfuge tube starting with the largest volume (usually H_2O) first and finishing with the smallest volume.

Distilled H_2O up to 25 μl
50 ng genomic DNA
2.5 μl 10× PCR buffer (200 m*M* Tris–HCL, pH 8.4, 500 m*M* KCl)
0.5 μl 10 m*M* dNTP (10 m*M* for each dNTP
0.75 μl 50 m*M* $MgCl_2$
2.5 μl 10 μM primer 1
2.5 μl 10 μM primer 2
0.125 μl *Taq* DNA polymerase (5 U/μl)

Mix thoroughly. Amplify for 35 cycles at 95°C for 30 sec, 52–60°C for 60 sec, and 70°C for 60 sec, followed by extension at 70°C for 5 min.

Gel Electrophoresis

Microsatellite PCR products can be analyzed on either polyacrylamide or high-resolution agarose gels. When optimized, both methods can resolve 1-bp size differences. The high-resolution agarose gels are made from low-melting-point agarose preparations (e.g., Metaphor, FMC). They are easy to prepare, require no PCR product labeling, and allow high-efficiency extraction of DNA from excised gel slabs using commercial purification columns (e.g., Qiaquick, Qiagen, Valencia, CA). However, polyacrylamide gels are more commonly used compared to agarose gels for microsatellite analysis because they offer better band resolution and higher sensitivity for faint bands when used with labeled PCR products.

High-resolution agarose gels: Unlabeled PCR products can be visualized after electrophoresis by staining with ethidium bromide or fluorescence dyes.

Polyacrylamide gels: The protocol is suitable for radiolabeled PCR products. For fluorescence-labeled PCR products, some modifications are required.

Prepare a batch of 6–8% polyacrylamide–urea gel in 0.6× TBE buffer. Formamide (30%) may be added to improve the resolution.

Apparatus for sequencing is usually used for microsatellite analysis.

Prerun the gel at 2000 V, 100 mA, and 8 W for 10–15 min.

Add loading buffer containing formamide (e.g., 95% formamide, 20 m*M* EDTA, 0.05% bromophenol blue, 0.02% xylene cyanole FF) to each PCR product and heat at 75°C for 3 min, then snap cool on ice.

Load 3 μl of each PCR sample onto the gel. Any unused sample can be stored at −20°C.

Run the gel at 2000 V, 100 mA, and 80W for 2–4 hr depending on the sizes of the markers.

After completing the electrophoresis, gels can be transferred on blotting papers, dried, and autoradiographed. For fluorescence-labeled products, analyze the gel/glass plate assembly on a scanner.

Interpretation of Microsatellite Analysis

To evaluate allelic size of microsatellites, PCR products can be electrophoresed on a polyacrylamide gel alongside a DNA sequencing reaction sample whose sequence is already known. The size of any A, C, G, or T dideoxy termination band of the sequencing reaction can be calculated by its distance from the sequencing primer; the size of any comigrating PCR product that is detected can be estimated by direct comparison.

To evaluate LOH in DNA from a sample of tumor cells or disease tissue, DNA from normal tissue of the same individual is usually used as an internal control. Only markers showing heterozygosity in the normal tissue can be evaluated. Visual inspection of PCR product bands is usually adequate, showing complete loss of one allele band. However, amplification of DNA from primary tumors with genomic deletion at one allele often shows a faint persistence of the deleted allele as a result of unavoidable contamination with normal cells. To compensate for this limitation, LOH is defined as a >50% diminution in the intensity ratio of the "lost" allele relative to the "retained" allele when comparing a test sample against a normal control. Densitometry can augment visual inspection to quantitate the ratios of allelic band intensities.

Occasionally, normal control DNA is unavailable for comparison with the disease sample (e.g., DNA from tumor cell lines). In such a solitary case, the presence of only one allele cannot determine LOH; however, the presence of only a single allele in a cohort of cases (e.g., a panel of tumor cell lines) at a higher frequency than that found in a cohort of normal DNA samples suggests that LOH is occurring in the case cohort. Likewise, a solitary DNA sample which shows a single allele at many adjacent, highly polymorphic microsatellites probably has LOH and a deletion over part of the corresponding chromosomal locus. The previous evaluations for a solitary DNA sample are statistical analyses for predicting LOH, and to optimize accuracy requires the use of a large number of closely localized, highly polymorphic microsatellites.

Troubleshooting: Modification of Protocols

Nonspecific PCR products may result in various unpredicted bands. Ideally, these can be eliminated by testing primer/marker conditions prior to assaying test samples. A first step would be to maximize the PCR annealing temperature without overly sacrificing the robustness of the desired reaction product. Alternatively, if the robustness of the desired reaction product is good, the number of cycles can be reduced (e.g., 35 → 30 → 25). If nonspecific bands still persist, hot start PCR can be employed. To streamline the application of hot start PCR to a large number of samples, Taq-start (Life Technologies, GIBCO-BRL, Gaithersburg, MD) can be added to the reaction mixtures prior to beginning PCR. Taq-start is a monoclonal antibody that binds *Taq* polymerase until reaction temperatures rise sufficiently to denature the antibody, thus preventing the premature activity of *Taq*. Addition of DMSO (0.5 μl/12.5-μl reaction) may help reduce the amplification of nonspecific products, particularly when the specific product is GC rich.

References

Anonymous (1992). A comprehensive genetic linkage map of the human genome. NIH/CEPH Collaborative Mapping Group. *Science* **258,** 67–86.

Beckman, J. S., and Weber, J. L. (1992). Survey of human & rat microsatellites. *Genomics* **12,** 627–631.

Boige, V., Laurenpuig, P., Fouchet, P., Flejou, J. F., Monges, G., Bedossa, P., Bioulacsage, P., Capron, F., Schmitz, A., Olschwang, S., and Thomas, G. (1997). Concerted nonsyntenic alleic losses in hyperploid hepatocellular carcinoma as determined by a high-resolution allelotype. *Cancer Res.* **57,** 1986–1990.

Dib, C., *et al.* (1996). A comprehensive genetic map of the human genome based on 5,264 microsatellites. *Nature* **380,** 152–154.

Emmert-Buck, M. R., Bonner, R. F., Smith, P. D., Chuaqui, R. F., Zhuang, Z., Goldstein, S. R., Weiss, R. A., Liotta, L. A. (1996). Laser capture microdissection. *Science* **274,** 998–1001.

Eshleman, J. R., and Markowitz, S. D. (1995). Microsatellite instability in inherited and sporadic neoplasms. *Curr. Opin. Oncol.* **7,** 83–89.

Fiallo, P., Williams, D. L., Chan, G. P., and Gillis, T. P. (1992). Effects of fixation on polymerase chain reaction detection of Mycobacterium leprae. *J. Clin. Microbiol.* **30,** 3095–3098.

Kosel, S., and Graeber, M. B. (1994). Use of neuropathological tissue for molecular genetic studies: Parameters affecting DNA extraction and polymerase chain reaction. *Acta Neuropathol.* **88,** 19–25.

Lindqvist, A. K. B., Magnusson, P. K. E., Balciuniene, J., Wadelius, C., Lindholm, E., Alarconriquelme, M. E., and Gyllensten, U. B. (1996). Chromosome-specific panels of tri- and tetranucleotide microsatellite markers for multiplex fluorescent detection and automated genotyping—Evaluation of utility in pathology & forensics. *Genome Res.* **6,** 1170–1176.

Mao, L., Lee, D. J., Tockman, M. S., Erozan, Y. S., Askin, F., and Sidransky, D. (1994). Microsatellite alterations as clonal markers for the detection of human cancer. *Proc. Natl. Acad. Sci. U.S.A.* **91,** 9871–9875.

Mao, L., Lee, J. S., Fan, Y. H., Ro, J. Y., Batsakis, J. G., Lippman, S., Hittelman, W., and Hong, W. K. (1996). Frequent microsatellite alterations at chromosomes 9p21 and 3p14 in oral premalignant lesions and their value in cancer risk assessment. *Nature Med.* **2,** 682–685.

Touitou, I., Rey, J. M., Dross, C., Dupont, M., Brun, O., Ciano, M., Demaille, J., Smaoui, N., Nedelec, B., Hamidi, L., Cattan, D., Mery, J. P., Prier, A., Cabane, J., Choukroun, G., Godeau, P., Delpech, M., Grateau, G., Faure, S., Prudhomme, J. F., Clepet, C., Weissenbach, J., Akopian, K., Kouyoumidjian, J. C., Amselem, S., *et al.* (1996). Localization of the familial Mediterranean fever gene (FMF) to a 250-kb interval in non-Ashkenazi Jewish founder haplotypes. *Am. J. Hum. Genet.* **59,** 603–612.

Weber, J. L., and May, P. E. (1989). Abundant class of human DNA polymorphisms which can be typed using the polymerase chain reaction. *Am. J. Hum. Genet.* **44,** 388–396.

23

REAL-TIME QUANTITATIVE PCR: USES IN DISCOVERY RESEARCH

P. Mickey Williams and Ayly L. Tucker

The recent explosion of genomic information is occurring concurrently with an explosion in technologies designed to harvest new information and analyze its meaning. Novel gene sequence is accumulating at the proverbial "exponential rate." Researchers are searching for rapid methods to utilize this blossom of new sequence data. Hallmark in the understanding of novel gene function is a study of the expression pattern. Many questions will need to be answered, such as In which tissues is the gene expressed? What cell types express the gene? and Is gene expression altered in response to biological stimuli of interest (e.g., disease, differentiation, and embryonic development)? Traditional methods for the study of gene expression, Northern blots and RNAse protection, are rapidly falling behind in their ability to produce necessary data in a timely and efficient manner.

Applications of the PCR continue to grow in all aspects of biological research. Several years ago a PCR application was described which allows for quantitation of gene expression—PATTY (Becker-Andre, 1989, 1991, 1993) or quantitative competitive reverse transcriptase (RT)-PCR (Piatak *et al.,* 1993). These methods were designed to address two major hurdles inherent in PCR that confound quantitative gene analysis: (i) the plateau of product accumulation and (ii) variable reaction efficiency in different samples under analysis. As most now know, a PCR amplification only continues to generate product exponentially for a finite time. Eventually, due to a multitude of potential reasons (product concentration, limiting substrates, etc.) the reaction ceases to produce further product. At this point the ability to discriminate quantitative results is difficult at best. Also, various inhibitors of PCR can contaminate samples introducing variability in amplification efficiencies. This also makes quantitation difficult. Therefore, the quantitative competitive approach was developed. This method uses a reaction spike as a normalization standard to monitor reaction efficiency and normalize the product accumulation, even into the plateau. By design this spike is validated to amplify with equal efficiency as the target gene of interest. Importantly, the spike must compete with the target for substrate and thus when plateau is reached both spike and target cease amplification regardless of the input quantities. Thus, a known amount of spike in a reaction serves to normalize the target molecule quantity. Quantitative competitive PCR has been successfully applied in a variety of arenas. It offers advantages to the traditional methods (Northern blot and RNAse protection) by allowing very small sample size for analysis, very reproducible quantitation, and somewhat greater sample throughput.

As the demand for quantitative gene expression analysis continues to grow, other methodologies are being developed. DNA arrays show promise for the analysis of many genes simultaneously, whereas screening methods such as stimulation proximity assays (Amersham) and branched DNA hybridization (Chiron) are well suited for analysis of a few genes and large numbers of samples. Another recently developed quantitative gene expression method, real-time or kinetic PCR (Heid *et al.*, 1996; Gibson *et al.*, 1996; Ririe *et al.*, 1997; Wittwer *et al.*, 1997), shows promise for utility in the study of multiple genes in multiple samples. Several different approaches have been described for detection of PCR product accumulation; TaqMan probe chemistry (Holland *et al.*, 1991; Heid *et al.*, 1996), double-stranded DNA binding dyes, and dual-probe energy transfer (Wittwer *et al.*, 1997). The remainder of this chapter will discuss development of TaqMan methods for quantitation of gene expression.

Real-time PCR using TaqMan chemistry is proving to be a reliable method which can be rapidly developed for new genes and samples of interest. The method has been described in detail previously (Heid *et al.*, 1996; Gibson *et al.*, 1996). Briefly, in addition to forward and reverse primers designed to amplify the gene of interest, an oligonucleotide hybridization probe is designed which will bind to the amplicon sequence. During primer extension, the 5′ nucleolytic activity of *Taq* polymerase degrades the hybridization probe (Holland *et al.*, 1991). Furthermore, the hybridization probe is dual labeled with two different fluorescent dyes: one is a reporter dye (usually FAM) and the other is termed the quenching dye (usually TAMRA). When the hybridization probe is intact, most of the emission energy of the reporter dye is efficiently transferred to the quenching dye. Upon hybridization probe degradation, the reporter dye is no longer spatially close enough to the quencher to permit energy transfer and thus an increase in reporter fluorescent emission intensity can be observed (Fig. 1). Instrumentation has been developed and discussed previously which permits on-line monitoring of PCR amplifications in real time (Heid *et al.*, 1996; Wittwer *et al.*, 1997).

Sample Preparation

Sample preparation is extremely critical to successful quantitative PCR. Contaminants from either the biological matrix (e.g., heme from blood samples) or the sample preparation reagents (e.g., phenol and salts) can significantly decrease PCR efficiency. Therefore, care should be taken to find sample preparation methods that yield relatively clean samples. For RNA purification, it is also important to remove contaminating genomic DNA. We utilized several methods for RNA purification and found that RNEasy (Qiagen) and PolyATract (Promega) are well suited to real-time reverse transcriptase (RT)-PCR.

Primer and Probe Design

In order to achieve successful PCR results in any complex sample, it is imperative that one begins with successful primer design. Primer design becomes even more critical when constrained by the need to

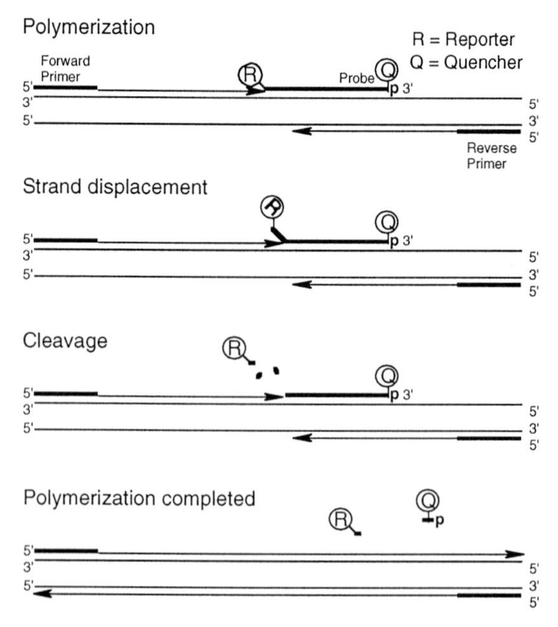

Figure 1 Real-time PCR with TaqMan chemistry. The PCR begins with a denaturation step, followed by an annealing step when the probe and primers hybridize. Extension of the primers begins (the probe is modified such that it will not serve as an extension primer). The nascent strand reaches the site of probe hybridization. The probe strand is displaced, followed by 5′ nucleolytic cleavage of the probe. At this point, the emission strategy of the reporter probe dye is no longer transferred to the quenching dye, and this results in an increase in reporter fluorescent emission intensity. The fluorescent emission intensity can be monitored in real time.

simultaneously design a hybridization probe. A variety of software is available to assist in this task (Oligo, National BioScience, and Primer Express, ABI–Perkin–Elmer). We have utilized both programs successfully for design of TaqMan primer and probe sets. The following are our rules of thumb for successful primer and probe design. Primers should be screened for potentially stable secondary structures and potential of primer dimer formation. Additionally, the 3′ end of the primer should be selected to have moderate to low G + C content, such that base mismatches are nonproductive in PCR preventing nonspecific target amplification. We routinely search for primers that have three or less G + C's in the last 5 bases of the 3′ end. Furthermore, it is probably best not to end on a string of three G + C's. Primers should be selected, their length adjusted if necessary, to have predicted T_m's within 2° of the mate. The hybridization probe is designed such that it binds to sequence within the amplicon and results in a predicted T_m approximately 10° higher than that of the primer pair selected. The maximum length of the probe should be no than 40 bases for efficient fluorescent energy transfer to occur. We have begun selecting primers which have T_m's predicted by Primer Express to be between 58 and 60°C and probe T_m's to be >68°; this permits multiple primer/probe sets to be analyzed simultaneously on one instrument. Recently, we had great success when maintaining the amplicon size in the range of 70–150 bp. This small amplicon size may allow very efficient amplification of targets and add to successful assay design.

Reaction Mixture

Many potential approaches are possible for RT-PCR, e.g., RT of the entire sample which is then split into aliquots for quantitative PCR or RT coupled to PCR of a sample aliquot in a single tube. RT can be performed with oligo-(dT) primer, random hexamers, the reverse PCR primer, and so on. We have chosen to use single-tube analysis, in which the RT and PCR amplification are performed in a single tube using the reverse PCR primer as the RT primer. This prevents the need to open tubes and decreases the amount of hands-on time necessary to achieve results. Additionally, many enzymes and enzyme sources are available for RT and PCR. We have chosen a system, developed by Perkin–Elmer and Roche Molecular Systems, using MuLV RT and AmpliTaq Gold DNA polymerase and a single buffer (Buffer A, supplied by Perkin–Elmer). One advantage of this system is that it utilizes a passive dye reference (ROX) for normalization of tube-to-tube fluorescent variability. We begin by using a final $MgCl_2$ concentration of 5 mM per reaction. This can be optimized in further experiments if necessary. Also 100 ng of total RNA is a good starting point, although often much less is necessary to detect moderate level expression.

10× Buffer A	1×
dNTPs (2.5 mM each)	200 μM
25 mM $MgCl_2$	5 mM (2.5–5.5)
Primer, forward	25 pmol
Primer, reverse	25 pmol

TaqMan probe	100 n*M*
RNAse inhibitor	20 U
MuLV	12.5 U
TaqGold	2.5 U
Water	Volume brought up to 40 μl
RNA sample	10 μl (100 ng total RNA)

In order to accommodate multiple primer/probe combinations on a single plate, we designed primers and probes such that they share similar predicted T_m's (see primer/probe design section). Having done this, the following is a general thermal cycling profile: 48°C for 30 min for RT; 95°C for 10 min for AmpliTaq Gold activation and RT inactivation; and 40 cycles of 95°C for 15 sec and 60°C for 1 min. Of course, these conditions can be modified and optimized for each project.

Quantitative Analysis

As outlined elsewhere (Heid *et al.*, 1996), amplification plots depicting the normalized fluorescent intensity of the reporter dye over time (cycles) are generated for each reaction tube analyzed. The time (in cycles) at which reporter dye fluorescent intensity increases over a threshold value is proportional to the amount of starting target molecules present in the sample. We call the time at which this increase occurs C_t (Fig. 2). Therefore, C_t values are reflective of the quantity of target in any sample. Standard curves generated with serial dilutions of a sample can be used for relative comparison of any unknown sample.

Quantitative competitive RT-PCR utilizes a spike of a known amount of the RNA competitor into each sample as a means of normalization of amplification efficiency. Real-time RT-PCR does not require the design and validation of such a spike approach (although this is still a valid method). An advantage that real-time RT-PCR offers is the use of another gene, which has been demonstrated to remain invariant in expression, for normalization of amplification efficiencies and also to add information on total mass of the sample analyzed.

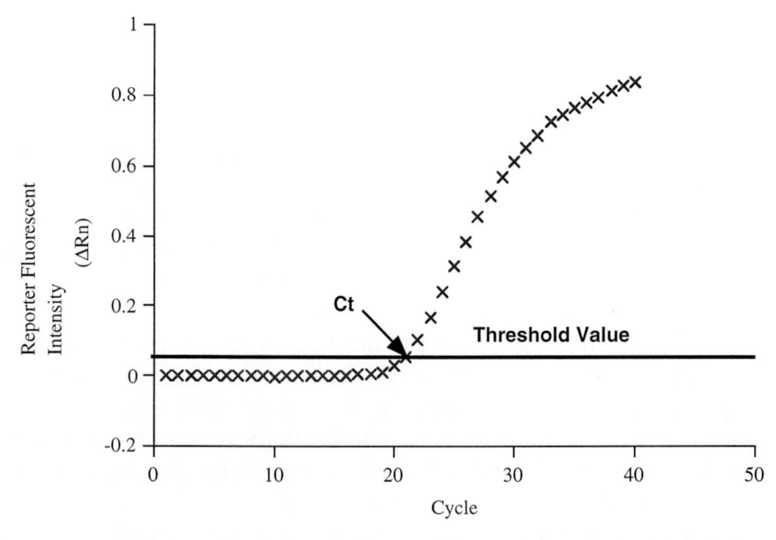

Figure 2 Amplification plot generated by Model 7700 Sequence Detector software. The fluorescent emission intensity is plotted on the *y* axis versus time (cycle numbers of PCR) on the *x* axis. A fluorescent threshold is calculated based on the fluorescent background of the initial cycles (generally cycles 3–15). The point at which the fluorescent intensity crosses the threshold is defined as C_t (i.e., threshold cycle). The C_t value is proportional to the starting target copy number (Heid *et al.*, 1996) and thus can be used as a quantitative comparison between samples. Lower C_t values correlate with greater target copies.

Example of Real-Time RT-PCR

Our goal was to develop a sensitive assay for detection of the presence of human type 1 interferon in a sample. We chose to develop a quantitative real-time RT-PCR assay monitoring the expression of Mx-α mRNA. Previous work has shown that cells treated with interferon-α (IFN-α) exhibit an increase in Mx-α expression (Aebi *et al.*, 1989); von Wussow *et al.*, 1990). Furthermore, Mx-α mRNA expression induction appears to be fairly specific to type 1 interferon treatment (i.e., IFN-α) because other cytokines do not seem to modulate its expression (von Wussow *et al.*, 1990). We chose to work with an established human tissue culture cell line, A549. The assay design was to add known amounts of IFN-α to A549 cells and monitor Mx-α expression. By adding serial dilutions of known amounts of IFN-α and measuring resultant gene expression, a standard curve can be generated. This standard curve would then be used to calculate the amounts of IFN-α present in serum samples. Our first objective was to monitor a "housekeeping gene," β-actin, to determine if expression levels remained invariant with IFN-α treatment. Cells were treated with various amounts of IFN-α in an overnight incubation. Approximately 18 hr after treatment, cytoplasmic RNA was harvested using the RNAeasy 96 kit (Qiagen). Insufficient RNA was recovered to spectroscopically determine RNA mass yields, so equal aliquots (reflecting equal volumes) of sample were analyzed for expression of β-actin. As shown in Table 1, the C_t values remained relatively constant regardless of the dose of IFN-α that was added (ranging from 17.7 to 18.5 C_t's). We believe that real-time RT-PCR has the ability to discriminate at least a twofold change, which is approximately 1 C_t unit. Our results demonstrated that β-actin expression was not altered upon IFN-α treatment and could be used to normalize Mx-α gene expression (i.e., β-actin C_t values for each sample will reflect RNA mass and reaction efficiency). Next, we analyzed the expression of Mx-α mRNA upon treatment of the cells with various doses of IFN-α. We used β-actin expression to normalize the Mx-α results. As mentioned previously, this corrects for any differences in sample load and also any changes in amplification

Table 1

Duplicate C_t Values for β-Actin Expression in A549 Cells Treated with Varying Amounts of IFN-α[a]

INF-α (U/ml)	β-Actin C_t	β-Actin C_t	Average
500	18.64	18.27	18.46
100	17.81	17.85	17.83
20	18.03	17.80	17.92
4	17.72	17.76	17.74
0.8	17.76	17.65	17.71
0.16	17.75	17.85	17.80
0.032	17.72	17.97	17.85
0	17.99	18.39	18.19

[a] The table demonstrates that β-actin expression is not regulated by IFN-α treatment in these cells. Therefore, β-actin expression will serve as a good housekeeping gene to normalize Mx-α data.

efficiency. As shown in Table 2, there is an IFN-α dose-dependent induction of Mx-α expression (C_t values for Mx-α decrease with increasing IFN-α dose, whereas β-actin remains relatively constant). To normalize the Mx-α C_t values to β-actin, we subtracted the β-actin C_t value from that of the Mx-α C_t (C_t values are log base values consistent with the exponential amplification of PCR). The normalized values are reported as ΔC_t in Table 2. The relative normalized (to β-actin) expression of Mx-α was correlated with the known amounts of IFN-α added to the A549 cells. From this data an IFN-α standard curve was derived (Fig. 3). Using this approach, serum samples containing unknown amounts of IFN-α can be incubated with A549 cells in culture. Mx-α and β-actin gene expression can be determined for these unknown sample wells. The delta Ct of an unknown sample can be plotted on the standard curve (Fig. 3), thus solving for the amount of IFN-α present in the serum sample.

Finally, Fig. 4 depicts β-actin normalized values for Mx-α expression in A549 cells treated with varying amounts of IFN-α. The data represent assays performed on several different days, indicating the assay is very reproducible. Furthermore, the assay detected IFN-α in a sample at a concentration as low as 0.16 U/ml, which we consider to be very sensitive compared to our typical ELISA assay.

This is just one example of how real-time PCR can be applied. We have had great success in implementing this method in a variety of other research projects. Certain advantages of this method make it extremely useful. First, minimal sample is necessary. We often work with samples derived from several hundred cells or less. Sec-

Table 2

Duplicate C_t Values for Mx-α β-Actin Expression in A549 Cells Treated with Varying Amounts of IFN-α

INF-α (U/ml)	MX C_t	MX C_t	Average	β-Actin C_t	β-Actin C_t	Average	ΔC_t
500	18.81	18.27	18.54	18.64	18.27	18.46	0.09
100	19.51	19.48	19.50	17.81	17.85	17.83	1.67
20	19.70	20.52	20.11	18.03	17.80	17.92	2.20
4	21.26	20.93	21.10	17.72	17.76	17.74	3.36
0.8	22.12	22.02	22.07	17.76	17.65	17.71	4.37
0.16	23.14	22.98	23.06	17.75	17.85	17.80	5.26
0.032	24.76	24.30	24.53	17.72	17.97	17.85	6.69
0	40.00	40.00	40.00	17.99	18.39	18.19	21.81

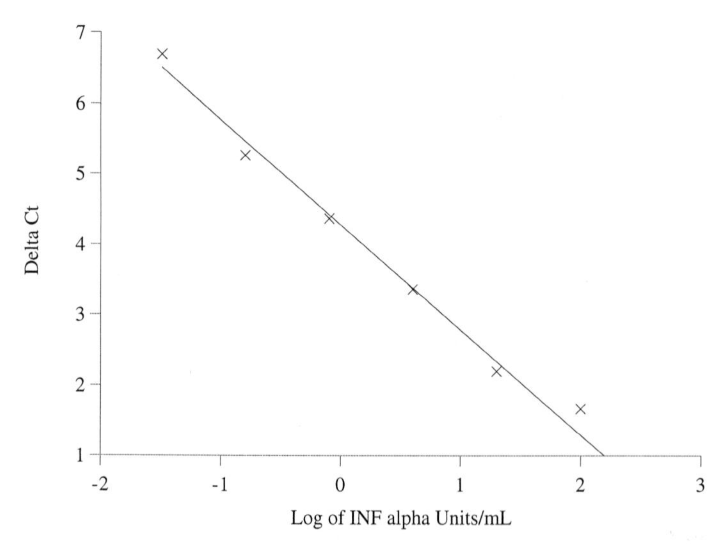

Figure 3 Standard curve of IFN-α induction of Mx-α gene expression. This is a plot of C_t values on the *y* axis versus the concentration of IFN-α added to the A549 cells. This standard curve can be used to calculate the amount of IFN-α in an unknown sample.

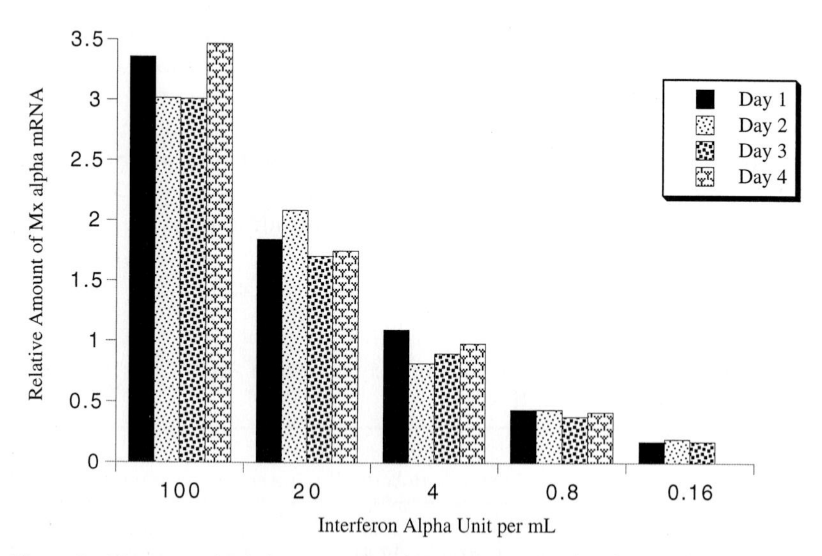

Figure 4 Histogram depicting normalized Mx-α expression levels in response to varying amounts of IFN-α treatment. C_t values were determined for MX-α and β-actin. Relative and normalized amounts of MXα mRNA were calculated by solving the equation:

$$2^{-(\text{Ct Mx}\alpha\ -\ \text{Ct }\beta\text{-actin})} \times 10.$$

This calculation assumes that the PCR is 100% efficient. The data represent replicate assays performed on different days.

ond, the time necessary for assay development is minimal compared to that for quantitative competitive PCR. Thus, we can apply this method for the analysis of many gene targets. Third, sample throughput is much greater than that of quantitative competitive PCR. Finally, no post-PCR manipulations are necessary which saves time and prevents potential laboratory PCR product contamination. We believe that this advance in methodology will find continued utility in the study of novel genes as we search for their function.

References

Aebi, M., Fah, J., Hurt, N., Samuel, C. E., Thomis, D., Bazzigher, L., Pavlovic, J., Haller, O., and Staeheli, P. (1989). cDNA structures and regulation of two interferon-induced human Mx proteins. *Mol. Cell. Biol.* **9,** 5062–5072.

Becker-Andre, M. (1989). Absolute mRNA quantification using the polymerase chain reaction (PCR). A novel approach by a PCR aided transcript titration assay (PATTY). *Nucleic Acids Res.* **17,** 9437–9446.

Becker-Andre, M. (1991). Quantitative evaluation of mRNA levels. *Methods Mol. Cell. Biol.* **2**(189), 189–201.

Becker-Andre, M. (1993). Absolute levels of mRNA by polymerase chain reaction-aided transcript titration assay. *Methods Enzymol.* **218,** 420–445.

Gibson, U. E. M., *et al.* (1996). A novel method for real time quantitative RT-PCR. *Genome Res.* **6,** 995–1001.

Heid, C., *et al.* (1996). Real time quantitative PCR. *Genome Res.* **6,** 986–994.

Holland, P. M., *et al.* (1991). Detection of specific polymerase chain reaction product by utilizing the 5′ → 3′ exonuclease activity of Thermus aquaticus DNA polymerase. *Proc. Natl. Acad. Sci. U.S.A.* **88**(16), 7276–7280.

Piatak, M. J., *et al.* (1993). Quantitative competitive polymerase chain reaction for accurate quantitation of HIV DNA and RNA species. *BioTechniques* **14,** 70–81.

Ririe, K. M., Rasmussen, R. P., and Wittwer, C. T. (1997). Product differentiation by analysis of DNA melting curves during the polymerase chain reaction. *Anal. Biochem.* **245**(2), 154–160.

von Wussow, P., Jakschies, D., Hochkeppel, H.-K., Fibich, C., Penner, L., and Deicher, H. (1990). The human intracellular Mx-homologous protein is specifically induced by type 1 interferons. *Eur. J. Immunol.* **20,** 2015–2019.

Wittwer, C. T., Ririe, K. M., Andrew, R. V., David, D. A., Gundry, R. A., and Balis, U. J. (1997). The LightCycler: A microvolume multisample fluorimeter with rapid temperature control. *BioTechniques* **22,** 176–181.

24

HOMOLOGY CLONING: A MOLECULAR TAXONOMY OF THE ARCHAEA

Anna-Louise Reysenbach and Costantino Vetriani

The recognition that the Archaea form a separate domain within the evolutionary tree of life has revolutionalized our view of prokaryote and eukaryote evolution. This discovery was largely a result of the sequence comparison of the evolutionarily conserved small-subunit rRNA genes (16S-like rDNAs), which placed the Archaea in a domain as distinct as the Bacteria are from the Eucarya (Woese *et al.*, 1990). Rooting this universal phylogenetic tree by comparing gene homologs that were duplicated early in the evolution of life [such as membrane ATPase α and β subunits (Gogarten *et al.*, 1989) and elongation factor EF-Tu (Iwabe *et al.*, 1989)], the Archaea appear to be more closely related to the Eucarya than the Bacteria are to the Eucarya. Based primarily on 16S rRNA phylogenies, the Archaea comprise two kingdoms, the Crenarchaeaota and the Euryarchaeota. Recently, a third kingdom has been proposed (the Korarchaeota) that is represented by sequences obtained from a thermal spring in Yellowstone National Park (Fig. 1) (Barns *et al.*, 1994, 1996). As more archaeal genes and genomes are sequenced, phylogenies based on other genes will continue to test the robustness of this domain and provide a molecular taxonomy for the Archaea. Brown

PCR Applications

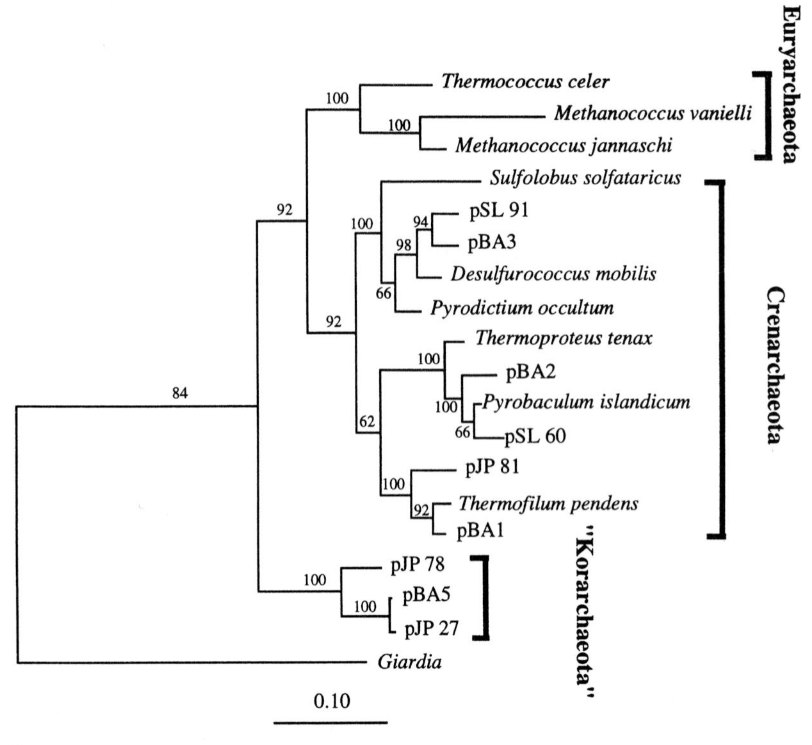

Figure 1 Archael phylogenetic tree of the small subunit rRNA molecule generated from maximum likelihood analysis (Felsenstein, 1981; fastDNAml is available through the RDP). Representatives of isolates and environmental 16S rRNA sequence ("phylotypes," e.g., pJP27) were used in this analysis. The tree was rooted with the Eucarya. Bootstrap values were generated from 100 boostrap resamplings. The bar represents the expected number of changes per sequence position.

and Doolittle (1997) provide an excellent review of the interdomain relationships implied from examining many different protein gene trees.

From genomic studies, the Archaea are more closely allied to the Eucarya with respect to informational characteristics of the cell, whereas most of the metabolic characteristics are shared by the Bacteria (Olsen and Woese, 1997) yet this latter relationship remains equivocal. Therefore, when designing primers for amplification of homologous genes from different Archaea, these observations may be taken into consideration. As more archaeal genomes are sequenced, the database for primer design will become larger, and robust primers will be easier to design. However, with more than 50% of the *Methanococcus jannaschii* genome as yet unaccountable for a specific function (Bult *et al.*, 1996), cloning of gene homologs that are evolutionarily less conserved may be a daunting task. Because the 16S rRNA genes are highly conserved, primer design is relatively easy. The 5′ end of the archaeal 16S rRNA is conserved within the archaeal and eukaryal domains and is universally conserved toward the 3′ end. Therefore, a primer set that will amplify almost the entire archaeal and eukaryal 16S rRNA gene has been designed (Reysenbach and Pace, 1995). However, in genes of interest that are less highly conserved, degenerate primers can be designed to obtain a portion of the gene [e.g., DNA polymerase (Uemori *et al.*, 1993) and glutamine synthetase genes (Brown *et al.*, 1994)], and the PCR fragment in turn can be used to screen a genomic DNA library and identify the full-length gene.

In this chapter, we describe the amplification of a very well-studied gene, namely, the 16S rRNA gene. We use this gene as the methodological framework, from which other genes can be amplified and cloned in a similar manner, to develop a molecular taxonomy of the Archaea. Because it has been estimated that perhaps only 1% of the microbial world has been grown in pure culture (Amann *et al.*, 1995), considering both archaeal isolates and archaeal diversity associated with naturally occurring communities is important to establish a molecular taxonomy of the domain. We will consider both approaches for cloning environmental and laboratory isolate DNAs.

Protocols

The general scheme for homology cloning is outlined in Fig. 2. In general, all cloning and probing procedures used are standard (Sambrook *et al.*, 1989). Detailed procedures used for molecular taxonomic studies are covered in Hillis *et al.* (1996). However, optimal results are dictated most by

The purity of the genomic DNA
The PCR primer design
The PCR conditions

Impurities copurified with genomic DNA during extraction proce-

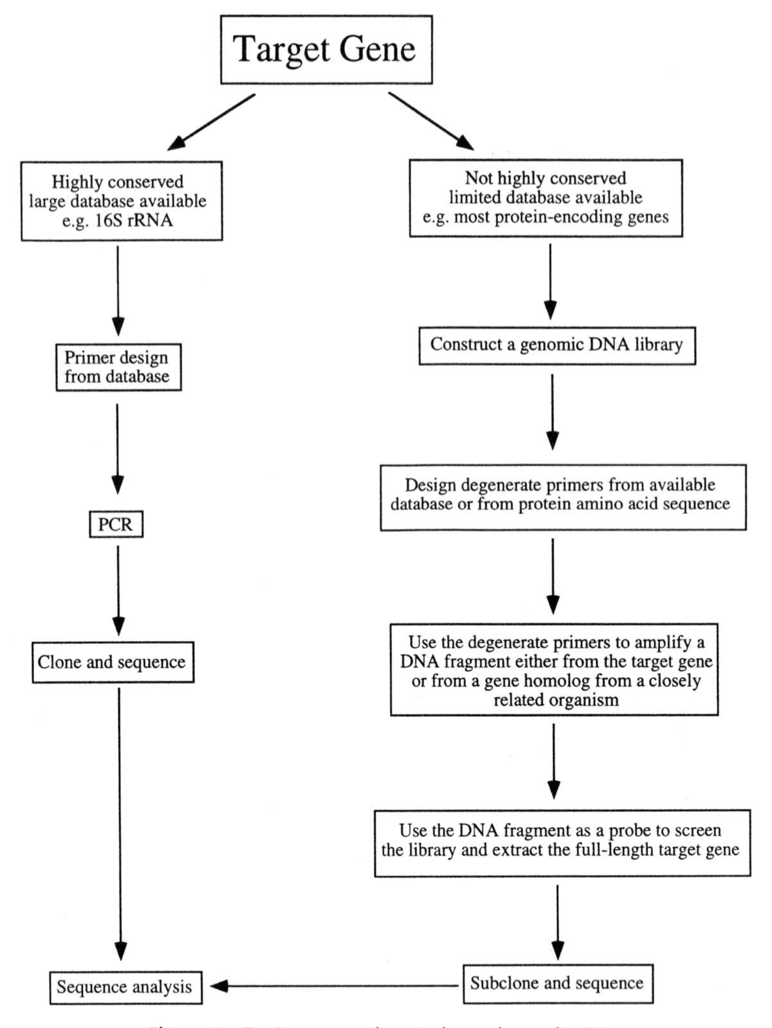

Figure 2 Basic approaches to homology cloning.

dures can be a serious problem. Some archaeal DNA may be difficult to amplify using the PCR; therefore, optimization of the appropriate PCR reaction conditions may be necessary.

DNA Extraction Procedure

DNA from Archaeal Isolates

If a pellet of a pure culture is used for genomic DNA extraction, most extraction procedures work well. However, some of the growth media for thermophiles may be a source of metals and sulfides that can affect the DNA purity and additional purification steps may be necessary. In some cases, purification of DNA is unnecessary, and the gene (e.g., 16S rRNA gene) can be amplified directly by using a dilution of the resuspended cell pellet and incorporating a hot start (5 min at 95°C) at the beginning of the PCR.

Environmental DNA Extraction

There are many different methods for extracting environmental DNA. Minimizing the environmental contaminants that may cause inhibition of the PCR is one of the goals of all extraction methods. Most methods for extracting DNA from soils use a protocol to remove humics from the preparation by using, for example, polyvinylpyrrolidone (Young *et al.*, 1993), and combining this with purification by low-melt agarose gel electrophoresis (Barns *et al.*, 1994). DNA extraction procedures from high-temperature environments, such as deep-sea hydrothermal vents or terrestrial thermal springs, often are inefficient due to the presence of metals such as iron. Methods for chelating the high concentrations of metals, such as increased EDTA concentrations (up to 100 m*M*) and/or the use of chelators such as Chelex 100 (Bio-Rad), have proved to be very efficient (Cary *et al.*, 1997). Every effort should be made to minimize the amount of possible genomic DNA shearing because this may result in chimeric PCR products (Robison-Cox *et al.*, 1995). For large fragment libraries, very gentle extraction procedures are required (Stein *et al.*, 1996).

PCR Primer Design

Primers are designed by aligning homologous sequences from as many taxa as possible. These sequences can be retrieved from large sequence databases such as Genbank (*http://www.ncbi.nlm.nih.gov/web/Genbank*). In selecting for homologs within the Archaea, primers specific for the archaeal domain are the most effective. However, it is not always possible to design primers that span the entire archaeal domain, and therefore primers may be restricted to a single group. Homologous regions that are conserved throughout the selected organisms serve as sites for primer design. In many cases these primers will only amplify a portion of the gene. For protein-encoding genes, identical amino acid sequence over a seven-amino acids stretch is optimal. Due to the degeneracy of the genetic code, the third position of the nucleotide in such a stretch may be highly variable. Thus, designing a primer that takes into account the twofold codons (amino acids that are encoded for by more than one codon), and varying the third nucleotide accordingly, reduces nonspecificity of the primer. Single mismatches in the internal part of the primer are generally not problematic. However, because the polymerase binds to the 3′ end, this region should be designed to perfectly match the template sequence. Additionally, internal sequence complementarity should be avoided to prevent formation of secondary structures within the primers. If possible, either a G or a C should be included at the 3′ end of the primer to enhance annealing efficiency. Generally, primers are designed to be between 18 and 39 nucleotides in length. Some primers are modified to facilitate cloning, such as the addition of polylinker tails that contain endonuclease restriction sites (Reysenbach and Pace, 1995) or dU-containing tails to allow cloning by the uracil DNA glycosylase (UDG) method (see Cloning).

Oligonucleotide Primers for the Archaeal 16S rRNA Gene

The Ribosomal Database Project (RDP; Maidak *et al.,* 1994; *http://www.cme.msu.edu/RDP*) provides a database of aligned rRNA sequences and services such as chimeric molecule checks, probe specificity checks, and preliminary phylogenetic analyses. Primers for 16S rRNA genes can be designed using the RDP database or the oligonucleotide probe database (OPD; Alm *et al.,* 1996; *http://www.cme.msu.edu/OPD/*), which provides a list of primers and oligonucleotides most relevant to archaeal 16S rRNA studies. Table 1 is a synthesis of some of the archael primers and probes for the 16S rRNA gene.

Oligonucleotide Primers for Other Archaeal Gene Homologs

The basic approach for cloning of archaeal gene homologs for which the sequence information is limited or nonexistent relies on the use of oligonucleotide probes to screen genomic DNA libraries and extract the full-length genes. The PCR fragment that serves as a probe may be generated many ways, including (i) designing PCR degenerate primers that anneal to regions of the gene that are relatively conserved throughout the Archaea (e.g., the glutamine synthetase gene from *Pyrococcus furiosus;* Brown *et al.,* 1994; Table 2); (ii) using a PCR primer with high specificity for the target gene together with a primer homologous to highly conserved regions throughout the entire gene family (e.g., the citrate synthase gene from *P. furiosus,* Muir *et al.,* 1995; Table 2); and (iii) using part of a homologous gene that has been amplified from a closely related organism (e.g., the glutamate dehydrogenase gene from *Thermococcus litoralis;* DiRug-

Table 1

Oligonucleotide Sequence for Archaeal 16S rRNA Gene Primers and Probes

Name (OPD Format)	Target Group	Sequence (5′ to 3′)	Position in *E. coli* Numbering	Reference
S-D-Arch-0021-a-S-20	Archaea	TTCCGGTTGATCCYGCCGGA	0021–0040	DeLong (1992)
S-D-Arch-0338-a-A-24	Archaea	GCGCCTGSTGCSCCCCGTAGGGCC	0338–0361	Giovannoni *et al.* (1990)
S-D-Arch-0344-a-A-20	Archaea	TCGCGCCTGCTGCICCCCGT	0344–0363	Raskin *et al.* (1994)
S-D-Arch-0915-a-A-20	Archaea	GTGCTCCCCCGCCAATTCCT	0915–0934	Amann *et al.* (1990)
S-D-Arch-0958-a-A-19	Archaea	YCCGGCGTTGAMTCCAATT	0958–0976	DeLong (1992)
S-K-Cren-0028-a-S-23	Crenarchaeota	AATCCGGTTGATCCTGCCGGACC	0028–0050	Schleper *et al.* (1997)
S-K-Cren-0457-a-A-22	Crenarchaeota	TTGCCCCCCGCTTATTCSCCCG	0457–0478	Schleper *et al.* (1997)
S-K-Cren-0499-a-A-18	Crenarchaeota	CCAGRCTTGCCCCCCGCT	0499–0515	Burggraf *et al.* (1994)
S-K-Cren-0667-a-A-15	Crenarchaeota	CCGAGTACCGTCTAC	0667–0681	DeLong *et al.* (1994)
S-K-Eury-0498-a-A-14	Euryarchaeota	CTTGCCCRGCCCTT	0498–0510	Burggraf *et al.* (1994)
S-K-Kora-0236-a-S-20	Korarchaeota	GAGGCCCCAGGRTGGGACCG	0236–0255	Burggraf *et al.* (1997)
S-K-Kora-1135-a-A-20	Korarchaeota	GTTTGCCCGGCCAGCCGTAA	1135–1154	Burggraf *et al.* (1997)
S-O-Cenar-0554-a-A-20	Group I marine Archaea	TTAGGCCCAATAATCMTCCT	0554–0573	Massana *et al.* (1997)
S-O-ArGII-0554-a-A-20	Group II marine Archaea	TTAGGCCCAATAAAAKCGAC	0554–0573	Massana *et al.* (1997)
S-O-Msar-0860-a-A-21	Methanosarcinales	GGCTCGCTTCACGGCTTCCCT	0860–0880	Raskin *et al.* (1994)
S-O-Mmic-1200-a-A-21	Methanomicrobiales	CGGATAATTCGGGGCATGCTG	1200–1220	Raskin *et al.* (1994)
S-F-Mbac-0310-a-A-22	Methanobacteriaceae	CTTGTCTCAGGTTCCATCTCCG	0310–0331	Raskin *et al.* (1994)
S-F-Mbac-0314-a-A-22	Methanobacteriaceae	GAACCTTGTCTCAGGTTCCATC	0314–0335	Raskin *et al.* (1994)
S-F-Mbac-1174-a-A-22	Methanobacteriaceae	TACCGTCGTCCACTCCTTCCTC	1174–1195	Raskin *et al.* (1994)
S-F-Mcoc-1109-a-A-20	Methanococcaceae	GCAACATAGGGCACGGGTCT	1109–1128	Raskin *et al.* (1994)
S-F-Msae-0825-a-A-23	Methanosaetaceae	TCGCACCGTGGCCGACACCTAGC	0825–0847	Raskin *et al.* (1994)
S-F-Msar-1242-a-A-22	Methanosarcinaceae	GGGAGGGACCCATTGTCCCATT	1242–1263	Raskin *et al.* (1994)
S-F-Msar-1414-a-A-21	Methanosarcinaceae	CTCACCCATACCTCACTCGGG	1414–1434	Raskin *et al.* (1994)

Note. R = A or G; Y = C or T; K = G or T; M = A or C; and S = C or G.

Table 2

Examples of Oligonucleotide Primers and Probes Used to Identify Gene Homologs within the Archaea

Target Gene	Organism	Sequence (5′ to 3′)	Primer Design and Use	Reference
Glutamine synthetase (*glnA*)	*Pyrococcus furiosus*	F: CATCATGA(AG)GT(TACG)GC(AT)AC(TCA) GC(AT)GG(TACG)CA R: AGCAGCAAAATGCTAGATACGGATT	Degenerate primers corresponding to *glnA* conserved regions II and V; used to amplify a 700-bp fragment	Brown *et al.* (1994)
Citrate synthase	*Pyrococcus furiosus*	F: GG(AT)CT(TC)GA(AG)GA(TC)GT(AT)TA(CT)AT (ATC)GA(CT)CA(GA)AC(AT)AA(CT)AT R: TACCC(TA)AA(AG)CC(TA)GT(AG)TC (CT)CA(TA)AT	Degenerate primers used to amplify a 760-bp fragment of the *P. furiosus* citrate synthase gene; the R primer sequence is highly conserved in all citrate synthase genes	Muir *et al.* (1995)
Glucose dehydrogenase	*Thermoplasma acidofilum*	GAA(G)CAA(G)AAA(G)GCA(CGT)ATA(CT)GT ACA(CGT)ATA(CT)GAC(T)ATGCCA(CGT)GA	Degenerate primers for the glucose dehydrogenase gene used to screen a genomic DNA library from *T. acidofilum*	Bright *et al.* (1993)
DNA polymerase	*Sulfolobus solfataricus*	F: GA(TC)CCNAAC(CT)T(CG)CA(AG)AA(CT)ATNCC R: (GT)A(CG)(CG)A(GT)(TC)TC(AG)TCGTGNAC (CT)TG	Degenerate primers conserved in all DNA polymerase I-like genes	Uemori *et al.* (1993)
Extracellular α-amylase (*amyA*)	*Pyrococcus furiosus*	F: AGCTAGCTTGGAGCTTGAAGAGGGAG R: ACTCGAGACCACAATAACTCCATACGGAG	Primers designed to amplify and clone the *P. furiosus amyA* gene	Dong *et al.* (1997)

giero *et al.*, 1997). Table 2 shows some examples of oligonucleotide primers and probes used to identify gene homologs within the Archaea.

PCR

The PCR may need to be optimized for a particular genomic sample, although several standard precautions can be made in advance to ensure successful amplification of a product. For example, it is our experience that some hyperthermophilic archaeal rRNA genes may be difficult to amplify due to their high GC content which may prevent both complete denaturation of the template and efficient binding of the primers. Addition of mild denaturants such as acetamide has been shown to increase amplification efficiency in some cases.

Some optimization procedures may include

Addition of denaturants such as acetamide (Reysenbach *et al.*, 1992) and/or inclusion of a hot start (Chou *et al.*, 1992)

Optimizing the PCR buffer (the $MgCl_2$ concentration can be optimized using an experimental matrix as described by Cobb and Clarkson, 1994): In some cases, the substitution of Tris with 300 m*M* tricine in the PCR buffer increases amplification efficiency.

Optimization of primer annealing temperature: Creating different temperatures for primer annealing in the same experiment using a thermal cycler such as the Robocycler from Stratagene is very convenient for this purpose.

Additional strategies for optimization of PCR conditions have been reviewed elsewhere (Innis and Gelfand, 1990; Ehrlich *et al.*, 1991).

The following is an example of a PCR mixture for amplification of archaeal 16S rRNA genes:

10 × reaction buffer	10 μl
1% Igepal (Sigma)	5 μl
50% acetamide (optional)	10 μl
dNTPs (each at 1.5 μ*M*)	10 μl
Forward primer (50 pmol/μl) (S-D-Arch-0021-a-S-20)	2 μl
Reverse primer (50 pmol/μl) (S-*-Univ-1492-a-A-19)	2 μl

DNA	10–100 ng
Taq DNA polymerase	1 U

Adjust final volume with sterile double-distilled water to 100 μl.

The PCR buffer contains 300 m*M* tricine (pH 8.4), 500 m*M* KCl, 15 m*M* $MgCl_2$. Igepal replaces detergents such as Nonidet P-40 that are often included in PCR mixtures. The primer choice is such that the forward domain-specific primer will select the archaeal gene and the reverse primer ensures that almost the entire 16S rRNA gene will be amplified.

The cycling conditions should be optimized for a particular template and primer set, but the following conditions generally work well for Archaea: denaturation of the DNA template at 94°C for 4 min, and 30 cycles of 92°C for 1.5 min, 50°C for 1.5 min, and 72°C for 2 min, where the primer extension is extended for 5 sec after each cycle. The resulting PCR products can be visualized on a preparative agarose gel and bands of the expected size can be gel purified (Sambrook *et al.*, 1989) and concentrated using columns such as those provided by Qiagen. If no archaeal PCR products are obtained, the following options to optimize the amplification reaction can be used:

The DNA template should be purified further

From environmental DNA the archaeal template may be low, so increasing the number of cycles for the reaction to 40 cycles may help

Using other PCR conditions such as "touchdown" PCR to prevent nonspecific annealing of the primer in the early amplification stages

Cloning

Standard cloning, screening, and sequencing procedures (Sambrook *et al.*, 1989) apply for archael genes. The PCR product can either be sequenced directly (if it is from a pure culture) by automated sequencing or cloned using standard cloning techniques and vectors. If the primers have polylinkers containing restriction endonuclease sites, the products should be cut with the endonucleases and cloned into a vector such as pBluescript (Stratagene). In this case the use of rare-cutting restriction endonucleases (e.g., *Not*I) will reduce the possibility of internal digestion of the PCR inserts. Another option is to include dU-containing 5′ tails in the primer sequence. Subsequent removal of the deoxyuracil bases by digestion of the PCR product with UDG will disrupt base pairing, exposing 3′ overhangs and facilitating cloning of the product in an appropriate plasmid vector (e.g., the pAMP cloning systems, GIBCO BRL). We routinely use this latter system, which provides a high rate of insert-containing clones. Amplified genes can be screened by restriction fragment-length polymorphisms analysis using tandem tetrameric restriction endonuclease pairs such as *Hae*III and *Msp*I (Moyer *et al.*, 1994). This is particularly important for screening of gene homologs from environmental-derived DNAs. The unique clones can then be purified for automated sequencing using standard protocols. The entire double-stranded 16S rRNA gene fragment can be sequenced using a number of internal reverse and forward primers (Reysenbach and Pace, 1995).

Analysis of Sequence

The sequences are aligned with a subset of related sequences obtained from sequence databases such as Genbank or RDP. An initial choice of apppropriate sequences can be made using the BLAST algorithm (software available at *http://www.ncbi.nlm.nih.gov*) that does a similarity search of a selected database. Additionally, the program Entrez is convenient for searching nucleotide and protein databases. Phylogenetic analyses are based on aligning homologous nucleotides, and a good analysis of the data requires an excellent alignment. Although this may appear to be simple to perform, it is deceiving and can be the most difficult and least understood component of a phylogenetic analysis (Swofford *et al.*, 1996). Due to the highly conserved nature of the 16S rRNA molecule, the folded secondary structure of the inferred rRNA sequence can be determined using an archaeal secondary structure template. This ensures that only homologous regions are compared in the alignments. Initial multiple alignments can be done by pairwise alignments using the program Clustal (available by anonymous ftp: *ftp.bio.indiana.edu* in the *molbio/align* directory). Refining these alignments can be done manually using, for example, the Genetic Data Environment program available from the RDP. This program is Unix based and interfaces with most phylogenetic analysis packages and databases. There are a multitude of programs and software packages available for conducting phylogenetic and population genetic analyses. Swofford *et al.* (1996) provide an excellent overview of these methods and list many of these programs. For additional discussion of analysis of archaeal protein gene trees, see Brown and Doolittle (1997).

Applications

Studies of gene homologs within the Archaea are of interest for many reasons. The evolution of the Archaea is pivotal to understanding the evolution of the Eucarya and Bacteria. Therefore, studies of gene evolution at the DNA sequence level have confirmed the small subunit rRNA-based phylogeny of life (Iwabe *et al.*, 1989). Additionally, generalizations that the basic archaeal cell functions, such as translation machinery, are more similar to those of the Eucarya, whereas metabolic functions are more closely shared with those of the Bacteria, can be tested and have been confirmed for a number of genes (e.g., RNA polymerases; Langer *et al.*, 1995). Therefore, studying the expression and regulation of these genes may have an important impact on our understanding of related fields in molecular, cellular, and developmental biology. Furthermore, cross-domain gene homolog comparisons have revealed that a significant amount of gene transfer has occurred during the evolution of the Bacteria and Archaea (Hilario and Gogarten, 1993; Brown and Doolittle, 1997). Lastly, the use of 16S rRNA phylogenetic techniques in assessing archaeal diversity has dramatically expanded our view of archaeal habitats. Many new lineages have been identified that are based on 16S rRNA gene sequence comparisons within the archaeal domain. The genetic diversity of these new lineages has yet to be realized. For example, the Crenarchaeota were traditionally thought to be restricted to high-temperature environments; however, new sequences have been obtained from soils around the world (Hershberger *et al.*, 1996) and open-ocean environments (DeLong, 1992; Fuhrman *et al.*, 1992) and have been associated with sponges (Preston *et al.*, 1996) and holothurians (McInerney *et al.*, 1995). The proposed new kingdom within the Archaea, the Korarchaeota, was identified from a study of the archaeal diversity based on the 16S rRNA gene sequences obtained from a Yellowstone National Park thermal spring (Barns *et al.*, 1994). A few studies are beginning to tap this novel diversity (Brown *et al.*, 1996; Stein *et al.*, 1996; Diversa Corp.). As more archaeal genes and genomes are sequenced, primers to explore this as yet uncultured archaeal diversity will continue to build on the molecular taxonomy for the Archaea.

References

Alm, E. W., Oerther, D. B., Larsen, N., Stahl, D. A., and Raskin, L. (1996). The oligonucleotide probe database. *Appl. Environ. Microbiol.* **62,** 3557–3559.

Amann, R. I., Krumholz, L., and Stahl, D. A. (1990). Fluorescent-oligonucleotide probing of whole cells for determinative, phylogenetic, and environmental studies in microbiology. *J. Bacteriol.* **172,** 762–770.

Amann, R. I., Ludwig, W., and Schleifer, K.-H. (1995). Phylogenetic identification and in situ detection of individual microbial cells without cultivation. *Microbiol. Rev.* **59,** 143–169.

Barns, S. M., Fundyga, R. E., Jeffries, M. W., and Pace, N. R. (1994). Remarkable archaeal diversity in a Yellowstone National Park hot spring environment. *Proc. Natl. Acad. Sci. U.S.A.* **91,** 1609–1613.

Barns, S. M., Delwiche, C. F., Palmer, J. D., and Pace, N. R. (1996). Perspectives on archaeal diversity, thermophily and monophyly from environmental rRNA sequences. *Proc. Natl. Acad. Sci. U.S.A.* **93,** 9188–9193.

Bright, J. R., Byrom, D., Danson, M. J., Hough, D. W., and Towner, P. (1993). Cloning, sequencing and expression of the gene encoding glucose dehydrogenase from the thermophile Archaeon *Thermoplasma acidophilum. Eur. J. Biochem.* **211,** 549–554.

Brown, J. R., and Doolittle, W. F. (1997). Archaea and the prokaryote-to-eukaryote transition. *Microbiol. Mol. Biol. Rev.* **61,** 456–502.

Brown, J. R., Masuchi, Y., Robb, F. T., and Doolittle, W. F. (1994). Evolutionary relationships of bacterial and archaeal glutamine synthetase genes. *J. Mol. Evol.* **38,** 566–576.

Brown, J. W., Nolan, J. M., Haas, E. S., Rubio, M. A., Major, F., and Pace, N. R. (1996). Comparative analysis of ribonuclease P RNA using gene sequences from natural microbial populations reveals tertiary structural elements. *Proc. Natl. Acad. Sci. U.S.A.* **93,** 3001–3006.

Bult, C. J., White, O., Olsen, G. J., Zhou, L., Fleischmann, R. D., Sutton, G. G., Blake, J. A., FitsGerald, L. M., Clayton, R. A., Gocayne, J. D., *et al.* (1996). Complete genome sequence of the methanogenic arachaeon, *Methanococcus jannaschii. Science* **272,** 1058–1073.

Burggraf, S., Mayer, T., Amann, R., Schadhauser, S., Woese, C. R., and Stetter, C. O. (1994). Identifying members of the domain Archaea with rRNA-targeted olignucleotide probes. *Appl. Environ. Microbiol.* **60,** 3112–3119.

Burggraf, S., Heyder, P., and Eis, N. (1997). A pivotal Archaea group. *Nature (London)* **385,** 780.

Cary, S. C., Cottrell, M. T., Stein, J. L., Camacho, F., and Desbruyeres, D. (1997). Molecular identification and localization of filamentous symbiotic bacteria associated with the hydrothermal vent annelid *Alvinella pompejana. Appl. Environ. Microbiol.* **63,** 1124–1130.

Chou, Q., Russell, M., Birch, D. E., Raymond, J., and Bloch, W. (1992). Prevention of pre-PCR mis-priming and primer dimerization improves low-copy-number amplifications. *Nucleic Acids Res.* **20,** 1717–1723.

Cobb, B. D., and Clarkson, J. M. (1994). A simple procedure for optimising the polymerase chain reaction (PCR) using modified Taguchi methods. *Nucleic Acids Res.* **22,** 3801–3805.

DeLong, E. F. (1992). Archaea in coastal marine environments. *Proc. Natl. Acad. Sci. U.S.A.* **89,** 5685–5689.

DeLong, E. F., Ying, Wu, K., Prezelin, B. B., and Jovine, R. V. M. (1994). High abundance of Archaea in Antarctic marine picoplankton. *Nature (London)* **371,** 695–697.

DiRuggiero, J., Tolliday, N, Borges, K. M., Veillerot, E., Vetriani, C., and Robb, F. T. (1997). Cloning and overexpression of the glutamate dehydrogenase from *Thermococcus litoralis.* Submitted for publication.

Dong, G., Vieille, C., Savchenko, A., and Zeikus, J. G. (1997). Cloning, sequencing and expression of the gene encoding extracellular α-amilase from *Pyrococcus furiosus* and biochemical characterization of the recombinant enzyme. *Appl. Environ. Microbiol.* **63,** 3569–3576.

Ehrlich, H. A., Gelfand, D., and Sninsky, J. J. (1991). Recent advances in the polymerase chain reaction. *Science* **252,** 1645–1651.

Felsenstein, J. (1981). Evolutionary trees from DNA sequences: A maximum likelihood approach. *J. Mol. Evol.* **17,** 368–376.

Fuhrman, J. A., McCallum, K., and Davis, A. A. (1992). Novel major archaebacterial group from marine plankton. *Nature (London)* **356,** 148–149.

Giovannoni, S. J., Britschgi, T. B., Moyer, C. L., and Field, K. G. (1990). Genetic diversity in Sargasso Sea bacterioplankton. *Nature (London)* **345,** 60–63.

Gogarten, J. P., Kibak, H., Dittrich, P., Taiz, L., Bowman, E. J., Bowman, B. J., Manolson, M. F., Poole, R. J., Date, T., Oshima, T., Konishi, J., Denda, K., and Yoshida, M. (1989). Evolution of the vacuolar H+ -ATPase: Implications for the origin of eukaryotes. *Proc. Natl. Acad. Sci. U.S.A.* **86,** 6661–6665.

Hershberger, K. L., Barns, S. M., Reysenbach, A.-L., Dowson, S. C., and Pace, N. R. (1996). Wide diversity of Crenarchaeota. *Nature (London)* **384,** 420.

Hilario, E., and Gogarten, J. P. (1993). Horizontal transfer of ATPase genes—The tree of life becomes a net of life. *BioSystems* **31,** 111–119.

Hillis, D. M., Moritz, C., and Mable, B. K. (1996). *Molecular Systematics,* 2nd ed. Sinauer, Sunderland, MA.

Innis, M. A., and Gelfand, D. H. (1990). Optimization of PCRs. In *PCR Protocols. A Guide to Methods and Applications* (M. A. Innis, *et al.,* Eds.), pp. 3–12. Academic Press, San Diego.

Iwabe, N., Kuma, K., Hasegawa, M., Osawa, S., and Miyata, T. (1989). Evolutionary relationship of archaebacteria, eubacteria, and eukaryotes inferred from phylogenetic trees of duplicated genes. *Proc. Natl. Acad. Sci. U.S.A.* **86,** 9355–9359.

Langer, D., Hain, J., Thuriax, P., and Zillig, W. (1995). Transcription in Archaea: Similarity to that in Eucarya. *Proc. Natl. Acad. Sci. U.S.A.* **92,** 5768–5772.

Maidak, B. L., Larsen, N., McCaughey, M. J., Overbeek, R., Olsen, G. J., Fogel, K., Blandy, J., and Woese, C. R. (1994). The Ribosomal Database Project. *Nucleic Acids Res.* **22,** 3485–3487.

Massana, R., Murray, A. E., Preston, C. M., and DeLong, E. F. (1997). Vertical distribution and phylogenetic characterization of marine planktonic. Archaea in the Santa Barbara channel. *Appl. Environ. Microbiol.* **63,** 50–56.

McInerney, J. O., Wilkinson, M., Patching, J. W., Embley, T. M., and Powell, R. (1995).

Recovery and phylogenetic analysis of novel arachael rRNA sequences from a deep-sea deposit feeder. *Appl. Environ. Microbiol.* **61,** 1646–1648.

Moyer, C. L., Dobbs, F. D., and Karl, D. M. (1994). Estimation of diversity and community structure through restriction fragment length polymorphism distribution analysis of bacterial 16S rRNA genes from a microbial mat at an active, hydrothermal vent system, Lohi Seamount, Hawaii. *Appl. Environ. Microbiol.* **60,** 871–879.

Muir, J. M., Russell, R. J. M., Hough, D. W., and Danson, M. J. (1995). Citrate synthase from the hyperthermophilic Archaeon, *Pyrococcus furiosus. Protein Eng.* **8,** 583–592.

Olsen, G. J., and Woese, C. R. (1997). Archael genomes: An overview. *Cell* **89,** 991–994.

Preston, C. M., Wu, K. Y., Molinsky, T. F., and DeLong, E. F. (1996). A psychrophilic Creanarchaeote inhabits a marine sponge: *Cenarchaeum symbiosum* gen. nov., sp. nov. *Proc. Natl. Acad. Sci. U.S.A.* **93,** 6241–6246.

Raskin, L., Stromley, J. M., Rittmann, B. E., and Stahl, D. A. (1994). Group-specific 16S rRNA hybridization probes to describe natural communities of methanogens. *Appl. Environ. Microbiol.* **60,** 1232–1240.

Reysenbach, A.-L., and Pace, N. R. (1995). Reliable amplification of hyperthermophilic archaeal 16S rRNA genes by PCR. In *Archaea: A Laboratory Manual. Thermophiles* (F. T. Robb and A. R. Place, Eds.), pp. 101–106. Cold Spring Harbor Laboratory Press, Cold Spring Harbor, NY.

Reysenbach, A.-L., Giver, L. J., Wickham, G. S., and Pace, N. R. (1992). Differential amplification of rRNA genes by polymerase chain reaction. *Appl. Environ. Microbiol.* **58,** 3417–3418.

Robison-Cox, J., Bateson, M. M., and Ward, D. M. (1995). Evaluation of nearest-neighbor methods of detection of chimeric small-subunit rRNA sequences. *Appl. Environ. Microbiol.* **61,** 1240–1245.

Sambrook, J., Fritsch, E. F., and Maniatis, T. (1989). *Molecular Cloning: A Laboratory Manual,* 2nd ed. Cold Spring Harbor Laboratory Press, Cold Spring Harbor, NY.

Schleper, C., Holben, W., and Klenk, H.-S. (1997). Recovery of crenarchaeotal ribosomal DNA sequences from freshwater-lake sediments. *Appl. Environ. Microbiol.* **63,** 321–323.

Stein, J. L., Marsh, T. L., Wu, K. Y., Shizuya, H., and DeLong, E. F. (1996). Characterization of uncultivated prokaryotes: Isolation and analysis of a 40-kilobase-pair genome fragment from a planktonic archaeon. *J. Bacteriol.* **178,** 591–599.

Swofford, D. L., Olsen, G. J., Waddell, P. J., and Hillis, D. M. (1996). Phylogenetic Inference. In *Molecular Systematics,* (D. M. Hillis, C. Moritz, and B. Mable, Eds.), 2nd ed. Sinauer, Sutherland, MA.

Uemori, T., Ishino, Y., Fujita, K., Asada, K., and Kato, I. (1993). Cloning of the DNA polymerase gene of *Bacillus caldotenax* and characterization of the gene product. *J. Biochem.* **113,** 401–410.

Woese, C. R., Kandler, O., and Wheelis, M. L. (1990). Towards a natural system of organisms: Proposal for the domains Archaea, Bacteria, Eucarya. *Proc. Natl. Acad. Sci. U.S.A.* **84,** 4576–4579.

Young, C. C., Burghoff, R. L., Keim, L. G., Minak-Bernero, V., Lute, J. R., and Hinton, S. M. (1993). Polyvinylpyrrolidone-agarose gel electrophoresis purification of polymerase chain reaction-amplifiable DNA from soils. *Appl. Environ. Microbiol.* **59,** 1972–1974.

25

CLONING MAMMALIAN HOMOLOGS OF *DROSOPHILA* GENES

Filippo Randazzo

The *Drosophila* model system is a powerful tool for mammalian gene discovery and for providing insights into mammalian gene function (Miklos and Rubin, 1996). *Drosophila* genes commonly have similar functions to their mammalian counterparts. By using genetically defined *Drosophila* genes as a starting point, mammalian genes with a wide range of biological functions can be identified. Hundreds of *Drosophila* mutants with well-characterized functions exist and many have been cloned or are in the process of being cloned. Furthermore, there are hundreds of *Drosophila* genes with important biological functions that have undiscovered mammalian counterparts. Therefore, *Drosophila* genetics may open up new classes of genes which have been difficult to discover in mammals by more conventional cloning approaches. This has been the case with many classes or subclasses of genes, including the segmentation (Joyner and Martin, 1987; Sasai *et al.*, 1992), the homeotic (Graham *et al.*, 1989; Duboule and Dolle, 1989), the neurogenic (Ema *et al.*, 1996; Weinmaster *et al.*, 1991; Conlon *et al.*, 1995), the voltage-gated potassium channel (Chandy and Gutman, 1996), and the Polycomb group (Pearce *et al.*, 1992; Stankunas *et al.*, 1998; Takihara *et al.*, submitted)

PCR Applications

and Trithorax group genes (Khavari *et al.,* 1993; J. Berger and F. Randazzo, manuscript in preparation).

A number of strategies are available for cloning mammalian homologs of *Drosophila* genes. The advent of the human and mouse genome projects have made expressed sequence tag (EST) database screening the method of choice. However, the public databases are incomplete, and thus more conventional cloning approaches, such as low-stringency hybridization, degenerate PCR, and intermediate species cloning, are important alternative strategies to database screening. Together, none of these strategies usually results in a full-length cDNA and thus full-length cDNA cloning approaches are a necessity.

Mammalian *Drosophila* Homologies

Gene homologs have been traditionally defined as genes with evolutionarily conserved functions. Without functional data, two genes with highly similar sequence can only be referred to as putative homologs. The most stringent definition for a mammalian homolog includes whether it has the ability to rescue function in *Drosophila.* If the mammalian gene is demonstrated to function in the same pathway as its *Drosophila* counterpart, then the genes can be considered homologs as well. Because the extent of sequence similarities can vary tremendously with different genes, deciding on whether two genes are potentially homologs in the absence of functional data can be difficult. Knowledge of the particular class of genes, their functional domains, and precedents for other homologs within the gene class can give clues as to whether a particular gene is a putative homolog.

The percentage identity and similarity between homologous *Drosophila* and mammalian genes varies tremendously depending on their conserved function(s). Sequences that are critical for function are conserved between divergent species and these sequences usually lie within the protein coding region of the gene. Because of third-base degeneracy and codon bias, nucleotide sequence identity is never as high as amino acid sequence identity. However, codon bias may actually work to increase simularity in some instances. Thus a highly conserved domain may be 90% identical at the amino acid level and only 70% identical at the nucleotide level.

How conserved are *Drosophila* and mammalian genes? Some housekeeping genes, such as structural genes and enzymes, tend to be highly conserved, whereas genes involved in developmental pathways can vary tremendously in their conservation. For instance, when comparing mammalian and *Drosophila* housekeeping proteins, β tubulin is 97% similar and 94% identical, and myosin light chain 1 is 65% similar and 41% identical.

Sequence homologies for developmental genes tend to be restricted to contiguous stretches rather than a general homology over the entire protein. These conserved stretches define functional domains. The murine *M33* gene is the homolog of the *Drosophila polycomb* gene and has only two small stretches of homology with *Drosophila* polycomb, one at the NH_2 terminus and one at the COOH terminus (Fig. 1A; Pearce *et al.*, 1992). Even with only limited sequence homology, *M33* knockout mice and *polycomb Drosophila* mutants have similarities in genetic phenotype (Core *et al.*, 1997). The similarity between *engrailed* in flies and mice relies in the homeobox region near the COOH terminus (Fig. 1B; Joyner and Martin, 1987). The central portion of the NH_2 half of the *engrailed* protein has only slight sequence similarity between the two species (Fig. 1B). On the other hand, high sequence similarity between *Drosophila numb* and murine *numb*, a signaling adapter protein, is restricted to the NH_2 half of the protein where the 148-amino acid putative SHC PTB domain is 70% identical (Fig. 1C; Verdi *et al.*, 1996). The *patched* gene has weaker and more widely dispersed sequence similarities (Fig. 1D; Goodrich *et al.*, 1996). In the case of *patched*, sequence similarities span the entire protein; however, the similarities are relatively weaker and are frequently interrupted by nonsimilar regions. Clearly, the patterns of sequence conservation can vary tremendously between different genes.

Not all *Drosophila* genes have mammalian counterparts, so failure to clone a homolog may not be a function of methodological problems. The *Drosophila sevenless* gene is one such example.

Database Searching

Database searching allows a researcher to electronically identify putative mammalian homologs of *Drosophila* genes. Banfi and coworkers (1996) identified ESTs with sequence similarities to 66 *Drosophila* genes with known functions. Many strategies are available for searching through the public EST databases. Once a cDNA clone

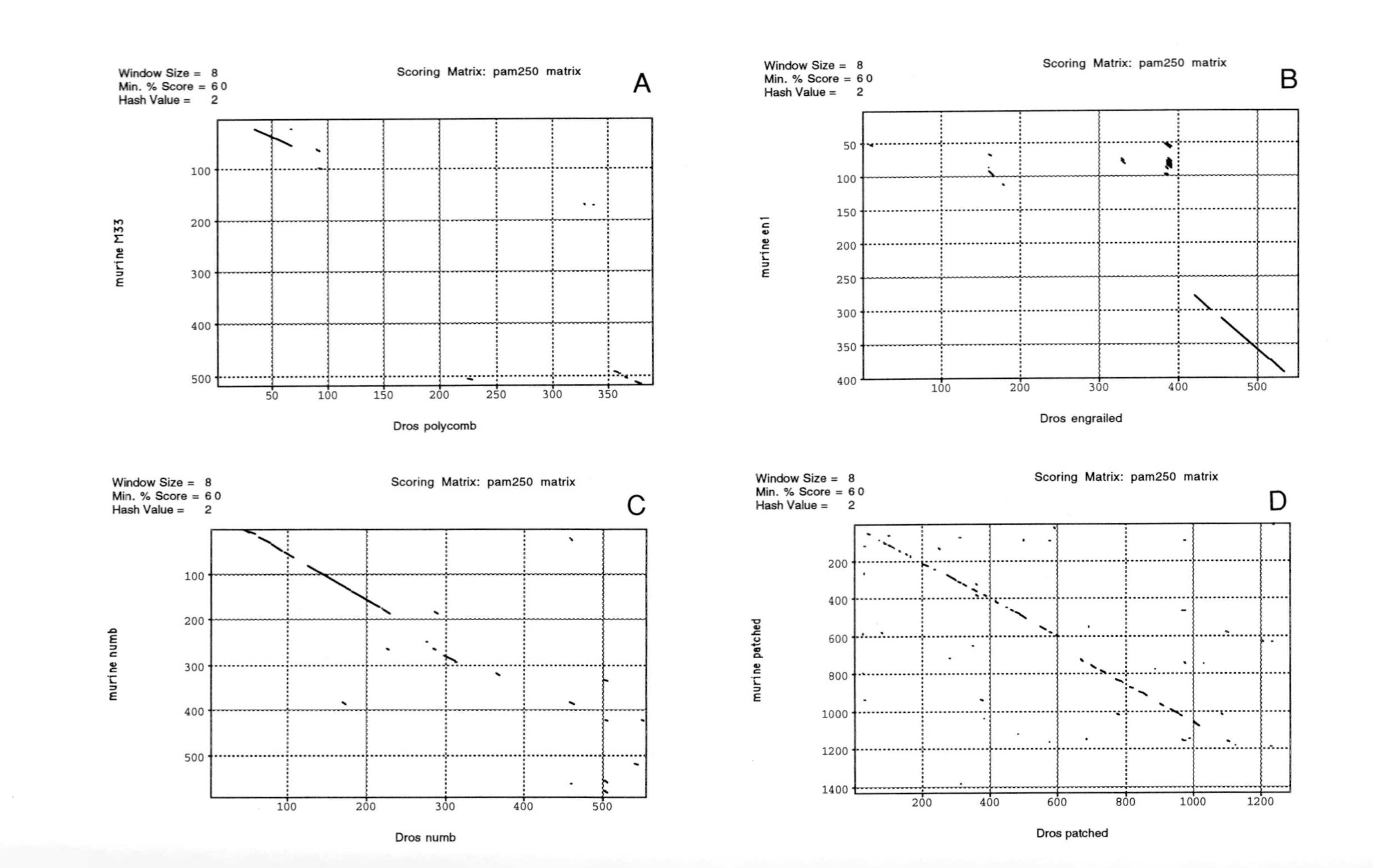
Window Size = 8
Min. % Score = 6 0
Hash Value = 2
Scoring Matrix: pam250 matrix
A
murine M33
Dros polycomb
Window Size = 8
Min. % Score = 6 0
Hash Value = 2
Scoring Matrix: pam250 matrix
B
murine en1
Dros engrailed
Window Size = 8
Min. % Score = 6 0
Hash Value = 2
Scoring Matrix: pam250 matrix
C
murine numb
Dros numb
Window Size = 8
Min. % Score = 6 0
Hash Value = 2
Scoring Matrix: pam250 matrix
D
murine patched
Dros patched

of interest has been identified, it can be purchased through the IMAGE Consortium.

One of the simplest search methods is the Basic Local Alignment Search Tool search, which is available on the world wide web (*http://www.ncbi.nlm.nih.gov/BLAST/*). This method allows one to search the EST databases using either *Drosophila* DNA or protein sequence as a template.

An alternative search tool is the FASTA algorithm, which can be found in many packages including the Genetics Computing Group package. Sensitivity can be further enhanced using the Smith–Waterman algorithm. However, robust hardware is necessary to run this algorithm. Using parallel implementation of the Smith–Waterman algorithm on a MASPAR computer, we find that protein backtranslated and searched against the ESTs, in some cases, will identify an EST clone when DNA/DNA searches fail.

Once a potential EST homolog has been identified, the length of the sequence can be expanded by contiging. Contiging is the process of identifying additional sequence data from a particular gene by systematic and sequential screening for additional cDNAs that overlap the ends of the parent molecule (Fig. 2). Contiging is necessary because EST sequence consists of a single sequence pass from the end of a cDNA. Thus, the cDNA may not be fully sequenced from end to end. Since the region of homology between the *Drosophila* gene and the mammalian gene can be limited to one region, contiging allows one to extend the sequence beyond the homology region by finding overlap with ESTs that have no homology to *Drosophila* but that do have 100% homology to the original EST sequence (Fig. 2). This method can potentially define an entire open reading frame for smaller genes. Full-length cDNAs, however, are not commonly found in the IMAGE consortium libraries because almost all the ESTs are derived from oligo-(dT)-primed cDNAs. Consequently, the libraries are biased in content toward the 3′ end of cDNAs.

Once contiging is complete, the longest cDNA clone can be ordered from the IMAGE consortium (*http://www.genomesystems.com/* or *http://www.resgen.com/online_catalog/*). DNA is then prepared,

---

Figure 1 McVector–Pustell matrix alignment of mouse proteins versus their *Drosophila* counterparts. (A), Murine *M33* vs *Drosophila polycomb;* (B), murine *en-1* vs *Drosophila engrailed;* (C), murine and *Drosophila numb;* (D), murine and *Drosophila patched.* Minimum percentage score, 60%; window size, 8; pam 250 matrix. Protein sequences were obtained from NCBI.

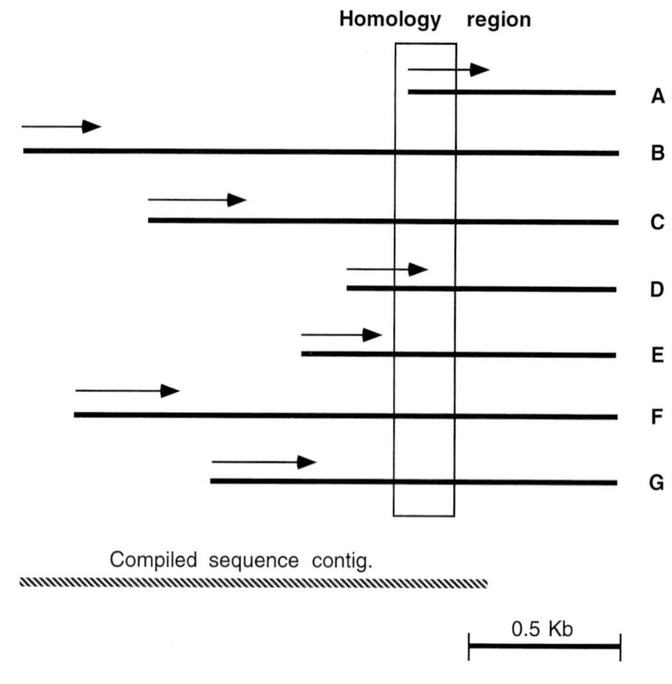

Figure 2 Theoretical contig of ESTs. Solid bars represent human EST cDNAs; right arrows indicate corresponding sequenced EST DNA. Rectangular region circumscribes the region of homology with *Drosophila* sequence. The stippled bar represents the compiled sequence contig.

and the ends of the cDNA are sequenced to confirm the identity of the clone. The clone is then fully sequenced. Splice variants are often uncovered and can be identified by restriction mapping and sequencing multiple EST clones.

A number of caveats may result in failure to identify an EST clone by database searching. Sequences that are critical for function are conserved between flies and mammals. Three prime untranslated (3′ UTR) sequences, however, are rarely conserved. If the mammalian gene has a long 3′ UTR sequence, then it is unlikely that it will be discovered by database searching since the EST cDNA clones are rarely >2 kb in length and are oligo-(dT) primed in origin. Even if the 3′UTR portion of the mammalian gene is short, the homology region with the *Drosophila* gene may reside solely at the 5′ end of the coding region. If the protein coding region is long, then it is unlikely that an EST sequence will be found with homology to that particular *Drosophila* gene.

Low-Stringency Hybridization

Some of the first cloned mammalian homologs of *Drosophila* developmental genes were the homeobox genes (McGinnis *et al.*, 1984; Levine *et al.*, 1984). These genes were cloned using standard low-stringency hybridization. This technique involves plating lambda libraries and then transferring the plaques onto duplicate nitrocellulose filters. Using radiolabeled *Drosophila* cDNA probe, library filters are then both hybridized and washed under low-stringency conditions. Lambda clones that remain hybridized to the probe on both filters are isolated. DNA is prepared and subsequently sequenced. A number of putative mammalian homologs have recently been cloned using low-stringency hybridization, including the Trithorax group gene *brg1* (Khavari *et al.*, 1993) and the calcium-activated potassium channel gene *slowpoke* (Pallanck and Ganetzky, 1994).

Degenerate PCR Cloning

Cloning mammalian genes by low-stringency hybridization using *Drosophila* cDNA probes is often unsuccessful. Frequently, the percentage similarity is too low or the homology is too dispersed. Thus, positive clones are difficult to identify because the signal-to-noise ratio is too low. PCR is a powerful and successful alternative to conventional library hybridization techniques. The design of degenerate PCR primers is discussed elsewhere (Compton, 1990).

Cloning a mammalian homolog by PCR requires identifying conserved regions from which to design primers. There are two approaches to identifying conserved sequences. If the *Drosophila* gene of interest has related family members with a common domain, primers can be designed from these sequences. Alternatively, homologs of the gene of interest can be cloned from species more closely related to *Drosophila* (intermediate species). Conserved sequences are then identified between species. Optimal primers are designed from seven-amino acid stretches of 100% identity. Choosing amino acid stretches with minimal base pair degeneracy further optimizes cloning.

The basic helix–loop–helix (bHLH) *MASH1* and *MASH2* genes were originally cloned by PCR as putative homologs of the *Drosophila* achaete-scute complex (AS-C) family of genes (Johnson *et al.*,

1990). Using the aligned amino acid sequences of *AS-C T5, T8, T3,* and *T4* regions, two short stretches of 100% amino acid identity, NARERNR and AVEIR, were chosen for designing fully degenerate oligos. Fourfold degenerate nucleotides were substituted with inosine. Restriction sites were added at their 5′ ends. Reverse transcriptase (RT)-PCR was performed from random-primed cDNA for 35 cycles with an annealing temperature of 45°C. PCR products were separated by agarose gel electrophoresis, subcloned, and sequenced. Of 24 clones sequenced, 16 were *MASH1* and 2 were *MASH2.* The mammalian bHLH gene homologs of the *Drosophila hairy* and *enhancer of split* genes were cloned in a similar manner (Sasai *et al.,* 1992).

Low-stringency hybridization cloning, combined with degenerate PCR cloning, resulted in successful cloning of eag, the mouse homolog of the *Drosophila* potassium channel gene *ether-a-go-go* (*eag*) (Warmke and Ganetzky, 1994). Attempts at cloning mammalian homologs directly using low-stringency hybridization and a *Drosophila eag* probe proved unsuccessful (Warmke and Ganetzky, 1994). In order to define a conserved domain, Warmke and Ganetzky screened for closely related *Drosophila* family members by low-stringency hybridization. The authors identified a new gene, eag-like K+ channel gene (*elk*), that was 49% identical to *eag* at the amino acid level.

Nested degenerate primers were created using amino acid sequence stretches that were 100% identical between *elk* and *eag* (Warmke and Ganetzky, 1994). The first set of primers were generated from the conserved ACIWY and TYCDL sequences, whereas the second set used the original ACIWY-derived forward primer and a nested internal ILGKGD-derived reverse primer. PCR reactions were performed for 35 cycles, 1 min at 95°C, 1 min at 42°C, and 1 min at 72°C using mouse cDNA as a template. The first PCR reaction yielded a predicted 850-nt fragment as shown by low-stringency Southern blot hybridization using the *Drosophila elk* probe. The remaining reaction mixture was separated by agarose gel electrophoresis and the 750–950 nt band was isolated. This DNA was used as template in the second PCR reaction along with the internal nested primer set. The products of this enriching PCR reaction were then subcloned and sequenced. A putative mouse *eag* homolog (*meag*) was isolated that was 65% identical to *Drosophila eag* at the amino acid level.

The use of intermediate species cloning is a proven method for cloning mammalian genes with highly dispersed sequence conservation (Goodrich *et al.,* 1996; Fig. 1d). Homologs of *Drosophila patched* were first cloned from more closely related species, mosquito (*Anopheles gambiae*), beetle (*Tribolium casteneum*), and butterfly (*Precis coenia*), by PCR and low-stringency hybridization (L. Goodrich, personal communication). Optimum degenerate PCR primers were designed using highly conserved regions between these species. Murine *patched* was subsequently cloned by RT-PCR (Goodrich *et al.*, 1996).

For a number of genes, the red flour beetle *Tribolium casteneum* has been shown to be more closely related to mammals than *Drosophila* (R. Denell, personal communication). Besides the cloning of *patched, Tribolium* has been used as an intermediate step in the cloning of other putative mammalian homologs (Ben-Arie *et al.*, 1996).

Successful cloning of mammalian *Notch* required intermediate species cloning approaches (Weinmaster *et al.*, 1991; Conlon *et al.*, 1995). A *Xenopus* homolog was first cloned using a *Drosophila Notch* probe and low-stringency hybridization to a *Xenopus* cDNA library (Coffman *et al.*, 1990). The *Xenopus* cDNA probe was subsequently used to clone, by hybridization, a murine *Notch* gene (Weinmaster *et al.*, 1991; Conlon *et al.*, 1995).

Full-Length cDNA Cloning

In order to extend the length of an existing cDNA and generate a full-length cDNA, a combination of PCR and the Gene Trapper cDNA positive selection system (GIBCO BRL) can be utilized. This is a powerful technique because an entire cDNA library can be screened in a few days (10^{12} clones). Thus, the chance of recovering a full-length cDNA is greatly increased. We have cloned numerous full-length cDNAs, including one 7.0-kb clone, using this approach (data not shown). The system works by annealing a biotinylated oligonucleotide probe, complementary to the desired cDNA, to a single-stranded cDNA library. We use a probe complementary to the 5′ region of our incomplete cDNA. Strepavadin-coated paramagnetic beads are then used to isolate the cDNA that binds to the specific gene probe. The target cDNA is then primed with a second probe specific to the target cDNA and the ssDNA is converted to dsDNA. The repaired cDNAs are then transformed into *Escherichia coli* and are ready for secondary screening by colony hybridization. Positive colonies are subsequently isolated, DNA is prepared, and clone identity is confirmed by DNA sequencing.

The power and efficiency of this technique is increased by the addition of two simple PCR steps. Before beginning the GeneTrapper procedure, a number of GeneTrapper plasmid cDNA libraries are prescreened by semiquantitative PCR to first determine if the desired cDNA exists in any of the libraries. Observing gene expression on a tissue Northern blot does not guarantee the presence of the cDNA in any particular library made from that tissue. In addition, libraries made from tissues that fail to show expression on a polyA$^+$ Northern blot have yielded positive clones.

The following is the protocol for semiquantitative evaluation of cDNA abundance in a GeneTrapper library:

1. Dilute an aliquot of the library at 1:50, 1:500, and 1:5000.
2. Use 5 μl of each dilution as template in a standard 50-μl PCR reaction.
   30 cycles: denaturation 94°C, 5 sec (30 sec, first cycle); annealing/extension 68°C
   Thin-walled PCR tubes (Perkin–Elmer 9600)
   KlenTac enzyme plus KlenTac buffer (Clonetech, Inc.)
3. Use a primer set near the 5′ end of the truncated cDNA for the PCR.
4. Run 5 μl of the completed PCR reaction on a mini-agarose gel.
5. The abundance of cDNA is evaluated by whether a band is observed at one of the three dilutions.
   1:5000 Abundant cDNA
   1:500 Moderately abundant cDNA
   1:50 Rare cDNA

Given a choice of libraries, the most abundant libraries are chosen for subsequent screening by GeneTrapper protocol.

Sometimes positive clones are not found at the colony screening step. In order to predetermine whether the bacterial transformation and colony hybridization steps will yield positive clones, we first perform PCR to determine if we have indeed trapped our cDNA. Using the PCR primer set described previously and the trapped DNA as template, we assay for the presence of trapped cDNA clones before we transform the DNA into *E. coli.*

Conclusions

Because of the great depth of the human and mouse EST databases, cloning mammalian homolog of *Drosophila* genes has become routine. In these cases, the primary challenge is cloning full-length cDNA. However, not all mammalian homologs can be discovered through this method because the public databases are incomplete and because of the nature of ESTs. Alternative approaches, such as low-stringency hybridization, degenerate PCR design, and intermediate species cloning, are necessary in those cases. Finally, failure to discover a mammalian homolog may not be a function of cloning problems but may be due to the fact that not all *Drosophila* genes have mammalian counterparts.

Acknowledgments

I thank Joel Berger for developing the PCR protocol modification for the GeneTrapper kit. I also thank Scott Chouinard, Rob Denell, Lisa Goodrich, Michael Innis, Steve Harrison, David Pot, and Mathew Scott for helpful comments.

References

Banfi, S., Borsani, G., Rossi, E., Bernard, L., Guffanti, A., Rubboli, F., Marchitiello, A., Giglio, S., Coluccia, E., Zollo, M., Zuffardi, O., and Ballabio, A. (1996). Identification and mapping of human cDNAs homologous to Drosophila mutant genes through EST database searching. *Nature Genet.* **13,** 167–174.

Ben-Arie, N., McCall, A. E., Berkman, S., Eichele, G., Bellen, H. J., and Zoghbi, H. Y. (1996). Evolutionary conservation of sequence and expression of the bHLH protein Atonal suggests a conserved role in neurogenesis. *Hum. Mol. Genet.* **5,** 1207–1216.

Chandy, K. G., and Gutman, G. A. (1996). Voltage-gated potassium channel genes. In *CRC Handbook of Receptors and Channels* (A. R. North, Ed.), pp. 1–71. CRC Press, Boca Raton, FL.

Coffman, C., Harris, W., and Kintner, C. (1990). Xotch, the Xenopus homolog of Drosophila notch. *Science* **249,** 1438–1441.

Compton, T. (1990). Degenerate primers for DNA amplification. In *PCR Protocols. A Guide to Methods and Applications* (M. A. Innis, D. H. Gelfand, J. J. Sninsky, and T. J. White, Eds.). Academic Press, San Diego.

Conlon, R. A., Reaume, A. G., and Rossant, J. (1995). Notch1 is required for the coordinate segmentation of somites. *Development* **121,** 1533–1545.

Core, N., Bel, S., Gaunt, S. J., Aurrand-Lions, M., Pearce, J., Fisher, A., and Djabali, M. (1997). Altered cellular proliferation and mesoderm patterning in Polycomb-M33-deficient mice. *Development* **124,** 721–729.

Duboule, D., and Dolle, P. (1989). The structural and functional organization of the murine HOX gene family resembles that of Drosophila homeotic genes. EMBO J. **8,** 1497–1505.

Ema, M., Morita, M., Ikawa, S., Tanaka, M., Matsuda, Y., Gotoh, O., Saijoh, Y., Fujii, H., Hamada, H., Kikuchi, Y., and Fujii-Kuriyama, Y. (1996). Two new members of the murine Sim gene family are transcriptional repressors and show different expression patterns during mouse embryogenesis. *Mol. Cell. Biol.* **16,** 5865–5875.

Goodrich, L. V., Johnson, R. L., Milenkovic, L., McMahon, J. A., and Scott, M. P. (1996). Conservation of the hedgehog/patched signaling pathway from flies to mice: Induction of a mouse patched gene by Hedgehog. *Genes Dev.* **10,** 301–312.

Graham, A., Papalopulu, N., and Krumlauf, R. (1989). The murine and Drosophila homeobox gene complexes have common features of organization and expression. *Cell* **57,** 367–378.

Johnson, J. E., Birren, S. J., and Anderson, D. J. (1990). Two rat homologues of Drosophila achaete-scute specifically expressed in neuronal precursors. *Nature* **346,** 858–861.

Joyner, A. L., and Martin, G. R. (1987). En-1 and En-2, two mouse genes with sequence homology to the Drosophila engrailed gene: Expression during embryogenesis [published erratum appears in *Genes Dev* 1987 Jul; **1**(5), 521]. *Genes Dev.* **1,** 29–38.

Khavari, P. A., Peterson, C. L., Tamkun, J. W., Mendel, D. B., and Crabtree, G. R. (1993). BRG1 contains a conserved domain of the SW12/SNF2 family necessary for normal mitotic growth and transcription. *Nature* **366,** 170–174.

Levine, M., Rubin, G. M., and Tjian, R. (1984). Human DNA sequences homologous to a protein coding region conserved between homeotic genes of Drosophila. *Cell* **38,** 667–673.

McGinnis, W., Hart, C. P., Gehring, W. J., and Ruddle, F. H. (1984). Molecular cloning and chromosome mapping of a mouse DNA sequence homologous to homeotic genes of Drosophila. *Cell* **38,** 675–680.

Miklos, G. L., and Rubin, G. M. (1996). The role of the genome project in determining gene function: Insights from model organisms. *Cell* **86,** 521–529.

Pallanck, L., and Ganetzky, B. (1994). Cloning and characterization of human and mouse homologs of the Drosophila calcium-activated potassium channel gene, slowpoke. *Hum. Mol. Genet.* **3,** 1239–1243.

Pearce, J. J., Singh, P. B., and Gaunt, S. J. (1992). The mouse has a Polycomb-like chromobox gene. *Development* **114,** 921–929.

Sasai, Y., Kageyama, R., Tagawa, Y., Shigemoto, R., and Nakanishi, S. (1992). Two mammalian helix–loop–helix factors structurally related to Drosophila hairy and Enhancer of split. *Genes Dev.* **6,** 2620–2634.

Stankunas, K., Berger, J., Ruse, C., Randazzo, F. M., and Brock, H. W. (1998). The *Enhancer of Polycomb* gene of *Drosophila* encodes a chromatin protein conserved in yeast and mammals. *Development* **125,** 4055–4066.

Takihara, Y., Tomotsune, D., Berger, J., Kyba, M., Shirai, M., Ohta, H., Matsuda, Y., Honda, B., Simon, J., Shimada, K., Brock, H., and Randazzo, F. Mouse Sex comb on midleg proteins are highly conserved, and interact with mph/rae28 *in vitro.* (Submitted for publication).

Verdi, J. M., Schmandt, R., Bashirullah, A., Jacob, S., Salvino, R., Craig, C. G., Program, A. E., Lipshitz, H. D., and McGlade, C. J. (1996). Mammalian NUMB is an evolutionarily conserved signaling adapter protein that specifies cell fate. *Curr. Biol.* **6,** 1134–1145.

Warmke, J. W., and Ganetzky, B. (1994). A family of potassium channel genes related to eag in Drosophila and mammals. *Proc. Natl. Acad. Sci. U.S.A.* **91,** 3438–3442.

Weinmaster, G., Roberts, V. J., and Lemke, G. (1991). A homolog of Drosophila Notch expressed during mammalian development. *Development* **113,** 199–205.

26

CLONING HUMAN HOMOLOGS OF YEAST GENES

Todd Seeley

Saccharomyces cerevisiae: A Model Organism in the Determination of Functional Relationships between Genes

The yeast *Saccharomyces cerevisiae* is widely exploited as a model eukaryote in studies relating to basic biology. This includes extensive analyses of genes relating to eukaryotic cell cycle, intracellular signaling, stress response, and other cellular processes. For many genes with functions defined in *S. cerevisiae,* structural homologs exist in other organisms. Functional models based on study of *S. cerevisiae* genes have proved their utility in the study of processes and pathways common to more complex organisms, including humans.

Recently, the complete DNA sequence of the *S. cerevisiae* genome became known (Dujon, 1996; Goffeau *et al.,* 1996; Mewes *et al.,* 1997). This heralded event allows us to define the contribution of DNA to the biology of this organism. For many of the genes which make up the genome of *S. cerevisiae,* a function has been defined

PCR Applications

by prior mutational analyses. For others, knowledge of the gene sequence allows us to infer relationships to genes of known function. The relationships between known and novel genes defined by this sequencing effort is certain to be a topic of study for many years. Knowledge of the *S. cerevisiae* genome will greatly accelerate the process of functional discovery in this organism, as will the proposed human genome sequencing effort upon its completion.

The "Power of Yeast Genetics"

Drawing on the power of yeast genetics, much can be learned by studying structural homologs of yeast genes. The fundamental requirements of yeast are common to all eukaryotic cells, including human cells. Why, then, have genetic studies so widely exploited yeast as a model eukaryote? The utility of genetic studies in yeast has classically been associated with the ability to freely propagate and study both haploid and diploid forms of this organism. In haploid cells, low genome copy number increases the probability that a random mutagenesis study will result in a strain exhibiting a useful phenotype. Relationships between various haploid yeast mutants can be studied by mating cells to form stable diploid strains. Through meiosis and sporulation, novel haploid strains harboring complex combinations of mutations can easily be constructed to determine the functional relationships between different genes. These relationships can be combined to obtain functional pathways.

In recent years, tools have been refined to allow functional analyses of yeast genes more directly, using strains transformed by libraries of DNA carried on episomal plasmid vectors and by direct replacement of genes with selectable marker DNAs. Hybridization and DNA sequence technologies have assisted in the rapid identification of relevant genes in functional tests. The functional relationships between these genes and existing genes identified by more classical methods can be determined. Newer strategies, including total genome scanning and use of miniaturized hybridization arrays (Shalon *et al.*, 1996; Smith *et al.*, 1996), have arisen to exploit the knowledge of the yeast genome further in defining gene function.

Human Homologs of Yeast Genes

The contribution of yeast genetic studies to the understanding of human gene function cannot be overstated. The ability to extensively manipulate the genome of *S. cerevisiae* has allowed yeast researchers to conduct experiments that were difficult or impossible to conduct in other eukaryotes. This included the early development of useful transformation vectors and the ability to perform gene replacements (Scherer and Davis, 1979; Struhl *et al.*, 1979). These technological breakthroughs combined with classical genetic analyses led to the ability to isolate, characterize, and determine the functions of large numbers of yeast genes.

Early molecular studies led to the realization that the yeast genome harbors close structural homologs of many genes from complex multicellular eukaryotes, including humans (Botstein and Fink, 1988). These studies described highly conserved yeast structural protein genes for actin and tubulin (Gallwitz and Seidel, 1980; Neff *et al.*, 1983; Ng and Abelson, 1980), for example. Later studies identified human homologs of *S. cerevisiae* genes known to be required for cell cycle progression, including *cdc28* and cyclin genes (Lorincz and Reed, 1984; Gautier *et al.*, 1988). Recently, structural homology was observed between the human ATM tumor suppressor gene and yeast mitotic checkpoint genes (Greenwell *et al.*, 1995; Morrow *et al.*, 1995; Savitsky *et al.*, 1995; Weinert *et al.*, 1994). This pattern of complementary experiments describing functionally related human and yeast genes has continued to evolve and has led to unified models describing various growth regulation and signal transduction pathways (Tugendreich *et al.*, 1994).

The substantial body of knowledge concerning the function of individual yeast genes has only begun to be exploited in the study of human gene function. A recent enumeration suggested that ~40% of all yeast genes are associated with an experimentally determined function. For ~35% of yeast genes, sequence similarities and motifs suggest additional functions. This leaves ~25% of yeast genes with no recognizable function, based on experiment or homology (Botstein and Cherry, 1997). Of these, many have mammalian structural homologs and constitute good candidates for future functional tests.

From Structure to Function

Functional knowledge gained from study of yeast genes can be useful in understanding pathways in other organisms, especially if structural homologs of yeast genes of interest can be demonstrated to exist. Direct comparison of *S. cerevisiae* and human genomes will require the completion of the human genome sequencing effort. In the meantime, sequence data from an increasing fraction of individual human cDNAs have become available, particularly since the advent of vast deposits of expressed sequence tag (EST) data. Knowledge of the human genome will soon progress to the point where all human structural homologs of yeast genes can be identified directly by database screening. As an illustration, a recent Serial Analysis of Gene Expression (SAGE) survey of expressed human transcripts suggested that of ~45,000 unique SAGE-tagged transcripts, ~50% could be identified as corresponding to known human Genbank mRNA and EST entries (Zhang *et al.*, 1997). This can be compared to current estimates of 60,000–100,000 total expressed human genes.

Examples exist in which human structural homologs of yeast genes have been identified directly by functional complementation of yeast mutants (Botstein and Fink, 1988). Yeast mutants transformed with libraries of human cDNA inserts in yeast expression vectors can be screened for plasmids which suppress (complement) a useful yeast mutant phenotype. In other cases, human cDNAs identified by independent means (e.g., positional cloning) with fortuitous homology to an interesting yeast gene can be expressed in yeast mutants as part of functional complementation studies. However, human structural homologs of yeast genes may fail these functional tests for a number of reasons. For example, structural conservation may be insufficient to allow interaction of human gene products with yeast proteins. Due to differences in preferred codon usage and other factors, human cDNAs may be relatively poorly expressed in yeast. To overcome these barriers, successful complementation studies can require substantial overexpression of a human transcript. In addition, successful complementation studies may utilize yeast mutants with a relatively mild phenotype (i.e., a relatively weak point mutant or temperature-sensitive mutant allele held at a semipermissive temperature). Negative results in these tests are uninformative and cannot be used to rule out that structural homologs function in similar pathways in their respective hosts. Ultimately, the activity of human genes must be tested directly by examining human cells for relevant phenotypes.

We have used computerized database screening strategies to identify human structural homologs of yeast genes of functional interest. We have found PCR-based techniques to be useful at multiple stages in the process of characterizing these genes, including the design of functional tests in human cells.

Selection of an Interesting Yeast Gene

The breadth and extended history of the yeast field has led to a broad knowledge base concerning the respective functions of individual yeast genes. Journal articles describing yeast gene functions can be identified through literature searches and from Genbank references. Literature references can be examined for descriptions of any known human structural homologs. The process of compiling a complete literature for a given yeast gene is complicated by multiple synonyms for many yeast genes. The same name may be assigned to unrelated genes from other organisms—another opportunity for confusion. In addition, the known functions of a yeast gene may bear little relationship to the manner in which a known human homolog was identified. For these reasons, the nomenclature by which related yeast and human genes are classified may give no clue to relatedness.

The Yeast Protein Database (YPD; *http://www.proteome.com/YPDhome.html*) is a useful resource in compiling a functional literature for yeast genes (Payne and Garrels, 1997). YPD is commercially maintained and is available without cost to academic researchers. Individual YPD entries can be extensive and include gene name synonyms. Known and predicted physical, functional, and genetic characteristics of each yeast ORF are cataloged. Amino acid sequence motifs (e.g., phosphorylation, glycosylation, and other posttranslational processing sites) are noted. Links to Genbank, PIR, Swissprot, literature references, and other functionally and structurally related genes are provided.

Database Searching

Once an interesting yeast gene has been identified, human homologs can be identified by computer-based search. In most labs, searches can be conducted via internet search. As a first effort, a search of the larger annotated databases (Genbank and PIR) for text strings corresponding to the name of your yeast gene and any synonyms is recommended. The common databases now include the results of automated homology searches, and annotations may include the name of your yeast gene. This is the fastest way to identify known homologs. Restrictive field descriptors can be added to searches to limit database output to human genes. Database matches from other organisms can be useful, however, and restricting searches in this manner is not generally recommended.

The starting sequence in a homology search is the predicted amino acid sequence of a yeast gene product rather than a DNA sequence. Due to differences between yeast and human preferred codon usage, direct DNA to DNA sequence searches may result in failure to detect important amino acid sequence similarities. Automated reverse translation from a submitted protein sequence is a common feature of most database searching resources, and amino acid sequences can be used to search nucleotide databases directly.

The choice of database search software depends on the user. Database searching resources available over the internet include the popular NCBI BLAST site (*http://www.ncbi.nlm.nih.gov/blast*). MPSRCH software, which runs on a MASPAR computer, is also highly recommended. MPSRCH uses a Smith–Waterman algorithm to score sequences on a massively parallel computer. This approach substantially improves the ability to pick up distantly related matches and is especially tolerant of small gaps and nucleotide sequence errors, commonly encountered when using yeast sequence to search human EST databanks. Yeast protein sequences can be used to search both protein and DNA databases. The nucleic acid databases are generally more complete, although unique entries can generally be identified in the protein databases as well.

Evaluating Search Scores

A search session produces a series of results accompanied by scores and alignments. An examination of the top scores should reveal a trend of highly clustered scores giving way to a few outlying high scores (Fig. 1). In an unrestricted search, outlying high scores will include entries corresponding to the yeast gene used as sequence probe. If human sequences lie within these outlying scores, the next part of the project can begin.

An excess of high scores can arise from domains of low amino acid complexity (e.g., polyalanine) or from a conserved domain present in a relatively large family of proteins. Both of these results can be informative with respect to the function of proteins. To reveal which of these cases has occurred, sequence alignments can be examined directly. In the cases of too many high scores or too few high scores (no outlying scores), additional searching may be required. Searches with smaller portions of the protein sequence can be used to omit domains of low complexity or excessive conservation. Large proteins can give relatively few high scores due to the fact that scores are averaged across the whole protein and conservation may be limited to smaller subdomains. Homologs of *S. cerevisiae* genes from closely

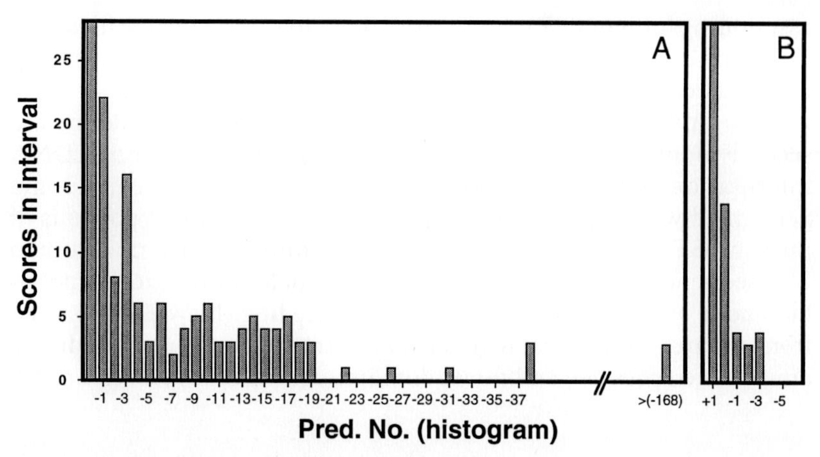

Figure 1 In *S. cerevisiae,* the CDC4 protein acts in a selective ubiquitin degradation pathway which targets cyclin/cdk complexes (Skowrya *et al.,* 1977; Feldman *et al.,* 1977). (A) A MASPAR search of the CDC4 protein sequence against protein data banks produced a closely related *Candida albicans* CDC4 sequence (Pred. No. 8.21e-168; Genbank: CACDC4). Both proteins include a conserved repeat structure identified in a large family of proteins (variously termed WD40, trp-asp, β-transducin repeats) and a second motif, termed the F-box (Bai *et al.,* 1996). The *S. cerevisiae* CDC4 F-box is known to bind SKP1. (B) A secondary search using the CDC4 F-box region alone (amino acids 255–321) identified the murine MD6 transcript from Genbank searches (Pred. No. 310e-01; Genbank: MDMD6RNA). MD6 retains the WD40 repeat motifs from CDC4. The relationship between CDC4 and MD6 was reported previously (Bai *et al.,* 1996).

related organisms (i.e., fungi) can be used to identify conserved subdomains of a protein for further searches.

EST Contig Assembly

Currently, the product of a typical search is a cluster of related ESTs or, with luck, a full-length mRNA sequence. Because most EST data have been collected from human cDNA libraries, database matches have a good chance of being human in origin. The significance of any human matches can be confirmed by matches to cDNAs from other multicellular organisms. Individual human matches are compiled and their overlap, if any, is determined. A crude map of the yeast gene and the approximate location of the EST matches can be constructed by hand using alignment data from the search. Alternatively, DNA sequence project software can be useful in aligning and assembling human EST contigs from downloaded sequence files (Fig. 2). In this manner, the number of unique human cDNAs represented by a cluster of related ESTs can be estimated.

Once one or more unique human cDNAs have been defined, a second round of searching begins with a goal of extending cDNA information as much as possible. New DNA to DNA searches are performed with sequence corresponding to the endmost portions of human EST contigs, with the goal of obtaining overlapping human EST sequences. The purpose is to gain as much information as possible about a potential mRNA before proceeding. New ESTs gained from a search may reveal regions of less strongly conserved predicted amino acid sequence and novel domains absent in a yeast homolog.

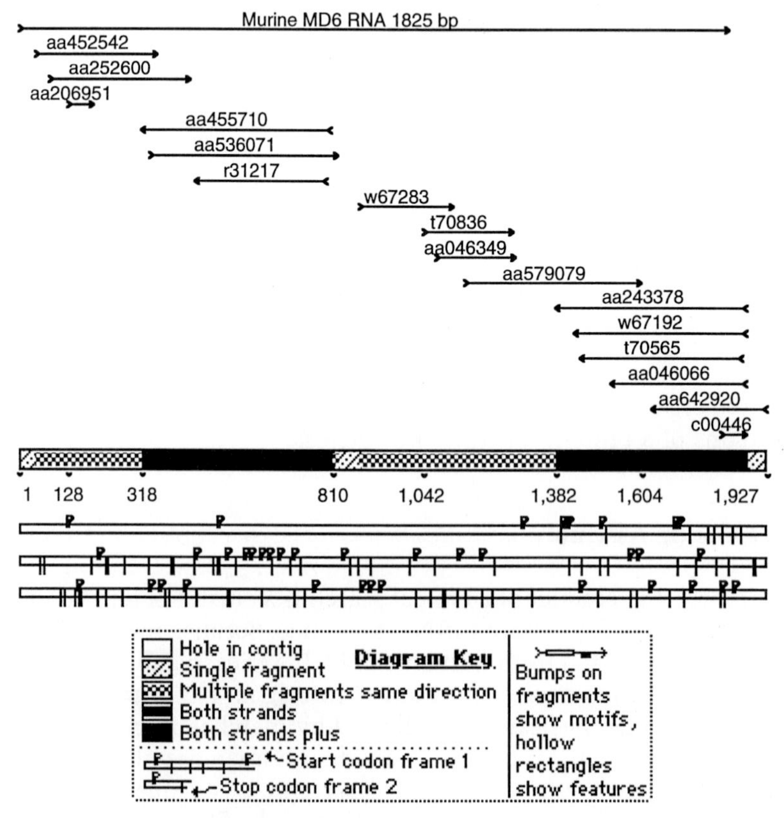

Figure 2 EST contig assembly. Close human EST matches to murine MD6 were retrieved and arrayed using Sequencher DNA sequence analysis software (Lifecodes). The result is an interrupted human EST contig.

Sequence errors are common in EST data. To reduce the chance that a corrupted EST sequence may be selected for later steps, all available ESTs should be compiled. Alternative splice junctions also complicate the analysis of EST data. Currently, automated EST contig resources are limited. Human EST contig data can be accessed via THC BLAST (*http://www.ncbi.nlm.nih.gov/cgi-bin/THCBlast/nph-thcblast*). Resources for acquiring up-to-date and accurate automated EST contig data should improve in the near future.

In compiling ESTs, we have noted individual entries which exhibit high levels of polymorphism when compared to closely related ESTs. Many of these ESTs can be judged to represent database entries of poor sequence quality rather than unique, separate genes. Artifacts of EST library preparation can also be observed, such as redundant ESTs which end abruptly at sequences of high GC content and artifacts likely caused by annealing of phased poly-dT primer to short internal poly-A tracts within an mRNA. As the redundancy level of EST data improves, these artifacts should become less of an issue.

Using yeast genes as homology probes, we have obtained single matches of murine EST origin. Using PCR primers designed from regions most highly conserved with yeast sequence, we have obtained small fragments of human cDNA sequence and used these data to isolate the corresponding human cDNA. In other examples, the complete sequence of a previously uncharacterized human cDNA was successfully assembled by EST contigging, with sufficient redundancy to have a high confidence in the predicted sequence. In another instance, a full-length murine mRNA match was used as a guide to compile human ESTs, producing a gapped human EST contig.

Success at this stage is in part related to the frequency of representation of relevant cDNAs within the databases, which is in turn related to the relative transcription level of genes in the tissues used to construct EST cDNA libraries. As EST data become more complete, the need to prepare EST contigs will diminish.

From the Database to the Bench

Once available EST database data have been cataloged, the goal shifts to obtaining any remaining cDNA sequence. Most EST sequence matches consist of a 200- to 350-bp sequence from the end of a cDNA clone of undetermined size. Additional sequence can be obtained by acquiring specific EST clones from the appropriate sources and determining the full sequence of cDNA inserts. Artifacts can occur in the process of cDNA cloning, and the sequence of multiple library clones should always be determined. EST clones are rarely full length and remaining sequence data must usually be acquired through some form of additional cDNA cloning.

Yeast researchers may be relatively unfamiliar with working with cDNAs. Introns are rare in yeast, and in virtually all applications moderately sized yeast genomic fragments can be used in place of cDNAs. The complete knowledge of the sequence of the yeast genome means that all sequence information required to initiate functional studies of a yeast gene already exists, without a need to acquire novel sequence data. In contrast, a newly characterized human cDNA fragment often requires further cloning and sequencing. The following sections include information on useful PCR-based cDNA techniques, including RACE (rapid amplification of cDNA ends).

PCR vs Library-Based Screening

With respect to the goal of obtaining remaining cDNA sequence, PCR-based approaches can be contrasted to classical library-based techniques. Both methods have evolved in a manner which improves the quality of data and shortens the time required for results. Both methods are valid and have specific advantages and disadvantages. In both cases, the size of the expected mRNA should be confirmed by Northern blots.

Classical library-based approaches using plaque or colony lift techniques can be time-consuming. However, new library-based approaches can dramatically improve the throughput of library screening compared to earlier approaches. These approaches combine library screening with oligonucleotide-based affinity enrichments (e.g., Genetrapper, GIBCO BRL). In library-based screening, success depends on the quality and representation of the library. A specific human cDNA sequence in a given library may be exceptionally rare, present in a rearranged form, or entirely absent due to artifacts of cDNA preparation and cloning. Relative inefficiency of cloning large inserts combined with inefficient extension of a poly-dT 3′ phasing primers in cDNA synthesis reactions can produce clones exhibiting truncated ends. Deletions or the absence of sequences can also result from toxicity of specific sequences when propagated in bacteria.

Given ready access to primers, PCR-based approaches can afford substantial time savings in the characterization of cDNAs. A series of PCR primers based on the sequence of EST contigs can be designed and used for various applications (Fig. 3). Primers which amplify an internal fragment from a cDNA pool can be used to generate DNA fragments for use as hybridization probes. Other primers can be used for RACE reactions to characterize the 3′ and 5′ ends of the human cDNA. For expression studies, a full-length cDNA or an ORF fragment can be directly amplified by PCR or, alternatively, can be assembled by subcloning from internal cDNA and RACE products.

For many, a PCR-based approach can be used throughout the process of characterizing a novel cDNA, with a potential for significant time savings compared to other approaches. To some extent, PCR can avoid artifacts introduced by library preparation or by instability of specific sequences in bacteria.

Primer Design

The success of a PCR-based approach depends entirely on the design and synthesis of useful PCR primers. The process of identification

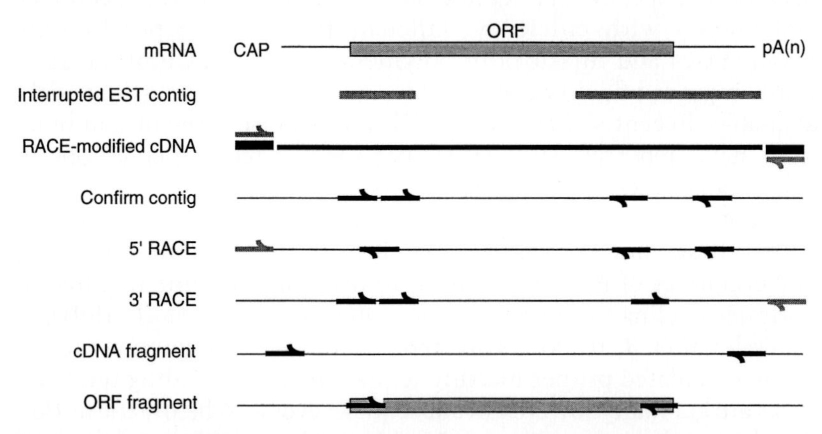

Figure 3 Primer design strategies for cloning human homologs of yeast genes. A hypothetical mRNA is indicated along with hypothetical EST contig data from a homology search. To obtain remaining sequence data, internal sequences can be amplified from cDNA pools. In turn, external sequence can be amplified from RACE-modified cDNA pools using primers designed from contig data. To ensure success, RACE primers should be paired in separate reactions with a series of gene-specific primers. Once a complete cDNA sequence has been determined, constructs for functional assays (including full-length ORF fragments) can be amplified.

of primers for PCR-based applications can be greatly aided by primer design software. In our experience, appropriate use of primer design software reduces the time required to optimize PCR protocols, increases the likelihood of obtaining desired products, and is highly recommended. Numerous software packages for primer design are available. Selection of specific packages will depend on user requirements because both simple and complex programs are available. Primer design software should evaluate all possible primers for uniqueness against a database compiled from the anticipated target DNA (e.g., human cDNA). Additional filters should exist for screening primers based on length, melting temperature, base composition, and other physical parameters. Filters should also eliminate primers exhibiting self- and/or cross-complementarity and those with potential secondary priming sites within the anticipated target.

PCR Conditions

Generally, high annealing temperature primers should be used whenever possible to minimize potential for PCR mispriming artifacts. Selection of available PCR enzyme–reagent combinations will depend on the specific application and the need to balance specificity and fidelity with efficiency. Efficient thermostable proofreading polymerases and mixtures of polymerases which exhibit substantially improved fidelity compared to earlier PCR reagents are widely available. Recently, many new PCR polymerase–reagent combinations have appeared, suggesting that specific recommendations in this area are likely to become rapidly outdated.

Cycling parameters depend on primer design. When used appropriately, "touchdown" cycling strategies can increase the specificity and efficiency of PCR, in addition to reducing the time required to optimize cycling parameters for new oligo pairs (Don *et al.*, 1991). In touchdown PCR, reactions are precycled at annealing temperatures above calculated primer melting temperatures. Annealing temperatures are gradually lowered in subsequent cycles to below the anticipated annealing temperature. In this manner, PCR amplification initiates at the highest possible annealing temperature for the highest possible specificity. The bulk of the cycling is performed at a lower temperature, resulting in high efficiency of product formation. Because a range of annealing temperatures is used, small deviations in the actual annealing temperatures of a given primer pair do not affect the success of the PCR. Product formation can be highly specific and efficient without the need to test a wide range of annealing temperatures, thus avoiding the time-consuming process of extensive PCR optimization.

Source of Target

To determine the length and relative tissue distribution of a human mRNA, an initial DNA fragment is required to serve as a hybridization probe in Northern blots. This fragment could come from a digest of an EST deposit clone or from a PCR fragment isolated from an existing cDNA pool. Unlike cloned library DNAs, cDNA pools have the advantage that DNA has not been passaged through bacteria, in which specific sequences may be lost or rearranged. cDNA pools are commercially available from a wide variety of human tissues. In initial experiments, the tissue origin of EST library matches can be used to select an appropriate cDNA pool. Alternatively, an inferred function may suggest expression in specific tissues. Genes associated with proliferation, for example, might be expected to be well represented in cDNA pools from proliferative tissue, including fetal, placental, and adult testis tissues.

To determine the relative representation of existing EST contigs, the length of the mRNA products should be determined using a Northern blot consisting of RNA from various tissues. Commercially produced multiple-tissue Northern blots can speed this step of an analysis. With luck, this blot will reveal relative mRNA tissue distributions and the mRNA length, which can be compared to EST contig lengths. Multiple-sized bands may indicate alternative promoters, mRNA splice sites, multiple termination sites, specific degradation products, or the presence of closely related genes.

Confirming an EST Contig by PCR

The contiguous nature of the ESTs represented in the contig under study can be confirmed by amplifying segments of the predicted cDNA using PCR primers designed from contig data. Conventional commercial cDNA pools can be utilized as PCR templates. Later RACE reactions, required to characterize 5′ and 3′ cDNA ends, will require modified cDNA pools (e.g., Clontech Marathon-Ready cDNA). For many applications, RACE-modified pools can be used in place of conventional cDNA pools, eliminating the need to purchase separate pools.

A contig consisting of a large number of unique, overlapping EST sequences provides strong evidence for a single common mRNA. When only a few ESTs are used to assemble a contig, the synthesis of an expected PCR product from a cDNA pool can confirm the notion that ESTs arrayed by homology searching represent a common cDNA. To resolve and/or confirm sequence polymorphisms derived from EST contigging, the DNA sequence of these PCR fragments should be determined. If discrete, single PCR products are produced, fragments can be submitted for direct sequencing. Alternatively, PCR products can be readily cloned into specialized PCR vectors, and the sequence of individual clone inserts can be determined. Multiple cloned products should always be sequenced to allow for the resolution of sequence polymorphisms, including potential PCR errors. Upon sequence determination of cloned fragments, a variety of small splice variants may be encountered. Variants with relatively short stretches of divergent sequence may not be detected when directly sequencing PCR products.

RACE

Homology searching may produce a single EST or EST contigs which clearly lack 5′ and/or 3′ end sequences. Northern blots can be used to estimate the extent of missing sequence information. To obtain the remaining sequences, DNA fragments can be generated from existing clones and used as hybridization probes in library screenings. RACE is a useful PCR-based alternative to isolating missing 5′ and/or 3′ ends of a cDNA. RACE can be performed in most laboratories and can provide substantial time savings compared to classical library screening methods. In current versions of this technique, a common primer is designed to anneal to an arbitrary adaptor sequence ligated to cDNA ends (Apte and Siebert, 1993; Edwards *et al.*, 1991). When a single gene-specific RACE primer is paired with the common primer, preferential amplification of sequences between the single gene-specific primer and the common primer occurs. Commercial cDNA pools modified for use in RACE are widely available.

The use of a single gene-specific primer in RACE can lead to the formation of unanticipated products. Abundant transcripts with a poor match to the primer may be preferentially amplified over the desired target, for example. The design at onset of three or more different RACE primers, spaced at intervals across the predicted cDNA, can increase the chance of obtaining useful products without the need for extensive PCR optimization (Fig. 3). "Nested" PCR strategies can also be used to increase the specificity of RACE. Nested PCR employs a secondary amplification step using primary product as template and new primers which bind to sequences internal to the expected primary product. The 5′ ends of RACE cDNA sequences are rarely discrete, a result likely related to the addition of extra nucleotides by reverse transcriptase; thus, 5′ ends of cDNAs should be confirmed by sequencing genomic DNA fragments.

Functional Validation

Understanding the function of human genes ultimately requires the design of appropriate tests in human cells. In practice, useful phenotypes result from examining cells in which the activity of a gene is increased (e.g., transcriptional overexpression) or decreased (e.g., dominant-negative mutations, targeted gene disruption, and antisense mRNA ablation). A highly specific phenotypic test is preferable to tests which monitor a relatively vague phenotype (e.g., "slow growth" and "toxicity"). For many phenotypic tests, PCR can be a convenient means of generating constructs for the functional validation of genes.

Assembly of a Complete ORF by PCR

For expression studies, ORF fragments can be directly amplified from cDNA pools using primers designed for this purpose (Fig. 3). Commercial PCR-ready eukaryotic expression vectors (e.g., pCR3.1, Invitrogen) allow direct cloning of PCR products. Transient transfections of full-length ORF expression constructs may be sufficient to produce useful phenotypes. Alternatively, stably transfected cell lines can be produced using vectors with selectable marker genes that are active in human cells. Transfected cell lines may exhibit unstable phenotypes, especially if the expressed protein has a detrimental effect on cell growth. In these cases, constructs which allow transcriptional regulation in human cells may be used. Tightly regulated expression systems currently in use, including tetracycline and ecdysone regulated vectors, may require specialized recipient cells, and this is an important consideration in experimental design.

Estrogen receptor (ER) fusion proteins have been described as an alternative means of regulating the activity of expressed proteins (Picard *et al.,* 1988). In the absence of ER receptor ligand, the ER domain can inhibit the expression of enzymatic or protein-binding activities of fused protein domains. In the presence of ER ligand, inhibition is relieved and the activity of the fused protein is rapidly expressed. ER and other protein fusions are readily produced by overlapping PCR or by subcloning techniques. Similar methods can be used to produce epitope-tagged proteins for immunoprecipitation and subcellular localization studies.

Dominant-Negative Phenotypes

Through self-association or by association with other proteins, a mutant form of a protein may inhibit the activity of a whole complex (a "dominant-negative" phenotype). In many cases, this phenotype has been achieved with simple protein truncations. Truncation variants can easily be produced from full-length ORF constructs by PCR or by subcloning. In instances in which these are novel genes from relatively well studied classes of proteins (e.g., kinases), prior mutational analyses may allow design of inactivated clones by site-directed mutagenesis, for which many PCR-based methods are available.

Antisense-mediated mRNA ablation is also well suited to the analysis of specific mRNAs in functional assays. In this strategy, antisense sequences are engineered to bind to a specific mRNA, which leads to destabilization and degradation of the mRNA. Complete inhibition of protein expression may be difficult or impossible to achieve by antisense strategies but may be substantial enough to produce useful phenotypes. Single members of highly conserved gene families can be targeted with constructs consisting of 5′ and 3′ untranslated (UTS) antisense sequences, which tend to be poorly conserved. Protein levels can be monitored using antibodies, if available. The extent of mRNA ablation can be validated using quantitative PCR, using primers directed at portions of an mRNA not represented by the ablation construct.

Antisense methods include synthesis of small oligonucleotide primers used in transient transfection assays. Successful inhibition may required the design and testing of a number of oligos. Antisense oligos must also be chemically modified to inhibit degradation and are required in large amounts. Alternatively, antisense cDNA expression constructs have been used to study gene function (Albert and Morris, 1994). Using PCR expression vectors, cDNA-based mRNA ablation constructs can be produced immediately from PCR reaction products. Because a larger sequence is targeted for ablation, this strategy may have a higher probability of success than oligo-based antisense methods. With either method, experiments should be designed to account for the possibility that proteins with long half-lives may be present long after their mRNA has been ablated. In the case of cDNA-based ablation constructs, the selection of stable cell lines may avoid this problem.

Radiation Hybrid Mapping

Knowledge of the chromosomal map position of a newly characterized gene may be useful in characterizing a known genetic disorder. In turn, the existence of known somatic or germline mutations can assist in understanding the function of a newly characterized gene. Many genetic disorders have been mapped to general regions of the genome without sufficient resolution to allow identification of specific genes. In somatic cells, recurring cancer-related mutations have been mapped to specific chromosomal regions, including gene amplifications, loss of heterozygosity, and gross karyotypic deletions and translocations.

Currently, a newly characterized EST contig is unlikely to already be associated with a chromosomal site. Large-scale efforts are under way to determine the map positions of all human cDNAs. Newly characterized sequences can be compared with NCBI BLAST against the sequence-tagged site (STS) database. Any STS matches can in turn be used to search the NCBI Entrez Genome feature (*http://www.ncbi.nlm.nih.gov/Entrez/Genome/org.html*), in which known map positions of STS sequences are cataloged. Relationships between specific genomic sites and human disease states can be determined by using On-line Mendelian Inheritance in Man (OMIM; *http://www3.ncbi.nlm.nih.gov/Omim*). Chromosomal sites of recurrent karyotypic alterations in cancer have been cataloged (Mitelman *et al.*, 1997).

A map position for a given sequence can be independently determined or confirmed in most laboratories by relatively straightforward PCR-based techniques, including radiation hybrid (RH) mapping (Walter *et al.*, 1994). RH mapping uses panels of genomic DNA from rodent–human hybrid cell lines that contain discrete human fragments. This genomic DNA is used as template in PCRs, and the pattern of retention of specific human DNAs within the panel is determined. Using a statistical analysis, the similarity between retention patterns of known markers is evaluated, and a map location is derived. Genomic DNAs for RH mapping are commercially available. The Stanford G3 panel (Research Genetics), for example, consists of DNA from 83 human–hamster hybrid cell lines, with sufficient fragmentation of the human chromosomes to map a gene within 500 kb. Other panels are available that allow physical map positions to be determined within 100 kb.

RH mapping primer sets may fail to produce informative data for a number of reasons. On genomic templates, large introns may separate PCR primers designed from cDNA sequence. For this reason, RH primers should be engineered relatively close together (~100–200 nt) on the cDNA sequence. Another common problem includes the synthesis of similarly sized PCR products from rodent control templates, resulting in unscorable PCRs. This may arise from poor technique (contamination with human DNA) or from highly conserved rodent sequences. Once poor technique has been ruled out, murine and other rodent sequences can be compiled with database searches to identify divergent regions for primer design. The 5′ and 3′ UTS sequences, which are more highly divergent, can be especially useful for this purpose.

Occasionally, an especially highly conserved ORF accompanied by particularly short UTS regions can interfere with the identification of useful RH primers. In these cases, flanking genomic DNA can be used to design new oligos. Genomic sequence is relatively divergent between rodent and human.

Genomic DNA by PCR

Understanding the regulation of a gene may require determining that the sequence of promoters, enhancers, and other regulatory sites is not represented within a cDNA sequence. For example, EST analyses and Northern blots may suggest tissue-specific expression patterns or alternative mRNA splice variants. A complete analysis of the surrounding genomic DNA may require obtaining sets of large overlapping BAC and/or YAC clones corresponding to your cDNA.

PCR is the accepted technique for identifying genomic clones from YAC or BAC libraries.

For smaller projects, single-sided PCR (Siebert *et al.*, 1995) can be used to characterize genomic fragments flanking a cDNA sequence. The strategy of single-sided PCR of genomic fragments is related to adaptor-mediated RACE. Synthetic adaptors ligated to the ends of genomic DNA restriction fragments can be used to provide common priming sites. Genomic sequence is obtained by combining a common primer with a cDNA-specific primer in PCR reactions. Because the distance between a gene-specific primer and the next genomic restriction site is usually unknown, a series of genomic templates constructed using different restriction enzymes is tested (e.g., Genomewalker, Clontech). In this manner, internal cDNA primers can be used to identify fragments with intron/exon boundaries. Primers designed close to the 5′ end of a cDNA can be used to obtain fragments encompassing the promoter regions of a gene.

Conclusions

An overview of specific and general considerations in the design of experiments aimed at the characterization of human homologs of yeast genes was provided. The large body of information concerning the functions of specific yeast genes can be readily exploited in the functional characterization of previously uncharacterized human cDNAs. Rapid technical progress in PCR techniques has simplified the process of isolating and characterizing human cDNAs. The use of database searching to identify targets for functional studies is widely employed and will continue to contribute to the understanding of gene function in humans in the future.

References

Albert, P. R., and Morris, S. J. (1994). Antisense knockouts: Molecular scalpels for the dissection of signal transduction. *Trends Pharmacol. Sci.* **15,** 250–254.

Apte, A. N., and Siebert, P. D. (1993). Anchor-ligated cDNA libraries: A technique for generating a cDNA library for the immediate cloning of the 5′ ends of mRNAs. *BioTechniques* **15,** 890–893.

Bai, C., Sen, P., Hofmann, K., Ma, L., Goebl, M., Harper, J. W., and Elledge, S. J.

(1996). SKP1 connects cell cycle regulators to the ubiquitin proteolysis machinery through a novel motif, the F-box. *Cell* **86,** 263–274.

Botstein, D., and Cherry, J. M. (1977). Yeast as a model organism. *Science* **277,** 1259–1260.

Botstein, D., and Fink, G. R. (1988). Yeast: An experimental organism for modern biology. *Science* **240,** 1439–1443.

Don, R. H., Cox, P. T., Wainwright, B. J., Baker, K., and Mattick, J. S. (1991). "Touchdown" PCR to circumvent spurious priming during gene amplification. *Nucleic Acids Res.* **19,** 4008.

Dujon, B. (1996). The yeast genome project: What did we learn? *Trends Genet.* **12,** 263–270.

Edwards, J. B., Delort, J., and Mallet, J. (1991). Oligodeoxyribonucleotide ligation to single-stranded cDNAs: A new tool for cloning 5′ ends of mRNAs and for constructing cDNA libraries by in vitro amplification. *Nucleic Acids Res.* **19,** 5227–5232.

Feldman, R. M. R., Correll, C. C., Kaplan, K. B., and Deshaies, R. J. (1977). A complex of Cdc4p, Skp1p, and Cdc53p/Cullin catalyzes ubiquitination of the phosphorylated CDK inhibitor Sic1p. *Cell* **91,** 221–230.

Callwitz, D., and Seidel, R. (1980). Molecular cloning of the actin gene from yeast *Saccharomyces cerevisiae. Nucleic Acids Res.* **8,** 1043–1059.

Gautier, J., Norbury, C., Lohka, M., Nurse, P., and Maller, J. (1988). Purified maturation-promoting factor contains the product of a *Xenopus* homolog of the fission yeast cell cycle control gene cdc2+. *Cell* **54**(3), 433–439.

Goffeau, A., Barrell, B. G., Bussey, H., Davis, R. W., Dujon, B., *et al.* (1996). Life with 6000 genes. *Science* **274,** 546, 563–567.

Greenwell, P. W., Kronmal, S. L., Porter, S. E., Gassenhuber, J., Obermaier, B., and Petes, T. D. (1995). TEL1, a gene involved in controlling telomere length in *S. cerevisiae,* is homologous to the human ataxia telangiectasia gene. *Cell* **82,** 823–829.

Lorincz, A. T., and Reed, S. I. (1984). Primary structure homology between the product of yeast cell division control gene CDC28 and vertebrate oncogenes. *Nature* **307,** 183–185.

Mewes, H. W., Albermann, K., Bahr, M., Frishman, D., Gleissner, A., *et al.* (1997). Overview of the yeast genome. *Nature* **387,** 7–65.

Mitelman, F., Mertens, F., and Johansson, B. (1997). *Nature Genet* (*Suppl.*), 417–474.

Morrow, D. M., Tagle, D. A., Shiloh, Y., Collins, F. S., and Hieter, P. (1995). TEL1, an *S. cerevisiae* homolog of the human gene mutated in ataxia telangiectasia, is functionally related to the yeast checkpoint gene MEC1. *Cell* **82,** 831–840.

Neff, N. F., Thomas, J. H., Grisafi, P., and Botstein, D. (1983). Isolation of the beta-tubulin gene from yeast and demonstration of its essential function in vivo. *Cell* **33,** 211–219.

Ng, R., and Abelson, J. (1980). Isolation and sequence of the gene for actin in Saccharomyces cerevisiae. *Proc. Natl. Acad. Sci. U.S.A.* **77,** 3912–3916.

Payne, W. E., and Garrels, J. I. (1997). Yeast protein database (YPD): A database for the complete proteome of *Saccharomyces cerevisiae. Nucleic Acids Res.* **25,** 57–62.

Picard, D., Salser, S. J., and Yamamoto, K. R. (1988). A movable and regulable inactivation function within the steroid binding domain of the glucocorticoid receptor. *Cell* **54,** 1073–1080.

Savitsky, K., Bar-Shira, A., Gilad, S., Rotman, G., Ziv, Y., *et al.* (1995). A single ataxia telangiectasia gene with a product similar to PI-3 kinase. *Science* **268,** 1749–1753.

Scherer, S., and Davis, R. W. (1979). Replacement of chromosome segments with altered DNA sequences constructed in vitro. *Proc. Natl. Acad. Sci. U.S.A.* **76,** 4951–4955.

Shalon, D., Smith, S. J., and Brown, P. O. (1996). A DNA microarray system for analyzing complex DNA samples using two-color fluorescent probe hybridization. *Genome Res.* **6,** 639–645.

Siebert, P. D., Chenchik, A., Kellogg, D. E., Lukyanov, K. A., and Lukyanov, S. A. (1995). An improved PCR method for walking in uncloned genomic DNA. *Nucleic Acids Res.* **23,** 1087–1088.

Skowyra, D., Craig, K. L., Tyers, M., Elledge, S. J., and Harper, J. W. (1997). F-Box proteins are receptors that recruit phosphorylated substrates to the SCF ubiquitin–ligase complex. *Cell* **91,** 209–219.

Smith, V., Chou, K. N., Lashkari, D., Botstein, D., and Brown, P. O. (1996). Functional analysis of the genes of yeast chromosome V by genetic footprinting. *Science* **274,** 2069–2074.

Struhl, K., Stinchcomb, D. T., Scherer, S., and Davis, R. W. (1979). High-frequency transformation of yeast: Autonomous replication of hybrid DNA molecules. *Proc. Natl. Acad. Sci. U.S.A.* **76,** 1035–1039.

Tugendreich, S., Bassett, D. E., Jr., McKusick, V. A., Boguski, M. S., and Hieter, P. (1994). Genes conserved in yeast and humans. *Hum. Mol. Genet.* **3,** 1509–1517.

Walter, M., Spillett, D., Thomas, P., Weissenbach, J., and Gooodfellow, P. N. (1994). A method for constructing radiation hybrid maps of whole genomes. *Nature Genet.* **7,** 22–28.

Weinert, T. A., Kiser, G. L., and Hartwell, L. H. (1994). Mitotic checkpoint genes in budding yeast and the dependence of mitosis on DNA replication and repair. *Genes Dev.* **8,** 652–665.

Zhang, L., Zhou, W., Velculescu, V. E., Kern, S. E., Hruban, R. H., *et al.* (1997). Gene expression profiles in normal and cancer cells. *Science* **276,** 1268–1272.

Part Four

GENOMICS AND EXPRESSION PROFILING

27

CELLULAR TRANSCRIPTOME ANALYSIS USING A KINETIC PCR ASSAY

John J. Kang and Michael J. Holland

With the emergence of complete genome sequences, it is possible to analyze cellular transcriptomes which determine, in large measure, the phenotype of a cell. Transcriptome analysis requires high-throughput methods capable of quantitating transcript levels over a wide intracellular abundance range. The transcript assay described here is kinetically monitored, reverse transcriptase initiated PCR (kRT-PCR). This assay is based on the principle that PCR reactions proceed as relatively simple base 2 exponentials, where the rate of double-stranded (ds) DNA product formation is quantitatively related to initial template concentration. The kRT-PCR assay is rapid and highly automated, and it utilizes unfractionated total cellular RNA as template. Most important, this assay is accurate for measuring transcript level differences in two or more physiological or genetic states within a factor of ±20%. Kinetic RT-PCR assays are accurate for measuring transcript copy number per cell in a single state within a factor of two. Finally, the kRT-PCR assay is capable of quantitating cellular transcripts differing in abundance over at least six orders of magnitude.

In kRT-PCR, a digital camera and computer monitor accumulation of dsDNA product at each PCR cycle using ethidium bromide fluorescence. Thus, the entire kinetic course of a PCR reaction is captured in a fully automated process, without postreaction handling, i.e., gels. Numerical analysis of the kinetic data permits quantitative comparison of initial template concentrations.

Protocols

The kRT-PCR Reaction Composition

Because kRT-PCR assays utilize an RNA template, reagents should be sterile and, where possible, treated with DEPC. A typical 100-μl reaction is assembled from the following components:

Stock solutions		Final concentration
4×	Reaction buffer	1×
200 m*M*	Tricine (pH 8.3)	50 m*M*
440 m*M*	Potassium acetate (pH 7.5)	110 m*M*
52%	Glycerol	13%
50×	dNTP mix	1×
15 m*M*	dATP	0.3 m*M*
15 m*M*	dcTP	0.3 m*M*
15 m*M*	dGTP	0.3 m*M*
2.5 m*M*	dTTP	0.05 m*M*
25 m*M*	dUTP	0.5 m*M*
500 μ*M*	Ethidium bromide (Sigma)	2.5 μ*M*
1 U/μl	UNG	2 U
2.5 U/μl	r*Tth* DNA polymerase	10 U
50 m*M*	Manganese acetate (pH 6.5)	2.4 m*M*
	Total cellular RNA template	0.12 μg
	Primers	0.25 μ*M*

Reactions are performed in capless MicroAmp tubes (Perkin–Elmer) using the standard Perkin–Elmer 9600 reaction plate. Reaction tubes are presorted on a UV transilluminator (covered with fresh plasticwrap to prevent RNase or template contamination) to eliminate sporadic tubes with high background fluorescence.

For ease and reproducibility, the reaction array is assembled by aliquoting primer pairs into the reaction array (50 μl of a 0.50 μ*M* dilution). The remaining components (including RNA template) are assembled as a 2x master mix. Typically, the RNA template and $Mn(oAc)_2$ are added as the last components of the master mix, and 50 μl of master mix is added to the reaction array. For vapor seal, 50 μl of mineral oil (Sigma) is placed over each reaction. To force out air bubbles that could interfere with fluorescence detection, the assembled reaction plate is centrifuged for 1 min at room temperature.

External standards to control for cycle-to-cycle variations in ethidium bromide fluorescence are prepared by adding 2x master mix to 50 μl of 10 m*M* EDTA instead of primers.

Primer Design and Quality Assurance

Primer pair design represents a critical, but controllable, variable for kRT-PCR. For successful kRT-PCR, primer pairs must produce transcript-specific product efficiently without producing significant competing product, in particular, template-independent products ("primer dimer"). To design yeast transcript-specific primer pairs, the open reading frame for a target gene product is retrieved from the Saccharomyces Genomic Database (*http://genome-www.stanford.edu/Saccharomyces/*). Sequence homologies that could lead to mispriming (e.g., overlapping open reading frames or gene families) are identified and eliminated by BLAST search of the entire genome sequence with the gene-specific target sequence. Primer pairs within the target sequence were designed using the Oligo software package (Version 5, National Biosciences, Inc.). In the first step, candidate primers on the upper strand and the lower strand are selected for minimal self-priming potential, balanced T_m, and optimal base composition. At the same time, potential primers with high annealing stability for the five nucleotides at the 3′ terminus are eliminated because of their intrinsic potential for mispriming. In particular, primers that contain at least one dA nucleotide homologous to template at their 3′ ends are selected. Inclusion of this common 3′ terminus reduces production and delays appearance of primer dimer products (R. Saiki, personal communication). For transcripts derived from closely related gene families, primers that selectively anneal to each specific gene family transcript are extracted from the list. In the second step, primer pairs spaced at 100- to 300-bp intervals are selected from the list of upper strand and lower strand primers and then analyzed for minimal intermolecular primer dimer potential.

Primers were synthesized and purified by standard methods. Reverse-phase chromatography of tritylated primers was used to eliminate incomplete oligonucleotide synthesis products that can contribute to template-independent PCR product formation. Primers are resuspended to a stock concentration of 100 μM in 10 mM Tris–HCl (pH 8.0) and 0.1 mM EDTA. Subsequent dilutions are made in DEPC-treated distilled water to prevent inhibition of the kRT-PCR reaction by excessive EDTA carryover.

Primer pair quality is assessed by kRT-PCR. Reactions performed with no added template provide an assessment of the PCR cycle value (C_t) at which very low-efficiency template-independent (primer dimer) signals are detected. Typically, these C_t values range from 30 to >50 PCR cycles. This latter value sets the lower limit of template-dependent signal that will provide quantitatively reliable information [i.e., the C_t value for a template-dependent signal should be less than (earlier) the template-independent C_t value]. In parallel reactions, the primer pair is analyzed by kRT-PCR with a series of total yeast RNA template dilutions over several orders of magnitude. The template-dependent signals are computationally assessed for curve spacing, slope, and plateau. For 10-fold serial dilutions, template-dependent curves should be parallel and spaced approximately 3.34 PCR cycles apart (since $2^{3.34} = 10$). The absence of proper spacing can indicate inefficient reactions or competition with template-independent product formation. Inhibitors decrease amplification efficiency and thereby reduce the slope of the kinetic curve. Plateau value is the fluorescence signal obtained when the reaction reaches the limit imposed by primer pair limitation. In productive reactions, this occurs when most of the primer has been incorporated into product. Thus, the plateau value varies with the starting primer concentration and amplicon size (ethidium bromide fluorescence yield per cycle is proportional to the length of the PCR product). For new primer pairs, the correspondence between kinetic data and a PCR product of the predicted size is confirmed by agarose gel electrophoresis. These quality assessment procedures serve to eliminate problematic PCR primers from further consideration. Utilizing this approach to primer pair design, approximately 85% of the primer pairs tested met the specifications described previously.

RNA Template Preparation

Total yeast RNA is extracted as previously described (Kang *et al.*, 1995). Yeast cells are grown to early log phase in a complex medium (YP; 2% yeast extract and 1% peptone) or a defined medium [0.67% YNB (yeast nitrogen base), Difco) supplemented with 2% glucose

(d) or 2% each glycerol plus 2% lactate (gl). Cells are harvested by centrifugation and disrupted by vortexing with glass beads in the presence of phenol and NNES buffer [50 m*M* sodium acetate (pH 5.0), 100 m*M* NaCl, 10 m*M* EDTA, and 0.5% SDS) preheated to 55°C. After vigorous vortexing, the hot suspension is plunged into dry ice/ethanol bath until the phenol begins to freeze. The sample thaws during centrifugation to separate phases. The aqueous layer is extracted with chloroform and nucleic acids precipitated by addition of ethanol. Residual DNA contamination is removed enzymatically with RNase-free DNase I (40 U/240 μg RNA, Promega) in the presence of placental RNase inhibitor (36 U/240 μg RNA, Pharmacia). DNase I is removed by extraction with NNES-saturated phenol, followed by chloroform extraction and ethanol precipitation. A final ethanol precipitation is performed with ammonium acetate to eliminate salt carryover. This procedure consistently yields several milligrams of high-quality RNA (8 mg per 100-ml culture wild-type cells in YPd) while eliminating the bulk of contaminating DNA. As shown later, 10^5 yeast cell equivalents of RNA (0.12 μg) (Sherman, 1991) are sufficient for the current assay; thus, each milligram of RNA provides template for more than 8000 reactions.

The RNA template dependence of kRT-PCR reactions can be validated by pretreatment of an aliquot of the template with RNase A. If the reaction is RNA template dependent, RNase A treatment will cause a substantial shift of the observed kinetic curve to a higher C_t value. Alternatively, the amount of DNA contamination in the preparation of total cellular RNA can be estimated by replacing $Mn(oAc)_2$ with $MgCl_2$. Since the reverse transcriptase activity of DNA polymerases is manganese dependent, this change eliminates RNA-templated products.

The kRT-PCR Reaction

The kRT-PCR reaction occurs in three phases in a single-reaction tube. In the first phase, a 2-min incubation at 50°C, the enzyme uracil-*N*-glycosylase (UNG) hydrolyzes any dU-containing PCR product carryover (Longo *et al.*, 1990; Mulder *et al.*, 1994; Udaykumar *et al.*, 1993). As indicated previously in the reaction composition, all kRT-PCR reactions are performed with a nucleotide mixture which contains a molar excess of dUTP over dTTP. During the second phase of the reaction template RNA is reverse transcribed to cDNA (30 min at 60°C). Reaction conditions [in particular, buffer, nucleotide concentration, temperature, and $Mn(OAc)_2$ concentration] are optimized for the reverse transcriptase reaction. Similarly, the reaction temperature (60°C) provides for efficient *Tth* DNA polymerase reverse transcriptase activity (Myers and Gelfand, 1991; Myers and Sigua, 1995) but reduces UNG activity. Thermostable *Tth* DNA polymerase is used in these reactions because this enzyme exhibits high levels of RNA-dependent and DNA-dependent DNA polymerase activity (Myers and Gelfand, 1991; Myers and Sigua, 1995). The reaction buffer has been formulated to provide the broadest Mn^{2+} optimum for the reverse transcriptase activity. Having created the DNA template by reverse transcription, the reaction enters the third phase, DNA amplification by thermal cycling with collection of kinetic data (95°C, 20 sec; 55°C, 30 sec; 72°C, 30 sec). As in conventional PCR, DNA-dependent DNA polymerase activity catalyzes cycle-dependent product accumulation. As it accumulates in the reaction tube, double-stranded DNA product is measured by detection of ethidium bromide fluorescence (Higuchi *et al.*, 1992).

The Kinetic Thermal Cycler

The experiments described here were performed with a kinetic thermal cycler (KTC) similar to the instrument previously described by Higuchi *et al.* (1993). Software has been improved, and refinements in full-field illumination make the optical path self-contained and compact (R. Higuchi and R. Watson, this volume). The optical components of the KTC have been built into a light-tight enclosure which fits over the sample block of a standard Perkin-Elmer GeneAmp 9600 thermal cycler platform. The thermal cycler's heated sample lid was removed to permit optical monitoring of the sample block. Mineral oil overlaid on the reactions provides a transparent vapor seal. A full-field UV light source (midrange, 300 n*M*) illuminates the sample block (96 reactions) through a dichroic mirror mounted at 45° directly above the sample block. The dichroic mirror also reflects ethidium bromide fluorescence from the reactions through a large Fresnel lens (which corrects optical distortion) to an eight-bit digital (charged-coupled device) camera mounted at the back of the enclosure. Toward the end of the anneal/extend phase of each PCR cycle, a software-controlled shutter opens to expose the sample block to UV excitation while the digital camera takes two exposures. A video-grabber board captures the sample images and stores them as TIF images on a PC-compatible microcomputer. Based on a user-determined grid over the sample block, the KTC software averages the images and integrates the fluorescence signal for each sample in the block. The integrated signals for each reaction well are stored in spreadsheet format at each PCR cycle, providing the raw data for analysis. The digital imaging system (video camera, video frame-grabber board, computer, and controlling software) are a modification of the IS1000 system from Alpha Innotech Corporation (San Leandro, CA).

After thermal cycling ends, the accumulated raw data are imported to a Microsoft Excel spreadsheet for a series of computational refinements. Each of the reaction wells is assigned a label to simplify identification of kRT-PCR reactions. The reaction array includes "external standards" (tubes containing ethidium bromide but incapable of amplification) placed at strategic positions in the array. From the average fluorescence of up to six external standards, a normalization factor is computed for each PCR cycle and applied to the entire reaction array to eliminate cycle-to-cycle variation due to lamp variation or other system fluctuations. During the first few cycles, the background fluorescence of external standards (and experimental reactions) usually decreases somewhat before stabilizing. The normalized data are then plotted for each reaction as relative, as opposed to absolute, increase in fluorescence versus cycle number (Higuchi *et al.,* 1993). For a typical 50-cycle run, each reaction generates 50 relative fluorescence data points that constitute a kinetic growth curve of dsDNA product accumulation during the PCR.

Kinetic curves from separate reactions can be compared by their threshold cycle value of C_t. The C_t is the fractional PCR cycle number at which the curve reaches an arbitrary fluorescence level (AFL) which is at least 10 standard deviations above the background fluorescence of the external standards. Since C_t values are inversely proportional to the log of the initial RNA template concentration, they are used to calculate transcript copy number. A summary table calculates C_t values for all kRT-PCR reactions in the array. For the analyses presented here, the AFL was set at 1.5.

Efficiency of the kRT-PCR Reaction

The overall efficiency of the kRT-PCR reaction can be affected by the efficiency of the reverse transcriptase step as well as by PCR cycle efficiency. It is well established that reverse transcriptase efficiency decreases with increasing length of the cDNA product. Similarly, the efficiency of the reverse transcriptase step in kRT-PCR could be influenced by the size of the sequence interval between the primer pairs (amplicon size). To assess reverse transcriptase as well as PCR cycle efficiency in the kRT-PCR reaction, four primer pairs for *ACT1* and eight primer pairs for *PGK1* were analyzed using the same RNA preparation as template. The sequence interval between the primer pairs varied from 205 to 716 bases. PCR cycle efficiency was assessed from the observed ΔC_t for 10^5 vs 10^3 cell equivalents of total yeast RNA template for each primer pair (Fig. 1). As expected for a highly processive DNA-dependent DNA polymerase such as *Tth* DNA polymerase, the PCR cycle efficiencies for all 12 primer pairs were similar and were not significantly affected by amplicon size.

In contrast, C_t values, from which transcript abundance is calculated, increased with increasing size of the sequence interval between the primer pairs (Fig. 1). For three *PGK1* primer pairs that generate products from 230 to 288 bps (average $C_t = 21.9 \pm 0.3$ cycles), three *PGK1* primer pairs that generate products from 451 to 527 bps (average $C_t = 23.1 \pm 0.6$ cycles), and *PGK1* primer pairs that generate 658- and 716-bp products (average $C_t = 25.3 \pm 0.7$ cycles), the C_t values increased with increasing sequence interval between the primer pairs. Taken together, these data strongly suggest that reverse transcriptase efficiency falls off with increasing amplicon size. Because of this consideration, primer pairs with amplicons between 100 and 300 bps were selected for kRT-PCR analysis. For four *ACT1* primer pairs that generate products between 205 and 318 bps, the average C_t was 23.1 ± 0.6, suggesting that the overall kRT-PCR efficiency of primer pairs specific for the same transcript varies within one PCR cycle.

Accuracy of kRT-PCR Assays

The accuracy of the kRT-PCR assay was assessed for two different kinds of analyses. The first measures the difference in transcript level between two or more states. In this kind of analysis, differences in overall kRT-CR efficiency for a given primer pair should be constant if the efficiency is reproducible from assay to assay. The second application is quantitation of transcript copy number per cell in

Primer pair	Product size	C_T @ 1E+4	C_T @ 1E+3 - C_T @ 1E+5
PGK1-A	230	22.2	7.4
PGK1-C	262	21.6	7.4
PGK1-D	288	21.8	7.0
	Average	21.9	7.3
	Std Dev	0.3	0.2
PGK1-B	469	23.5	6.4
PGK1-E	527	23.3	7.7
PGK1-F	451	22.4	6.8
	Average	23.1	7.0
	Std Dev	0.6	0.7
PGK1-G	716	25.8	7.3
PGK1-H	658	24.8	7.0
	Average	25.3	7.2
	Std Dev	0.7	0.2
ACT1-A	285	23.2	6.9
ACT1-B	318	23.8	6.5
ACT1-C	205	22.5	7.1
ACT1-D	238	22.7	7.4
	Average	23.1	7.0
	Std Dev	0.6	0.4

Figure 1 Assessment of kRT-PCR efficiency differences among different primer pairs for the same transcript. Four different *ACT1* primer pairs and eight different *PGK1* primer pairs with the indicated product sizes (amplicon size) were evaluated in kRT-PCR assays using total cellular RNA template isolated from strain S173-6B grown in YP medium containing glucose. PCR cycle efficiencies were determined from the difference in C_t values in assays utilizing 10^3 (1E + 3) versus 10^5 (1E + 5) cell equivalents of cellular RNA template. A value of 6.7 PCR cycles is expected for 100% PCR cycle efficiency. The observed C_t values at an AFL of 1.5 for each primer pair using 10^4 (1E + 4) cell equivalents of RNA template are indicated together with averages and standard deviations.

a single state. For this latter measurement, primer pair-dependent differences in kRT-PCR efficiency could introduce an error in the determination.

To assess the ability of the kRT-PCR assay to reproducibly determine small differences in transcript levels between two states, *ACT1* (actin) and *CYC1* (iso-1-cytochrome c) mRNAs were quantitated in total cellular RNA isolated from cultures of strain S173-6B harvested in late log phase (culture S173-6B-a) versus early log phase (culture S173-6B-c) as well as a congenic strain harvested in early log phase (culture CPY1-b). Strains were grown in YNB medium containing 2% glycerol plus 2% lactate as carbon source. *ACT1* mRNA levels ranged from 0.71 to 1.48, whereas *CYC1* transcript levels ranged from 0.32 to 2.29 in total cellular RNA preparations from the three cultures (Fig. 2). To confirm the accuracy of differences measured

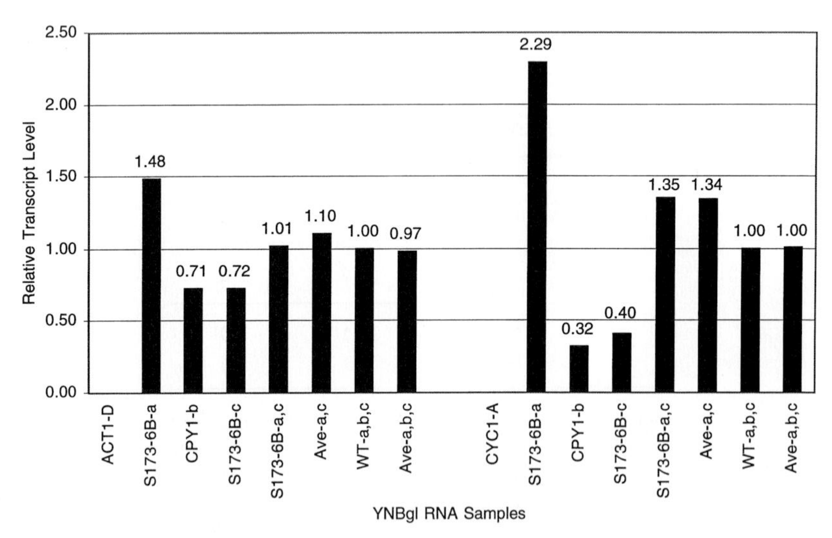

Figure 2 Accuracy of the kRT-PCR assay for determining transcript level differences in multiple states. Total cellular RNA was isolated from two independent cultures of strain S173-6B harvested in late log phase (culture a) or early log phase (culture c) and a congenic strain CPY1 harvested in early log phase (culture b). Yeast strains were grown in YNB medium containing 2% glycerol plus 2% lactate as carbon source (YNBgl). *ACTI* and *CYC1* transcript levels were determined by kRT-PCR assays using 0.12 μg of each RNA preparation as well as a mixture of equal amount of RNA from cultures a and c (S173-6B-a and -c). Relative transcript levels were normalized (1.00) to the kRT-PCR assay for a mixture of equal amounts of each RNA preparation (WT-a,b,c) and are indicated above each bar. Mathematical averages were computed from the kRT-PCR data for individual RNA preparations (Ave-a,c and Ave-a,b,c).

by the kRT-PCR assay, total cellular RNA was mixed for the two most disparate preparations as well as for all three preparations, and kRT-PCR assays were performed to measure relative transcript levels in the mixtures. The results were compared to the arithmetic averages for the kRT-PCR determinations for each separate RNA preparation. The experimental data obtained by kRT-PCR assays of the mixtures were within 10% or less of the arithmetic averages (Fig. 2). These results show that the overall kRT-PCR efficiencies for the *ACT1* and *CYC1* primer pairs are reproducible. Consequently, the kRT-PCR assay provides highly accurate transcript level differences between states, which could include different physiological, developmental, or genetic states.

To further evaluate the intrinsic reproducibility of the assay, kRT-PCR assays were performed with 21 different yeast mRNA-specific primer pairs using 10^5 cell equivalents of total cellular RNA isolated from cultures grown in YP medium with 2% glycerol plus 2% lactate as template. Three independent kRT-PCR reaction arrays were evaluated. Two analyses were performed sequentially in the same kinetic thermal cycler and the third was performed in a different instrument. The average standard deviation for the three analyses of all 21 primer pairs was 0.17 PCR cycles (Fig. 3). These experiments confirm the reproducibility of primer pair-specific C_t values. These observations, together with those described in Fig. 2, demonstrate that kRT-PCR-based quantitation of transcript level differences in two or more states is accurate within $\pm$20%.

To assess the accuracy of the kRT-PCR assay for determining transcript copy number per cell, kRT-PCR assays were performed with 10^5 cell equivalents of total cellular RNA isolated from strain S173-6B grown in YP medium containing 2% glucose as carbon source and primer pairs specific for *ENO2* (enolase 2), *ACT1, GPD2* (glycerol phosphate dehydrogenase 2), and *PCK1* (phosphoenolpyruvate carboxykinase) mRNAs (Fig. 4). The observed C_t values at an AFL of 1.5 for *ENO2, ACT1, GPD2,* and *PCK1* mRNAs were 17.46, 19.89, 23.55, and 29.79, respectively. Conversion of these C_t values to transcript copy number per cell requires knowledge of C_t values for known amounts of a test transcript. An extensive kRT-PCR analysis of known amounts of HIV mRNA performed by Roche Molecular Systems, Inc. showed that the average C_t value observed at an AFL of 1.5 for 10^5 HIV mRNA templates was 25.5. Utilizing this calibration factor, transcript copy numbers per cell were as follows: *ENO2,* 263; *ACT1,* 49; *GPD2,* 3.9; and *PCK1,* 0.05. These values are in close agreement with those obtained independently by

Gene	Primer pair	C_T@1.5 Exp 1 KTC 1	Exp 2 KTC 1	Exp 3 KTC 2	Average C_T	Std Dev
Glucokinase	GLK1-A	20.38	20.64	20.40	20.47	0.15
Hexokinase 1	HXK1-A	22.18	22.39	22.36	22.31	0.11
Hexokinase 2	HXK2-A	22.41	22.66	22.64	22.57	0.14
Phosphoglucoisomerase	PGI1-A	22.68	22.63	22.80	22.70	0.09
Phosphofructokinase-subunit 1	PFK1-A	23.06	23.33	23.13	23.17	0.14
Phosphofructokinase-subunit 2	PFK2-A	22.64	22.64	22.41	22.56	0.13
Fructosebisphosphate aldolase	FBA1-A	19.12	19.13	18.90	19.05	0.13
Triose phosphate dehydrogenase 2	TDH2-A	19.67	20.27	19.73	19.89	0.33
Triose phosphate dehydrogenase 3	TDH3-A	17.34	17.41	17.06	17.27	0.19
Triose phosphate dehydrogenases	TDHx-A	17.59	17.52	17.51	17.54	0.05
Phosphoglycerate kinase	PGK1-D	19.96	20.29	19.95	20.07	0.19
Glycerol phosphate mutase	GPM1-A	19.45	19.64	19.28	19.46	0.18
Enolase 1	ENO1-D	20.17	20.13	19.82	20.04	0.19
Enolase 2	ENO2-D	20.19	20.18	20.02	20.13	0.09
Pyruvate kinase	PYK1-A	19.75	20.10	19.54	19.80	0.28
PEP carboxykinase	PCK1-A	20.64	20.90	20.80	20.78	0.13
Fructose bis-phosphatase	FBP1-A	23.56	24.19	23.77	23.84	0.32
Alcohol dehydrogenase 1	ADH1-A	18.35	18.82	18.63	18.60	0.24
Alcohol dehydrogenase 2	ADH2-A	20.21	20.35	20.48	20.35	0.13
Actin	ACT1-D	21.12	21.54	21.45	21.37	0.22
Iso-1-cytochrome C	CYC1-A	24.00	23.86	24.20	24.02	0.17

Figure 3 Reproducibility of kRT-PCR assays. To evaluate run to run variability for kRT-PCR assays, three separate reaction arrays were generated using 21 different primer pairs specific for the indicated yeast gene products and 10^5 (1E5) cell equivalents of cellular RNA isolated from strain 6B grown in YP medium containing 2% glycerol plus 2% lactate as carbon source (YPgl). Two of the reaction arrays were analyzed sequentially using the same thermal cycler (KTC 1) and the third array was analyzed using a different thermal cycler (KTC 2). C_5 values are tabulated at an AFL of 1.5 for each assay reaction as is the average and standard deviation for the three determinations for each primer pair. The average standard deviation for all the determinations was 0.17 PCR cycles, which is <20% variation.

SAGE analysis of the yeast transcriptome in log phase cultures grown in YP medium containing 2% glucose (Velculescu *et al.*, 1997); *ENO2*, 289; *ACT1*, 60, *GPD2*, 3.7, and *PCK1*, not detected. Approximately 20,000 tags were sequenced in the SAGE analysis of a log phase yeast culture. Consequently, *PCK1* mRNA may not have been detected in the SAGE analysis because of the low abundance of this transcript. The accuracy of kRT-PCR for determining transcript copy number per cell depends on the variation in overall efficiency of transcript-specific primer pairs. Based on the results shown in Fig. 1, this variation is within one PCR cycle, suggesting that the kRT-PCR-based determination of transcript copy number per cell will be accurate within a factor of 2.

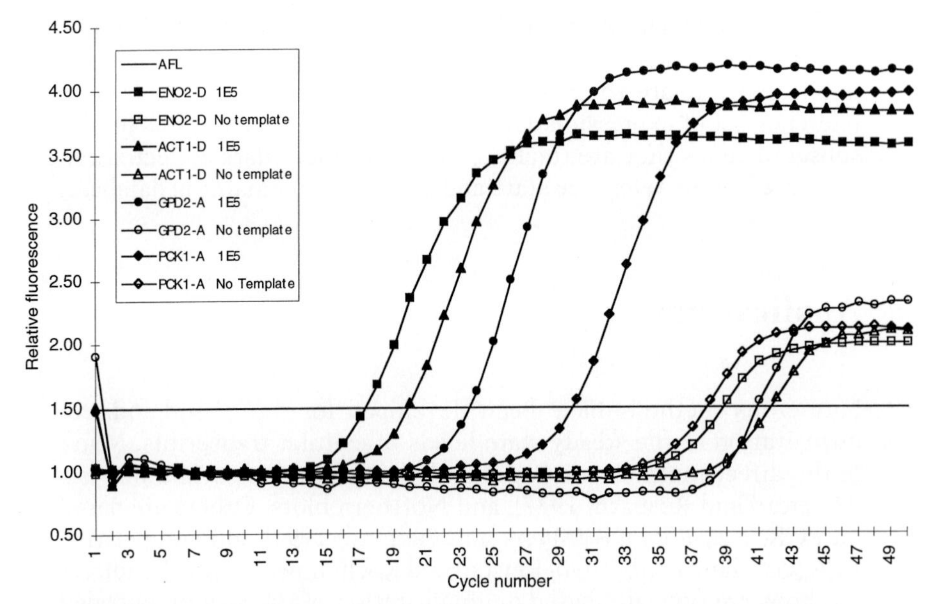

Figure 4 Accuracy of kRT-PCR assays for determining transcript copy number per cell. Kinetic RT-PCR assays were performed using primer pairs specific for yeast *ENO2* (enolase 2), *ACT1* (actin), *GPD2* (glycerol phosphate dehydrogenase 20, and *PCK1* (phosphoenolpyruvate carboxykinase) mRNAs and total cellular RNA isolated from log phase cultures of strain S173-6B grown in YP medium containing 2% glucose. Kinetic curves are shown for reactions containing 10^5 cell equivalents of RNA (1E5) and no template for each primer pair. Relative fluorescence is plotted against PCR cycle number. C_t values were measured at an AFL of 1.5.

The results shown in Fig. 4 also illustrate the dynamic range of the kRT-PCR assay for quantitating transcript that differs in relative abundance by almost four orders of magnitude. The wide dynamic range of the assay derives from the high intrinsic sensitivity of PCR as well as the fact that the output fluorescence signal of the kRT-PCR assay is independent of transcript abundance; only PCR cycle number varies.

Because of the high degree of reproducibility of kRT-PCR transcript level determinations, it will not be necessary to repeat a complete analysis for any physiological or genetic reference state. Standardization of data sets for a two-state comparison is accomplished by inclusion of a few well-characterized reference transcripts in each reaction array. It is important to emphasize that the data set for any state comparison can be computationally normalized to any transcript or group of transcripts that were analyzed. Thus, the data sets are robustly relational and can accommodate incremental additions of new data for additional transcripts in any state. For example, investigators studying expression of a subset of yeast genes can computationally relate their data set to all other sets by collecting a reference set of expression data in one of the reference states for the subset of genes they are tracking. All subsequent data collected can be related to the reference state and thereby the remaining database.

Applications

Numerous methods have been developed for direct and indirect quantitation of the steady-state levels of cellular transcripts. Many of the direct methods are based on hybridization, e.g., Rot analysis (Hereford and Rosbash, 1977), and Northern blots. Others are based on PCR, e.g., RT-PCR (Myers and Gelfand, 1991) and competitive PCR (Gilliland *et al.*, 1990; Reischl and Kochanowski, 1995). Indirect methods are typically based on quantitation of the protein encoded by the mRNA, e.g., Western blotting, ELISA, and enzymatic activity of the natural product or of a reporter gene product. The accuracy and relative simplicity of the kRT-PCR assay offers some significant advantages with respect to time and effort compared to each of these established methods.

It is likely that genomics scale analyses will involve not only an assessment of large numbers of transcripts in a few states but also analysis of large numbers of states or cell types. For *S. cerevisiae,* comprehensive genetic studies of gene expression for subsets of coordinately regulated genes, including epistasis analysis, will require transcript-level determinations in large numbers of different genetic states. The kRT-PCR assay readily accommodates multiple-state comparisons using total cellular RNA isolated from cells in each state. Most important, a single kRT-PCR reaction array can include any customized combination of transcript-specific primer pairs with template RNAs corresponding to any state. The intrinsic flexibility of the kRT-PCR reaction array offers some significant advantages over DNA microarray assays, which require separate hybridizations with each state-specific hybridization probe.

A major application for technologies that monitor transcript levels is cell typing (Lander, 1996). Cell typing is particularly important for studying development and complex populations of cells in multicellular organisms. Cell type-specific transcript markers will be extremely valuable in this regard. Such markers could be used to assess cell populations in organs and the effects of drugs or cytokines on cell population dynamics. It seems likely that cell-specific transcript markers will correspond to transcripts over a wide range of relative abundance. A particularly attractive feature of kRT-PCR is the ability of this methodology to quantitate transcripts over a wide range of abundance.

The wide dynamic range intrinsic to the kRT-PCR assay should be useful for quantitation of transcripts from known and computationally annotated ORFs in the yeast genome, for example. For yeast where large numbers of mutations in transcriptional regulatory genes are available, comprehensive identification of coordinately regulated gene groups is possible. On a more global scale, a genetic dissection of the yeast transcriptional regulatory network is approachable with the kRT-PCR assay. Such analysis clearly requires the ability to quantitatively relate transcript levels with genetic state. For this latter application, the reproducibility of the kRT-PCR assay and, consequently, the highly relational nature of the data sets, offers significant advantages compared to other technologies.

Acknowledgments

We thank Robert Watson, Mary Fisher, David Gelfand, Russell Higuchi, John Sninsky, and Randy Saiki at Roche Molecular Systems, Inc., for invaluable advice and assistance.

References

Gilliland, G., Perrin, S., Blanchard, K., and Bunn, H. F. (1990). Analysis of cytokine mRNA and DNA: Detection and quantitation by competitive polymerase chain reaction. *Proc. Natl. Acad. Sci. U.S.A.* **87,** 2725–2729.

Hereford, L. M., and Rosbash, M. (1977). Number and distribution of polyadenylated RNA sequences in yeast. *Cell* **10,** 453–462.

Niguchi, R., Dollinger, G., Walsh, P. S., and Griffith, R. (1992). Simultaneous amplification and detection of specific DNA sequences. *Biotechnology* **10,** 413–417.

Higuchi, R., Fockler, C., Dollinger, G., and Watson, R. (1993). Kinetic PCR analysis: Real-time monitoring of DNA amplification reactions. *Biotechnology* **11,** 1026–1030.

Kang, J. J., Yokoi, T. J., and Holland, M. J. (1995). Binding sites for abundant nuclear factors modulate RNA polymerase I-dependent enhancer function in Saccharomyces cerevisiae. *J. Biol. Chem.* **270,** 28723–28732.

Lander, E. S. (1996). The new genomics: Global view of biology. *Science* **274,** 536–539.

Longo, M. C., Berninger, M. S., and Hartley, J. L. (1990). Use of uracil DNA glycosylase to control carry-over contamination in polymerase chain reactions. *Gene* **93,** 125–128.

Mulder, J., McKinney, N., Christopherson, C., Sninsky, J., Greenfield, L., and Kwok, S. (1994). Rapid and simple PCR assay for quantitation of human immunodeficiency virus type 1 RNA in plasma: Application to acute retroviral infection. *J. Clin. Microbiol.* **32,** 292–300.

Myers, T. W., and Gelfand, D. H. (1991). Reverse transcription and DNA amplification by a *Thermus thermophilus* DNA polymerase. *Biochemistry* **30,** 7661–7666.

Myers, T. W., and Sigua, C. L. (1995). Amplification of RNA: High temperature reverse transcription and DNA amplification with *Thermus thermophilus* DNA polymerase. In *PCR Strategies* (M. A. Innis, D. H. Gelfand, and J. J. Sninsky, Eds.), pp. 55–68. Academic Press, San Diego.

Reischl, U., and Kochanowski, B. (1995). Quantitative PCR. A survey of the present technology. *Mol. Biotechnol.* **3,** 55–71.

Sherman, F. (1991). Getting started with yeast. *Methods Enzymol.* **194,** 3–21.

Udaykumar, Epstein, J. S., and Hewlett, I. K. (1993). A novel method employing UNG to avoid carry-over contamination in RNA-PCR. *Nucleic Acids Res.* **21,** 3917–3918.

Velculescu, V. E., Zhang, L., Zhou, W., Vogelstein, J., Basrai, M. A., Bassett, D. J., Hieter, P., Vogelstein, B., and Kinzler, K. W. (1997). Characterization of the yeast transcriptome. *Cell* **88,** 243–251.

28

PARALLEL ANALYSIS WITH BIOLOGICAL CHIPS

Mark Schena and Ronald W. Davis

Complete genome sequences provide the information required for functional analysis of whole genomes. Systematic examination of 10^4 or 10^5 genes poses a major technical challenge that is best addressed with parallel approaches utilizing emerging microarray or biological "chip" technologies. Microarray assays have a myriad of applications, including gene expression monitoring (Schena *et al.*, 1995, 1996; Shalon, 1996; Lockhart *et al.*, 1996; DeRisi *et al.*, 1996; Heller *et al.*, 1997), DNA resequencing (Chee *et al.*, 1996), mutation detection (Cronin *et al.*, 1996; Kozal *et al.*, 1996; Hacia *et al.*, 1996), and many other aspects of genome analysis (Sapolsky and Lipshutz, 1996; Shalon *et al.*, 1996; Shoemaker *et al.*, 1996). Chips will play a central role in biological research by providing a high-capacity link between sequence information and function (Southern, 1996; Schena, 1996; Lander, 1996; Strauss and Falkow, 1997; Fodor, 1997). The impact of microarrays may be as significant as that of recombinant DNA and PCR.

Background

The development of solid-surface assays for gene expression arose out of our basic research on transcription factors. My (MS) earliest thinking on the subject dates back to the beginning graduate school in 1985. During that period, Keith Yamamoto (UCSF) described an experiment in which Ivarie and O'Farrell (1978) used two-dimensional gel electrophoresis to analyze protein extracts from glucocorticoid-treated mammalian cells. Comparisons of treated and control samples revealed that as many as 1% of all mammalian genes (~1000) are subject to steroid regulation. Although the identity of the proteins and the corresponding hormone-regulated genes was not readily determinable, these early experiments underscored the fact that a single inducer (dexamethasone) working through its cognate transcription factor (glucocorticoid receptor) could cause specific, widespread changes in gene expression. It also suggested the need for assays capable of whole genome expression analysis.

The need for a high-capacity expression monitoring platform gained momentum during the early 1990s with our work on plant transcription factors. Working at Stanford University, Schena and Davis (1992) identified a large number of homeobox genes in higher plants and demonstrated that ectopic expression of these regulators could trigger global developmental changes (Schena *et al.*, 1993). These observations further illustrated the importance and complexity of transcription factors and indicated that a detailed view of such regulators would require sophisticated assays for gene expression analysis.

Coincident with our work on gene expression, Steve Fodor and colleagues (Affymetrix) and Dari Shalon and Patrick Brown (Stanford University) were developing DNA microarray technologies based on light-directed combinatorial chemistry (Fodor *et al.*, 1991) and mechanical microspotting (Shalon *et al.*, 1996), respectively. These microarray technologies enabled the automated production of high-density arrays of biologically active nucleic acids. Working with Affymetrix and Shalon and Brown, we devised the first chip-based assays for biological analysis (Schena *et al.*, 1995, 1996; Shalon, 1996; Lockhart *et al.*, 1996; DeRisi *et al.*, 1996; Heller *et al.*, 1997).

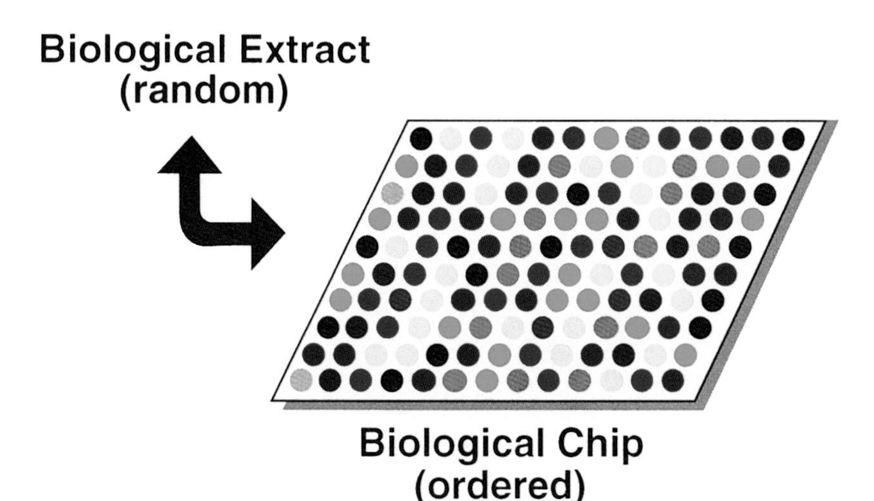

Fig. 1. Microarray concept. Shown is the basic concept of a biological chip assay. Biological extracts are deciphered by reacting these mixtures with an ordered array of biological molecules (biological chip). The extent of binding of labeled products to their cognate array elements provides a quantitative readout of important cellular processes such as gene expression. Parallelism allows precise comparisons to be made between all the genes or gene products represented in the array. Pseudocolor representations simplify data analysis.

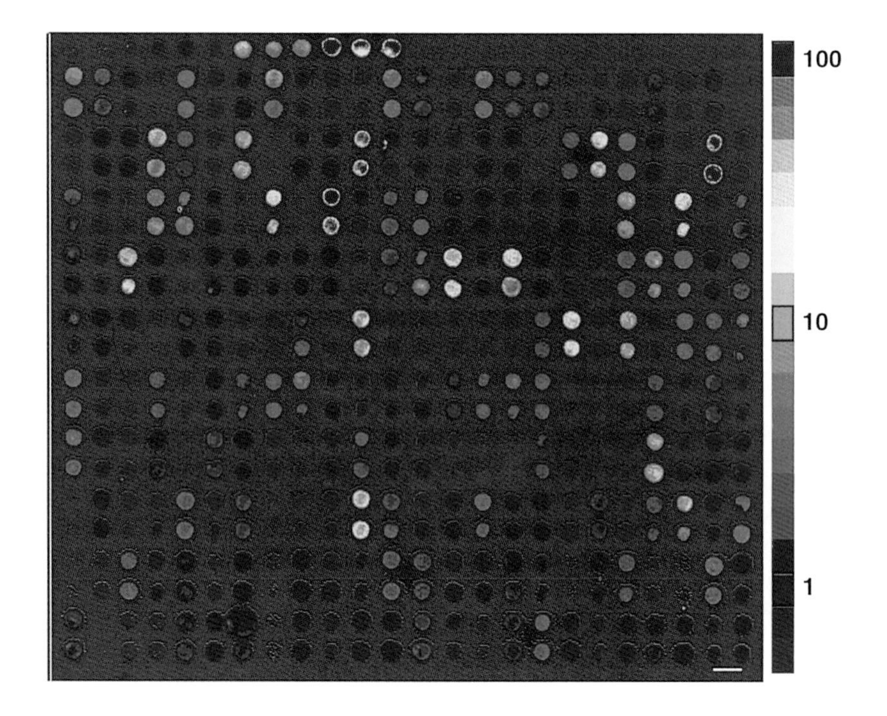

Fig. 3. Detection technology. Human cDNAs were amplified by PCR, purified with an ArrayIt kit (TeleChem), and spotted in duplicate on modified glass microscope slides (CEL Associates) with a robotic spotting device. The cDNAs were attached covalently, denatured, and hybridized to a fluorescent probe prepared by reverse transcription of total human mRNA. The detection device from Norgen Systems (*rick@isl.stanford.edu*) included a xenon light source and a charge-coupled device camera. The color bar (right), which indicates gene expression levels (mRNAs per 100,000), was calibrated using control features spotted in triplicate (top line). Scale bar = 200μm.

Basic Concept

The complexity of biology derives mainly from the fact that biological systems (i) contain a large (but finite) amount of genetic information and (ii) are largely refractory to analysis except as biochemical extracts which are randomized collections of molecules. The complexity and disorder of biological systems and extracts can be deciphered by ordering the information on a solid support (Fig. 1, see color insert). This is the central concept of the microarray approach (Fig. 1).

Unlike molecules in an extract, the position and identity of each analytical element in a microarray is known or is easily determinable. Under the appropriate experimental conditions, biochemical reactions with chips can provide a quantitative "readout" of the identity and concentration of important molecules such as gene transcripts. The solid surfaces used in chip-based approaches enable rapid-reaction kinetics and a high degree of assay sensitivity and precision that cannot be achieved with conventional methods. The parallel format allows meaningful comparisons to be made between each of the genes or gene products represented in the array; moreover, miniaturization allows large numbers of determinations, ultimately entire genomes, to be made simultaneously.

Theory

Biological chip experiments can be expedited by following formal methodological guidelines. The so-called "twelve rules of parallel gene analysis" (Table 1) were created to provide an epistemological and theoretical scaffold on which to design experiments, develop protocols, and optimize methods and enabling technologies for parallel biological studies. The first eight rules (Table 1) define the experimental cycle and, using predicate logic, lay out a contiguous string of commands to ensure that the experimental data accurately reflect the biological question. The last four rules focus mainly on data interpretation and on the theoretical aspects of biological systems.

The 12 rules mainly address *what* needs to be done and *why*, as opposed to describing *how* a particular step should be performed (Table 1); accordingly, the rules are intended to be timeless and

Table 1

Twelve Rules of Parallel Gene Analysis[a]

Number	Rule
1	The assay cycle contains five essential components: biological question, sample preparation, biochemical reaction, detection, and data analysis and modeling.
2	Manipulations of biological systems must precisely reflect the biological question.
3	Biological samples must precisely reflect the biological specimen.
4	All gene analyses must be performed in parallel.
5	Technologies for parallel gene analysis must be amenable to miniaturization and automation.
6	Parallel formats must provide a precise and ordered reflection of the biological sample.
7	Detection systems must allow the precise acquisition of data from the parallel format.
8	Data from the detection system must be precisely manipulated and modeled.
9	Comparisons of two or more parallel data sets shall be subject to the limitations inherent in comparing separate experiments.
10	Absolute relational hierarchies can only be constructed from parallel data sets assembled from experiments that singularly interrogate a complete set of system variables.
11	A universal parallel format is one that contains analytical elements for a complete set of system variables for which all of the intrinsic and extrinsic properties have been delineated.
12	Parallel gene analysis for a biological system is said to be complete when a four-dimensional data set has been assembled for all of the system variables in each system module.

[a] The rules provide a formal architecture for chip-based biological analysis. A detailed definition of the terms and theory is available electronically (*http://cmgm.stanford.edu/~schena/;http://www.technologymentors.com/*).

should thus survive the rapid evolution of techniques and technologies expected in the genomics field during the next two decades.

Assay Cycle

The microarray assay cycle (Fig. 2) contains five essential components: biological question, sample preparation, biochemical reaction,

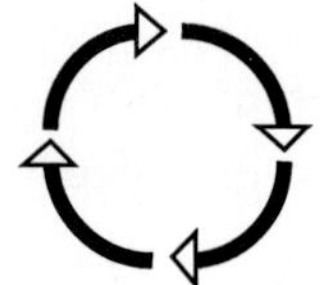

Figure 2 Experimental lifecycle. Shown are the basic components of the microarray assay cycle (boxes) together with the protocols, methods, and enabling technologies (parentheses) that were used in the first gene expression experiments (Schena *et al.*, 1995). Circular arrows denote the cyclic nature of the experimental process.

detection, and data analysis and modeling (Table 1, rule 1). The assay cycle can be thought of as a "life cycle" in that each round of analysis is usually followed by subsequent rounds that proceed in an orderly progression. Each step in the cycle embodies a large number of methods and protocols that are chosen on the basis of their compatibility with particular enabling technologies and research applications. The first assay with a biological chip, that of gene expression monitoring in plants (Schena *et al.*, 1995), illustrates how a specific set of experimental protocols fit into the assay cycle (Fig. 2). The structure of the life cycle provides a experimental foundation on which to build increasingly powerful tools for parallel gene analysis.

Biological Questions

One key difference between chip experiments and conventional approaches is the amount of time required for data analysis and modeling. Unlike an RNA blot, for example, the data generated from a single chip often requires weeks or even months to analyze. Sound experimental design is thus key to experimental success with microoarrays and each experiment should begin with a biological question. Questions need to be tailored to a specific biological problem and posed within the context of a given set of tools. Evaluation of a biological question can be made by asking some basic questions:

Is the question appropriate for the experimental system?
Is the question too broad or too narrow?
What are the possible artifacts?
Is the experiment reproducible?
Have the right controls been included?

Sample Preparation

The second step in the experimental cycle is sample preparation (Fig. 2). The preparation of high-quality biological samples is as difficult as it is important. Care must be take when manipulating cells, tissue samples, or whole organisms in the context of a particular biological question (Table 1, rule 2). Slight alterations in temperature, hormonal and nutritional environment, genetic background, tissue composition, and so forth can lead to significant changes in the gene expression readout. An experiment designed to test the biological effects of dexamethasone should take note of the temperature of the incubator, steroid contaminants in dexamethasone, solvents used to dissolve the hormone, the clonality and growth phase of the cells, and so forth.

Biological samples should reflect the biological specimen as closely as possible (Table 1, rule 3). Prior to preparing a fluorescent complementary DNA (cDNA) mixture for expression analysis, for example, one might consider the following pitfalls:

Enzymatic and chemical degradation of the messenger RNA (mRNA)
Sequence-specific labeling effects
Hybridization artifacts associated with fluorescent moieties
Skewing of the mRNA population during amplification

A labeling protocol for total mRNA that reproducibly yields high-quality fluorescent samples is performed as follows:

Mix 5.0 μg mRNA, 5.0 ng control mRNA cocktail, 4.0 μg oligo-dT 21 mer, and 17.0 μl water (treated with 0.1% diethyl pyrocarbonate).
Incubate the 27-μl mixture at 65°C for 3 min to denature the

mRNA and then at 25°C for 10 min to allow annealing of the oligo-dT.

To the 27-μl mixture, add 10 μl 5× First Strand Buffer [250 m*M* Tris–HCl (pH 8.3), 375 m*M* KCl, 15 m*M* $mgCl_2$], 5.0 μl DTT (0.1 *M*), 1.5 μl RNase block (20 U/μl), 1.0 μl dNTP cocktail (25 m*M* each of dATP, dGTP, and dTTP), 2.0 μl of dCTP (1 m*M*), 2.0 μl Cy3-dCTP (1 m*M*), and 1.5 μl SuperScript II reverse transcriptase (200 U/μl).

Incubate the 50-μl reaction at 37°C for 2 hr.

Biochemical Reaction

The third step in the experimental cycle is the biochemical reaction (Fig. 2). This step involves the preparation and use of biological chips (Table 1, rules 4–6) and is thus extremely complex and undergoing rapid technological evolution. The fundamental goal is to carry out the biochemical reaction in such a way that the parallel format (chip) provides a precise reflection of the biological sample. The main components of the biochemical reaction are the following:

Biological chip technologies (e.g., photolithography, microspotting, and "ink-jetting")

Chip substrate selection (e.g., cDNAs, oligonucleotides, and proteins and other biomolecules?)

Surface considerations (e.g., array element density and purity, substrate attachment, biological activity, and specific versus nonspecific binding)

Biochemical interactions (e.g., DNA–DNA hybridization, RNA–DNA hybridization, and protein–protein recognition?)

The first published gene expression experiments utilized microarrays of cDNAs amplified by PCR and deposited into known locations on glass using a mechanical spotting device (Schena *et al.*, 1995). The biochemical reaction in this case was a hybridization between single-stranded fluorescent cDNA (0.5–2.5 kb) in solution and denatured cDNA (0.5–2.5 kb) attached to the glass surface. Hybridization reactions with cDNA microarrays are typically carried out for 6–12 hr at 60–65°C at a total nucleic acid concentration of 0.1–0.5 mg/ml nucleic acid in buffers containing ~1.0 molar salt (5X SSC or 6X SSPE) and 0.1 or 0.2% SDS or an equivalent detergent. Reactions are typically performed under glass coverslips in volumes of 1.0–1.5 μl per $cm.^2$

Hybridization signals are usually linear up to 24 hr, after which time elevated nonspecific hybridization may be observed. Increases in background fluorescence are also observed when hybridization temperatures fall below 55°C. Microarray wash steps are typically performed at ambient temperature (25°C) with buffers containing 10–100 m*M* salt (0.1–1.0X SSC or SSPE) and 0.1 or 0.2% SDS. Experiments with oligonucleotides are similar to cDNA chip protocols except that hybridization temperatures are typically 25–42°C (Lockhart *et al.*, 1996; Chee *et al.*, 1996; Cronin *et al.*, 1996; Kozal *et al.*, 1996; Hacia *et al.*, 1996; Sapolsky and Lipshutz, 1996; Shoemaker *et al.*, 1996).

Detection

The fourth step in the assay cycle is detection (Fig. 2). Above all, the detection system must allow accurate acquisition of the microarray data (Table 1, rule 7). Some of the key considerations in designing chip-based detection systems are the following:

Speed
Resolution
Sample alteration (e.g., photobleaching)
Dynamic range
Multiplexing
Data distortion (e.g., mechanical and electronic noise)

The first published microarray expression experiments utilized a detection system based on confocal fluorescence scanning (Schena *et al.*, 1995). In this system, the chip was positioned on a motorized translational stage perpendicular to an upward-facing confocal microscope lens, moving in a raster pattern at ~10 cm/sec. Excitation of the substrate was achieved with a multiline gas Argon/Krypton laser, which supplied ~5 mW beam power per line to the glass surface. The emission from separate 488- and 568-nm lines was detected using two photomultiplier tubes fitted with optical filters designed for fluorescein and lissamine detection. Signals from the photomultiplier tubes were read into a personal computer using an analog-to-digital conversion board which provided a final pixel spacing of 10–20 μm.

Recent detection system designs utilize combinations of single-line gas lasers for excitation or continuous wavelength light sources such as those obtained with xenon lamps. The use of charge-coupled device cameras, as an alternative to confocal scanners and photomultiplier tubes, has been demonstrated recently (Fig. 3, see color insert).

Data Analysis and Modeling

Three of the 12 rules (Table 1, rules 8–10) pertain directly to the fifth phase of the experimental cycle—data analysis and modeling (Fig. 2). It is vital that the data from the detection system are manipulated and modeled with precision (Table 1, rule 8). The following are some key considerations in analyzing data from chip experiments:

Suitability of controls
Comparing control and experimental data
Interpreting multiplex microarray formats
Comparing conventional data and chip-based results

Controls for nucleic acid-based gene expression experiments typically take the form of mRNAs added to the reverse transcription reaction as a dilution series that spans the partial concentrations of mRNAs present in the biological sample (e.g., 10^{-5}, 10^{-4}, and 10^{-3} mole per mole of total polyA$^+$ mRNA). mRNAs added to the reverse transcription reaction provide controls for the labeling, hybridization, and detection steps of the assay. Controls can also be added during the RNA isolation step to correct for RNA degradation. In most experiments, quantitation is made by comparing the fluorescence signals from the experimental portions of the microarray to the controls. Data for multiple biological samples obtained from multiplex experimental formats such as those utilizing two-color fluorescence are directly comparable, providing that the two labeling schemes reproducibly yield 1 : 1 results with the same mRNA sample.

Chip-to-chip comparisons are inherently more complicated than data from single microarrays and are subject to the limitations inherent in comparing separate experiments (Table 1, rule 9). The main pitfalls in comparing two or more parallel data sets arise from discrepancies in

Chip manufacturing
Biochemical reaction (e.g., hybridization) conditions
Sample preparation

Detection

Data analysis

The availability of high-quality microarrays, coupled with careful experimentation at each of the downstream steps, allows for meaningful data to be acquired and compared from independent experiments. This capability is crucial in high-throughput settings, particularly those involving diagnostics and other clinical applications in which a large number of patient samples are examined.

Applications

Similar to PCR and other important advances in medical research, microarray-based assays have a vast number of applications, many of which probably have yet to be realized. Virtually any problem in which massive, parallel analysis affords a distinct advantage over current methods will probably see biological chips employed as the platform of choice. The application areas currently most active include

Gene expression monitoring

DNA resequencing

Point mutation analysis

Genotyping applications

Biological chips have immediate applications in both academic and commercial settings, allowing accelerated data acquisition for drug discovery, disease diagnosis and prognosis, and basic research.

Many biological applications would benefit from the availability of microarrays that allow genomewide analysis (Table 1, rule 10). This necessitates technologies compatible with data sets containing information for ~100,000 genes. Continued advances in photolithography, mechanical spotting, piezoelectric, and other arraying techniques together with continued detailed structural and biochemical studies of single genes and gene products will ultimately provide "genome chips" for parallel expression analysis (Table 1, rule 11). This will allow the assembly of enormous databases of biological information, including whole organism gene expression profiles with spatial and temporal coordinates for all the genes and cells in an organism (Table 1, rule 12). Biological chips used in massive, parallel assays will assist in accelerating the information age of biology.

References

Chee, M., Yang, R., Hubbell, E., Berno, A., Huang, X. C., Stern, D., Winkler, J., Lockhart, D. J., Morris, M. S., and Fodor, S. P. A. (1996). Accessing genetic information with high-density DNA arrays. *Science* **274,** 610–614.

Cronin, M. T., Fucini, R. V., Kim, S. M., Masino, R. S., Wespi, R. M., and Miyada, C. G. (1996). Cystic fibrosis mutation detection by hybridization to light-generated DNA probe arrays. *Hum. Mutat.* **7,** 244–255.

DeRisi, J., Penland, L., Brown, P. O., Bittner, M. L., Meltzer, P. S., Ray, M., Chen, Y., Su, Y. A., and Trent, J. M. (1996). Use of a cDNA microarray to analyze gene expression patterns in human cancer. *Nat. Genet.* **14,** 457–460.

Fodor, S. P. A. (1997). Massively parallel genomics. *Science* **277,** 393–395.

Fodor, S. P. A., Read, J. L., Pirrung, M. C., Stryer, L., Tsai Lu, A., and Solas, D. (1991). Light-directed, spatially addressable parallell chemical synthesis. *Science* **251,** 767–773.

Hacia, J. G., Brody, L. C., Chee, M. S., Fodor, S. P. A., and Collins, F. S. (1996). Detection of heterozygous mutations in BRCA1 using high density oligonucleotide arrays and two-colour fluorescence analysis. *Nature Genet.* **14,** 441–447.

Heller, R. A., Schena, M., Chai, A., Shalon, D., Bedilion, T., Gilmore, J., Woolley, D. E., and Davis, R. W. (1997). Discovery and analysis of inflammatory disease-related genes using cDNA microarrays. *Proc. Natl. Acad. Sci. U.S.A.* **94,** 2150–2155.

Ivarie, R. D., and O'Farrell, P. H. (1978). The glucocorticoid domain: Steroid-mediated changes in the rate of synthesis of rat hepatoma proteins. *Cell* **13,** 41–55.

Kozal, M. J., Shah, N., Shen, N., Yang, R., Fucini, R., Merigan, T. C., Richman, D. D., Morris, D., Hubbell, E., Chee, M., and Gingeras, T. R. (1996). Extensive polymorphisms observed in HIV-1 clade B protease gene using high-density oligonucleotide arrays. *Nature Med.* **2,** 793–799.

Lander, E. S. (1996). The new genomics: Global views of biology. *Science* **274,** 536–539.

Lockhart, D. J., Dong, H., Byrne, M. C., Follettie, M. T., Gallo, M. V., Chee, M. S., Mittmann, M., Wang, C., Kobayashi, M., Horton, H., and Brown, E. L. (1996). Expression monitoring by hybridization to high-density oligonucleotide arrays. *Nature Biotechnol.* **14,** 1675–1680.

Sapolsky, R. J., and Lipshutz, R. J. (1996). Mapping genomic library clones using oligonucleotide arrays. *Genomics* **33,** 445–456.

Schena, M. (1996). Genome analysis with gene expression microarrays. *BioEssays* **18,** 427–431.

Schena, M., and Davis, R. W. (1992). HD-Zip proteins: Members of an *Arabidopsis* homeodomain protein superfamily. *Proc. Natl. Acad. Sci. U.S.A.* **89,** 3894–3898.

Schena, M., Lloyd, A. M., and Davis, R. W. (1993). The *HAT4* gene of *Arabidopsis* encodes a developmental regulator. *Genes Dev.* **7,** 367–379.

Schena, M., Shalon, D., Davis, R. W., and Brown, P. O. (1995). Quantitative monitoring of gene expression patterns with a complementary DNA microarray. *Science* **270,** 467–470.

Schena, M., Shalon, D., Heller, R., Chai, A., Brown, P. O., and Davis, R. W. (1996). Parallel human genome analysis: Microarray-based expression monitoring of 1000 genes. *Proc. Natl. Acad. Sci. U.S.A.* **93,** 10614–10619.

Shalon, D. (1996). PhD thesis, Stanford University.

Shalon, D., Smith, S. J., and Brown, P. O. (1996). A DNA microarray system for analyzing complex DNA samples using two-color fluorescent probe hybridization. *Genome Res.* **6,** 639–645.

Shoemaker, D. D., Lashkari, D. A., Morris, D., Mittmann, M., and Davis, R. W. (1996). Quantitative phenotypic analysis of yeast deletion mutants using a highly parallel molecular bar-coding strategy. *Nature Genet.* **14,** 450–456.

Southern, E. M. (1996). DNA chips: Analysing sequence by hybridization to oligonucleotides on a large scale. *Trends Genet.* **12,** 110–115.

Strauss, E. J., and Falkow, S. (1997). Microbial pathogenesis: Genomics and beyond. *Science* **276,** 707–712.

29

HIGH-DENSITY cDNA GRIDS FOR HYBRIDIZATION FINGERPRINTING EXPERIMENTS

Armin O. Schmitt, Ralf Herwig, Sebastian Meier-Ewert, and Hans Lehrach

Although the speed of sequencing DNA has accelerated enormously in the past decade, rapid screening of genetic material remains indispensable for many purposes. For example, it would be infeasible to monitor the changes in the mRNA expression patterns in young organisms by sequencing completely all expressed mRNA, even if the speed of sequencing were orders of magnitudes higher than it is currently. Other examples in which detailed information can be replaced by an approximate knowledge about the mRNA expression pattern include discrimination between cell and tissue types, between normal and diseased cells, or between different organisms.

Such "approximate knowledge" about mRNA could also help avoid unnecessary sequencing efforts since large-scale sequencing projects are entering a stage in which the danger of sequencing DNA multiple times is no longer negligible.

Hybridization fingerprinting experiments with short oligonucleotides as probes provide a method to screen efficiently and at relatively low costs huge quantities of genetic material without losing too much information (Poustka *et al.*, 1986; Lehrach *et al.*, 1990; Drmanac *et al.*, 1991). Hybridization fingerprinting of clones with oligonucleotides combines advantages of two other common techniques: the robustness against repeats of gel fingerprinting approaches in which restriction digest patterns serve as fingerprints (Coulson *et al.*, 1986) and the high data rates of hybridization experiments with pooled probes (Evans and Lewis, 1989).

It could be shown that the analysis of whole genomes with this technique is viable using automated protocols at various steps (Craig *et al.*, 1990; Hoheisel *et al.*, 1991; Maier *et al.*, 1992). Rough estimates indicate that fingerprints derived from about 100–200 such hybridization cycles with nonhomologous octamers should be sufficient to discriminate all the estimated 100,000 genes in mammals (Meier-Ewert *et al.*, 1994).

Essentially, a fingerprint of a clone is a series of M numbers (signals from an imager) which can be considered as a typical signature of the gene or gene family to which this clone belongs. For simplicity's sake, we assume that the numbers can only take on the values 0 and 1. A 1 at a given position j of the fingerprint (jth component when interpreted as a mathematical vector) means that the jth oligonucleotide from a given set of oligonucleotides hybridizes to that clone; on the other hand, a 0 means that it does not (see Table 1). The order of the M oligonucleotides is arbitrary but fixed for one experiment. To achieve a high throughput of such experiments the hybridization procedure has to be massively parallelized. Robots for picking and spotting clones onto membranes in high-density grids have been devised to this end (Meier-Ewert *et al.*, 1993; Maier *et al.*, 1994).

In the following sections, we present a protocol for the production of such high-density grids, discuss a number of built-in controls to judge their quality, and discuss a number of possible applications.

Protocols

Construction of cDNA Libraries

About 100,000 individual cDNA sequences constituting a representative cDNA library of a given tissue or organism are inserted in the plasmid pSPORT1 carried by *Escherichia coli* and grown as individual colonies on an agar plate. Picking colonies and inoculating microtiter

Table 1
Scheme of a Fingerprint

Clone	AAGAAGAA	ATCATCAT	ACATCATG	CAGAGGAG	CTCTCCAC	...	Binary Fingerprint
MPImouse_9day_16P23	0	1	1	0	0	...	01100 ...
MPImouse_9day_58N22	0	0	0	1	0	...	00010 ...
MPImouse_9day_40K4	1	0	1	0	0	...	10100 ...
MPImouse_9day_173N1	0	1	0	1	0	...	01010 ...
MPImouse_9day_50L10	...	...	...	...	...	...	...
...							

The fingerprint of a clone carries a 1 if it hybridizes to a given oligonucleotide; otherwise, a 0 is assigned.

plates is a very laborious task when carried out manually. A picking robot was devised to perform this task which accelerates the manual rate by about an order of magnitude (Maier *et al.,* 1994). Its image processing software is able to identify and locate the individual colonies on the 22 × 22-cm agar plate. Each colony is picked by 1 of the 96 pneumatically controlled pins of a mobile picking head and transferred into a 384-well microtiter plate (Genetix Ltd., Christchurch, Dorset, UK). With the picking rate being about 3500 clones per hour, a library comprising 100,000 clones can be picked in <30 hr. A stack of 36 such high-capacity microtiter plates can be processed without human supervision in about 4 hr. The obtained set of microtiter plates serves as a reference set and is stored at −70°C. These reference filters are made available to laboratories worldwide (*http://www.rzpd.de/;* Zehetner and Lehrach, 1994).

PCR Amplification

In principle, *in situ* hybridization experiments can be carried out with the plasmid–host cell complex (Hoheisel *et al.,* 1994). The use of very short oligonucleotides as probes, however, requires that the cDNA be purified away from the host cell and vector DNA to suppress signals due to hybridization. The most effective way to accomplish this is massive and specific amplification by means of the PCR. To perform a large number of reactions simultaneously a robot was developed (Meier-Ewert *et al.*, 1993) which cycles a stack of 120 plastic sealed microtiter plates (Genetix Ltd.) 30 times between three water tanks of 94°C (10 sec), 72°C (1 sec), and 65°C (3 min). Figure 1 is a photograph of our PCR robot.

In such a way, 46,080 (= 120 × 384) clones can be amplified to a very high copy number of about 10^{11} (about 1 p*M*) from an initial number of 10,000 cells. A 40-μl reaction is carried out with 200 nl *E. coli* culture, 5 pmol each of forward and reverse primer, 0.001% gelatine, 1 U *Taq* polymerase, 0.1 m*M* each dNTP, and PCR buffer. PCR buffer contains 1.5 m*M* $MgCl_2$, 10 m*M* Tris–HCl (pH 8.55), and 50 m*M* KCl. A 3-min extension at 65°C was done after thermocycling. Testing is done on a 1% agarose gel using *Bst*EII-digested λ DNA as size standard. The PCR is estimated to succeed at a rate of 75–80%, producing clones of 1.4-kbp length on average.

Figure 1 The photograph displays the PCR robot which enables us to process 120 plates holding 384 clones each simultaneously. The three water baths contain 250 liters each and are kept at 94, 72, and 65°C.

Production of High-Density cDNA Grids

To subject the PCR products to efficient probing with oligonucleotides, they are spotted onto 22 × 22-cm Hybond N^+ nylon membranes (Amersham), also referred to as "filters." The spotting procedure is performed automatically by basically the same robot—with the picking head being replaced by a spotting head—as was used for picking the clones (Meier-Ewert *et al.*, 1993; Lehrach *et al.*, 1997). Up to 250,000 PCR products can be arrayed in about 90 min without human intervention. A filter is subdivided into 2304 (= 48 × 48) square blocks, each carrying 25 cDNA products. Thus, a filter can host a grid of 57,600 cDNA clones which corresponds to a density of 1.2 clones per mm^2. Since 3 of the 25 spots are used for control purposes (1 so-called guide dot and one pair of clones with known sequence) and since all cDNA products are spotted in duplicates (resulting in 12 different clones per block and 27,648 different clones in total), the effective density is lowered by a factor of about 2. Several dozen copies are produced from each library. The ensemble of all copies of a filter constitutes a set, and four sets are necessary to accommodate a library of almost 100,000 clones when clones are spotted in duplicates.

The filters are probed with radioactively labeled oligonucleotides in a 600 m*M* NaCl, 60 m*M* Na citrate, 7.2% Sarkosyl solution at 8°C overnight and washed at 5°C in the same buffer. The amount of radioactivity in a clone–oligonucleotide duplex is quantified by a phosphor imager (Molecular Dynamics, Sunnyvale, CA). After each hybridization cycle, filters are washed in the buffer twice at 65°C for 20 min to remove the probe. About 15 hybridization cycles can be carried out with one filter before it has to be replaced by a fresh copy. In order to measure the amount of PCR product present in each dot, the first hybridization of each filter is carried out with one of the primers since the primer should hybridize to all clones. This procedure is called background hybridization.

Choice of Probes

The *n*-mer composition of DNA sequences does not deviate much from that of random sequences (Schmitt and Herzel, 1997); therefore, a selection of randomly synthesized oligonucleotides performs well as a set of probes. Nevertheless, sets designed especially for hybridization experiments with a specific type of cDNA outperform random sets by about 20%, *i.e.*, the same amount of information can be gathered from an experiment with a set which is 20% smaller than a random selection. In view of the costs that can be saved by reducing the number of redundant hybridization cycles, it is worthwhile to perform experiments with tailored sets of probes (Kel *et al.*, 1996).

Quality Control Procedures

The immediate result of one hybridization experiment is stored as a file in the TIF format. Such files can be viewed with the program XDigitise written by Huw Griffith (Imperial Cancer Research Institute, London, UK); an example of the image produced by XDigitise is given in Figure 2.

Nine bright guide dots can be recognized in the enlarged section in the top right corner of Fig. 2, together with several fainter hybridization signals. These guide dots play a crucial role in the software, which is used to determine the coordinates of dots on the filter and to quantify intensities (Hartelius, 1996; R. Schattevoy, personal communication). The series of all intensity arrays are finally assembled into an $N \times M$ matrix representing the fingerprints of all N clones of a library, with M being the number of different oligonucleotides with which experiments have been carried out.

It is obvious that such a multistep procedure line is very error prone. Therefore, several control procedures are incorporated to assess and ensure the quality of data.

Background Hybridization

The background hybridization enables us to exclude clones from the analysis whose background signal indicates that the PCR amplification has failed. Figure 3 shows that these clones can be discriminated clearly from those with a considerable amount of PCR product. The area below the first maximum confirms the estimated failure rate of about 20% in the PCR. Furthermore, the background signal can be used to normalize for the varying amount of PCR product in each dot.

Duplicate Spotting

The spotting of PCR product in duplicates allows for another type of control. Ideally, both dots of a duplicate pair would contain exactly the same amount of PCR material, a situation which can only be approximated in real experiments. This coherence between duplicates can be quantified mathematically by the empirical correlation coefficient

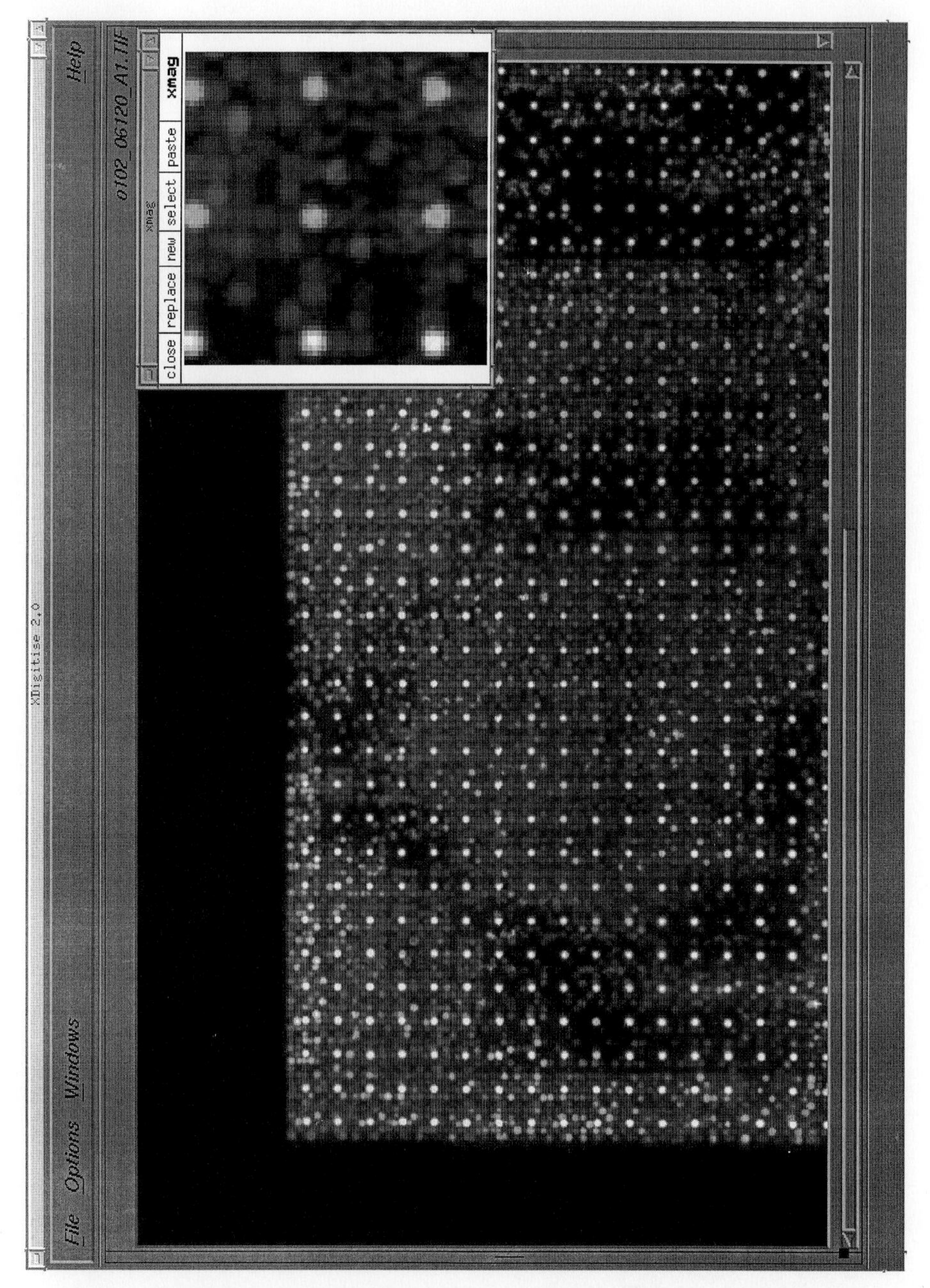

Figure 2 The window produced by the software package XDigitise 2.0 used to visualize hybridiation filters. In the magnified insert nine guide dots are clearly visible. A guide dot consists of salmon sperm and produces a clear signal for any oligo. It occupies the center of a block of 5 × 5 dots and is used for correction purposes in the image processing.

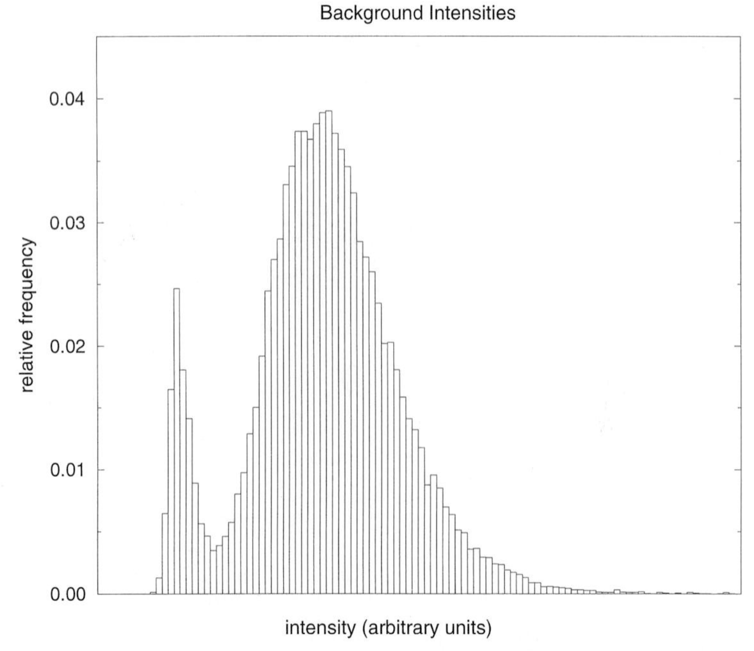

Figure 3 Probing the clones with the primers (background hybridization) is a means of determining the rate of failed PCR reactions. A small amount of PCR product (failed reactions) binds only a small amount of radioactivity resulting in a weak signal (left peak). About 80% of all reactions produce a sufficient amount of clones which can then undergo probing (right peak).

$$r = \frac{\sum_i (x_i - \overline{x})(y_i - \overline{y})}{\sqrt{\sum_i (x_i - \overline{x})^2 \sum_i (y_i - \overline{y})^2}}$$

where x_i and y_i are the intensities of a pair of duplicates, $\overline{x}$ and $\overline{y}$ are the respective average intensities, and the sum extends over all pairs of intensities from a filter. The correlation coefficient is 1 (perfect linear correlation) if $x_i = ay_i + b$ and $a > 0$ for all pairs i; it is -1 (perfect linear anticorrelation) if $x_i = ay_i + b$ and $a < 0$ holds; and it is 0 (noncorrelation) if there is no linear relation on average. Figure 4 shows two scatter plots of duplicate intensities, *i.e.*, one intensity of each pair is plotted against the other. The intensities shown in the left plot exhibit a high correlation of $r = 0.98$ and those in the right plot a poor correlation of $r = 0.31$. We discard filters whose correlation coefficient drops below 0.6 (about 10% of all filters) since we believe that a reliable quantification of their intensities cannot be guaranteed. Visual inspection of

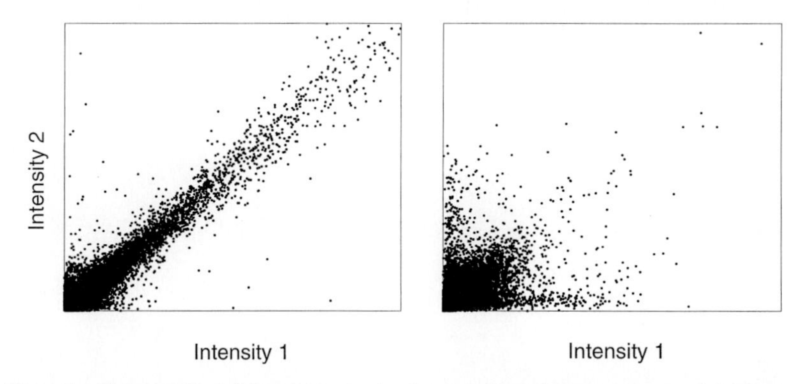

Figure 4 The spotting of the clones in duplicates allows for assessment of the overall quality of a filter and of the image processing steps. The second value of each pair of intensities is plotted against the first value. A high correlation coefficient r (see text) between the two intensity values of a duplicate pair (left, $r = 0.98$) indicates that we can trust the data, whereas a low correlation (right, $r = 0.31$) warns us not to accept such values and to locate the source of error.

low-correlation filters suggest that at some point in the production line a mistake must have occurred.

CONTROL SEQUENCES

We previously mentioned clones with known sequence. These clones are 500- to 1500-bp long segments of the human major histocompatibility complex sequence (Beck *et al.*, 1992). Hybridizations of the probes with these MHC clones allow for assessing the intensities of the signals, *i.e.*, they permit conclusions about the distribution of signal intensities if a stretch complementary to the probe sequence is present in a clone and if no such stretch is present. The first of these distributions describes the intensities of true positive signals and the latter of true negative signals. Figure 5 shows these two distributions for one oligonucleotide. It can be seen that the two distributions can be discriminated with high security, *i.e.*, intensities in the medium or high range are likely caused by true hybridizations, whereas the low-range intensities can in most cases be classified as nonhybridizations. The average value of the true signal intensities constitutes a threshold above which an intensity has likely been generated by a perfectly matching oligonucleotide sequence.

Clustering of Fingerprints

The essential idea of the fingerprinting approach is that identical or similar cDNA clones entail identical or similar fingerprints. Realisti-

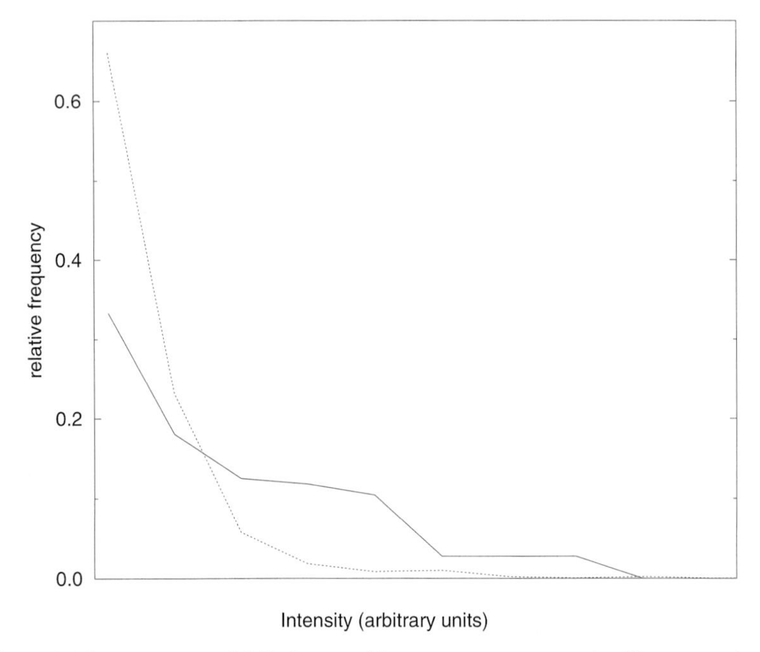

Figure 5 The presence of 712 clones of known sequence on the filters permits us to study the intensities which are generated by matching and nonmatching oligos. The solid line displays the frequency distribution of intensities from full matches (true positives) of the oligo CTGGAGGA, *i.e.,* at least one copy of the oligonucleotide sequence (or its complementary strand) is present in the clone sequence; the dotted line displays the intensities of nonperfect hybridizations (true negatives) of that oligonucleotide against the control sequences.

cally, we cannot expect that two fingerprints become absolutely identical even if they are derived from identical clones; there are too many sources of noise and other influences corrupting the data. The clustering of experimental data such as hybridization fingerprints, *i.e.,* the finding of groups whose members are relatively similar to each other but dissimilar to the members of all other groups, is a great computational challenge (Jain and Dubes, 1988; Mirkin, 1996). The clustering technique we apply (Gerst, 1997) is a very fast and sensitive approach based on Bayesian codeword analysis. Fine-tuning of parameters controlling the connectivity of clusters is done by clustering a reference set of approximately 2300 fingerprints from 18 different genes for which we know the correct sequences.

Results and Discussion

The following results refer to a library of 35,635 mouse clones, 30,471 of which were taken from 9-day-old mice and 5164 from 12-day-old mice. The clones were probed with 125 8-mers, and the derived fingerprints were attributed to 4323 different clusters ranging in size from 2 to 438 members. These were 12,801 fingerprints classified as singletons, *i.e.,* they could not be assigned to any cluster and are also significantly different from each other. Figure 6 depicts a histogram of the size distribution in logarithmic scale. About 10 clusters contain more than 80 members and are not visible in the truncated histogram. Figure 7 shows the fingerprints of a cluster of 23 members with the intensities represented in four shades of gray.

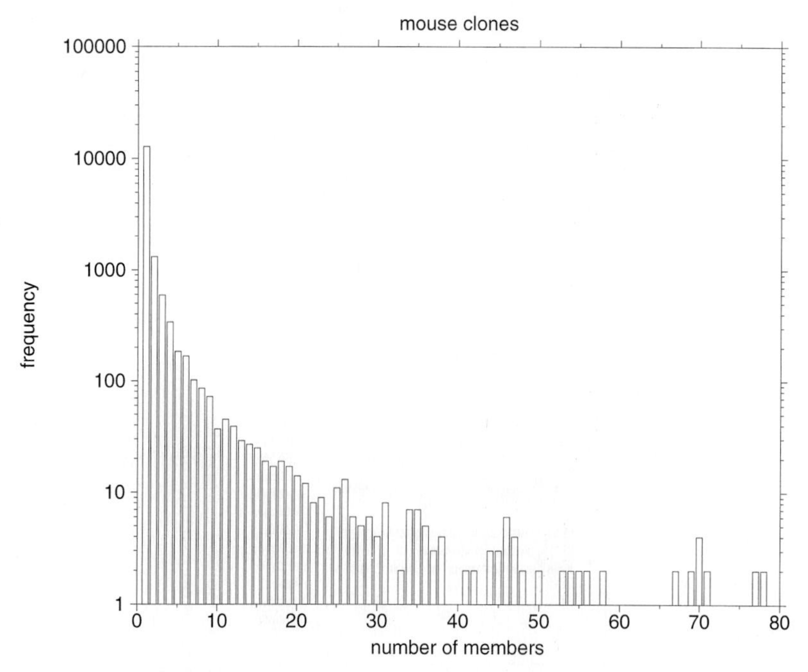

Figure 6 The distribution of the sizes of clusters found in the fingerprint data of about 35,000 clones from 9-day-old (about 30,000) and 12-day-old (about 5000) mice (linear-log scale). About one-third of the fingerprints remain as singletons; very few clusters contain more than 80 members.

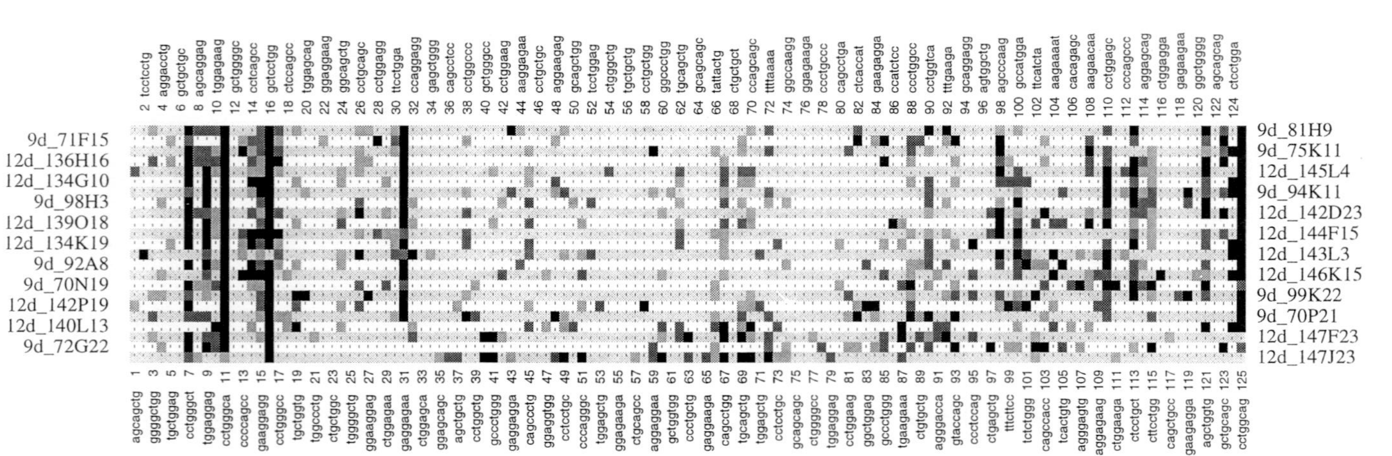

Figure 7 A graphical representation of the fingerprints of a cluster. The darkness of the squares symbolizes the intensity measured in the hybridization experiments. The clone names are given on the left and right and the oligonucleotide sequences on the top and bottom.

An interesting question regarding the individual clusters is whether the DNA sequence of its member clones is already known and available in public databases. To answer this question, the fingerprints in the cluster must be compared with the theoretical fingerprints of a nonredundant representative database. We assembled such a database from all the rodent coding sequences stored in the Genbank and EMBL databases. Where necessary, these sequences were trimmed from the 3′ end to a maximum of 2000 bp in order to fit them to the experimental conditions. Redundant sequences were removed by means of the program CLEANUP (Grillo *et al.*, 1996). The obtained collection of sequences, C_{seq}, comprises almost 13,000 entries. The generation of a theoretical fingerprint from a sequence is straightforward: Assign a 1 to the component of the theoretical fingerprint if the sequence (or its reverse complement) of the respective oligonucleotide is contained in the DNA sequence and a 0 otherwise. We term the obtained collection of theoretical fingerprints C_{tfp}.

In order to identify the theoretical fingerprint $f_{best\ theo}$ which coincides best with a given experimental fingerprint f_{obs} we use the following combinatorial approach. For each theoretical fingerprint from C_{tfp} we determine the number of congruent matches n (*i.e.*, the number of times that the theoretical and experimental fingerprints have oligonucleotide matches at the same positions—the so-called simple-matching coefficient), the expected number of congruent matches μ if the matching oligos were located at random positions in the fingerprints, and the standard deviation σ of μ.

The sequence from C_{seq} whose theoretical fingerprint maximizes the quantity $Q = \frac{n - \mu}{\sigma}$, i.e., the theoretical fingerprint with the number of congruent matches which is highest in terms of standar deviations above expectation value, is considered as the most likely representative of f_{obs} in C_{seq}. Table 2 shows the database matching results for a cluster comprising 23 members. The fingerprints with the highest Q values in that cluster are hitting (α or β) globin sequences (sequence length about 500 bp).

Xanthine dehydrogenase (sequence length about 130 bp) is erroneously selected in three cases because its fingerprint contains only two matching oligonucleotides, whose positions are congruent with those in the experimental fingerprint.

Further analysis with standard tools shows that synaptotagmin (sequence length about 4000 bp) has a 40% homology with globin sequences over a stretch of about 500 bp. Therefore, the fingerprints

Table 2

Matching Experimental against Theoretical Fingerprints

Clone	n	m_{obs}	m_{theo}	μ	σ	Q	ACC	Description
12d_142P19	6	11	7	0.58	0.78	7.55	L75940	Mus musculus alpha-globin mRNA, complete cds. 1/96
12d_145L4	7	11	10	0.84	0.84	7.28	M26894	Mouse beta-H1-globin mRNA, 3' end. 3/90
12d_147J23	6	14	6	0.64	0.74	7.22	M10524	Mouse beta-globin mRNA, partial cds. 3/88
9d_71F15	7	13	10	0.99	0.91	6.59	M26894	Mouse beta-H1-globin mRNA, 3' end. 3/90
9d_94K11	8	19	10	1.45	1.07	6.10	M26894	Mouse beta-H1-globin mRNA, 3' end. 3/90
12d_134K19	7	15	10	1.14	0.97	6.03	M26894	Mouse beta-H1-globin mRNA, 3' end. 3/90
9d_70N19	7	19	8	1.16	0.97	6.03	D00920	Rat mRNA. 4/92
12d_144F15	6	11	10	0.84	0.84	6.01	M26894	Mouse beta-H1-globin mRNA, 3' end. 3/90
12d_142D23	6	11	10	0.84	0.84	6.01	M26894	Mouse beta-H1-globin mRNA, 3' end. 3/90
9d_98H3	2	9	2	0.12	0.33	5.58	X75139	M.musculus (129/Sv) gene for xanthine dehydrogenase, exon 21. 7/95
12d_143L3	7	14	12	1.28	1.02	5.58	X78545	M.musculus MCP-8 mRNA for serine protease. 6/94
9d_81H9	6	13	10	0.99	0.91	5.49	M26894	Mouse beta-H1-globin mRNA, 3' end. 3/90
12d_136H16	6	13	10	0.99	0.91	5.49	M26894	Mouse beta-H1-globin mRNA, 3' end. 3/90
9d_92A8	6	14	10	1.06	0.94	5.23	M26894	Mouse beta-H1-globin mRNA, 3' end. 3/90
9d_75K11	6	15	10	1.14	0.97	4.99	M26894	Mouse beta-H1-globin mRNA, 3' end. 3/90
12d_134G10	2	12	2	0.16	0.37	4.90	X75139	M.musculus (129/Sv) gene for xanthine dehydrogenase, exon 21. 7/95
12d_139O18	2	12	2	0.16	0.37	4.90	X75139	M.musculus (129/Sv) gene for xanthine dehydrogenase, exon 21. 7/95
12d_140L13	5	16	7	0.85	0.85	4.90	M13750	Rat prolactin-like protein A (rPLP-A) mRNA, complete cds. 9/88
12d_146K15	9	19	17	2.46	1.36	4.81	U10355	Mus musculus BALB/c synaptotagmin 4 (syt4) mRNA, complete cds. 8/94
9d_70P21	5	17	7	0.90	0.89	4.71	Z66540	M.auratus mRNA (440 bp). 12/95
9d_72G22	6	21	8	1.28	1.01	4.67	D00920	Rat mRNA. 4/92
12d_147F23	12	26	23	4.56	1.74	4.27	M33863	Mouse 2'-5' oligo A synthetase mRNA, complete cds. 6/93
9d_99K22	6	19	10	1.45	1.07	4.24	M26894	Mouse beta-H1-globin mRNA, 3' end. 3/90

The names of clones in a cluster and the description of database entries from GenBank or the EMBL database which are most compatible with them. ACC, accession number; m_{obs} and m_{theo}; number of matching oligonucleotides in the experimental and theoretical fingerprints, respectively; n, number of congruent matches. The Q value quantifies the security of identification of a database entry. For μ and σ, see text.

of these two clones resemble each other. These last examples indicate the limits of the database matching algorithm in its current form.

Another problem that we identified is the splitting off of true clusters into two or more clusters on account of the varying clone lengths (Meier-Ewert *et al.*, 1997). Although the average clone length can be estimated to be about 1400 bp, the individual clone lengths are subject to considerable variation of about 500 bp, which is reflected in missing or additional signals in the fingerprints of some of the clones. Endeavors are under way to generalize the similarity measure applied during the clustering procedure in order to treat this phenomenon appropriately.

Despite some shortcomings from which the technique still suffers, we believe that hybridization fingerprinting is a powerful method to very efficiently extract valuable information about large quantities of cDNA material. Attaining a survey of the mRNA expression pattern of a cell in as little as 6 weeks (including production of library, hybridization, and data analysis) is an achievement which is unrivaled by other experimental approaches.

Acknowledgments

We thank H. Gerst, J. O'Brien, and J. Freund for valuable contributions and the Novartis Pharma Ltd. for financial support.

References

Beck, S., Kelly, A., Radley, E., Khurshid, F., Alderton, R. P., and Trowsdale, J. (1992). DNA sequence analysis of 66kb of the human MHC class II region encoding a cluster of genes for antigen processing. *J. Mol. Biol.* **228,** 433–441.

Coulson, A., Sulston, J., Brenner, S., and Karn, J. (1986). Toward a physical map of the nematode *Caenorhabditis elegans. Proc. Natl. Acad. Sci. U.S.A.* **83,** 7821–7825.

Craig, A. G., Nizetic, D., Hoheisel, J. D., Zehetner, G., and Lehrach, H. (1990). Ordering of cosmid clones covering the herpes simplex virus type I (HSV-I) genome: A test case for fingerprinting by hybridisation. *Nucleic Acid Res.* **18,** 2653–2660.

Drmanac, R., Lennon, G., Drmanac, S., Labat, I., Crkvenjakov, R., and Lehrach, H. (1991). Partial sequencing by oligohybridization: Concept and applications in genome analysis. *In Proceedings of the First International Conference on Electrophoresis, Supercomputing and the Human Genome* (C. Cantor and H. Lim, Eds.), pp. 60–75. World Scientific, Singapore.

Evans, G. A., and Lewis, K. A. (1989). Physical mapping of complex genomes by cosmid multiplex analysis. *Proc. Natl. Acad. Sci. U.S.A.* **86,** 5030–5034.

Gerst, H. (1997). High performance, high speed code analysis based on the Bayes theorem. *In* Proceedings of the EMBL conference "Automation in Mapping and Sequencing," Heidelberg.

Grillo, G., Attimonelli, M., Liuni, S., and Pesole, G. (1996). CLEANUP: A fast computer program for removing redundancies from nucleotide sequence database. *CABIOS* **12,** 1–8.

Hartelius, K. (1996). Analysis of irregularly distributed points. PhD thesis, Institut før Matematisk Modellering, Danmarks Tekniske Universitet, Lyngby, Denmark.

Hoheisel, J., Lennon, G., Zehetner, G., and Lehrach, H. (1991). Use of high coverage reference libraries of *Drosophila melanogaster* for relational data analysis. *J. Mol. Biol.* **220,** 903–914.

Hoheisel, J., Ross, M. T., Zehetner, G., and Lehrach, H. (1994). Relational genome analysis using reference libraries and hybridization fingerprinting. *J. Biotechnol.* **35,** 121–134.

Jain, A. K., and Dubes, R. C. (1988). *Algorithms for Clustering Data.* Prentice Hall, Englewood Cliffs, NJ.

Kel, A. E., Philipenko, M., Babenko, V., and Kolchcanov, N. A. (1996). Genetic algorithm for the selection of oligonucleotides for identification by hybridisation of genomic DNA fragments possessing gene potential. *In Proceedings of the German Conference on Bioinformatics (GCB'96)* (R. Hofestädt, T. Lengauer, M. Löffler, and D. Schomburg, Eds.), pp. 238–240. Universität Leipzig, Leipzig.

Lehrach, H., Drmanac, R., Hoheisel, J., Larin, Z., Lennon, G., Monaco, A. P., Nizetic, D., Zehetner, G., and Poustka, A. (1990). Hybridization fingerprinting in genome mapping and sequencing. *In Genome Analysis Volume 1: Genetic and Physical Mapping* (K. E. Davies and S. Tilghman, Eds.), Cold Spring Harbor Laboratory Press, pp. 39–81. Cold Spring Harbor, NY.

Lehrach, H., Bancroft, D., and Maier, E. (1997). Robotics, computing, and biology—An interdisciplinary approach to the analysis of complex genomes. *Interdisciplinary Sci. Rev.* **22,** 37–44.

Maier, E., Meier-Ewert, S., Ahmadi, A., Curtis, J., and Lehrach, H. (1994). Application of robotic technology to automated sequence fingerprint analysis by oligonucleotide hybridisation. *J. Biotechnol.* **35,** 191–203.

Maier, E., Hoheisel, J., McCarthy, L., Mott, R., Grigoriev, A., Monaco, A., Larin, Z., and Lehrach, H. (1992). Complete coverage of the *Schizosaccharomyces pombe* genome in the yeast artificial chromosomes. *Nature Genet.* **1,** 273–277.

Meier-Ewert, S., Maier, E., Ahmadi, A., Curtis, J., and Lehrach, H. (1993). An automated approach to generating expressed sequence catalogues. *Nature* **361,** 375–376.

Meier-Ewert, S., Rothe, J., Mott, R., and Lehrach, H. (1994). Establishing catalogues of expressed sequences by oligonucleotide fingerprinting of cDNA libraries. *In Identification of Transcribed Sequences* (U. Hochgeschwender and K. Gardiner, Eds.), pp. 253–260. Plenum, New York.

Meier-Ewert, S., Lange, J., Gerst, H., Herwig, R., Schmitt, A., Freund, J., Elge, T., Mott, R., Herrmann, B., and Lehrach, H. (1997). Comparative gene expression profiling by oligo fingerprinting. Submitted for publication.

Mirkin, B. (1996). *Mathematical Classification and Clustering.* Kluwer, Dordrecht.

Poustka, A., Pohl, T., Barlow, D. P., Zehetner, G., Craig, A., Michiels, F., Ehrich, E., Frischauf, A. M., and Lehrach, H. (1986). Molecular approaches to mammalian genetics. *Cold Spring Harbor Symp. Quant. Biol.* **51,** 131–139.

Schmitt, A. O., and Herzel, H. (1997). Estimating the entropy of DNA sequences. *J. Theor. Biol.* **188,** 369–377.

Zehetner, G., and Lehrach, H. (1994). The Reference Library System—Sharing biological material and experimental data. *Nature* **367,** 489–491.

30

COMPARATIVE GENOMIC HYBRIDIZATION

Koei Chin and Joe W. Gray

Comparative genomics hybridization (CGH) is a "genome scanning" technique that allows entire test genomes to be analyzed for changes in relative DNA sequence copy number (Kallioniemi *et al.*, 1992). In CGH, total genome DNA from a test sample (e.g., tumor DNA) and from a reference sample (typically normal genomic DNA) are labeled independently with different fluorochromes and cohybridized to normal chromosome preparations along with excess unlabeled *Cot-1* DNA (DeVries *et al.*, 1995; Kallioniemi *et al.*, 1994b). The general strategy is illustrated in Fig. 1. The ratio of the amounts of the two genomes that hybridize to each location on the target chromosomes is an indication of the relative copy number of the two DNA samples at that point in the genome. Typically, the test DNA is labeled with a green fluorescing dye (e.g., fluorescein) and the reference DNA is labeled with a red fluorescing dye (e.g., Texas red). In this case, regions of increased relative copy number gain in the tumor will have a higher green to red fluorescence ratio than average, and regions of reduced copy number will appear to have a lower green to red fluorescence ratio. The chromosomes are also usually stained with a blue fluorescing dye such as DAPI to elicit bands along the target chromosomes, thereby allowing chromosome

PCR Applications

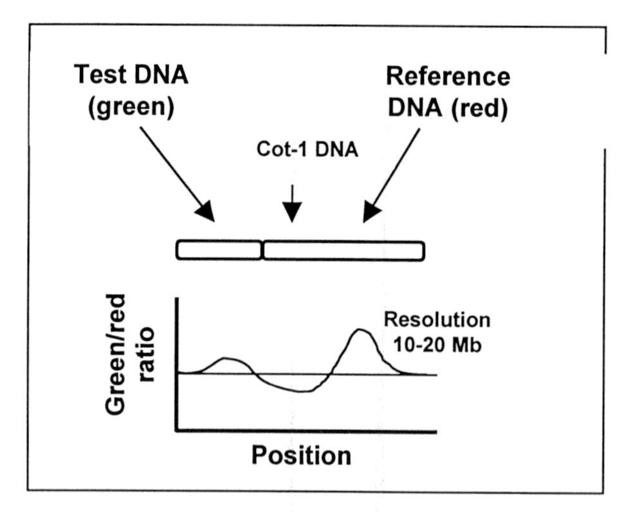

Figure 30.1 Schematic diagram of the steps in comparative genomic hybridization. Test and reference genomic DNA samples labeled so that they fluoresce green and red, respectively, are hybridized along with unlabeled *cot*-1 DNA to normal metaphase chromosomes. Green:red fluorescence ratios are measured along each chromosome as an indication of relative DNA sequence copy number.

identification. Alternately (or in addition), the chromosomes may be stained using fluorescence *in situ* hybridization with one or more chromosome-specific probes to aid in identification (Shi *et al.*, 1997). This may be particularly useful when analyzing mouse chromosomes (Donehower *et al.*, 1995). Important aspects of CGH include the size distribution of labeled test and references genomes after labeling, the quality of metaphase spreads to which hybridization is performed, and the efficiency with which test and reference genomes are labeled and hybridized (DeVries *et al.*, 1995; Du Manior *et al.*, 1995b; Kallioniemi *et al.*, 1994b). These aspects of CGH are discussed in detail in the following protocol.

CGH hybridizations are usually analyzed using digital imaging microscopy (Du Manoir *et al.*, 1995a,b; Piper *et al.*, 1995). In this process, three images of blue, green, and red fluorescence are acquired. The blue DAPI image is used for chromosome identification (often after contrast enhancement), whereas the green and red images are used to calculate profiles along each chromosome showing the ratios of green to red fluorescence. The green : red profiles are calculated after normalization of the green and red images to unit intensity and subtraction of the background fluorescence that underlies each chromosome image. Green : red fluorescence ratios are then calculated and displayed from pter to qter along each chromosome. Typically, about 10 chromosomes are analyzed per test and the results are combined to show the mean (±1 standard deviation). Often, regions of abnormal increased relative copy number are defined as those in which the average green to red ratio is >1.25 and regions of high-level copy number increase are defined as regions in which the ratio is >1.5. Regions of reduced relative copy number are defined as those in which the mean ratio is <0.75. However, the actual thresholds should be based on replicate analyses of normal samples (DeVries *et al.*, 1995; Kallioniemi *et al.*, 1994b). In situations in which the consequences of an error are high, statistical procedures based on the *t* test may be used to define regions that are significantly different from normal (Moore *et al.*, 1997). Aberrations involving centromeric regions and chromosomes 1p32-pter, 16p, 19, and 22 should be interpreted with caution since erroneous excursions of the ratio above or below commonly established thresholds sometimes occur in these regions (DeVries *et al.*, 1995; Kallioniemi *et al.*, 1994b). In addition, users should understand that chromosome-based CGH cannot detect single-copy losses or gains unless the extent of the region of loss/gain is greater than about 10 Mb. Moreover, the CGH ratio may not be a quantitative measure of the number of copies lost or gained unless the involved region is much greater than 10 Mb in extent.

The principal advantages of CGH are that it maps changes in copy number in complex genomes onto normal metaphase chromosomes so they can be easily related to physical maps and/or physically mapped genes, and it employs genomic DNA so that cell culture is not required. As a result, CGH has proven especially useful for analysis of genetic changes in solid tumors that contain maker chromosomes, physical deletions, homogeneously staining regions, and double-minute chromosomes of unknown origin (Kellieniemi *et al.*, 1994a). In addition, CGH has proven useful for detection of segmental aneusomies during pre- or neonatal diagnosis (Bryndorf *et al.*, 1995), especially in samples that are difficult to grow in culture.

Basic Protocol

The basic protocol is divided into five steps:

Step A: Preparation of metaphase spreads
Step B: Genomic DNA labeling

Step C: *In situ* hybridization
Step D: Image acquisition
Step E: Image analysis and profile calculation

Step A: Preparation of Metaphase Spreads

The quality of the normal metaphase spreads to which reference and test DNA samples are hybridized during CGH strongly influences the reliability and sensitivity of CGH analyses. Metaphase spreads are prepared from synchronized cultures of peripheral blood cells from a normal healthy male donor. T lymphocytes in RPMI 1640 medium are stimulated with phytohemagglutinin (PHA) and cultured for 3 or 4 days. The cells are then synchronized by successive treatment with methotrexate (MTX) to inhibit DNA replication and thymidine to release the cells more or less synchronously from the MTX-induced block. Cells in the synchronized cultures are accumulated in mitosis by treatment with colcemid. The mitotic cells are then swollen in hypotonic and fixed in methanol/acetic acid. Cells in fixative can be stored at 4°C for several days until needed for metaphase spread preparation. Metaphase spreads are prepared by dropping cells in fixative in one or two spots on precleaned microscope slides and air-dried under standard conditions (we recommend 45% humidity and 25°C). Temperature and humidity must be carefully controlled to obtain high-quality metaphase spreads for CGH. The exact conditions may vary between laboratories. Even with careful control, some slide preparations that appear morphologically of high quality (i.e., appear well spread and exhibit high-resolution Giemsa banding) may not be suitable for CGH. Thus, representative slides from each preparation should be tested for CGH using test and reference DNA probes from samples with known CGH characteristics. In particular, the test sample should carry known regions of aberration over which one copy of the genome has been gained or lost.

Equipment and Reagents

Vacutainer containing sodium heparin (Becton–Dickinson)

RPMI 1640 medium with 10% fetal bovine serum (FBS), 10 ml/liter penicillin–streptomycin (Life Technology)

PHA stock (Life Technology) reconstituted in 10 ml sterile deionized water (store at 4°C)

10 μM MTX: Dissolve 0.5 mg MTX (Sigma) in 100 ml Hank's

balanced salt solution and sterilize by filtration through a 0.22-μm filter (store at -20°C)

1 m*M* thymidine: Dissolve 25 mg thymidine (Sigma) in 100 ml Hank's balanced salt solution and sterilize by filtration through a 0.22-μm filter (store at -20°C)

10 μg/ml colcemid (Life Technology)

75 m*M* KCl; 0.56 g in 100 ml H_2O (prepare fresh)

Fixative; 3:1 (v/v) methanol/glacial acetic acid (prepare fresh)

T75 tissue culture flask (Corning)

15- and 50-ml sterile disposable conical polypropylene centrifuge tubes (Falcon)

Pasteur pipet

Premium microscope slides (Fisher Scientific)

Thermotron (A Venturedyne, Ltd.)

Sorvall RT7 centrifuge (Sorvall, Inc., Du Pont Instruments)

Method

A1. Collect peripheral blood by venipuncture into Vacutainers containing sodium heparin.

A2. Inoculate 2 ml of the whole blood obtained in step A1 into one T75 flask containing 25 ml RPMI 1640/10% FBS medium. Add 1 ml of reconstituted PHA in 10 ml.

A3. Incubate 3 days in 5% CO_2 incubator at 37°C. Agitate flask one or two times per day during incubation.

A4. Add 250 μl of 10^{-5} M MTX (10^{-7} M final concentration) 1 day prior to harvest (e.g., at 4–6 p.m.). Incubate 14–16 hr (overnight).

A5. Transfer into a 50-ml centrifuge tube, centrifuge 10 min at 1000 rpm, room temperature, and remove supernatant by aspiration.

A6. Resuspend cells in 23 ml of RPMI medium and 250 μl of 1 m*M* thymidine (10^{-5} M final concentration) on the following day (e.g., 6–10 a.m.), return to the T75 flask, and incubate 4–6 hr at 37°C.

A7. Add 200 μl of 10 μg/ml colcemid and incubate for 20 min at 37°C.

A8. Divide into two 15-ml centrifuge tubes and spin for 10 min at 1000 rpm at room temperature. Remove supernatant by aspiration.

A9. Add 10 ml of 75 m*M* KCl at 37°C and gently resuspend cells. Incubate 30 min in water bath at 37°C and add 0.5 ml of fixative. Note that the solution color will change from red to very dark red.

A10. Centrifuge 10 min at 1000 rpm at room temperature. Aspirate supernatant.

A11. Add 10–12 ml of fresh fixative and mix well using a Pasteur pipet. Incubate for 20 min at room temperature. Repeat centrifugation and fixation two to four times.

A12. Check the quality of the metaphase spreads under a phase-contrast microscope by test dropping onto precleaned premium slides (e.g., in the Thermotron at 45% humidity, 25°C). When preparations of well-spread, reasonably straight metaphase chromosomes have been obtained, continue slide-making.

A13. Place the slides at room temperature for 1 or 2 weeks after preparation. Test their quality by CGH analysis of DNA samples of known quality and chromosomal composition. Store at −20°C in nitrogen-filled sealed plastic bags until needed for CGH experiments.

Step B: DNA Labeling

Approximately 0.5–1 μg of test and reference DNA is labeled by nick translation. Male or female reference DNA is prepared from healthy donor's peripheral blood. Test DNA is isolated from fresh tissue or cultured cells, archived frozen tissue, or paraffin embedded tissue as desired. Test and reference DNA are labeled by nick translation with fluorescein–12-dUTP and Texas red–5-dUTP, respectively. The amount of DNA polymerase I, 10× BioNick enzyme mix, and/or incubation time may have to be adjusted slightly so that the labeled DNA fragments range in size from 500 to 2000 bp.

Equipment and Reagents

10× nucleotide mix
- 5 μl of 10 m*M* dATP (200 μ*M* final concentration)
- 5 μl of 10 m*M* dCTP (200 μ*M* final concentration)
- 5 μl of 10 m*M* dGTP (200 μ*M* final concentration)
- 125 μl 1 M Tris–Cl, pH 7.2 (500 μ*M* final concentration)
- 12.5 μ*M* $MgCl_2$ (200 μ*M* final concentration)

2.5 μl 10 mg/ml bovine serum albumin (100 μg/ml final concentration)
93.3 μl H_20 (final volume 250 μl)

Fluorescein–12-dUTP (Du Pont, NEN)
Texas red–5-dUTP (Du Pont, NEN)
DNA polymerase I (Life Technologies)
BioNick 10× enzyme mix (Life Technologies)
15 and 73°C water baths
1% agarose gel: 1 g agarose in 100 ml 1× TAE buffer
1× TAE buffer: 40 m*M* Tris–acetate, 1 m*M* EDTA

METHOD

B1. Add the following reagents to a microcentrifuge tube on ice:
 5 μl 10× nucleotide mix (20 μ*M* each dATP, dCTP, and dGTP final concentration)
 1 μl fluorescein–12-dUTP or Texas red–5-dUTP (1 n*M* final concentration)
 1 μg test or reference DNA
 1 μl DNA polymerase I
 3 μl BioNick 10× enzyme mix
 H_2O to bring volume to 50 μl

B2. Mix well and incubate 60–70 min at 15°C. Incubation time should be adjusted to produce the desired probe size (500–2000 bp).

B3. Stop the enzyme reaction by heating 10 min at 70°C.

B4. Electrophorese 5 μl of the reaction on 1% agarose gel in 1× TAE buffer along with a size marker, stain DNA with ethidium bromide, and check size of labeled DNA fragments under UV illumination.

Step C: *In Situ* Hybridization

Approximately 200 ng each of FITC-labeled test and Texas red-labeled reference DNA samples is hybridized to normal metaphase spreads along with 20 μg of unlabeled human *Cot-1* DNA. The *Cot-1* DNA is added to block hybridization of labeled, interspersed repeated DNA sequences. Labeled probes are ethanol precipitated, air-dried, resuspended in master hybridization solution, and denatured by heating to 73°C. Probes are applied to denatured metaphase spreads, coverslipped and incubated in a moist chamber at 37°C for 2 or 3 days.

After hybridization, spreads are washed and counterstained with DAPI in antifade solution.

Equipment and Reagents

20 μg/ml fluorescein-labeled test probe DNA (see step B)
20 μg/ml Texas red-labeled reference probe DNA (see step B)
0.5–1.0 μg/ml human *Cot-1* DNA (Life Technologies)
3 M sodium acetate, pH 5.2
75, 85, and 100% ethanol
Master hybridization solution
  5 ml formamide
  1.0 ml 20× SSC, pH 7.0
  1 g dextran sulfate (Sigma)
  Heat 2 or 3 hr at 70°C, vortezing periodically to dissolve solids
  Check pH with pH paper; adjust to 7.0 if necessary
  Bring final volume to 10 ml with H_2O
  Store in 1-ml aliquots at −20°C
Denaturation solution (70% formamide/2× SSC)
  70 ml formamide
  10 ml of 20× SSC, pH 7.0
  Bring final volume to 100 ml with H_20 and adjust pH 7.0
  Store at 4°C
20× SSC stock
  Dissolve 1753 g NaCl and 882 g sodium citrate in 8 liters H_2O
  Adjust pH to 6.3 with 1 *M* HCl
  Bring final volume to 10 liters with H_2O
Metaphase spreads (see step A)
Diamond pen
37°C slide warmer
45 and 75°C water baths
18-mm² coverslips
Rubber cement
37°C moist chamber
Wash buffer (50% formamide/2× SSC, pH7)
  75 ml formamide
  15 ml of 20× SSC, pH 7.0
  Adjust pH to 7.0 if necessary with HCl
  Bring final volume to 150 ml with H_2O
  Store at 4°C.

PN buffer

Stock solution

Solution A: 0.1 *M* $Na_2HPO_47H_2O$/0.1% Nonidet P-40

Solution B: 0.1 *M* $NaH_2PO_4H_2O$/0.1% Nonidet P-40

A working solution is prepared by slowly adding solution B to solution A to bring the pH to 8.0. The final ratio of solution B to solution A should be approximately 5:1 (v/v). This may be stored at room temperature.

Antifade mounting medium

Dissolve 100 mg *p*-phenyleneediamine dihydrochrolide (Sigma) in 10 ml H_2O

Adjust to pH 8.0 with approx 10 ml of 0.5 *M* bicarbonate buffer (0.42 g $NaHCO_3$ in 10 ml H_2O, pH 9.0 with NaOH)

Filter through 0.22 μm to remove undissolved particulates

Add 90 ml glycerol

DAPI in antifade mounting medium

1 m*M* DAPI stock solution

Dissolve 1 mg DAPI in 2.86 ml H_2O

A working solution is prepared by mixing 0.4 μl of 1 m*M* DAPI stock solution with 1 ml antifade mounting medium. This should be stored in the dark at −20°C.

Method

C1. Add the following reagents to a microcentrifuge tube

20 μl human *Cot-1* DNA

10 μl 20 ng/μl fluorescein-labeled test probe DNA

10 μl 20 ng/μl Texas red-labeled reference probe DNA

4 μl (one-tenth vol) 3 *M* sodium acetate (pH 5.2)

100 μl (2.5× vol) 100% ethanol

C2. Mix well and microcentrifuge at 14,000 rpm at 4°C.

C3. Discard supernatant and allow pellet to air-dry for 10 min.

C4. Add 10 μl of master hybridization solution and resuspend pellet gently. Store mixture in the dark at room temperature until needed.

C5. Select a slide carrying high-quality metaphase spreads (see step A) and mark the location(s) of the metaphase spreads on the back of the slide using a diamond pen.

C6. Incubate slides in 2× SSC in a Coplin jar at 37°C in a water bath.

C7. Dehydrate slides by incubating them in 70, 85, and 100% ethanol in Coplin jars at room temperature (2 min each). The slides should be stacked vertically on blotting paper and allowed to air-dry.

C8. Prewarm slides 2 min on a slide warmer at 37°C and immerse for 2.5–10 min in denaturation solution in a Coplin jar prewarmed to 73°C in a water bath.

C9. Dehydrate slides by incubating them in 70% ethanol at −20°C and then in 85 and 100% ethanol at room temperature. All incubations should be in a Coplin jar for 2 min each. The slides should be stacked vertically on blotting paper and allowed to air-dry. Place slides on slide warmer at 37°C.

C10. Denature probe mix from step C4 by heating for 5 min at 73°C, and then immediately place denatured probe mix onto the inscribed area of each slide. Cover with 18 mm^2 coverslip and seal coverslip edges with rubber cement.

C11. Incubate in moist chamber in the dark for 2 or 3 days at 37°C.

C12. Peel off rubber cement and remove coverslips gently.

C13. Wash slides by immersing successively in Coplin jars at 45°C water bath for 10 min each in three changes of hybridization wash buffer and then for 10 min in 2× SSC.

C14. Wash slides by immersing successively for 10 min each in two changes of PN buffer at room temperature and then successively for 5 min each in two changes of distilled water.

C15. The slides should be stacked vertically on blotting paper and allowed to air-dry.

C16. Apply 10 μl of 0.4 μM DAPI in antifade to each inscribed area on slides and cover with a 24 × 50-mm coverslip. Store in a refrigerator at 4°C until image acquisition.

Step D: Image Acquisition and Analysis

Blue (DAPI), green (FITC), and red (Texas red) images can be acquired using any of several different image acquisition systems. Images should be acquired for 8–10 metaphase spreads for each hybridization. Three-color images are analyzed to determine green and red fluorescence intensity ratios along each chromosome. Analysis typically entails intensity normalization of green and red images, chromosome segmentation, background subtraction, medial axis calculation, integration of fluorescence intensity in bands perpendicular to the medial axis across each chromosome, and calculation of green : red ratios along each medial axis. These ratios indicate relative DNA sequence copy number at each point in the test genome. Approximately 10 metaphases should be analyzed per hybridization and averaged.

Equipment and Reagents

Several systems for image acquisition and CGH analysis are commercially available. Vendors include
Vysis, Inc. (Downers Grove, IL)
Perceptive Scientific Instruments, Inc. (League City, TX)
Applied Imaging (Santa Clara, CA)

Method

D1. Assess the quality of the hybridization by visual examination in a fluorescence microscope using filters appropriate for DAPI, fluorescein, and Texas red. Spreads are typically viewed using a 63× oil-immersion objective lens. Test (green) and reference (red) hybridizations should be smooth with uniform, low background. Test and references hybridization intensities should be very low at the chromosome centromeres. The CGH hybridization images should be similar for the two homologous chromosomes in each metaphase spread. DAPI fluorescence intensity should be high and bands should be visible.

D2. Acquire digital images of blue (DAPI), green (fluorescein), and red (Texas red) from 8–10 metaphases.

D3. Analyze the resulting three-color images to determine hybridization intensity, hybridization homogeneity, and signal-to-background ratio.

D4. Identify chromosomes based on DAPI staining and banding pattern, and eliminate overlapping or sharply bent chromosomes.

D5. Calculate green : red fluorescence ratios along the chromosome medial axis in each chromosome (after background correction and normalization).

D6. Calculate mean green : red fluorescence ratio profiles for each chromosome type by combining the individual ratio profiles from multiple images. It is helpful to display the mean ± 1 standard deviation for the combined measurements.

D7. Identify regions of increased or decreased relative DNA copy number. This may be accomplished by identifying regions where the ration mean exceeds or falls below preselected thresholds (often 1.25 and 0.75, respectively) or using statistical approaches as described previously.

References

Bryndorf, T., Kirchhoff, M., Rose, H., Maahr, J., Gerdes, T., Karhu, R., Kallioniemi, A., Christensen, B., Lundsteen, C., and Philip, J. (1995). Comparative genomic hybridization in clinical cytogenetics. *Am. J. Hum. Genet.* **57,** 1211–1220.

DeVries, S., Gray, J. W., Pinkel, D., Waldman, F. M., and Sudar, D. (1995). *Comparative Genomic Hybridization. Current Protocols in Human Genetics.* Suppl. 6, Unit 4.6, pp. 1–18. Wiley, New York.

Donehower, L. A., Godley, L. A., Aldaz, C. M., Pyle, R., Shi, Y.-P., Pinkel, D., Gray, J., Bradley, A., Medina, D., and Varmus, H. E. (1995). Deficiency of p53 accelerates mammary tumorigenesis in *Wnt-I* transgenic mice and promoted chromosomal instability. *Genes Dev.* **9,** 882–895.

Du Manoir, S., Kallioniemi, O.-P., Lichter, P., Piper, J., Benedetti, P., Carouthers, A., Fantes, J., Garcia-Sagredo, J., Gerdes, T., Giollant, M., Hemery B., Isola, J., Maahr, J., Morrison, L., Perry, P., Stark, M., Sudar, D., van Vliet, L., Verwoerd, N., and Vrolijk, J. (1995a). Hardware and software requierments for quantitative analysis of comparative genomic hybridization. *Cytometry* **19,** 4–9.

Du Manior, S., Schrock, E., Bentz, M., Speicher, M., Joos, S., Ried, T., Lichter, P., and Cremer, T. (1995b). Quantitative analysis of comparative genomic hybridization. *Cytometry* **19,** 27–41.

Kallioniemi, A., Kallioniemi, O.-P., Sudar, D., Rutovitz, D., Gray, J., Waldman, F., and Pinkel, D. (1992). Comparative genomic hybridization for molecular cytogenetic analysis of solid tumors. *Science* **258,** 818–821.

Kallioniemi, A., Kallioniemi, O.-P., Piper, J., Tanner, M., Stokke, T., Chen, L., Smith, H. S., Pinkel, D., Gray, J. W., and Waldman, F. (1994a). Detection and mapping of amplified DNA sequences in breast cancer by comparative genomic hybridization. *Proc. Natl. Acad. Sci. USA* **91,** 2156–2160.

Kallioniemi, O., Kallioniemi, A., Piper, J., Isola, J., Waldman, F., Gray, J., and Pinkel, D. (1994b). Optimizing comparative genomic hybridization for analysis of DNA sequence copy number changes in solid tumors. *Genes Chromosomes Cancer 10,* 231–243.

Moore, D. H., II, Pallavicini, M., Cher, M. L., and Gray, J. W. (1997). A t-statistic for objective interpretation of comparative genomic hybridization (CGH) profiles. *Cytometry* **28,** 183–190.

Piper, J., Rutovitz, D., Sudar, D., Kallioniemi, A., Kallioniemi, O.-P., Waldman, F., Gray, J., and Pinkel, D. (1995). Computer image analysis of comparative genomic hybridization. *Cytometry* **19,** 19–26.

Shi, Y.-P., Naik, P., Dietrich, W. F., Gray, J. W., Hanahan, D., and Pinkel, D. (1997). DNA copy number changes associated with characteristic LOH in islet cell carcinoma of transgenic mice. *Genes Chromosomes Cancer* **19,** 104–111.

31

GENETIC FOOTPRINTING AND FUNCTIONAL MAPS OF THE YEAST GENOME

Tracy Ferea, Barbara Dunn, David Botstein, and Patrick Brown

Upon completion of the *Saccharomyces cerevisiae* genome it became clear that many of the genes obtained by the sequencing effort did not have a function inferable from their sequence. Approximately 44% of the genes have been characterized either experimentally or by their strong similarity to proteins or protein families with known functions (Cherry *et al.,* 1998; Mewes *et al.,* 1997). The majority of the genome currently consists of functionally uncharacterized open reading frames (ORFs). In *S. cerevisiae,* as with other organisms, gene function has typically been inferred by generating mutations in an individual gene and then testing fitness under a variety of conditions to detect a phenotype that suggests a function. Genetic footprinting is a highly parallel genomic approach that was developed to meet the challenge of determining gene function more efficiently (Smith *et al.,* 1996).

Genetic footprinting employs an insertional mutagenesis strategy (Fig. 1). A thorough mutagenesis is accomplished by transiently driving the expression of the Ty1 transposon from a regulatable promoter. The mutated population is then tested for fitness under a variety of selective growth conditions. Fitness is determined for each ORF retrospectively using PCR to compare the Ty1 insertion pattern at

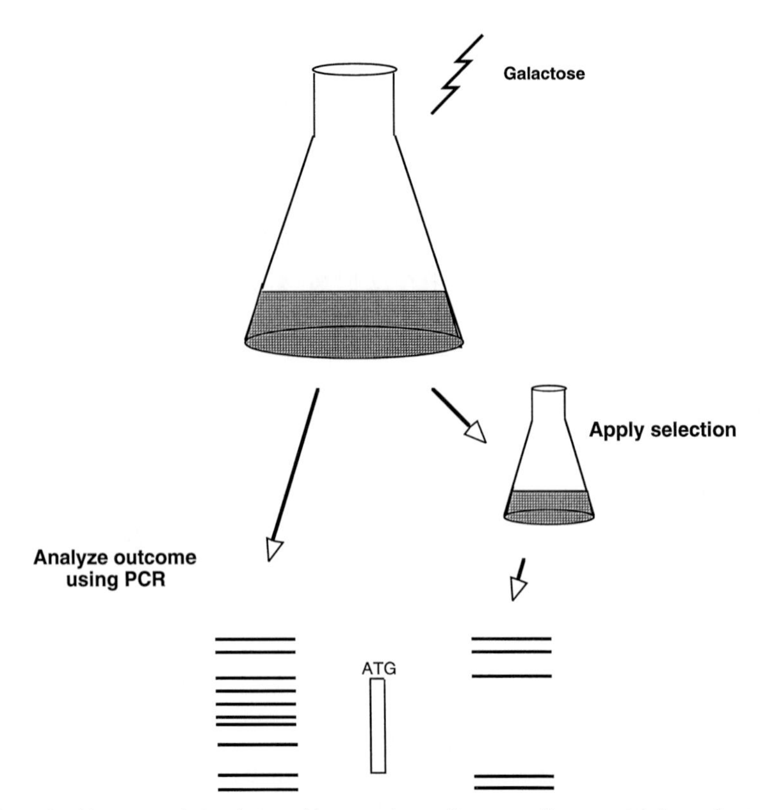

Figure 1 Mutagenesis is triggered by growing cells on medium containing galactose. Galactose induces the expression of the Ty1 transposon from the GAL1 promoter and subsequent insertion into the genome. For each gene, PCR is employed to compare the Ty1 insertion patterns both before and after growth on selective medium. Insertion into the coding region of a gene important for growth on selective medium results in a loss of cells containing those insertions from the growing population and a concomitant loss of PCR products.

each gene in the mutated population both before and after selection. When insertion in a gene results in a growth disadvantage, cells carrying insertions in that gene become underrepresented as the population expands. The relative depletion of these cells from the population results in a concomitant loss of their PCR products. PCR products from the selected population and the original mutagenized population are then compared. A growth disadvantage upon selection is seen as a "footprint" or loss of PCR products resulting from insertion within regions of the gene essential for function or expression. The pattern of PCR products from outside that region remains intact.

Genetic footprinting is efficient because both the mutagenesis and the selection processes are performed on large populations of cells. Therefore, these processes need to be performed only once and then PCR can be employed to analyze every gene for which a unique primer has been designed. When performed at the genome scale such an analysis is the equivalent of a saturation mutagenesis. The comprehensive nature and efficiency of the genetic footprinting method should expedite the genomic scale mapping of gene function.

Plasmid and Strain Construction

The plasmid pPBTy1 was constructed from two plasmids, pVIT41 and pJEF1562, obtained from Jef Boeke (Johns Hopkins, Boston). A 7.5-kb *Apa*I/*Xho*I fragment of pVIT41 was isolated. This fragment contained the Ty1 transposable element and a unique sequence complementary to the PBTy1R1 oligonucleotide (5′ AGAGCTCCCGG-GATCCTCTACTAAC 3′). This unique sequence serves as a molecular marker distinguishing the transposable element from endogenous transposons. After endonuclease digestion a 6.5-kb *Xho*I/*Apa*I fragment of pJEF1562 was isolated. The key features of the 6.5-kb fragment are the 2-μm-based plasmid backbone for high copy number, the *URA3* gene for use as a selectable marker, and the regulatable *Gal1* promoter. Ligation of the two fragments generated the plasmid pPBTy1.

A uracil auxotroph of the haploid yeast strain X2180-1A, a derivative of S288C, was selected on 5-fluoroorotic acid after transformation and homologous integration of a linear *Hin*dIII fragment of the *URA3* gene with an internal deletion of the 250-bp *Eco*RV/*Stu*I fragment. Effective induction of Ty1 transposition requires the strain to be GAL^{+}; therefore, the galactose permease gene (*GAL2*) which is defective in many strains of the S288C background was replaced by two-step gene replacement (Guthrie and Fink, 1991). Briefly, a YIp5 vector containing a 2.5-kb *Eco*RI/*Hin*dIII fragment of the *GAL2* gene was linearized within the *GAL2* coding region with *Bgl*II. The linear fragment with *GAL2* ends, for integration, also contained the *URA3* gene for selection. Integration at the *gal2* locus generated a duplication with a wild-type copy of the gene, flanked by the mutant copy. Plasmid excision was selected on 5-fluoroorotic acid and colonies were screened for the wild-type *GAL2* phenotype on

YP galactose medium containing antimycin A (10 μg/ml). Transformation with pPBTY1 completed the construction of the strain DBY7282 employed in the mutagenesis procedure.

Mutagenesis

In *S. cerevisiae,* the mutagenesis is performed by transiently expressing a marked Ty1 transposable element under the control of the regulatable *GAL1* promoter (Boeke *et al.*, 1985), from the high copy number plasmid pPBTy1. Exponential growth on SC galactose medium (Rose *et al.*, 1990) at 25°C for 20 generations induces sufficient transposition to obtain footprints for >97% of genes (Smith *et al.*, 1996). The efficiency of mutagenesis can be monitored by determining the number of canavanine-resistant mutants produced by insertion into the arginine permease gene *CAN1* (Rose *et al.*, 1990). Typically, a population with a measured mutant frequency of approximately 2×10^{-5} *can1*^{R} mutants/cell indicates sufficient transposition to obtain good footprinting results for other genes in the genome.

After a large-scale mutagenesis, DNA is isolated from at least 10^{12} cells. This cell number should yield enough DNA to analyze the effectiveness of the mutagenesis for nearly every gene in the genome. The remaining cells are frozen as glycerol stocks for future selection experiments in aliquots of $\geq 2 \times 10^{8}$ cells.

Selection

Typically a selection is initiated and maintained with $\geq 2 \times 10^{8}$ mutagenized cells to help prevent stochastic loss of mutants from the population (i.e., by random genetic drift). Cells are grown under the selective condition for 15–60 generations. The cells containing insertions into a gene important for growth under a particular selective conditions are depleted from the population. For example, when the overall population has doubled 15 times, a mutant growing at a rate of 80% of wild-type will have doubled only 12 times. Therefore, at 15 generations the mutant will be represented at $2^{12}/2^{15}$ or one-eighth of its original abundance in the mutagenized population. Thus, even more subtle growth defects can be detected by extending the number of generations the population is grown under selective pressure.

As with any selection the parameters of the experiment must be carefully designed. Therefore, pH, temperature, and oxygenation of the culture should be carefully controlled. Careful review of the components of the medium is also essential. The population should be maintained at a sufficient number of cells to avoid stochastic loss of individual mutations during the selection process. Unless the purpose of the experiment is to test the effects of mutations on cell viability during various phases of growth, exponential growth should be maintained throughout the selection process.

Large-Scale Yeast DNA Isolations

Large-scale DNA preparations from 1×10^{12} cells can be performed by scaling up the yeast DNA isolation procedure described by Rose *et al.*, (1990). Only slight modifications to the procedure are necessary.

1. Grow 10 liters of cells to approximately 1×10^{8} cells/ml. Centrifuge in a Sorvall GS3 rotor at 3000 rpm for 5 min (3 liters at a time). Discard the supernatant.
2. Wash cells by resuspending pellets in a total of 100 ml of 1.0 *M* sorbitol. Pool cells and centrifuge in GS3 rotor at 3000 rpm for 5 min. Discard the supernatant.
3. Resuspend the cells in 225 ml of 1.0 *M* sorbitol, 0.1 *M* Na_2 EDTA (pH7.5), and 14 m*M* β-mercaptoethanol.
4. Add 7.5 ml of a 2.5 mg/ml solution of zymolyase-100T (ICN Immunobiologicals, P. O. Box 1200, Lisle, IL 60532) dissolved in 1.0 *M* sorbitol, 0.1 *M* Na_2 EDTA (pH 7.5), and 14 m*M* β-mercaptoethanol. Incubate in a 2-liter flask at 37°C for 60 min, shaking gently (50 rpm).
5. Centrifuge the cells in a GSA rotor for 5 min at 3000 rpm. Discard the supernatant.
6. Resuspend the cell pellet in a total of 375 ml of 50 m*M* Tris–Cl (pH 7.4) and 20 m*M* Na_2 EDTA.
7. Add 37.5 ml of 10% SDS and mix.

8. Incubate at 65°C for 30 min in a 1-liter flask at 100 rpm.
9. Add 112.5 ml of 5 *M* potassium acetate and store on ice for 60 min.
10. Centrifuge in a GSA rotor at 11,000 rpm for 10 min. Transfer supernatant to a fresh tube and repeat centrifugation.
11. Transfer the supernatant to a fresh GSA tube and add 2 vol of 95% ethanol at room temperature. Mix and centrifuge at 11,000 rpm for 15 min at room temperature.
12. Discard the supernatant. Dry the pellet and then resuspend in 225 ml of 10 m*M* Tris–Cl (pH 7.4) and 1 m*M* EDTA. This may take several hours.
13. Add 1.25 ml of a 10 mg/ml solution of RNase A and incubate at 37°C for 30 min.
14. Add 1 vol of 100% isopropanol and gently mix. Remove the precipitate by stiring with a glass Pasteur pipet. The precipitate should wind around the pipet like a loose "cocoon" of fibers. Air-dry cocoon.
15. Wash with 70% ETOH, followed with a 95% ETOH wash. Resuspend in 10 m*M* Tris–Cl (pH 7.4) and 1 m*M* EDTA to a concentration of 500 ng/μl. Clarify the DNA solution by centrifugation in a Sorvall SS-34 rotor at 10,000 rpm for 15 min.

PCR

To examine a representative population of cells for each gene 1 μg of DNA (approximately 10^8 genomes) is used per PCR reaction. One primer is an oligonucleotide (PBTy1R1) complementary to a marked region of the *Ty1* element, that is not present in the endogenous element. The other is a gene-specific oligonucleotide that is fluorescently labeled at the 5′ end with FAM. A nearly complete set of primers for the *S. cerevisiae* genome is available from Research Genetics (2130 Memorial Parkway, Huntsville, AL 35801; *www.resgen.com*). The gene-specific primers were chosen to typically survey approximately 700 bp of coding region with a calculated T_M of 72°C by using the Design Primers program (*http://genome-www.stanford.edu/Saccharomyces/*). As a control for spurious products PCR analyses are also performed on DNA isolated from cells

(DBY7282) that have not been induced for Ty1 transposition, i.e., they were grown on SC glucose medium.

The PCR reaction volume is 50 μl and reactions are set up in 96-well plates.

1 μg DNA
1.5–2.5 m*M* MgCl
250 μ*M* each dNTP
50 m*M* KCl
10 m*M* Tris–HCl,pH 8.3
2 U *Taq* polymerase
0.64 μ*M* gene-specific oligonucleotide
0.38 μ*M* Ty1-specific oligonucleotide

The 50-μl PCR reaction mix is centrifuged to the bottom of the plate in a Beckman GS6 centrifuge set at 22°C, 600 rpm, for 1 min. The plate is then loaded into a preheated 93°C Perkin–Elmer Model 9700 thermocycler.

Thermocycling parameters
Denature at 93°C for 1 min
10 cycles of 92°C for 30 sec, 67°C for 45 sec, 72°C for 2 min
20 cycles of 92°C for 30 sec, 62°C for 45 sec, 72°C for 2 min

Analysis of PCR Products

PCR products are analyzed on Applied Biosystems DNA sequencing machines, Models 373 or 377, equipped with GeneScan DNA analysis software and with the XL upgrade. The advantage of the XL upgrade is that it expands the gel capacity to 64 samples per run. Notched microtrough gel plates with a 12-cm well-to-read distance for the ABI 373 and 6 cm for the ABI 377 are recommended (CBS Scientific, P. O. Box 856, Del Mar, CA 92014). The notches increase sample volume capacity, facilitating loading. A denaturing polyacrylamide gel matrix with slightly different compositions is used depending on the model of ABI machine: 4% Gene-Page Plus for the ABI 377 and 6% for the ABI 373 (Amresco, Inc., 30175 Solon Industrial Parkway, Solon, OH 44139).

Five microliters of each PCR product and 2 μl of loading buffer are combined in wells of a 96-well plate. Loading buffer is made beforehand and stored at −20°C in covered 96-well plates. Centrifuge at 1000 rpm to spin samples to bottom of plate and then denature at 95°C for 10 min. The denatured sample, now approximately 1 μl in volume, is loaded directly from the hot block onto the polyacrylamide gel using an eight-channel Hamilton syringe. Loading is staggered by loading every other lane and running the gel 2–5 min prior to the second loading; this staggers the molecular weight markers to facilitate lane tracking. Running parameters are 16 W for 8 hr for the ABI 373 or 640 V for 4 hr for the ABI 377.

Loading buffer

2 ml loading dye: 30 mg/ml dextran blue and 50 m*M* EDTA (pH 8)

1.8 ml GeneScan-2500 TAMRA molecular weight markers (Applied Biosystems, 850 Lincoln Center Dr., Foster City, CA 94404)

10 ml formamide

Interpretation

ABI GeneScan analysis software is employed to display electropherograms of the PCR products. Lane tracking is adjusted, and then the local Southern method (Southern, 1979) is used to examine the reciprocal relationship between mobility and fragment length to determine the molecular weights of PCR products relative to the GeneScan 2500 molecular weight standards. The accuracy of automated size calling should always be visually assessed. Although some PCR-to-PCR variability is seen in the pattern produced by genes that have relatively few insertions, the complex patterns of insertions found for most genes are well conserved even after selective growth. For example, electropherograms from the *MET10* gene (which encodes a protein necessary for methionine prototrophy) are shown in Fig. 2. The mutagenized cells (time 0) have a dense pattern of PCR products resulting from multiple independent insertions into the gene within the analyzed population. Growth on medium containing the nonfermentable carbon source lactate results in a pattern similar to that of the mutagenized population, suggesting that this gene is not essential for growth under this condition. However,

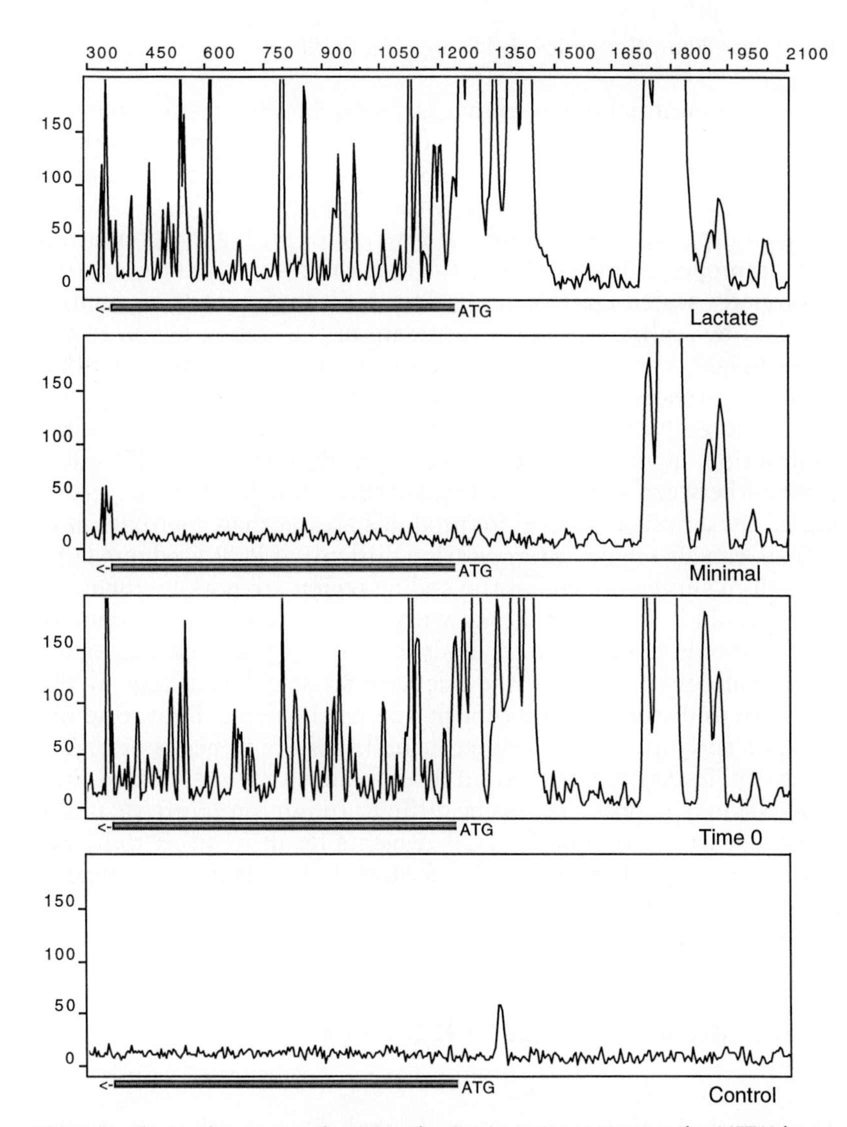

Figure 2 Electropherograms depicting the Ty1 insertion pattern at the *MET10* locus. Relative fluorescent intensity is indicated on the *y* axis and molecular weight on the *x* axis. Each electropherogram contains PCR products from cells grown as follows: the initial mutagenized population (time 0), unmutagenized cells (control), or cells from the mutagenized population after 21 population doublings on lactate or minimal selective medium. The area of the trace emcompassing the start codon (ATG) and first 1246 bases of the 3107-bp *Met10* coding region is depicted with a bar under each electropherogram. Note that peaks representing insertions in the *MET10* coding region are depleted after growth on minimal medium. The lack of significant peaks in the control indicates the specificity of the oligonucleotide primers.

growth on minimal medium that lacks methionine results in a loss of PCR products from both the coding region and the immediate upstream vicinity of the gene. This "genetic footprint" indicates that this gene is essential for growth in minimal medium.

The footprinting data should be interpreted with appropriate caution. Improper interpretation of the data can result when comparing improperly scaled electropherograms. It is important to normalize the data by scaling the peaks upstream of the coding region to the same height, the underlying assumption being that insertions far enough upstream of the coding region do not cause a selective growth disadvantage. Normalization compensates for differences in baseline subtraction, quantities loaded, and slight differences in PCR effectiveness between samples. Inadequate extension during PCR, resulting in low-amplitude signal for products greater than approximately 500 bp, can also make interpretation difficult. If PCR products large enough to survey outside of the coding region are not obtained, the PCR should be repeated. Peaks with relative fluorescent intensity of <50 should not be used for analysis or scaling because increasing their scale also increases the background signal resulting in the inability to distinguish a footprint among the noise. Repeating the PCR, or designing a new primer, usually results in peaks of higher intensity. Peaks that are too intense because they are located near integration hot spots can also be difficult to interpret. Intense peaks can cause exhaustion of the PCR reagents resulting in partially extended products that can mask a footprint. This is easily remedied by repeating the PCR with fewer cycles.

Application to Other Organisms

The genetic footprinting strategy is applicable to other organisms with minimal requirements: (i) a suitable insertional mutagen that can be used to generate enough independent mutations to saturate the genome; (ii) DNA sequence information sufficient to choose specific primers; (iii) a genome small enough that a population of mutagenized cells, sufficient to include a representative sample of mutations in each gene, can be present within every PCR reaction. Growth as a haploid or homozygous diploid is desirable if recessive phenotypes are to be analyzed. With the explosion of microbial sequencing projects, these requirements could easily be met for an increasing number of microorganisms. Modifications of the procedure permit footprinting of viral genomes or cloned DNA at a high resolution (Singh *et al.,* 1997).

Functional Maps

The ability to survey an entire genome for the contributions of each gene to cellular fitness under a variety of selective conditions makes it possible to generate comprehensive functional maps for an organism. The entire repertoire of genes essential for a specific regulatory mechanism, signal transduction, or biochemical pathway can be mapped with appropriately designed experiments. We initiated this process by examining several selective conditions that have been widely used in analysis of *S. cerevisiae* gene function. This has resulted in a functional map for the 268 genes that reside on chromosome V of *S. cerevisiae* (Smith *et al.,* 1996). A detectable mutant phenotype was found for 62% of these genes after analyzing only seven selections. It is only a matter of time before a complete functional map of the genome is made. With 31% of yeast genes having mammalian homologs (Botstein *et al.,* 1997) among the still rapidly growing Genbank entries, it is easy to imagine the impact of a eukaryotic functional map on the understanding of cognate genes in other organisms.

Acknowledgments

We thank Rebecca Koskela, Sandra Metzner, Edward Chung, and Katja Schwartz for critical reading of the manuscript.

References

Boeke, J. D., Garfinkel, D. J., Styles, C. A., and Fink, G. R. Ty elements transpose through an RNA intermediate. *Cell* **40**(3), 491–500.

Botstein, D., Chervitz, S. A., and Cherry, J. M. (1997). Yeast as a model organism. *Science* **277**(5330), 1259–1260.

Cherry, J. M., Adler, C., Ball, C., Chervitz, S. A., Dwight, S. S., Hester, E. T., Jia, Y., Juvik, G., Roe, T., Schroeder, M., Weng, S., and Botstein, D. (1998). SGD: Saccharomyces Genome Database. *Nucleic Acids Res.* **26**(1), 73–79.

Guthrie, C., and Fink, G. R. (1991). *Methods in Enzymology: Guide to Yeast Genetics and Molecular Biology,* Vol. 194, Academic Press, San Diego.

Mewes, H. E., Albermann, K., Bahr, M., Frishman, D., Gleissner, A., Hani, J., Heumann, K., Kleine, K., Maieri, A., Oliver, S. G., Pfeiffer, F., and Zollner, A. (1997). Overview of the yeast genome. *Nature,* **387** (Suppl.), 7–65.

Rose, M. D., Winston, F., and Hieter, P. (1990). *Methods in Yeast Genetics: A Laboratory Course Manual.* Cold Spring Harbor Laboratory Press, Plainview, NY.

Singh, I. R., Crowley, R. A., and Brown, P. O. (1997). High-resolution functional mapping of a cloned gene by genetic footprinting. *Proc. Natl. Acad. Sci. U.S.A.* **94**(4), 1304–1309.

Smith, V., Botstein, D., and Brown, P. O. (1995). Genetic footprinting: A genomic strategy for determining a gene's function given its sequence. *Proc. Natl. Acad. Sci. U.S.A.* **92**(14), 6479–6483.

Smith, V., Chou, K. N., Lashkari, D., Botstein, D., and Brown, P. O. (1996). Functional analysis of the genes of yeast chromosome V by genetic footpringing. *Science* **274**(5295), 2069–2074.

Southern, E. M. (1979). Measurement of DNA length by gel electrophoresis. *Anal. Biochem.* **100**(2), 319–323.

32

MOLECULAR ANALYSIS OF MICRODISSECTED TISSUE: LASER CAPTURE MICRODISSECTION

Nicole L. Simone, Jeffrey Y. Lee, Mary Huckabee, Kristina A. Cole, Rodrigo F. Chuaqui, Chetan Seshadri, Lance A. Liotta, Bob Bonner, and Michael R. Emmert-Buck

With the advent of PCR, and the development of high-throughput, automated microhybridization arrays, and mutation screening methods, DNA or RNA can be extracted from tissue biopsies and analyzed with a parallel panel of hundreds or even thousands of genetic markers. However, even the most sophisticated genetic testing methods will be of limited value when they are applied to tissue if the input DNA, RNA, or protein is not derived from pure populations of cells or is contaminated by the wrong cells. At the microscopic level, tissues are composed of complicated interacting and interdependent cell populations, which are regulated by the local extracellular matrix. Consequently, the task of analyzing critical gene expression patterns in development, normal function, and disease progression depends on the extraction of specific cells from their complex tissue milieu. Laser capture microdissection (LCM) (Emmert-Buck *et al.*, 1996; Bonner *et al.*, 1997) has been developed to provide a rapid, reliable method to procure pure populations of selected cells from specific microscopic regions of tissue sections for molecular analysis.

Several methods have been developed to microdissect stained tissue sections and amplify the procured material with PCR (Fearon *et al.*, 1987; Radford *et al.*, 1993; Emmert-Buck *et al.*, 1994). The procurement tool can be a pipet, pointed probe, fine needle, or blade (Shibata *et al.*, 1992) either hand-held or connected to a micromanipulator arm (Whetsell *et al.*, 1992). Boehm *et al.*, (1997) placed the tissue section on an ultrathin film and then used a UV laser to cut out the tissue field of interest. LCM, a new concept in microdissection, employs a transfer surface which is focally bonded to the targeted tissue (Emmert-Buck *et al.*, 1996) (Fig. 1). The operator views the tissue in a standard microscope through a glass slide, selects microscopic clusters of cells for analysis, and then pushes the button which activates a laser within the microscope optics. The pulsed laser beam is absorbed by the transfer film immediately above the targeted cells. At this precise location the film melts and fuses with the underlying cells of choice. When the film is removed, the chosen cell(s) remains bound to its undersurface while the rest of the tissue is left behind. Only the targeted cells are captured with a precision of transfer which can approach 1 μm. The remaining tissue on the slide remains fully accessible for further capture so that comparative

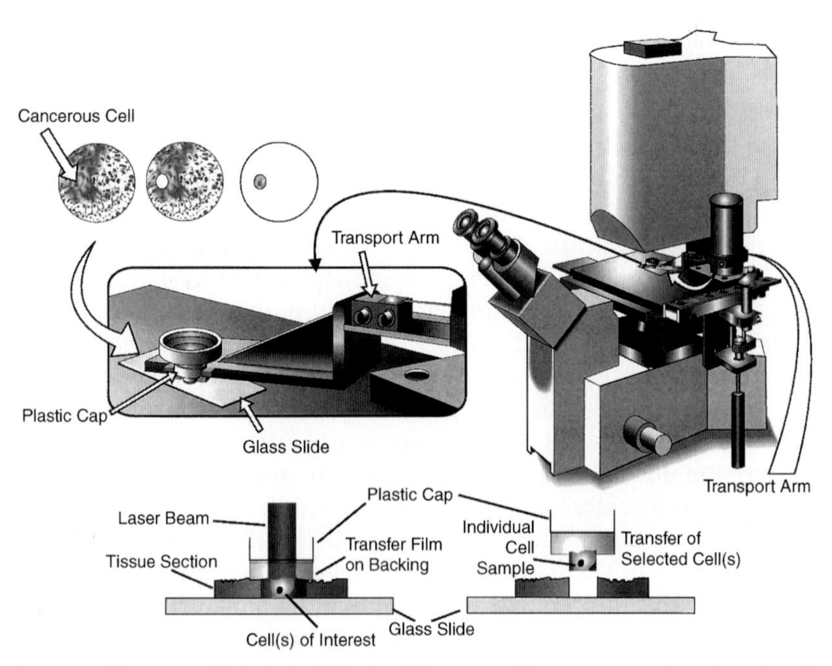

Figure 1 Elements of laser capture microdissection. The commercial version Pixcell developed by Arcturus, Inc. is depicted.

molecular analysis can be performed on adjacent cells. No micromanipulation is required. The distinct morphology of the procured cells is retained and held on the transfer film, providing a diagnostic record of the cells chosen for molecular analysis. Because of the thermoplastic bonding principle of LCM, chemical reactions which can cross-link biological molecules in the tissue and alter subsequent molecular analysis are avoided. Focusing the laser diode beam using the microscope objectives can focally activate the polymer in spots as small as 4 μm. The cells transferred to the film are firmly embedded but are also fully accessible to aqueous solutions necessary for the extraction of DNA, RNA, and proteins.

LCM has been successfully applied to DNA, RNA, and protein analysis from stained microdissected tissue cells (Fig. 2, see color insert). LCM microscopes and capture film technology are commercially available from Arcturus Engineering, Inc. (Santa Clara, CA). Readers interested in the details of the LCM microscope are referred to the manufacturer's instructions. Protocols for preparing the tissue before microdissection and conducting the PCR amplification after microdissection are provided here.

Protocols

Tissue Preparation

Histological slides are an ideal substrate for microdissection: They allow microscopic visualization and two-dimensional manipulation of a single cell layer obtained from a tissue specimen. Traditionally, fixation prior to embedding is performed to preserve the morphology of the tissue. While the routine fixative formalin preserves the tissue morphology, it cross-links RNA and DNA to protein, limiting the analysis of nucleic acids (Mies, 1994). Alcohol-based fixatives maintain the integrity of the nucleic acids and transfer well by LCM (Foss, *et al.*, 1994).

Once fixed, the tissue is embedded in paraffin. Paraffin processing can be performed via a routine overnight or accelerated cycle in an automated tissue processor. Once embedded, thin sections are mounted onto glass slides. Consistent LCM transfers have been demonstrated from 5- to 10-μm-thick paraffin-embedded tissue sections.

For a successful LCM transfer, the strength of the bond between polymer film and the targeted tissue must be stronger than that between the tissue and the underlying glass slide. Therefore, techniques which increase the adhesion of the tissue section to the glass slide should be avoided. These include the sectioning of tissue onto charged or positive glass slides and the use of an adhesive in the water bath prior to tissue mounting.

To prevent cross-contamination while sectioning, paraffin and tissue fragments should be wiped from the area with xylenes between each slide. A fresh microtome blade should be used for each block. The prepared tissue slides should be stored at room temperature and in a low-humidity environment until required for LCM transfer. Frozen tissues embedded in OCT are cut in a −20°C cryostat at 5–10 μm.

Staining

Staining should be performed as near as possible to the scheduled LCM transfer time using solution baths that are replaced regularly. Staining should be performed as follows (staining of frozen sections begins at step 4):

1. Xylenes to deparaffinize the slides (5 min ×2)
2. 100% ethanol wash
3. 95% ethanol wash
4. 70% ethanol wash
5. Purified water wash
6. Mayer's hematoxylin (1 or 2 min)
7. Purified water wash
8. Bluing reagent (30–60 sec)
9. 70% ethanol wash
10. 95% ethanol wash
11. Eosin Y (1 or 2 min)
12. 95% ethanol wash (×2)
13. 100% ethanol wash
14. Xylene wash for 1 min
15. Air-dry for at least 2 min to allow the xylene to evaporate completely

Poor transfers may result if the slide is not fully dehydrated (i.e., the 100% ethanol becomes hydrated after repeated use) or if the xylenes do not completely deparaffinize the slides. The final xylene

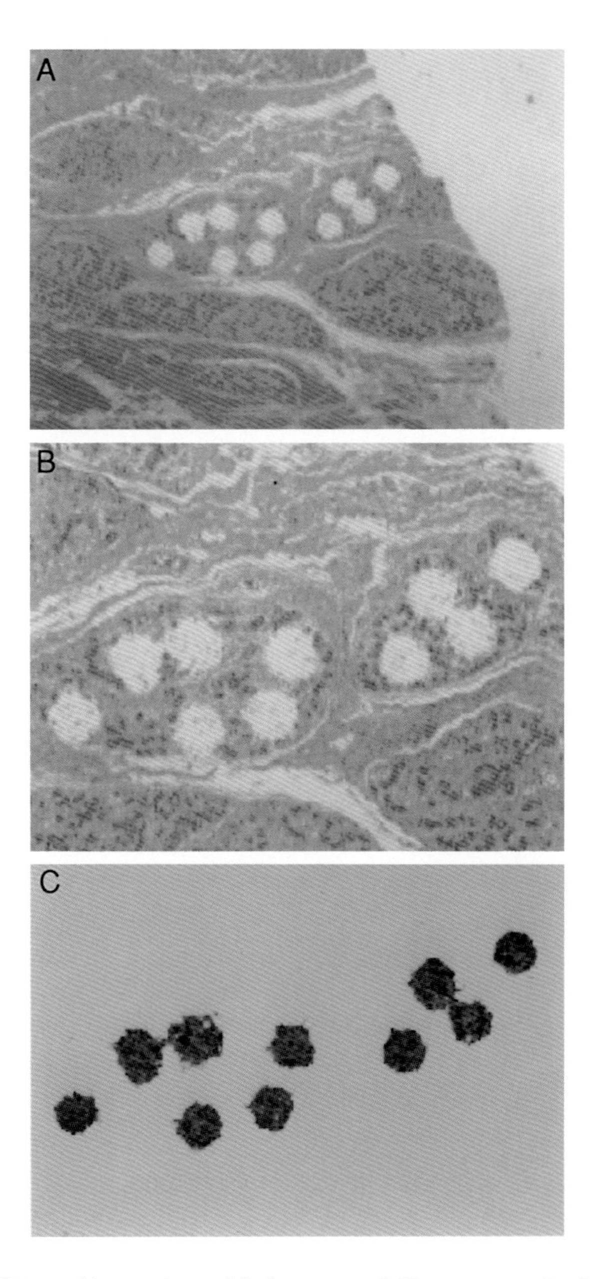

Fig. 2. (A) Normal human breast lobules surrounded by stroma, stained with hematoxylin and eosin (magnification 100×). (B) Sixty-micrometer diameter circular transfer voids left after LCM procurement from selected lobules (magnification 400×). (C) Cellular transfers adherent to polymer transfer film (magnification 400×).

rinse facilitates the efficiency of transfer with LCM. If a tissue section does not transfer well, a longer xylene rinse may help. While other staining protocols can be used, the slides should be dehydrated in a final xylene step.

The tissue section must be dessicated and not coverslipped for effective LCM transfer. The staining appears darker and more granular due to light scattered from the irregular air–tissue interface. The tissue on which the polymer melts and bonds after laser activation appears lighter and resembles a coverslipped slide due to the replacement of the air in the tissue with the polymer. This phenomenon is called index matching or "polymer wetting."

LCM

Follow the protocol from the LCM manufacturer. Consistent transfers (>95% transfer rate) have been demonstrated using a 60-μm beam diameter at 30 mW and a 30-μm beam diameter at 20 mW.

Due to the extreme sensitivity of PCR, much of the LCM procedure has been designed to reduce investigator contact with the tissue. For example, Arcturus has attached the transfer polymer to the undersurface of a vessel cap to simplify handling. The film is bonded to an optically clear plastic cap that can be accurately manipulated by a mechanical arm and fits into a 0.5-ml Eppendorf microfuge tube.

Molecular Analysis

PCR protocols are not altered for LCM, although the decreased starting template may require more cycles or the use of radioactive incorporation for improved detection. The extraction and isolation procedures are scaled down. For example, when using proteinase K for DNA extraction, decreasing the buffer volume will result in a final DNA PCR template concentration comparable to that of the full-scale procedures.

DNA Extraction

After microdissection, the cap is inserted into an Eppendorf tube containing digestion buffer [50 μl buffer containing 0.04% proteinase K, 10 m*M* Tris–HCl (pH 8.0), 1 m*M* EDTA, and 1% Tween-20]. The tube is placed in a 37°C oven to equilibriate. It is then placed upside down so that the digestion buffer contact the tissue on the cap. The incubation continues overnight at 37°C. The tube is centrifuged for

5 min and the cap is removed. The reaction is heated to 95°C for 8 min to inactivate the proteinase K. It can then be used directly as template for PCR. An example of a 10.0-μl PCR reaction is as follows: 1.0 μl 10X PCR reaction buffer, 1.5 μl DNA template, 0.8 μl 25 μM dNTP, 0.2 μl 10 μM 3′ primer, 0.2 μl 10 μM 5′ primer, 0.2 μl 5 U/μl *Taq* DNA polymerase, 0.1 μl 20 μCi/μl ^{32}P-labeled dCTP, and 6.0 μl DEPC-treated H_2O. Thermal cycler parameters are 94°C for 1 min, 35 cycles of 45 sec at 94°C, 45 sec at the annealing temperature, 2 min at 72°C, followed by a final extension for 2 min at 72°C.

RNA Analysis

The cap is inserted in a 0.5-ml Eppendorf tube containing 200 μl guanidinium isothiocyanate and 1.6 μl β-mercaptoethanol. Since this buffer rapidly lyses cells, the tissue only needs to be exposed to it for a few minutes. The RNA is extracted by adding 0.1X vol 2M sodium acetate (pH 4.0), 1X vol water-saturated phenol, and 0.3X vol chloroform-isoamyl alcohol. The aqueous and organic phases are separated by incubation on ice for 15 min, followed by centrifugation at 4°C for 30 min. The upper aqueous layer is placed into a fresh tube and the RNA is precipitated at −80°C for 30 min in 1X vol isopropanol containing 10 μg glycogen carrier. The RNA is then pelleted by spinning at 14,000 rpm for 30 min at 4°C.

Microdissected RNA often contains contaminating DNA. It is therefore important to DNAase treat the RNA for certain protocols [e.g., reverse transcriptase (RT)-PCR] with primers that would amplify DNA or library production requiring blunt-end ligation. It is important that the DNAase is certified as RNAase free.

DNAse treatment: To the pellet add 15 μl DEPC-H_2O, 1 μl 20 U/μl RNAse inhibitor, and 2 μl 10X reaction buffer. Once the pellet is resuspended, add 2 μl 10 U/μl DNAse I. Incubate at 37°C for 2 hr and then reextract and pellet the RNA as described previously.

Sample random-primed RT-PCR: Resuspend the pellet in 24 μl DEPC–H_2O with 1 μl 20 U/μl RNAase inhibitor. Aliquot 12 μl RNA into two tubes for the (+) and (−) RT reactions. Add 4 μl 5X RT buffer, 2 μl 250 μM dNTP, and 1 μl 20 μM random hexamers. Incubate for 5 min at 65°C and then 10 min at 25°C. Add 1 μl 100 U/μl MMLV to the (+) RT tubes only. Add 1 μl DEPC–H_2O to the (−) RT tubes. Continue the incubation for an additional 10 min at 25°C. Heat for 50 min at 37°C followed by 5 min at 95°C. If possible, perform the 10-μl PCR reaction described previously while the cDNA is fresh.

PCR amplification of cDNA from the (−) RT reaction would indicate that contaminating DNA is being amplified.

Applications

DNA obtained from cells procured by LCM from clinical specimens has been analyzed by a variety of methods, including loss of heterozygosity (Vocke *et al.*, 1996), single-stranded conformational polymorphism, dideoxy fingerprinting, clonal analysis, and direct sequencing (Chandrasekharappa *et al.*, 1997).

Similarly, LCM has been used to study expression differences in various tissues using RNA-based techniques, such as cDNA library construction (Krizman *et al.*, 1996), microarray hybridization, and differential gene expression (Chuaqui *et al.*, 1997).

References

Boehm, M., *et al.* (1997). Microbeam MOMeNT: Non-contact laser microdissection of membrane mounted native tissue. *Am. J. Pathol.* **151.**

Bonner, R. F., *et al.* (1997). Laser capture microdissection: Molecular analysis of tissue. *Science* **278.**

Chandrasekharappa, S. C., *et al.* (1997). Positional cloning of the gene for multiple endocrine neoplasia-type 1. *Science* **276,** 404–407.

Chuaqui, R. F., *et al.* (1997). Identification of a novel transcript up-regulated in a clinically aggressive prostate carcinoma. *Urology* **50,** 302–307.

Crisan, D., and Mattson, J. C. (1993). Retrospective DNA analysis using fixed tissue specimens. *DNA Cell Biol.* **12,** 455–464.

Emmert-Buck, M. R., *et al.* (1994). Increased gelatinase A (MMP-2) and cathepsin B activity in invasive tumor regions of human colon cancer samples. *Am. J. Pathol.* **145,** 1285–1290.

Emmert-Buck, M. R., *et al.* (1996). Laser capture microdissection. *Science* **274,** 998–1001.

Fearon, E. R., Hamilton, S. R., and Vogelstein, B. (1987). Clonal analysis of human colorectal tumors. *Science* **238,** 193–197.

Foss, R. D., *et al.* (1994). Effects of fixative and fixation time on the extraction and polymerase chain reaction amplification of RNA from paraffin-embedded tissue. Comparison of two housekeeping gene mRNA controls. *Diagn. Mol. Pathol.* **3,** 148–155.

Krizman, D. B., *et al.* (1996). Construction of a representative cDNA library from prostatic intraepithelial neoplasia. *Cancer Res.* **56,** 5380–5383.

Mies, C. (1994). Molecular biological analysis of paraffin-embedded tissues. *Hum. Pathol.* **25,** 555–560.

Radford, D. M., *et al.* (1993). Allelic loss on a chromosome 17 in ductal carcinoma in situ of the breast. *Cancer Res.* **53,** 2947–2949.

Schena, M., *et al.* (1995). Quantitative monitoring of gene expression patterns with a complementary DNA microarray. *Science* **270,** 467–470.

Shibata, D., *et al.* (1992). Specific genetic analysis of microscopic tissue after selective ultraviolet radiation fractionation and the polymerase chain reaction. *Am. J. Pathol.* **141,** 539–543.

Vocke, C. D., *et al.* (1996). Analysis of 99 microdissected prostate carcinomas reveals a high frequency of allelic loss on chromosome 8p12–21. *Cancer Res.* **56,** 2411–2416.

Whetsell, L., *et al.* (1992). Polymerase chain reaction microanalysis of tumors from stained histological slides. *Oncogene* **7,** 2355–2361.

33

AMPLIFIED FRAGMENT LENGTH POLYMORPHISM: STUDIES ON PLANT DEVELOPMENT

Emmanuel Liscum

During recent years tools of molecular genetics have been been used to answer questions about nearly every aspect of plant biology. Plant developmental biology is clearly one area that has benefited greatly from this "boom" in the use of molecular genetic techniques (Chory *et al.*, 1996, Meyerowitz, 1997a,b; Taylor, 1997). New technologies are being used to address problems in plant development almost as fast as they are developed. The amplified fragment length polymorphism (AFLP) technology[1] discussed in this chapter represents one such development. AFLP was designed (Zabeau and Vos, 1993; Vos *et al.*, 1995) as a nucleic acid fingerprinting method to exploit molecular genetic variations existing between closely related genomes in the form of restriction fragment length polymorphisms. As a DNA fingerprinting method for broad-application comparative genome analysis, AFLP likely has no rival, with the obvious exception of

[1] AFLP technology was developed and patented by Keygene n.v. (Agrobusiness Park 90, P. O. Box 216, NL-6700 AE Wageningen, The Netherlands). Perkin–Elmer (Applied Biosystems Division, Foster City, CA) has exclusive rights to this technology. Life Technologies, Inc. (Gathersberg, MD) also markets a research kit for AFLP fingerprinting of plant DNAs (under license from Keygene n.v.).

large-scale DNA sequencing. The number of fingerprints observed is 10- to 100-fold greater with AFLP than with other commonly employed methods. Furthermore, this wealth of information is derived essentially without the shortcomings commonly associated with other methods (Thomas *et al.*, 1995; Vos *et al.*, 1995; Bachem *et al.*, 1996; Habu *et al.*, 1997).

With respect to studies of plant development, the two most obvious applications of AFLP technology are in the generation of high-resolution genetic maps for use in positional cloning strategies (Thomas *et al.*, 1995; Cnops *et al.*, 1996; Büschges *et al.*, 1997; Huala *et al.*, 1997; Liscum and Oeller, 1999) and as a means to identify differentially expressed genes (Bachem *et al.*, 1996; Habu *et al.*, 1997). Although no one conventional methodology would be used to address such seemingly unrelated tasks, AFLP provides a superior method for both and requires only minor modifications of the basic protocol described here (see Variations).

Method

The basic AFLP protocol (Vos *et al.*, 1995) can be separated into four sequential steps (Fig. 1; see also Protocol). During the first step DNAs (either genomic DNAs or cDNAs depending on the question being addressed) are isolated from the sources one wishes to compare and then digested with appropriate restriction endonucleases (REs). For most plant DNAs (or cDNAs), two REs are used in the digestion—one a rare-cutter having a 6-bp recognition site (e.g., *Eco*RI) and the other a frequent cutter with a 4-bp recognition site (e.g., *Mse*I).

In the second step of AFLP, specific double-stranded (ds) oligodeoxynucleotide adapters are ligated to the ends of the digested DNAs to generate chimeric molecules of known and unknown sequence. Each adapter has one end that is homologous to the cohesive end of the target DNA but contains a single base pair change at the recognition site such that ligation can be done in the presence of the REs used to digest the DNAs (step 1). The importance of the ligation step will be discussed later.

The third step of AFLP is the PCR phase. The chimeric target DNA:adapter DNA fragments are first subjected to one or more

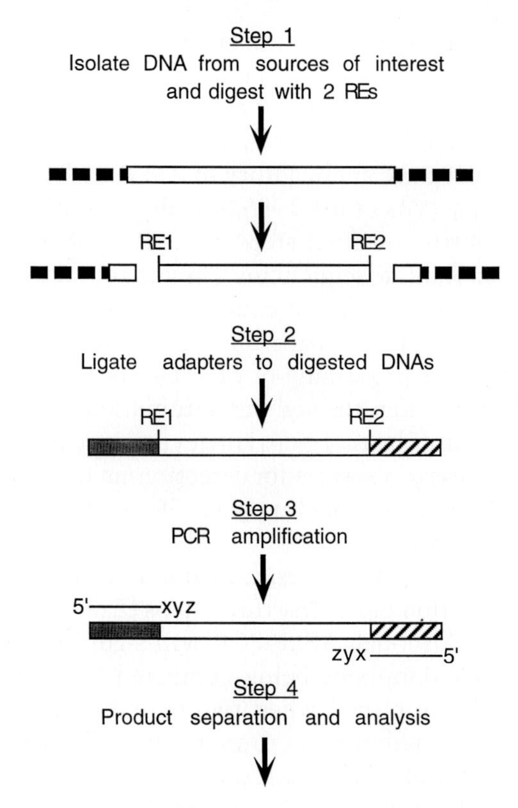

Figure 1. Flow diagram for the basic AFLP protocol.

rounds of PCR "preamplification" to ensure that the investigator has sufficient template DNA for fingerprinting PCRs. Preamplifications can also be used to reduce the complexity of the DNAs (see Variations). After preamplification the template DNAs are selectively PCR amplified (hereafter referred to as AFLP-PCR) using oligonucleotide primers that are homologous to the ds adapter and RE recognition site sequences but having extensions of one to three bases on their 3′ ends (AFLP primers). Although these extensions can vary in length and sequence from one primer to another, they are of a defined single length and sequence within any single primer. Because of the relative inability of *Taq* DNA polymerase to extend a DNA molecule if mismatches occur between a template and the 3′ end of a oligonucleotide primer (Kwok *et al.*, 1990; Newton *et al.*, 1989), only template DNAs with sequences internal to the adapter/RE site that are complementary to the primer extensions will be amplified. Hence, the use of AFLP primers allows the investigator to "fingerprint" a subset of the target genome in any given reaction in a selective and reproducible manner. Furthermore, longer extensions provide greater selectivity, ensuring the applicability of AFLP with genomes of any size or complexity (Vos *et al.*, 1995; see also Variations).

During the fourth and final step of AFLP, PCR products are resolved on polyacrylamide sequencing gels and subsequently analyzed for fingerprints of interest. Most commonly one of the primers used in AFLP-PCR is end labeled with some molecule that can be easily detected after proper processing of gels. Use of fluorescent dyes as end labels allows for high-throughput automation of both processing and analysis (Perkin-Elmer, 1995). However, since not all investigators will have access to a system for detection/analysis of fluorescent signals, radioisotopic end labeling (e.g., ^{32}P or ^{33}P) can be used as a cost-effective alternative.

The real power of AFLP is derived from a "synergistic" effect of the serial combination of its first three steps (Zabeau and Vos, 1993; Vos *et al.*, 1995). Although AFLP-PCR will significantly reduce the numbers of DNA templates being compared (e.g., only 1 in 256 molecules will be amplified when two base 3′ end extensions are used, assuming *Taq* polymerase cannot tolerate mismatches at the 3′ end), the number of DNA molecules being amplified will still be too great in most cases to be feasibly analyzed. This is especially true when examining plant nuclear genomes which are generally on the order of a few hundred to hundreds of thousands of mega-base pairs in size (Bennet and Letch, 1995). However, under the standard conditions described by Vos *et al.* (1995), the complexity of the DNA being amplified via AFLP-PCR is actually lower than that predicted by virtue of the *Taq* polymerase and primer selectively alone. This is apparently because of the higher annealing temperature of the *Eco*RI site/adapter primer compared to that of the *Mse*I site/adapter primer and the presence of an inverted repeat at the ends of *Mse*I–*Mse*I fragments which might form a stem-and-loop structure and compete with primer annealing (Vos *et al.*, 1995). Both of these factors promote the amplification of *Eco*RI–*Eco*RI- and *Eco*RI–*Mse*I-ended fragments while repressing amplification of the predominant *Mse*I–*Mse*I-ended fragments. End labeling of the primer directed against the 6-bp recognition site-specific adapter can further reduce complexity at the visualization stage since fragments having 6-bp recognition site-specific adapters on one or both ends will represent a limited subset of the total amplified DNA fragments. In the simplest sense, the digestion and ligation steps result in the generation of an extremely large number of distinct "molecular genetic windows," of which only a few are "opened and looked into" by any single AFLP-PCR.

The protocol presented in the following section represents the basic AFLP procedure we have used to generate molecular genetic markers for the positional cloning of multiple loci affecting phototropism in *Arabidopsis thaliana* (Huala *et al.*, 1997; R. Hausman, A. V. Motchoulski, E. Stowe-Evans, and E. Liscum, unpublished results) and in studies of differential gene expression (K. Lease and E. Liscum, unpublished results). With only minor modifications of this protocol, a variety of questions involving comparative genome analysis can be addressed (see Variations and Applications).

Protocol

Reagents

Reagents for DNA isolation	(See below)
Digestion/ligation buffer	50 m*M* Tris–acetate (pH 7.5) 50 m*M* Mg–acetate 250 m*M* K–acetate 25 ng/μl bovine serum albumin
Restriction enzymes	*Eco*RI *Mse*I
Adapter preparation buffer	100 m*M* Tris–acetate (pH 7.5) 100 m*M* Mg–acetate 500 m*M* *K*–acetate
ATP	10 m*M*
T4 DNA ligase	
Adapter oligonucleotides	*Eco*RI.Adapt1: 5′-CTCGTAGACTGCGTACC *Eco*RI.Adapt2: 5′-AATTGGTACGCAGTC

	*Mse*I.Adapt1: 5′-GACGATGAGTCCTGAG
	*Mse*I.Adapt2: 5′-TACTCAGGACTCAT
10× PCR buffer	500 m*M* KCl
	100 m*M* Tris–HCl (9.0)
	1.0% Triton X-100
TE	10 m*M* Tris (pH 7.5)
	1 m*M* EDTA
$MgCl_2$	25 m*M*
dNTP mix	dATP, dCTP, dTTP, dGTP (5 m*M* each)
[γ^{33}P]ATP	3000 Ci/mmol
Taq DNA polymerase	5 U/μl
T4 polynucleotide kinase	
*Eco*RI.Core+Adapt primer	5′-CTCGTAGACTGCGTACCAATTC
*Mse*I.Core+Adapt primer	5′-GACGATGAGTCCTGAGTAA
*Eco*RI–AFLP primers	5′-AGACTGCGTACCAATTCxyz
*Mse*I–AFLP primers	5′-GATGAGTCCTGAGTAAxyz
2× formamide buffer	98% (v/v) formamide
	10 m*M* EDTA (pH 8.0)
	0.025% (w/v) xylene cyanol FF
	0.025% (w/v) bromophenol blue
5× TBE buffer	450 m*M* Tris–borate
	10 m*M* EDTA (pH 8.0)
Acrylamide/ bisacrylamide	19:1
Urea	
Glacial acetic acid	

Step 1: DNA Isolation and Digestion

1. Isolate genomic DNA by any good mini(micro)preparation. A "good" preparation is one that yields RE-digestible DNAs. The most common problem associated with plant DNA mini(micro)preparations is contamination with starch and/or other complex carbohydrates. Hence, we suggest dark-adapting

plants of interest for 2 to 3 days prior to tissue harvest to deplete sugar reserves.

2. For each individual DNA sample, combine the following:

   0.5 μg DNA
   8 μl digestion/ligation buffer
   5 U each of *Eco*RI and *Mse*I
   dH_2O to a total volume of 40 μl

3. Incubate samples at 37°C for ~3 hr. It is essential that the digestion proceed to completion to ensure fidelity of subsequent AFLP-PCR. However, since it is necessary to have active enzymes present during the ligation step, digestions should not be done for significantly longer than 3 hr.

Step 2: Generation of Chimeric Template DNAs

1a. Prepare *Eco*RI-specific ds adapter by mixing the following:

   1.7 μg *Eco*RI.Adapt1
   1.5 μg *Eco*RI.Adapt2
   3 μl adapter preparation buffer
   dH_2O to a total volume of 60 μl

1b. Prepare *Mse*I-specific ds adapters by mixing the following:

   16 μg *Mse*I.Adapt1
   14 μg *Mse*I.Adapt2
   3 μl adapter preparation buffer
   dH_2O to a total volume of 60 μl

   The final *Eco*RI- and *Mse*I-specific adapter concentrations in these mixes are 5 and 50 pmol/μl, respectively. To prevent adapter self-ligation, adapters should not be phosphorylated.

2. Heat adapter mixes to 95°C and allow to cool to room temperature slowly. There will be enough of each ds adapter for 60 ligation reactions.

3. Add the following to each 40-μl digestion mix:

   1 μl ds *Eco*RI-specific adapter
   1 μl ds *Mse*I-specific adapter
   1 μl ATP
   4 μl digestion/ligation buffer
   1 U T4 DNA ligase
   dH_2O to a total volume of 50 μl

4. Incubate ligation reactions at 37°C for ≥3 hr. Although additional RE can be added to the ligation reactions to prevent re-ligation of restricted template DNAs, this is generally not necessary if digestions did not significantly exceed 3 hr. As discussed under Method, adapters are irreversibly ligated to template DNAs because of a single-bp change at the RE recognition site.

Step 3: PCR

The x, y, and *z'*s on the AFLP primers designate the 3′-end extensions that can vary in length and sequence between primers (see Method). We standardly use primers with two base extensions in studies with *Arabidopsis* genomic DNAs or cDNAs. Hence, we have 256 possible primer pair combinations which can be used for AFLP-PCRs.

1. Set up preamplification reactions by mixing 1 μl of chimeric template DNA (cut/ligated) with the following:
   0.5 μl *Eco*RI.Core+Adapt primer (at 50 ng/μl)
   0.5 μl *Mse*I.Core+Adapt primer (at 50 ng/μl)
   0.8 μl dNTPs
   2 μl 10× PCR buffer
   1.2 μl $MgCl_2$
   0.4 U *Taq* DNA polymerase
   dH_2O to a total volume of 20 μl
   The "Core+Adapt" primers are homologous to the RE recognition sites and adapter sequences only, thus promoting amplification of all chimeric molecules in a nonselective manner.
2. Amplify the reactions with the following thermocycle profile:
   a. 1 min at 94°C
   b. 30 sec at 94°C
   c. 1 min at 56°C
   d. 1 min at 72°C
   e. Repeat steps b–d 19 times
   f. 2 min at 72°C
   g. Hold at 4°C
   Since nonphosphorylated adapters were used in the ligation step, each adapter is only actually ligated to one strand of

target DNA. Hence, a "hot start" should never be performed during a preamplification since one strand of the adapter would be lost to solution during a quick denaturation in the absence of *Taq* polymerase. Use of the thermocycle profile given previously allows the recessed 3′ ends to be filled in by *Taq* polymerase during ramping to the first denaturation. An aliquot of preamplification reaction can be separated on a 1% (w/v) agarose TBE gel to check integrity and quality of the PCR. Products should resolve into a smear of relatively low-molecular-weight molecules (<1000 bp).

3. Dilute preamplification products 1:10 with TE, and store at 4°C (short term) or −20°C (long-term) until needed for AFLP-PCR.
4. Just prior to their use in AFLP-PCR, *Eco*RI–AFLP primers are end labeled with γ-^{33}P in the following mix:

   10 μl *Eco*RI–AFLP primer (at 50 ng/μl)

   10 μl[γ-^{33}P]ATP

   5 μl adapter preparation buffer

   5–10 U of T4 polynucleotide kinase

   dH_2O to a total volume of 50 μl

   Other end-labeling reagents may be used; however, visualization of final PCR products may require gel handling and processing different from that described here.
5. Incubate reaction mix at 37°C for 30 min to stimulate kinase activity and then at 70°C for 10 min to inactivate the kinase. Primers will be at a final concentration of 10 ng/μl.
6. Set up each AFLP-PCR by combining 1 μl of diluted preamplified template DNA with the following:

   0.125 μl labeled *Eco*RI–AFLP primer

   0.15 μl unlabeled *Mse*I–AFLP primer (at 50 ng/μl)

   0.2 μl dNTPs

   0.5 μl 10× PCR buffer

   0.3 μl $MgCl_2$

   0.1 U *Taq* DNA polymerase

   dH_2O to a total volume of 5 μl

   AFLP-PCRs can be scaled up or down to match investigators needs. We use 5-μl reactions in order to save on costs of radioisotope since <2 μl of any given reaction is typically used for analysis.

7. Amplify AFLP reactions with the following thermocycle profile:
   a. 2 min at 94°C
   b. 30-sec denaturation at 94°C
   c. 30-sec annealing at 65°C
   d. 1-min extension at 72°C
   e. Repeat steps b–d 12 times, reducing the annealing temperature 0.7°C per cycle
   f. 30 sec at 94°C
   g. 30 sec at 56°C
   h. 1 min at 72°C
   i. Repeat steps f–h 25 times
   j. 2 min at 72°C
   k. Hold at 4°C

Step 4: Gel Separation and Analysis of AFLP-PCR Products

1. Add an equal volume of 2× formamide buffer to each AFLP-PCR and incubate at 95°C for 10–15 min to denature the DNAs.
2. Load denatured samples on a 5% polyacryamide gel (made with 7.5 *M* urea and 1× TBE) and separate by electrophoresis at a constant current of 40–50 W using 0.5× TBE as the running buffer. Stop electrophoresis when the xylene cyanol dye front is about 2 or 3 cm from the bottom of the gel. Use of 1× TBE in the gel matrix promotes sharp banding, whereas use of 0.5× TBE as the running buffer allows for complete electrophoresis in ~3 hr without compromising banding.
3. Fix gel for 30 min in 10% (v/v) acetic acid, and then dry onto filter paper (with a vacuum/heat gel drier).
4. Expose dried gel to Kodak BioMax film (or equivalent from another manufacturer) to generate an autoradiograph of the labeled DNA fingerprints. Use of a single emulsion film (e.g., BioMax) that is more sensitive to low-energy β-particle emitters allows for the generation of clear high-resolution autoradiographs on overnight exposure to a standard AFLP gel generated as described here.
5. Examine autoradiograph for fingerprints of interest (i.e., bands exhibiting a desired level of genetic linkage when constructing genetic maps or bands exhibiting differential abundance when examining gene expression). The analysis of fingerprint patterns

can be done by eye or via computer using digitized images of the autoradiographs.

Variations

Although the protocol presented here provides a basic working version of AFLP technology, it can be modified in various ways to accommodate particular investigator wants and/or needs. Several such modifications are discussed in this section, especially where relevant to the generation of high-resolution genetic maps and studies of gene expression in plants. These and other "tailoring" steps are discussed in detail at our laboratory web page, *www.biosci.missouri.edu/liscum/LiscumLabPage.html.*

The most obvious modifications that can be made to the basic AFLP protocol described here relate directly to what type of data the investigator wishes to generate. For example, if the fingerprints to be generated will be analyzed for genetic linkage to a locus of interest, the template DNAs for AFLP will be derived from genomic DNAs isolated from segregant individuals or pools of individuals. However, if one is trying to identify genes that are up- or downregulated in response to a specific condition(s), target DNAs will be cDNAs derived from polyA$^+$ RNA isolated from plants grown under control and test conditions. By virtue of the questions being asked by these two examples, the manner by which the resultant fingerprints are analyzed will also be different. Readers are directed elsewhere for detailed examples of such data analyses (Bachem *et al.*, 1996; Liscum and Oeller, 1999) since they are beyond the scope of this chapter.

A majority of the modifications to the basic AFLP protocol are likely to be made to the preamplification(s). For example, although a nonselective preamplification (see Protocol, step 3) can be used to generate an essentially unlimited supply of template DNAs, it may not be suitable when the initial target DNAs (genomes) are of high complexity, such as plant nuclear genomes. In such cases, a single selective PCR (i.e., AFLP-PCR) may not reduce the complexity adequately to allow manageable data analysis, and thus a selective preamplification may be employed using a primer (or primers) with a

3′-end extension(s). The number of selective bases on the 3′ end, and whether one or both primers contain equal extensions, depends on the initial as well as desired final complexity of the target and product DNAs, respectively. Hence, with greater complexity it makes sense to consider a preamplification with greater selectivity. For instance, use of two single-base extension primers in a preamplification will effectively reduce the complexity of the initial chimeric template DNAs by at least 93.75%. With careful choice of selective amplification(s) (preamplification plus AFLP-PCR), similar levels of manageable product formation can be attained with any genome, regardless of size and/or complexity (Vos *et al.*, 1995).

It should be noted that not only does a selective preamplification reduce the complexity of template DNAs from single sources but also a similar complexity-reducing effect is observed with DNAs pooled from several sources. Consequently, bulked segregant analysis (Michelmore *et al.*, 1991) can be used in the generation of genetic linkage data (Thomas *et al.*, 1995; Vos *et al.*, 1995; Büschges *et al.*, 1997). In our experience, the nonselective preamplifications do not easily allow for use of bulked segregants (Liscum and Oeller, 1999). However, even when single-segregant analysis must be done, AFLP is generally more robust than other methods which can be used in conjunction with bulked segregants.

Applications

Although this chapter has focused on AFLP technology as a tool to address important questions in plant developmental biology, use of this technology is certainly not limited to such studies. For example, AFLP has been used for bacterial and pathogen genotyping (Huys *et al.*, 1996; Janssen *et al.*, 1996; Lin *et al.*, 1996; Picardeau *et al.*, 1997; Rosendahl and Taylor, 1997), for immunological and epidemiological studies (Dijkshoorn *et al.*, 1996; Janssen and Dijkshoorn, 1996; Schreiner *et al.*, 1996; Kuhn *et al.*, 1997), and for ecological, evolutionary, and biodiversity studies of organisms from a wide range of taxa (Karp *et al.*, 1996; Travis *et al.*, 1996; Hongtrakul *et al.*, 1997; Keim *et al.*, 1997; Pakniyat *et al.*, 1997). This amazing variety of uses for AFLP technology is a testament to its inventors, who proposed many of the broad uses in their original patent proposal (Zabeau and Vos, 1993), well before AFLP had been put to the tests. It is likely that this technology will be a major player in answering a wide range of questions being asked in biology, including those in plant developmental biology.

References

Bachem, C. W. B., van der Hoeven, R. S., de Bruijn, S. M., Vreugdenhil, D., Zabeau, M., and Visser, R. G. F. (1996). Visualization of differential gene expression using a novel method of RNA fingerprinting based on AFLP: Analysis of gene expression during potato tuber development. *Plant J.* **9,** 745–753.

Bennett, M. D., and Letch, I. J. (1995). Nuclear DNA amounts in angiosperms. *Ann. Bot.* **76,** 113–176.

Büschges, R., Hollricher, K., Panstruga, R., Simons, G., Wolter, M., Frijters, A., van Daelen, R., van der Lee, T., Diergaarde, P., Groenendijk, J., Töpsch, S., Vos, P., Salamini, F., and Schulze-Lefert, P. (1997). The barley *Mlo* gene: A novel control element of plant pathogen resistance. *Cell* **88,** 695–705.

Chory, J., Chatterjee, M., Cook, R. E., Elich, T., Fankhauser, C., Li, J., Nagpal, P., Neff, M., Pepper, A., Poole, D., Reed, J., and Vitart, V. (1996). From seed germination to flowering, light controls plant development via the pigment phytochrome. *Proc. Natl. Acad.Sci. U.S.A.* **93,** 12066–12071.

Cnops, G., den Boer, B., Gerats, A., Van Montagu, M., and Van Lijsebettens, M. (1996). Chromosomal landing at the *Arabidopsis TORNADO1* locus using an AFLP-based strategy. *Mol. Gen. Genet.* **253,** 32–41.

Dijkshoorn, L., Aucken, H., Gernersmidt, P., Janssen, P., Kaufmann, M. E., Garaizar, J., Ursing, J., and Pitt, T. L. (1996). Comparison of outbreak and nonoutbreak *Acinetobacter baumannii* strains by genotypic and phenotypic methods. *J. Clin. Microbiol.* **34,** 1519–1525.

Habu, Y., Fukada-Tanaka, S., Hisatomi, Y., and Iida, S. (1997). Amplified restriction fragment length polymorphism-based mRNA fingerprinting using a single restriction enzyme that recognizes a 4-bp sequence. *Biochem. Biophys. Res. Commun.* **234,** 516–521.

Hongtrakul, V., Huestis, G. M., and Knapp, S. J. (1997). Amplified fragment length polymorphisms as a tool for DNA fingerprinting sunflower germplasm—Genetic diversity among oilseed inbred lines. *Theor. Appl. Genet.* **95,** 400–407.

Huala, E., Oeller, P. W., Liscum, E., Han, I.-S., Larsen, E., and Briggs, W. R. (1997). Arabidopsis NPH1: A protein kinase with a putative redox-sensing domain. *Science* **278,** 2120–2123.

Huys, G., Coopman, R., Janssen, P., and Kersters, K. (1996). High-resolution genotypic analysis of the genus *Aeromonas* by AFLP fingerprinting. *Int. J. Syst. Bacteriol.* **46,** 572–580.

Janssen, P., and Dijkshoorn, L. (1996). High resolution DNA fingerprinting of *Acinetobacter* outbreak strains. *FEMS Microbiol. Lett.* **142,** 191–194.

Janssen, P., Coopman, R., Huys, G., Swings, J., Bleeker, M., Vos, P., Zabeau, M., and Kersters, K. (1996). Evaluation of the DNA fingerprinting method AFLP as a new tool in bacterial taxonomy. *Microbiology* **142,** 1881–1893.

Karp, A., Seberg, O., and Buiatti, M. (1996). Molecular techniques in the assessment of botanical diversity. *Ann. Bot.* **78,** 143–149.

Keim, P., Kalif, A., Schupp, J., Hill, K., Travis, S. E., Richmond, K., Adair, D. M.,

Hughjones, M., Kusker, C. R., and Jackson, P. (1997). Molecular evolution and diversity in *Bacillus anthracis* as detected by amplified fragment length polymorphism markers. *J. Bacteriol.* **179,** 818–824.

Kuhn, I., Albert, M. J., Ansaruzzaman, M., Bhuiyan, N. A., Alabi, S. A., Islam, M. S., Neogi, P. K. B., Huys, G., Janssen, P., Kersters, K., and Mollby, R. (1997). Characterization of *Aeromonas* spp isolated from humans with diarrhea, from healthy controls, and from surface water in Bangladesh. *J. Clin. Micobiol.* **35,** 369–373.

Kwok, S., Kellogg, D. E., McKinney, N., Spasic, D., Goda, L., Levenson, C., and Sninsky, J. J. (1990). The effects of primer-template mismatches on the polyermase chain reaction human immunodeficiency virus type 1 model studies. *Nucleic Acids Res.* **18,** 999–1006.

Lin, J. J., Kuo, J., and Ma, J. (1996). A PCR-based DNA fingerprinting technique—AFLP for molecular typing of bacteria. *Nucleic Acids Res.* **24,** 3649–3650.

Liscum, E., and Oeller, P. W. (1999). AFLP: DNA fingerprinting to positional cloning. In *Analytical Biotechnology: Genome Analysis* (P. Oefner, Ed.). CRC Press, Boca Raton, FL., in press.

Meyerowitz, E. M. (1997a). Genetic control of cell division patterns in developing plants. *Cell* **88,** 299–308.

Meyerowitz, E. M. (1997b). Plants and the logic of development. *Genetics* **145,** 5–9.

Michelmore, R. W., Paran, I., and Kesseli, R. V. (1991). Identification of markers linked to disease resistance genes by bulked segregant analysis: A rapid method to detect markers in specific genomic regions using segregating populations. *Proc. Natl. Acad. Sci. U.S.A.* **88,** 9828–9832.

Newton, C. R., Graham, A., Heptinstall, L. E., Powell, S. J., Summers, C., Kalsheker, N., Smith, J. C., and Markham, A. F. (1989). Analysis of any point mutation in DNA the amplification refractory mutation system ARMS. *Nucleic Acids Res.* **17,** 2503–2516.

Pakniyat, H., Powell, W., Baird, E., Handley, L. L., Robinson, D., Scrimgeour, C. M., Nevo, E., Hackett, C. A., Caligari, P. D. S., and Forster, B. P. (1997). AFLP variation in wild barley (*Hordeum spontaneum* C. Koch) with reference to salt tolerance and associated ecogeography. *Genome* **40,** 332–341.

Perkin-Elmer (1995). AFLP plant mapping kit protocol, Publ. No. 402083.

Picardeau, M., Prodhom, G., Raskine, L., Lepennec, M. P., and Vincent, V. (1997). Genotypic characterization of five subspecies of *Mycobacterium kansasii*. *J. Clin. Microbiol.* **35,** 25–32.

Rosendahl, S., and Taylor, J. W. (1997). Development of multiple genetic markers for studies of genetic variation in arbuscular mycorrhizal fungi using AFLP. *Mol. Ecol.* **6,** 821–829.

Schreiner, T., Prochnowcalzia, H., Maccari, B., Erne, E., Kinsler, I., Wolpl, A., and Wiesneth, M. (1996). Chimerism analysis after allogeneic bone marrow transplantation with nonradioactive RFLP and PCR-AFLP using the same DNA. *J. Immunol. Methods* **196,** 93–96.

Taylor, C. B. (1997). Plant vegetative development: From seed and embryo to shoot and root. *Plant Cell* **9,** 981–988.

Thomas, C. M., Vos, P., Zabeau, M., Jones, D. A., Norcott, K. A., Chadwick, B. P., and Jones, J. D. G. (1995). Identification of amplified restriction fragment length polymorphism (AFLP) markers tightly linked to the tomato *Cf-9* gene for resistance to *Cladosporium fulvum*. *Plant J.* **8,** 785–794.

Travis, S. E., Maschinski, J., and Keim, P. (1996). An analysis of genetic variation in

Astragalus cremnophylax var Cremnophylax, a critically endangered plant, using AFLP markers. *Mol. Ecol.* **5,** 735–745.

Vos, P., Hogers, R., Bleeker, M., Reijans, M., van de Lee, T., Hornes, M., Frijters, A., Pot, J., Peleman, J., Kuiper, M., and Zabeau, M. (1995). AFLP: A new technique for DNA fingerprinting. *Nucleic Acids Res.* **23,** 4407–4414.

Zabeau, M., and Vos, P. (1993). Selective restriction fragment amplification: A general method for DNA fingerprinting. *Eur. Patent Appl.,* Publ. No. 0534858A1.

34

A Fluorescent, Multiplex Solid-Phase Minisequencing Method for Genotyping Cytochrome P450 Genes

Tomi Pastinen, Ann-Christine Syvänen, Catherine Moberg, Gisela Sitbon, and Jörgen Lönngren

The sequence variations of the human genome ranges from mutations causing disease or predisposing for multifactorial disorders to neutral polymorphisms that have no influence on the phenotype of the individuals. Moreover, the genome contains variations that can be considered as normal, but they cause differences between individuals in various metabolic pathways. The genes encoding drug-metabolizing enzymes are examples of this type of genetic variation.

The hepatic cytochrome P450 system comprises enzymes that are responsible for drug oxidation and, thus, usually for the rate of drug clearance. There are profound interindividual and interethnic differences in drug metabolism which influence the disposition and effects of the drug. This is an important factor for therapy with many classes of drugs, in particular those which have a narrow therapeutic range. Some individuals are homozygous for defect P450 alleles, causing no enzyme to be produced. These subjects will, at ordinary dosage, reach too high plasma concentrations of the drug, with possible side effects as a consequence (Ingelman-Sundberg *et al.*, 1994; Linder *et al.*, 1997). Conversely, for drugs that are activated by the enzyme, too low plasma levels will be experienced.

PCR Applications

In humans, mainly four different P450 enzymes (i.e., *CYP2D6, CYP2C19, CYP2E1,* and *CYP3A4*), having different but partially overlapping substrate specificities, account for the primary metabolism of clinically important drugs. Of these, *CYP2D6* and *CYP2C19* are polymorphically distributed in the populations. The majority of mutations causing the genetic defects are known (Sachse *et al.*, 1997; deMorais *et al.*, 1994).

Regarding *CYP2D6,* more than 30 commonly used drugs, including β-receptor blockers, tricyclic antidepressants, SSRIs, and neuroleptics, are substrates for the enzyme. Due to unique structural properties of the active site, most *CYP2D6* substrates are not metabolized by any other P450 enzyme. The capacity to hydroxylate a *CYP2D6* substrate may be measured using the model substances debrisoquine or dexmethorphan. The compound is given orally and the ratio [metabolic ratio, (MR)] between the parent compound, e.g., debrisoquine and its metabolite 4-OH-debrisoquine, is determined in the 8-hr urine. Individuals who show a MR higher than 12.6 are referred to as poor metabolizers (PMs). About 5–10% of Caucasians and 1% of Orientals show this defect and are homozygous for defect *CYP2D6* alleles. Individuals with a MR lower than 12.6 are classified as efficient metabolizers (EMs), whereas individuals with duplicated or amplified *CYP2D6* genes are ultrarapid metabolizers (UMs).

The molecular basis for the poor metabolizer phenotype is due to deficient alleles of the *CYP2D6* gene that are distributed in the Caucasian population: *CYP2D6*3* and **6* have base deletions: *CYP2D6*4,* the most commonly defect allele with a frequency of about 23%, in which a mutation in the intron 3/exon 4 junction causes a splicing defect; and *CYP2D6*5,* in which the whole gene has been deleted. Individuals homozygous for the mutations are PMs, whereas subjects heterozygous for any of the mutations have an impaired rate of metabolism but are not defined as poor metabolizers. In the Caucasian population about 35% carry a heterozygous *CYP2D6* genotype.

The *CYP2C19* drug metabolism in humans is associated with the 4′-hydroxylation of *S*-mephenytoin. This enzyme metabolizes several important drugs, e.g., certain barbiturates, antimalaria drugs, SSRIs, and antidepressants. As for the *CYP2D6,* specific genetic alterations lead to the PM phenotype. The PM phenotype occurs in about 5% of Caucasians and 20% of Orientals. The major genetic defect in PMs is a single G → A substitution within exon 5 which creates an aberrant splice site. This allele is referred to as m_1 or *2 (deMorais *et al.*, 1994). This allele accounts for about 75% of the PMs. In addition, the PMs among Orientals may carry other deficient alleles (Xiao *et al.*, 1997). Regarding *CYP2D6*, PMs may risk adverse drug reactions due to too low (in case of metabolic activation by *CYP2C19* enzyme) or too high plasma levels of drug.

It is obvious that in many cases, in particular for drugs with a narrow therapeutic range, it would be of considerable importance to know the metabolic status of the individuals that are going to be treated with the drug. This can facilitate an individualized therapy which avoids unnecessary side effects or other adverse medical complications that may be experienced by PMs and also individuals heterozygous for mutations in case they are treated with neuroleptics, for example. One may also predict that the effect of *CYP2D6* inhibitors of the metabolism would be more pronounced in heterozygous EMs.

The method described in this chapter allows for simultaneous or individual determination of deficient alleles in both *CYP2D6* and *2C19*. For *CYP2D6* the method allows for more than 95% confidence in the assignment of the phenotype as derived from the genotype (i.e., the *CYP2D6*3, -*4, -*5,* and *-*6* alleles are characterized). For the *CYP2C19* the *2 allele is the major deficient allele known to confer the PM phenotype in Caucasians; this allele accounts for about 75% of the PMs.

Most of the currently used methods for detecting single nucleotide variations are based on amplification of the target DNA by the PCR technique, which allows both sensitive and specific analysis of the target DNA sequence by a variety of methods, such as hybridization with sequence-specific oligonucleotide probes or with the aid of nucleic acid modifying enzymes, such as DNA ligases or DNA polymerases. The different reaction principles for distinguishing between nucleotide sequence variants can be combined with solid-phase formats into assays that are easily amenable to automation (Syvänen and Landegren, 1994).

In the solid-phase minisequencing method, sequence variants are distinguished in PCR products immobilized on a solid support by mediation of the biotin–avidin interaction using single nucleotide primer extension reactions (Syvänen *et al.*, 1990). The variable nucleotide(s) is detected by specific extension of a primer that anneals immediately adjacent to the variable site with a single labeled nucleotide that is complementary to the nucleotide to be detected using a DNA polymerase. The method was previously set up for a large number of individual mutations causing genetic disorders (Syvänen, 1994) and it is currently used in routine diagnostics on a large scale (Hietala *et al.*, 1996).

The specificity of the solid-phase minisequencing method originates from the fidelity of the nucleotide incorporation catalyzed by the DNA polymerase, whereas the primer annealing reaction is performed at nonstringent conditions. Consequently, all sequence variants can be detected at the same reaction conditions, irrespective of the flanking nucleotide sequence. Therefore, the solid-phase minisequencing method is particularly well suited for the simultaneous detection of multiple sequence variants in each sample.

This chapter describes the application of a multiplex, solid-phase minisequencing method for simultaneous detection of four polymorphisms (**3, *4, *5, *6*) in *CYP2D6* and **2* in *2C19* genes using fluorescently labeled ddNTPs. It should be noted that the *CYP2D6*5* allele is a complete gene deletion. Homozygousity for this deletion is obvious by the abscence of a PCR product but should be corroborated by Southern blot analysis.

Principle of the Method

In the multiplex minisequencing method, multiple primers of various sizes that anneal immediately adjacent to the variable sites in the templates are extended with single-fluorescent ddNTPs by a DNA polymerase, followed by detection of the extended primers on a DNA autosequencing instrument. The size of the extended primer defines the position of the analyzed polymorphism, and the incorporated fluorescein–ddNTP gives the identity of the nucleotide at each site (Fig. 1). Because the ALF instrument that is used is based on the use of a single label, separate reactions must be performed for each of the four nucleotides. The use of streptavidin-coated comb-shaped manifold supports (Lagerkvist *et al.*, 1994) for capturing the amplified templates and for transferring them to the minisequencing reaction mixtures allows practical handling of large numbers of samples. The manifolds have been designed to fit the slots of the sequencing gel of the ALF instrument, and thus the complete procedure, including loading of the gel, can be carried out without pipetting steps.

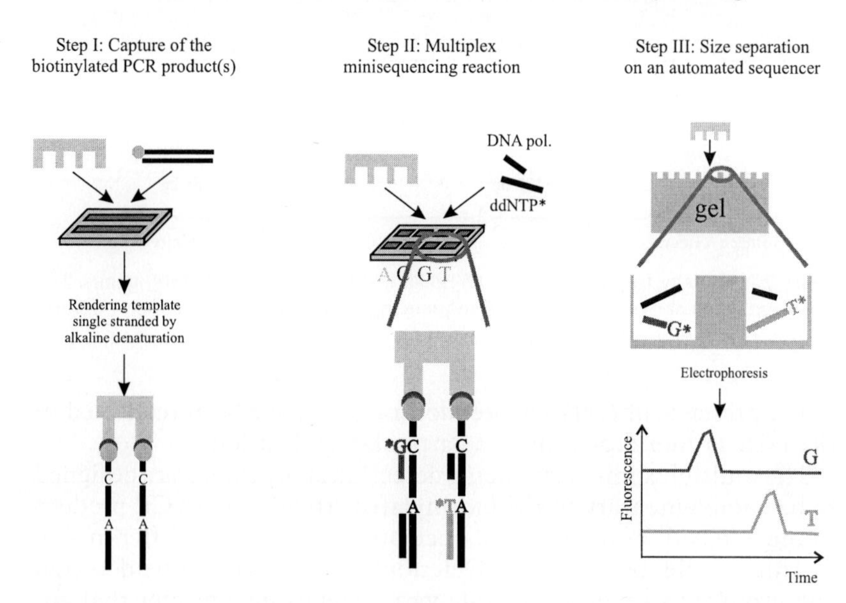

Figure 1. Principle of fluorescent, multiplex minisequencing method illustrated by the detection of two polymorphic nucleotides.

Design and Synthesis of Primers

A prerequisite for setting up the solid-phase minisequencing method is that nucleotide sequence information of the gene of interest is available for designing primers for PCR and for the minisequencing primer extension reactions. PCR primers and one minisequencing detection primer per variable nucleotide position are required. During the amplification, the PCR primers should produce fragments that are as small as possible and of equal size to ensure efficient and equal binding to the manifold support. The PCR primers should be 20–23 nucleotides long and have similar melting temperatures and noncomplementary 3′ ends (Dieffenbach *et al.*, 1993). One of the PCR primers is biotinylated in its 5′ end during its synthesis using a biotinyl phosphoramidite reagent. The biotinylated primer can be used without purification given that the efficiency of the biotinylation reaction on the oligonucleotide synthesizer has been close to 90%.

In this application three PCR reactions are performed (for *CYP2D6*4* and **6, CYP2D6*3,* and *CYP2C19,* respectively; due

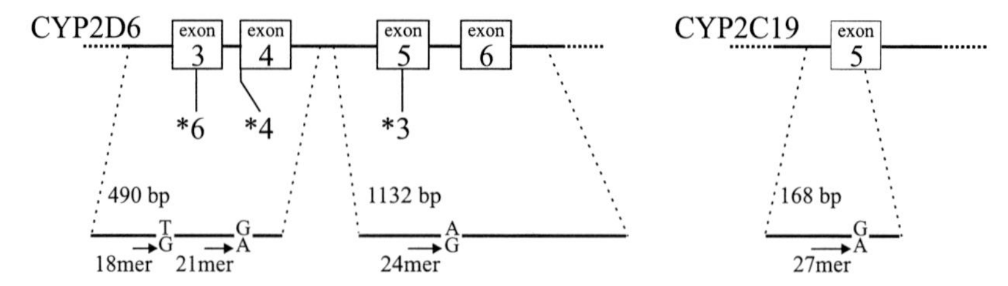

Figure 2. Strategy for typing of the *CYP2D6*3, -*4, -*6,* and *CYP2C19m$_1$* genes. The bases indicated above and below the sequencing primers indicate the incorporation for the *wt* and *mut* genotypes, respectively.

to the presence of *CYP2D6* pseudogenes, we have been restricted in the PCR primer design in this particular application).

The multiplex minisequencing detection step primers are designed to be complementary to the biotinylated strand of the PCR product immediately 3′ of the variable nucleotide position and differ in size by three nucleotides (Fig. 2). Deletions (insertions) can be detected analogously to single nucleotide variations using a primer that anneals immediately adjacent to the deletion (insertion) breakpoint. The first nucleotide within the deletion in the normal allele and the first nucleotide following the deletion in the mutant allele are detected in the minisequencing reactions.

Protocols

Sample Preparation

Equipment and Reagents

QIAamp Blood Kit (Qiagen, Hilden, Germany)

Method

DNA is isolated as described by the kit producer.

PCR Amplification

Equipment and Reagents

PCR instrument Perkin–Elmer GeneAmp 9600 (Perkin–Elmer, Norwalk, CT)

GNA 100 electrophoresis instrument (Pharmacia Biotech, Uppsala, Sweden)
2 and 3% NuSieve 3:1 agarose (FMC BioProducts, Rockland, ME)
Sizemarker VIII (Boehringer-Mannheim, Mannheim, Germany)
1 × TBE buffer (100 m*M* Tris base, 83 m*M* boric acid, 1 m*M* EDTA, pH 8.0)
10 × PCR buffer without $MgCl_2$ (Perkin–Elmer)
25 m*M* $MgCl_2$ (Perkin–Elmer)
AmpliTaq Gold (5 U/μl) (Perkin–Elmer)
2.5 m*M* dNTPs (Pharmacia Biotech)
DMSO (in AutoLoad kit, Pharmacia Biotech)
Primer 2D6/3/51: 5′GCG GAG CGA GAG ACC GAG GA
Primer 2D6/3/31: 5′CCG GCC CTG ACA CTC CTT CT (biotin labeled)
Primer 2D6/46/51: 5′TGG TGG ATG GTG GGG CTA AT
Primer 2D6/46/31: 5′CAG AGA CTC CTC GGT CTC TC (biotin labeled)
Primer 2C19/*2/51: 5′AAT TAC AAC CAG AGC TTG GC
Primer 2C19/*2/31: 5′TAT CAC TTT CCA TAA AAG CAA G (biotin labeled)

Method

The PCR method involves three PCR reactions:

1. The *CYP2D6*3* is amplified as a 1132-bp product.
2. The *CYP2D6*4* and **6* are amplified as a 490-bp product.
3. The *CYP2C19*2* is amplified as a 168-bp product.

Reaction 1: *CYP2D6*3* Allele

Reagent	Final concentration
10× PCR buffer without $MgCl_2$	1×
$MgCl_2$ (25 m*M*)	1 m*M*
dNTP (2.5 m*M* each)	250 μ*M* (each)
DMSO	10%
Primer 2D6-3-51 (10 μ*M*)	0.6 μ*M*
Primer 2D6-3-31 (10 μ*M*)	0.3 μ*M*
AmpliTaq Gold (5 U/μl)	0.03 U/μl
H_2O	

The components for an adequate number of reactions are mixed and DNA (0.06–0.2 μg) is added to a final volume of 50 μl.

Reaction 2: *CYP2D6*4* and *6 Alleles

Reagent	Final concentration
10× PCR buffer without $MgCl_2$	1×
$MgCl_2$ (25 m*M*)	1.5 m*M*
dNTP (2.5 m*M* each)	250 μ*M* (each)
DMSO	5%
Primer 2D6-46-51 (10 μ*M*)	0.6 μ*M*
Primer 2D6-46-31 (10 μ*M*)	0.3 μ*M*
AmpliTaq Gold (5 U/μl)	0.03 U/μl
H_2O	

The components for an adequate number of reactions are mixed and DNA (0.03–0.1 μg) is added to a final volume of 30 μl.

Reaction 3: *CYP2C19*2 Allele*

Reagent	Final concentration
10× PCR buffer without $MgCl_2$	1×
$MgCl_2$ (25 m*M*)	3 m*M*
dNTP (2.5 m*M* each)	250 μ*M* (each)
Primer 2C19-2-51 (10 μ*M*)	1 μ*M*
Primer 2C19-2-31 (10 μ*M*)	0.3 μ*M*
AmpliTaq Gold (5 U/μl)	0.03 U/μl
H_2O	

The components for an adequate number of reactions are mixed and DNA (0.03–0.1 μg) is added to a final volume of 30 μl.

The reactions are run on a Perkin–Elmer GeneAmp PCR System 9600 as follows:

PCR program for reaction 1	PCR program for reactions 2 and 3
10 min, 94°C: initial denaturation	10 min, 94°C: initial denaturation
20 sec, 95°C: denaturation	20 sec, 95°C: denaturation
35× 20 sec, 52°C: annealing	35× 20 sec, 58°C: annealing
90 sec, 72°C: elongation	60 sec, 72°C: elongation
5 min, 72°C: final extension	5 min, 72°C: final extension

The PCR products are visualized by 2 or 3% agarose/EtBr gel electrophoresis (8–10 V/cm for 45 min).

Affinity Capture of the Biotinylated PCR Products

Equipment and Reagents

Eight-tooth streptavidin-coated sequencing combs (in AutoLoad kit, Pharmacia Biotech)

Ten-well plates for the PCR product capture and denaturation step (in AutoLoad kit, Pharmacia Biotech)

Capturing buffer (2 *M* NaCl, 10 m*M* Tris–HCl, 1 m*M* EDTA, pH 7.5)

1× TE buffer (10 m*M* Tris–HCl, 1 m*M* EDTA, pH 7.5)

100 m*M* NaOH

Method

1. Fifteen microliters of *3** allele, 6 μl of *4** and *6** alleles, and 3 μl of *2 allele PCR product are transferred to each well of the 10-well plates; 60 μl of capturing buffer and 36 μl of distilled water are added to a final volume of 120 μl per well.
2. Four teeth of the sequencing combs are inserted per well in a 10-well plate. The combs are moved gently up and down two or three times and the plate is placed on a wet tissue paper and covered with plastic wrap. Alternatively, a humid chamber may be used.
3. The plate is incubated for a minimum of 1 hr at room temperature.
4. The sequencing combs are transferred to another 10-well plate containing 100 μl of 100 m*M* NaOH per well.
5. The plate is incubated for 10 min at room temperature.
6. The sequencing combs are rinsed for 2 × 2 min using a minimal volume of 100 ml 1× TE buffer.

It is of great importance that the sequencing combs do not become dry during the procedure.

The high biotin-binding capacity of the sequencing combs in combination with the detection sensitivity of the sequencing instrument should permit capture and analysis of at least 20 biotinylated PCR products. If the size of the PCR products differ considerably the relative amounts in the wells need to be balanced accordingly.

Fluorescent, Multiplex Minisequencing Reactions

Equipment and Reagents

40-Well plates for sequencing reactions (in AutoLoad kit, Pharmacia Biotech)

Heating blocks shaped for plates (Pharmacia Biotech).

Minisequencing primers

*2D6*6* 5′ AAG AAG TCG CTG GAG CAG (18-mer)

*2D6*4* 5′ TAC CCG CAT CTC CCA CCC CCA (21-mer)

*2D6*3* 5′ act GAT GAG CTG CTA ACT GAG CAC (24-mer)

*2C19*2* 5′ tag cta CAC TAT CAT TGA TTA TTT CCC (27-mer)

Note: lowercase letters are "random" sequence for making all 21 nucleotides CYP-specific sequences (except for 18-mer).

ThermoSequenase DNA polymerase E79000Y (Amersham)

10× ThermoSequenase buffer (260 m*M* Tris–HCl, 65 m*M* $MgCl_2$, pH 9.5)

Fluorescein-labeled dideoxynucleoside triphosphates (Renaissance Nucleotide Analogues, F-ddATP, NEL-402; F-ddCTP, NEL-400; F-ddGTP, NEL-403; F-ddUTP, NEL-401; NEN/DuPont, Boston)

10 μM ddNTPs (Pharmacia Biotech)

METHOD

1. Four master mixtures are prepared for the multiplex minisequencing reactions by combining 2 μl of 10× ThermoSequenase buffer; 2 μl of a 15 μ*M* solution of each detection step primer, except for the 2D6*6 primer for which 4 μl is used; 1 μl of a 5 μ*M* solution of F-ddATP, 1 μl of a 10 μ*M* solution of F-ddCTP, 1 μl of a 1 μ*M* solution of F-ddGTP, or 1 μl of a 10 μ*M* solution of F-ddUTP; 1 μl of a 10 μ*M* solution of the three other unlabeled ddNTPs; 0.26 U of ThermoSequenase DNA polymerase; and distilled water to a final volume of 20 μl per reaction.
2. Twenty-microliter aliquots of the four reaction mixtures are transferred to four adjacent wells in the 40-well plate.
3. The combs carrying the denatured PCR products are inserted into the wells of the plate. The plate is incubated for 5 min at 55°C and washed with 1× TE buffer, and the combs are transferred to another 40-well plate filled with 1× TE buffer.
4. The plate is stored on ice if the electrophoresis on the DNA sequencer will be performed during the same day. The plates may be stored at −20°C for longer periods.

Analysis of the Extended Minisequencing Primers

EQUIPMENT AND REAGENTS

Fluorescent DNA autosequencing instrument (ALF DNA autosequencer, Pharmacia Biotech)

Glass plates, spacers, and combs for short gels (18 × 27 cm) for the ALF sequencer

50% Hydrolink polyacrylamide (Long Ranger, FMC BioProducts)

Urea (ALF grade, Pharmacia Biotech).

10× TBE (1.0 *M* Tris base, 0.83 *M* boric acid, 10 m*M* EDTA, pH 8.0)

Stop solution (100% deionized formamide containing 5% mg/ml of Dextran Blue 2000)

METHOD

1. A 18 × 27-cm polyacrylamide gel containing 10% Hydrolink polyacrylamide and 8 *M* urea in 1.2 × TBE buffer is prepared according to the instructions in the ALF DNA sequencer manual.
2. 1.2 × TBE buffer is used for the electrophoresis. The gel is preheated to 45°C, the wells are rinsed with buffer, and stop solution is added to the wells.
3. The combs carrying the products of the multiplex minisequencing reactions are inserted into the wells of the gel.
4. The combs are left in the gel for 10 min at 45°C and then carefully removed.
5. The electrophoresis is carried out for 70 min at 1200 V, 45 mA, and 25 W. The gel can be reloaded twice.

The procedure can easily be adapted to other models of auto-sequencers working according to the same detection principle. The AutoLoad kit, however, is only compatible with the ALF instruments.

Interpretation of the Data

The results of the electrophoretic run saved in the computer of the sequencing instrument are interpreted by direct visual inspection of the electropherograms. It is of importance that the running conditions are such that the separation times between peaks are sufficient to allow for unambiguous peak identification. A suitable software package may also be used (e.g., ALF Fragment Manager). In samples from individuals homozygous for the analyzed polymorphism, only one peak is detected at the time point corresponding to the size of the primer, whereas in heterozygous samples there are two peaks at the corresponding time point. Figure 3 shows examples from typing of *CYP* genes as described in this chapter.

Variants of the Multiplex, Minisequencing Method

Here we provide a specific protocol for typing four polymorphisms in *CYP2D6* and *2C19* genes. The number of polymorphic sites that can be typed per multiplex minisequencing reaction can be increased significantly without increasing the amount of work required to carry out the assay by adding more primers to each reaction. In the previous protocol we used primers that differed in size by three bases, and these were clearly resolved by rapid electrophoresis on a short gel. Primers differing in size by one or two bases could be used if required. The number of PCR-amplified templates analyzed per reaction can also be increased. The method is generally applicable for detecting variations in any other gene employing the same reaction conditions. We have previously set up the method using a similar protocol for multiplex genotyping of nine polymorphic sites in the *HLA-DQA1* and *DRB1* genes (Pastinen *et al.*, 1996).

Fluorescent multiplex minisequencing methods, in which the primers are extended using ddNTPs labeled with four different fluorophores followed by multicolor detection in one lane using an Applied Biosystems DNA sequencer, have been reported (Shumaker *et al.*, 1996; Tully *et al.*, 1996). In principle, the use of four fluorophores reduces the number of multiplex minisequencing reactions per sample from four to one. However, since the separation of primers extended with four fluorophores with different electrophoretic mobility in one lane requires longer running times, the theoretical fourfold increase in electrophoretic capacity compared to the ALF-based system is not achieved. Moreover, the same gel may be reused in the ALF instrument, which also saves time.

Conclusion

All reagents and equipment required for the fluorescent, multiplex minisequencing method described previously are available from

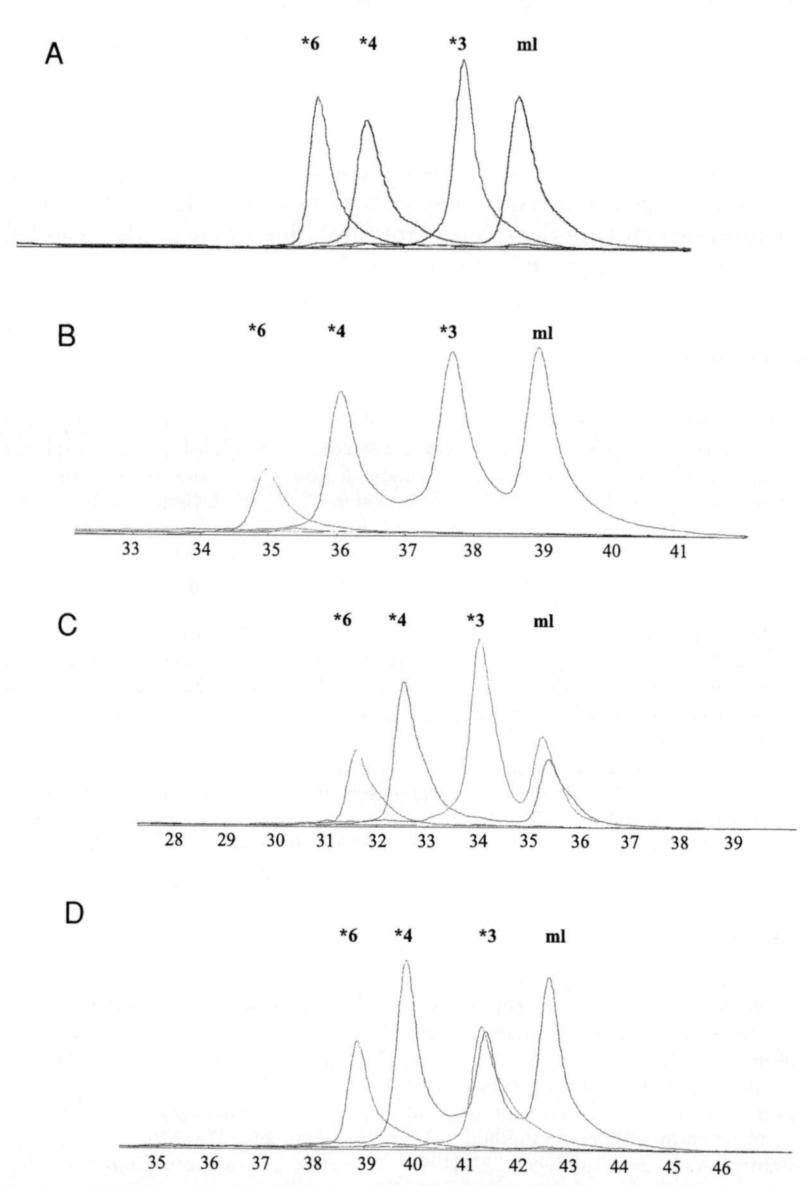

Figure 3. Electropherograms from *CYP2D6* and *2C19* multiplex, minisequencing. **3, *4,* and **6* are polymorphic alleles for *CYP2D6,* and **2* is a polymorphic allele in *CYP2C19.* A, Homozygous wt for all alleles (EM, extensive metabolizer); B, homozygous wt for alleles **6, *4,* and **2,* homozygous mutant for **3* allele (PM, poor metabolizer for 2D6); C, homozygous wt for **6, *4,* and **3* heterozygous for **2* allele (EM); D, homozygous wt for **6, *4,* and the **2* allele; heterozygous for the **3* allele (EM).

common suppliers of molecular biological reagents. Therefore, the method is easy to set up both in the research laboratory and for routine analysis of previously known sequence variants. The procedure involves simple and robust treatment procedures. In particular, the manifold format of the minisequencing step, facilities handling procedures, and interpretation of the results are simple. Furthermore, the method is highly flexible and mutant alleles may easily be added or subtracted to a given procedure as desired.

References

deMorais, S. M. F., Wilkinson, G. R., Blaisdell, J., Nakamura, K., Meyer, U. A., and Goldstein, J. A. (1994). The major genetic defect responsible for the polymorphism of *S*-mephenytoin metabolism in humans. *J. Biol. Chem.* **269,** 15419–15411.

Dieffenbach, C. W., Lowe, T. M. J., and Dveksler, G. S. (1993). General concepts for PCR primer design. *PCR Methods Appl.* **3,** S30–S37.

Hietala, M., Aula, P., Syvänen, A.-C., Isoniemi, A., Peltonen, L., and Palotie, A. (1996). DNA based carrier screening in primary health care: Screening for AGU mutations in maternity health offices. *Clin. Chem.* **42,** 1398–1404.

Ingelman-Sundberg, I., Johansson, I., Persson, I., Oscarsson, M., Hu, Y., and Bertilsson, L. (1994). Genetic polymorphism of cytochrome P450 Functional consequences and possible relationship to disease and alkohol toxicity. In *Towards a Molecular Basis of Alkohol Use and Abuse* (B. Jansson *et al.*, Eds.), p. 207. Birkhäuser–Verlag, Basel.

Lagerkvist, A., Stewart, J., Lagerström-Fermer, M., and Landegren, U. (1994). Manifold sequencing: Efficient processing of large sets of sequencing reaction. *Proc. Natl. Acad. Sci. U.S.A.* **91,** 2245–2249.

Linder, M. W., Prough, R. U., and Valdes, R., Jr. (1997). Pharmacogenetics: A laboratory tool for optimizing therapeutic efficiency. *Clin. Chem.* **42,** 254–266.

Pastinen, T., Partanen, J., and Syvänen A.-C. (1996). Multiplex, fluorescent solid-phase minisequencing for efficient screening of DNA sequence variation. *Clin. Chem.* **42,** 1391–1397.

Sachse, C., Brockmöller, J., Bauer, S., and Roots, I. (1997). Cytochrome P450 2D6 variants in a Caucasian population: Allele frequencies and phenotypic consequences. *Am. J. Hum. Genet.* **60,** 284–295.

Shumaker, J. M., Metspalu, A., and Caskey, T. C. (1996). Mutation detection by solid phase primer extension. *Hum. Mutat.* **7,** 346–354.

Syvänen, A.-C. (1994). Detection of point mutations in human genes by the solid-phase mini-sequencing method. *Clin. Chim. Acta* **226,** 225–236.

Syvänen, A.-C., and Landegren, U. (1994). Detection of point mutations by solid-phase methods. *Hum. Mutat.* **3,** 172–179.

Syvänen, A.-C., Aalto-Setälä, K., Harju, L., Kontula, K., and Söderlund, H. (1990). A primer-guided nucleotide incorporation assay in the genotyping of apolipoprotein E. *Genomics* **8,** 684–692.

Syvänen, A.-C., Sajantila, A., and Lukka, M. (1993). Identification of individuals by analysis of biallelic DNA markers using PCR and solid-phase minisequencing. *Am. J. Hum. Genet.* **52,** 46–59.

Tuly, G., Sullivan, K. M., Nixon, P., Stones, R. E., and Gill, P. (1996). Rapid detection of mitochondrial sequence polymorphism using multiplex solid-phase fluorescent minisequencing. *Genomics* **34,** 107–127.

Xiao, Z.-S., Goldstein, J. A., Xie, H.-G., Blaisdell, J., Wang, W., Jiang, C.-H., Yan, F.-X., Nan, N., Huang, S.-L., Xu, Z.-H., and Zhou, H.-H. (1997). Differences in the incidence of the CYP2C19 polymorphism affecting the S-mephenytoin phenotype in Chinese Han and Bai populations and identification of a new rare CYP2C19 mutant allele. *J. Pharmacol. Exp. Thera.* **281,** 604–609.

35

THE CLEAVASE I ENZYME FOR MUTATION AND POLYMORPHISM SCANNING

Mary Ann D. Brow

The Cleavase[1] I enzyme is a thermostable structure-specific endonuclease that cleaves at the junctions between single- and double-stranded DNA, on the 5′ side of the duplexed regions. The enzyme consists of the 5′ nuclease domain of the DNA polymerase from *Thermus aquaticus* (Lawyer *et al.*, 1989; Brow *et al.*, 1996), one of the eubacterial DNA polymerases that have been shown to cleave at similar DNA structures, provided such structures are somewhat close to the unpaired 5′ end (Lyamichev *et al.*, 1993). The Cleavase I enzyme is an engineered variant that does not retain any polymerase activity and that can cleave DNA without requiring a free 5′ end (M. A. Brow, unpublished data).

Single strands of DNA commonly contain short regions of self-complementarity and, in nondenaturing conditions, can fold on themselves to form partially base-paired secondary structures. These structural conformations are highly reproducible but are very sensitive to changes in the sequence of the strands. A single nucleotide substitution in a short (e.g., 300 nt) DNA fragment can change the resulting conformation enough to alter its electrophoretic mobility,

[1] Cleavase and CFLP are registered trademarks of Third Wave Technologies, Inc.

as is observed in single-strand conformation polymorphism (SSCP) analysis (Orita *et al.*, 1989). The method described here uses the structure-forming behavior of DNA in a different way. The Cleavase I enzyme recognizes and cleaves the folded strands, producing a set of fragments that, when resolved by electrophoresis, constitute "structural fingerprints" for identifying the specific DNA. The same conformational changes that alter the fragment mobility in a gel in SSCP are reflected as changes within the fingerprint pattern. Because analysis of the cleavage products does not require the structures to be resolved intact, this Cleavase-based analysis can be routinely applied to much larger fragments than can SSCP. The method has been applied to fragments from 100 to 2500 bp in length. Furthermore, the products can be resolved by rapid electrophoresis through short denaturing polyacrylamide gels for accurate sizing of the cleavage products. This method is called Cleavase fragment length polymorphism (CFLP) analysis. It has been applied successfully to mutation detection in human gene systems and to bacterial and viral typing (Brow *et al.*, 1996; Marshall *et al.*, 1997; Rossetti *et al.*, 1997). It can also be used for rapid screening of plasmids for misincorporation-induced sequence changes in cloned PCR fragments (M. A. Brow, unpublished data).

Analysis of PCR-generated fragments (Saiki *et al.*, 1985; Mullis and Faloona, 1987) by the CFLP method comprises the following steps: (i) separation of DNA strands by heating, (ii) formation of intrastrand structures on cooling, (iii) rapid enzymatic cleavage of these structures before they are disrupted by reannealing of the complementary strands, (iv) separation through denaturing polyacrylamide, and (v) visualization of the resulting fingerprint. The procedures in this chapter include instruction on preparing PCR-generated DNA for CFLP analysis, choosing optimal Cleavase I digestion conditions for new amplicons, and troubleshooting the most common problems with the procedure.

Protocols

Sample Preparation Following PCR Generation of DNA Fragments

The CFLP method can be used with a wide range of end-label moieties, including radioisotopes, biotin, digoxigenin, and fluorescent dyes, allowing visualization by autoradiography, chemiluminescence, fluorescence scanning, or colorimetric development of transfer membranes. Alternatively, the products of cleavage may be internally labeled or may be stained after electrophoresis (e.g., with silver stain or fluorescent stains such as SYBR Green) to visualize the fragments that are not seen when an end label is used (Brow and Fors, 1997). It should be noted that the patterns visualized from labeled DNAs will be quite different from those viewed by staining because the former show the concentration of labeled ends at a given gel position, whereas the latter show the mass of DNA at that site.

The quality and reproducibility of the CFLP pattern is dependent on the cleanliness of the starting material. For example, CFLP assays performed on amplification reactions that have a high background of unintended product will have background bands that are unlikely to be consistent from sample to sample and that will interfere with interpretation of the specific pattern. For this reason it is very important to assess the quality of the amplicon after the PCR and to take any steps that may be required to isolate the desirable material.

Step 1: Check the Specificity of the PCR

The PCR reaction products should always be checked on a gel to ensure the absence of multiple amplification (double-stranded) products. This verification is typically done by standard native agarose or polyacrylamide gel electrophoresis (Sambrook *et al.,* 1989; Saiki, 1990).

If multiple amplification products exist, either the desired product should be isolated from a denaturing polyacrylamide gel by standard methods or the PCR should be refined to produce only the desired amplicon.

Note: Use of a urea-containing gel for purification gives much cleaner final product.

If the PCR shows a single band, proceed to step 2.

Step 2: Remove Salts, Proteins, and Primers

Even for PCRs that show a single product, some form of sample cleanup is necessary before digestion with the Cleavase I enzyme in order to

Remove primers and extraneous single-stranded products generated during PCR reactions, both of which can interfere with the CFLP band patterns.

Remove excess salt, which can interfere with the CFLP reaction.

Removal of Primers, Primer Dimers, and Extraneous Single Strands

Analysis of labeled PCR products by electrophoresis follwed by detection of the specific label (e.g., by autoradiography or blotting) is much more sensitive than staining with ethidium bromide and usually reveals the presence of many truncated polymerization products. The use of a denaturing gel increases this type of lower molecular weight background, suggesting that some partially extended products are bound to templates even after extended incubation at 72°C at the end of the PCR. These products contribute an undesirable background to the CFLP pattern. The following procedures are effective in removing primers and some kinds of contaminating products. They are particularly effective when used in combination.

Treatment with Escherichia coli Exonuclease I

This enzymatic purification step is particularly effective at removing primers and other single-stranded contaminants, regardless of size.

Note: If uracil *N*-glycosylase (UNG) is used in the PCR, it must be removed by column purification (e.g., High Pure PCR Product Purification Kit) or by organic extraction before this incubation. Residual UNG activity will remove uracil bases during the *Exo*I treatments, which can cause background in the CFLP patterns due to structure-independent strand breakage at the abasic sites.

For each 100 μl of a completed PCR:

1. Warm the reaction to 70°C for 5–10 min. Lower the temperature to 37°C and incubate for 5 min. This incubation prior to enzyme addition melts any secondary structures that may have formed and enhances the ability of the exonuclease to digest single-stranded molecules;
2. If necessary, supplement the reaction to bring the $MgCl_2$ up to a concentration of at least 1.5 m*M* (most PCR reaction mixes contain 1.5–4 m*M* $MgCl_2$).
3. Add 100 U of *E. coli* Exonuclease I (Amersham, No. E 70073Z) directly to the PCR reaction tube.

4. Incubate at 37°C for 30 min and then inactivate the enzyme by heating to 70°C for 30 min.
5. Clean and desalt the remaining DNA in one of two ways:
   Ethanol precipitate as described below. Dissolve the DNA in 75% of the original starting volume, or at about 100 fmol per microliter, in either dH_2O or 10 m*M* Tris–Cl (pH 8.0) and 0.1 m*M* EDTA ($T_{10}E_{0.1}$).
   Use the High Pure PCR Product Purification Kit as described below.

High Pure PCR Product Purification Kit (Boehringer Mannheim, No. 1732668)

The High Pure PCR Product Purification Kit uses DNA-binding spin columns for the fractionation and desalting of the PCR-generated DNA. It is particularly effective at removing primers and low-molecular-weight (<100 bp) contaminating products, whether single or double stranded. The DNA is bound to glass fibers within the columns in the presence of a chaotropic salt, washed, and eluted in a low-salt buffer. To use the High Pure PCR Product Purification Kit, the manufacturer's protocol may be followed with the following alterations:

1. Use dH_2O or $T_{10}E_{0.1}$ instead of the elution buffer supplied.
2. For a PCR having an average yield, elute the DNA in 0.5–0.75× the volume of the PCR loaded.

Removal of Salts

The methods described above provide DNA in low-salt buffers suitable for direct use in CFLP analysis. On occasion, an amplicon is sufficiently pure (i.e., has no lower molecular weight background when resolved on a denaturing gel and visualized by one of the sensitive methods described above) that these methods can be omitted. Nonetheless, the DNA must still be thoroughly desalted before use in the CFLP reaction. The mono- and divalent cations present in the PCR buffer favor reannealing of the two DNA strands present in the reaction and decrease the formation of the intrastrand secondary structures that are the target of the Cleavase I endonuclease. In addition, both KCl and $MgCl_2$ have been found to slow the rate of cleavage in CFLP reactions.

Standard precipitation from 0.3 *M* sodium acetate by the addition of 2–4 vol of ethanol is sufficient for desalting; 1–4 μg of tRNA or

2 μg of glycogen added to the labeled DNA as carrier will not interfere with the CFLP reaction.

Even when PCR products have been purified by one of the methods cited previously, uncut amplicon should be included as a control for each CFLP reaction to check the final purity of the product. Any bands appearing in the uncut lane must be discounted from the CFLP pattern. One innocuous artifact seen with purified DNA is a faster migrating band or smear composed of incompletely denatured, full-length DNA. This usually appears only in the lanes for no-enzyme controls (see Fig. 1, far right lane) when a high concentration of full-length DNA is loaded in the well. Such bands typically do not appear in the corresponding cleaved samples and thus do not interfere with the CFLP pattern.

Digestion by the Cleavase I Enzyme

Materials

Sample DNA Considerations

The amount of DNA required will depend on the detection system used and the specific activity of the labeled DNA (Table 1).

As much as CFLP analysis can detect base changes, it is also sensitive to base modifications such as labeled nucleotides, methylation, deaza–purine analogs, or the presence of dU in place of dT. CFLP is fully compatible with the UNG-based methods of controlling PCR carry-over (Longo *et al.*, 1990), wherein dUTP must be substituted for TTP in PCR amplifications. However, because of the sensitivity of the method to any modifications in the DNA, samples to be compared should be generated with the same dye or base analog content.

CFLP Reaction Reagents

Cleavase I endonuclease, 25 U/μl

10× CFLP buffer

- 100 m*M* MOPS, pH 7.5
- 0.5% Tween 20
- 0.5% NP-40

2 m*M* $MnCl_2$

10 m*M* $MgCl_2$

CFLP stop solution

- 95% deionized formamide
- 10 m*M* EDTA, pH 8.0

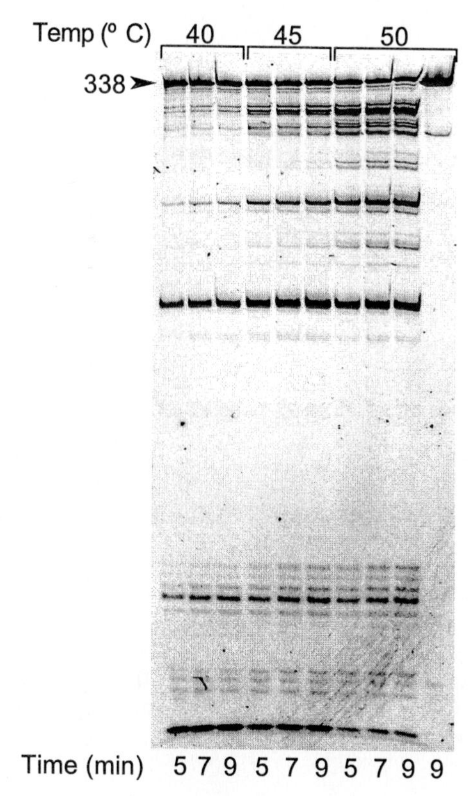

Figure 1 Hepatitis C 5′ UTR optimization panel; reactions with $MgCl_2$. A 338-nt fragment of the 5′ UTR of hepatitis C virus, type 1a, was amplified by PCR from a plasmid containing the sequence. The sense and antisense primers were 5′-CCTGATGGGGGC-GACACTCCACC-3′ and 5′ CTCATGGTGCACGGTCTACGAGA-3′, respectively. The antisense primer was labeled at the 5′ end with tetrachlorofluorescein. The resulting DNA fragment was first treated with exonuclease I as described, then purified using the High Pure PCR product purification system. The DNA was eluted from the High Pure column in dH_2O, and approx 200 fmol was used in each 20-μl cleavage reaction. The optimization panel of cleavage reactions was assembled as described and included 1 m*M* $MgCl_2$. The reactions were incubated at the temperatures indicated above each lane and for the times indicated below. The lane on the far right shows the no-enzyme control, which was incubated at 50°C for 9 min. The reactions were terminated by the addition of 16 μl of CFLP stop solution having 0.02% methyl violet as the indicator dye. Ten microliters of each reaction was loaded on a 20× 20 cm × 0.5 mm 10% polyacrylamide gel (19:1 cross-link) with 7 *M* urea in 0.5× TBE. Aliquots of CFLP stop with xylene cyanol and bromophenol blue were loaded into the two outermost empty lanes to monitor the electrophoresis. The gels were run at 20 W until the bromophenol blue reached the bottom and were then scanned with an Hitachi FMBIO fluorescence image analyzer using a 585-nm filter. Full-length fragment size (338 nt) is indicated on the left. Based on the distribution of band sizes and signal intensity, 8 min at 50°C was chosen as the best combination for Cleavase I digestion of this type of amplicon.

Table 1

Recommended Amounts of DNA for CFLP Analysis

End Label	Detection Method	DNA per 20-μl Reaction
Biotin and digoxigenin (each used with an appropriate alkaline phosphatase conjugate)	Chemiluminescent*[a]	
	CDP-*Star*	150–200 fmol
	Lumi-Phos 530	200+ fmol
	Colorimetric[a]	
	(NBT/BCIP)	200+ fmol
Fluorescent dye (will vary with dye/instrument compatibility)	Fluorescence imager:	
	Hitachi FMBIO	150–200 fmol
	ABI 373 Sequencer	250–300 fmol
Radioactivity (^{32}P)	Autoradiography	30 fmol or approx 100,000 counts
None	Direct staining techniques	
	Silver	100–200 ng
	SYBR Green	100–200 ng

[a] Chemiluminescent and colorimetric detection require transfer of the DNA from the gel to a nylon membrane. Contact blotting procedures are available from Tropix and Boehringer Mannhim Biochemical. The NBT/BCIP indicated for the colorimetric detection is a combination of the nitro blue tetrazolium/5-bromo-4-chloro-3-indolyl-phosphate substrates for alkaline phosphatase.

- Tracking dyes
  - 0.05% each of xylene cyanol and bromophenol blue
  - or
  - 0.02% methyl violet (for fluorescence scanning)
- Distilled water

Determination of Optimal Time and Temperature for the Cleavase I Reaction

The CFLP reaction is a partial digest producing a nested set of cleavage products from each 5′ end-labeled DNA fragment. The ideal CFLP pattern has a full range of band representation, from short products a few nucleotides long to residual undigested material, and it has uniform distribution of signal across the pattern. When single-base variants of a gene are compared, the patterns appear largely the same, with the differences revealed as changes in one or a few bands within a region of the variant patterns. These changes may include mobility shifts, the appearance of a new band(s), the disappearance of a band(s), or a combination of these effects.

The cleavage reaction can be optimized for each type of amplicon (e.g., variants of a gene of interest) by varying the temperature in 5° increments and using several different reaction times. All subsequent samples of this type of amplicon can then be analyzed under the established optimal conditions.

The following example of an optimization panel identifies the best CFLP reaction conditions for most unknowns. In general, underdigestion (e.g., too much uncut material remaining) indicates that a lower temperature is needed and moderate overdigestion (little full-length DNA remaining) is remedied by increased temperature. For further information, see Troubleshooting.

1. Prepare 10 DNA reaction tubes, labeled 1–10, with each containing

   DNA fragment (see Table 1 for recommended amount)

   dH_20 to 13 μl

   (Nine will be digested and one will be the no-enzyme control. An oil overlay is usually not necessary but may provide more consistent results with some DNAs.)

2. For each DNA reaction tube prepare the following volumes of these enzyme reaction mixes:

	+Enzyme (9)	−Enzyme (1)
10× CFLP buffer	2 μl	2 μl
2 m*M* $MnCl_2$	2 μl	2 μl
10 m*M* $MgCl_2$ (optional)[a]	2 μl	2 μl
25 U/μl Cleavase I enzyme	1 μl	0 μl
dH_2O	0 μl	1 μl
Total	7 μl	7 μl

[a]The addition of $MgCl_2$ at this concentration (1 m*M*) slows the cleavage rate of the enzyme approximately fivefold. Slowing the reaction can be desirable either for convenience in handling (allowing more time to process multiple samples) or to control cleavage of fragments that tend to easily overdigest to produce results that are less likely to vary with small variations in the reaction handling.

3. Place DNA tubes 1–3 in the thermal cycler. Denature at 95°C for 15 sec.

4. Bring the temperature down to 45°C. Add 7 μl of (+) enzyme mix.

5. While at reaction temperature, stop the reactions at 1, 3, and 5 min by adding 16 μl of CFLP stop solution. If $MgCl_2$ is used in the reactions, stop them at 5, 7, and 9 min. Remove the samples to room temperature.
6. Repeat steps 3–5 at 50 and 55°C, including the (−) enzyme tube with the 55°C sample set. (The temperature ranges tested may be varied depending on the behavior of the DNA, e.g., 40, 45, and 50°C were used for the samples in Fig. 1).
7. Prior to gel loading, heat the samples to 80°C for 2 min. Load 5–10 μl per lane. Resolve the products by electrophoresis through small-scale (10–20 cm × 0.5 mm) denaturing polyacrylamide gels (6–10% polyacrylamide with 19:1 cross-linking, 7 *M* urea, and 0.5–1 × TBE). As with sequencing reactions, the best resolution is achieved if the samples are fully denatured (i.e., run at sufficient wattage to warm the gels). The remainder of the samples may be stored at −20°C for later analysis.
8. Visualize according to the labeling method used. Select the combination of time and temperature that gives the best pattern of bands, as described above (Fig. 1).

CFLP ANALYSIS FOR DETECTION OF MUTATIONS AND POLYMORPHISMS

1. Prepare two DNA reaction tubes for each sample: one for the digest and one for a no-enzyme control:

   DNA fragment (see Table 1 for recommended amount)

   dH_20 to 13 μl

   (An oil overlay is usually not necessary but may provide more consistent results with some DNAs.)
2. For each DNA reaction tube prepare the following volumes of these enzyme reaction mixes:

	+Enzyme	−Enzyme
10× CFLP buffer	2 μl	2 μl
2 m*M* $MnCl_2$	2 μl	2 μl
10 m*M* $MgCl_2$ (optional)	2 μl	2 μl
25 U/μl Cleavase I enzyme	1 μl	0 μl
dH_2O	0 μl	1 μl
Total	7 μl	7 μl

3. Denature at 95°C for 15 sec.
4. Bring the temperature down to the previously determined reaction temperature.

5. At reaction temperature, add 7 μl of corresponding enzyme mix to (+) and (−) enzyme DNA reaction tubes. Incubate for the time selected after optimization.
6. While at reaction temperature, stop the reactions by adding 16 μl of CFLP stop solution. Remove to room temperature.
7. Prior to gel loading, heat the samples to 80°C for 2 min. Load 5–10 μl per lane. Resolve the products by electrophoresis as described above.
8. Visualize according to the labeling method used. Compare the patterns of the experimental (suspected mutant) DNAs to the wild-type DNA patterns to identify samples with sequence changes (Fig. 2).

Troubleshooting

Problem	Possible solution
Overdigestion (little full-length material left, weak signal at top of gel lane)	Temperature too low Increase temperature in 5°C increments. Too much cleavage activity Use the $MnCl_2/MgCl_2$ mixture in the cleavage reaction. and/or Reduce the amount of Cleavase I enzyme to 5–10 U per reaction. and/or Reduce incubation time to 0.5–1 min. Also see "No signal"
No signal (weak or blank gel lane)	DNA was not labeled, or specific activity of labeled DNA was too low. 5′ end label was "nibbled" off by the duplex-dependent 5′ to 3′ exonuclease activity of the Cleavase I enzyme Use the $MnCl_2/MgCl_2$ mixture in the cleavage reaction. and/or Reduce the amount of Cleavase I enzyme to 5–10 U per reaction. and/or Reduce incubation time to 0.5–1 min.

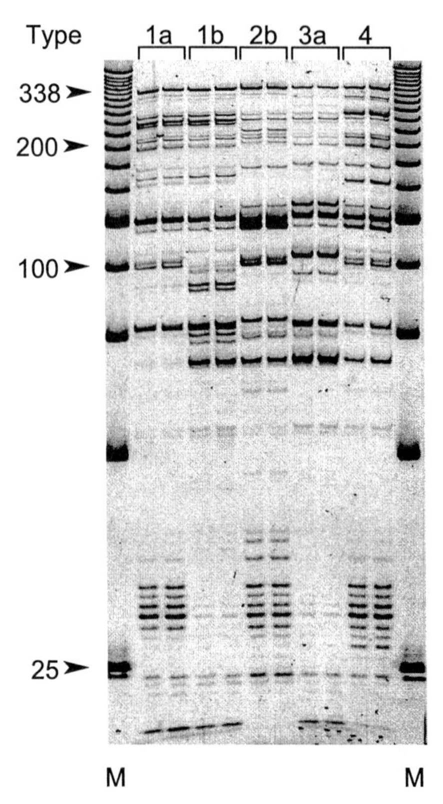

Figure 2 CFLP analysis of the hepatitis C 5′ UTR from several viral types. The 338-nt region of the 5′ UTR of hepatitis C virus was amplified by PCR from a set of plasmids containing the sequences for viral types 1a, 1b, 2b, 3a, and 4. The primers listed in the legend to Fig. 1 were used. The resulting DNA fragments were purified as described in the legend to Fig. 1. Single PCR isolates of each viral type were analyzed in duplicate CFLP reactions, and approx 120 fmol was used in each 20-μl reaction. The cleavage reactions were assembled as described and included 1 mM $MgCl_2$. The reactions were incubated for 8 min at 50°C and were terminated by the addition of 16 μl of CFLP stop solution having 0.02% methyl violet. The products were resolved and visualized as described in the legend to Fig. 1. The marker lanes, each containing a 25-bp ladder, are indicated by an "M" below each lane, and the sizes (nt) of selected marker bands and the full-length HCV amplicon are indicated on the left. Comparison of the patterns from each of the viral types shows that there are numerous distinctive pattern features that are associated with each specific viral type.

Underdigestion (weak pattern, too much uncut DNA)	High temperature may be melting out most or all structure Reduce temperature in 5 or 10°C increments to find optimum. Ionic strength too high—Salt speeds the reannealing of the strands and blocks CFLP digestion Precipitate the DNA from 200 m*M* NaCl (final aqueous concentration) with 2.5 vol of ethanol, and wash with 70% EtOH to eliminate any excess salt. Insufficient manganese available for the Cleavase I enzyme (the manganese optimum is affected by salt and EDTA; these will reduce the cleavage efficiency) DNA should be desalted and resuspended in DNA dilution buffer, water, or a Tris–HCl buffer with 0.1 m*M* or less EDTA. Insufficient enzyme activity Use the "no $MgCl_2$" reaction conditions to stimulate the cleavage. Incubation time not long enough Increase duration of incubation at reaction temperature to 4 or 5 min.
A very strong band dominates the pattern (sometimes with little or no pattern above it in the gel lane)	A strong structure is forming that is favored for cleavage. Test the full range of reaction temperature: lower temperatures can increase the stability of other structures, allowing them to compete more successfully for cleavage. In other cases, the use of high temperature (>65°C) to destabilize the dominant structure has proven effective. CFLP analysis may be tried on the complementary strand; the structures on this strand often exhibit different behavior in these reactions.

Acknowledgments

I thank Mary Oldenburg and Marianne Siebert for sharing data and protocols and Laura Heisler for critical reading of the manuscript and discussions.

References

Brow, M. A. D., and Fors, L. (1997). Cleavase fragment length polymorphism analysis for mutation scanning. *Biomed. Products* **22**(9), 22–24.

Brow, M. A. D., Oldenburg, M. C., Lyamichev, V., Heisler, L. M., Lyamicheva, N., Hall, J. G., Eagan, N. J., Olive, D. M., Smith, L. M., Fors, L., and Dahlberg, J. E. (1996). Differentiation of bacterial 16S rRNA genes, intergenic regions and *Mycobacterium tuberculosis katG* genes by structure-specific endonuclease cleavage. *J. Clin. Microbiol.* **34**(12), 3129–3137.

Lawyer, F. C., Stoffel, S., Saiki, R. K., Myambo, K., Drummond, R., and Gelfand, D. H. (1989). Isolation, characterization, and expression in *Escherichia coli* of the DNA polymerase gene from *Thermus aquaticus. J. Biol. Chem.* **264,** 6427–6437.

Longo, M. C., Berninger, M. S., and Hartley, J. L. (1990). Use of uracil DNA glycosylase to control carry-over contamination in polymerase chain reactions. *Gene* **93,** 125–128.

Lyamichev, V., Brow, M. A. D., and Dahlberg, J. E. (1993). Structure-specific endonucleolytic cleavage of nucleic acids by eubacterial DNA polymerases. *Science* **260,** 778–783.

Marshall, D. J., Heisler, L. M., Lyamichev, V., Murvine, C., Olive, D. M., Ehrlich, G. D., Neri, B. P., and Arruda, M. D. (1997). Determination of hepatitis C virus enotypes in the United States by Cleavase fragment length polymorphism analysis. *J. Clin. Microbiol.* **35**(12), 3156–3162.

Mullis, K. B., and Faloona, F. A. (1987). Specific synthesis of DNA in vitro via a polymerase-catalyzed chain reaction. *Methods Enzymol.* **155,** 335–350.

Orita, M., Suzuki, Y., Sekiya, T., and Hayashi, K. (1989). A rapid and sensitive detection of point mutations and genetic polymorphisms using polymerase chain reaction. *Genomics* **5,** 874–879.

Rossetti, S., Englisch, S., Bresin, E., Pignatti, P. F., and Turco, A. E. (1997). Detection of mutations in human genes by a new rapid method: Cleavage fragment length polymorphism analysis (CFLPA). *Mol. Cell. Probes* **11,** 155–160.

Saiki, R. K. (1990). Amplification of genomic DNA. In *PCR Protocols: A Guide to Methods and Applications* (M. A. Innis, D. H. Gelfand, J. J. Sninsky, and T. J. White, Eds.), pp. 13–20. Academic Press, San Diego.

Saiki, R. K., Scharf, S. J., Faloona, F., Mullis, K. B., Horn, G. T., Erlich, H. A., and Arnheim, N. (1985). Enzymatic amplification of β-globin genomic sequences and restriction site analysis for diagnosis of sickle cell anemia. *Science* **230,** 1350–1354.

Sambrook, J., Fritsch, E. F., and Maniatis, T. (1989). *Molecular Cloning: A Laboratory Manual.* Cold Spring Harbor Laboratory Press, Cold Spring Harbor, NY.

INDEX